DISCARDED
FROM
UNH LIBRARY

D1222430

Electronic Absorption Spectra and Geometry of Organic Molecules

An Application of Molecular Orbital Theory

Electronic Absorption Spectra and Geometry of Organic Molecules

An Application of Molecular Orbital Theory

HIROSHI SUZUKI

Department of Chemistry
College of General Education
University of Tokyo
Tokyo, Japan

1967

ACADEMIC PRESS NEW YORK · LONDON

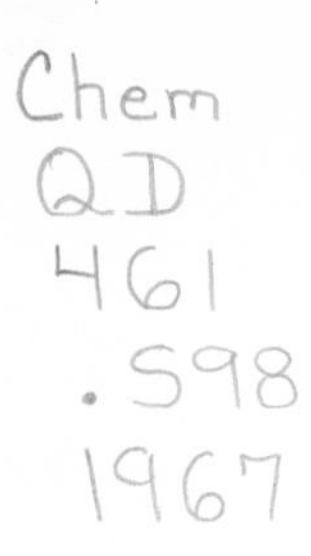

ACADEMIC PRESS INC.
111 Fifth Avenue, New York, New York 10003

United Kingdom Edition published by
ACADEMIC PRESS INC. (LONDON) LTD.
Berkeley Square House, London W.1

LIBRARY OF CONGRESS CATALOG CARD NUMBER: 66-30823

PRINTED IN THE UNITED STATES OF AMERICA

Preface

The main subject of this book is the relation between electronic absorption spectra of organic compounds and geometry of molecules. In most works which deal with electronic absorption spectroscopy this relationship is discussed rather briefly as a special topic. I believe, however, that an extensive discussion not only will afford important information about molecular geometry but also will facilitate the understanding of the nature of electronic states and electronic absorption spectra of molecules.

The book is intended primarily for research workers and graduate-level chemistry students. Hopefully, it will also be useful to beginners in the field.

The book treats the following subject areas: (1) electronic absorption spectroscopy, (2) molecular orbital (MO) theory, (3) the interpretation of electronic absorption spectra of organic compounds, especially of organic conjugated compounds, by the use of MO theory, and (4) the relation of the spectra to the geometry of molecules.

Chapters 1 and 4–6 present various important aspects of electronic absorption spectroscopy. While the vibrational absorption (in the infrared regions) of a compound is a property that may be attached primarily to definite parts of the molecule, the electronic absorption (in the visible and ultraviolet regions) of a compound, especially of a conjugated compound, is a property of the molecule as a whole. For this reason, the electronic absorption spectra are rather difficult to interpret; at the same time, if they are adequately interpreted, they can afford an important clue to the electronic structure and geometry of molecules.

Molecular orbital theory in its LCAO approximation forms the basis for the interpretation of the spectra given in this volume. In Chapters 2, 3, 7–11, 19, and 20, I have presented important principles and concepts of the theory. All essential facts are given herein and difficult mathematical details are omitted.

Chapter 2 is an introductory exposition of MO theory. Symmetry considerations based on group theory are important in the application of MO theory and in the understanding of molecular spectra. In Chapter 3 I have presented principles and basic aspects of group theory in a usable form insofar as they are believed necessary for a practical performance of calculation by MO theory and for an understanding of electronic spectra.

A detailed exposition of methods of calculating molecular electronic states within the framework of MO theory are presented. The overall aim is to enable the reader to make his own calculations and applications.

Chapters 7–9 are devoted to a presentation of the so-called simple LCAO MO methods. Here, the term *simple* is used to indicate methods which neglect electronic interaction, using the so-called one-electron approximation. On the other hand, methods which take account of electronic interaction are called *advanced* methods. The simple MO methods are particularly useful in the description of molecules in the ground state; in interpretation of electronic spectra they are useful for correlating the position and intensity of certain types of bands in a restricted class of compound. However, they alone cannot give an overall correct picture of an electronic spectrum; for this it is necessary to include electronic interaction. In Chapter 10 the usefulness and limitations of the simple MO methods in interpreting electronic spectra are discussed, and in Chapter 11 a discussion of advanced MO methods, especially of a semiempirical advanced LCAO MO method called the Pariser–Parr–Pople method, or more simply, the P-method, which is the most popular of the advanced methods, is presented.

The methods of building up molecules from fragments or of composite-molecule methods have proved very useful for the understanding of electronic structure as well as of electronic spectra of conjugated molecules. In particular, they are very useful for discussion of the main subject of this work. Because they are so important and are so frequently referred to herein, it seemed desirable to name them as simply as possible. Thus they are called the simple or the advanced composite-molecule methods according to whether they neglect or take into account electronic interaction. In Chapters 19 and 20 these methods are expounded in considerable detail.

In the remaining chapters electronic absorption spectra and data of various representative types of organic compounds are presented, discussed, and interpreted on the basis of the theoretical foundations set forth in the preceding chapters; results of calculation by various MO methods are given, with special emphasis placed on the relation of the spectra to the geometry of molecules. By the geometry of a molecule is meant the geometrical structure of a molecule, that is, the three-dimensional arrangement of the constituent atomic nuclei in a molecule. The selection of material has, of necessity, been severely limited and inevitably reflects my own work and interests. However, an effort has been made to include typical and important examples which are believed suitable for illustrating the various aspects of the relation between the spectra and the molecular geometry.

Chapters 12–15 are, to some extent, revised versions of papers I had published from 1959 to 1962. In the other chapters some of the interpretations of spectra and of the correlations of spectra with the molecular geometry are original and some are extensions drawn from the literature.

The systems of notation for electronic spectral bands and electronic transitions and conventions used are summarized in the appendix; other prevailing systems of notation and classification of bands and transitions are given a brief explanation too. A list of general references, arranged by topics, is also included.

In writing this volume, I have made reference to many articles, reviews, and books, some of which are listed in the General References, and especially to those written by A. E. Gillam and E. S. Stern; G. Herzberg; H. H. Jaffé and M. Orchin; J. N. Murrell; H, Eyring, J. Walter, and G. E. Kimball; R. Daudel, R. Lefebvre, and C. Moser; A. Streitwieser, Jr.; and the review written by S. F. Mason. I feel much indebted to these authors. I am also indebted to Professors K. Shiomi, O. Simamura, S. Nagakura, and M. Takahasi of the University of Tokyo for their helpful suggestions and encouragement during the research on the subject. Finally, it gives me great pleasure to acknowledge the understanding encouragement I have received from my patient family during the considerable length of time devoted to writing this book.

HIROSHI SUZUKI

Tokyo, Japan
May, 1967

Contents

Chapter 19. Simple Composite-Molecule Method and a Classification of π–π^* Transitions

Chapter 20. Advanced Composite-Molecule Method

Chapter 21. Carbonyl Compounds

Chapter 22. Nitrobenzene, Benzoic Acid, Aniline, and Related Compounds

Chapter 23. Azobenzenes and Related Compounds

Chapter 1

Introduction

1.1. Energy Levels of a Molecule and Molecular Spectra

1.1.1. Energy Levels of a Molecule

A molecule can exist only in discrete energy states, and the total energy E of a molecule in an energy state, aside from the translational energy, can be expressed as

$$E = E_{el} + E_{vib} + E_{rot}, \tag{1.1}$$

where E_{el}, E_{vib}, and E_{rot} represent the electronic, vibrational, and rotational energies, respectively. All of these three types of energy are quantized. With each electronic state there are associated a number of vibrational states, and with each vibrational state there are associated a number of rotational states.

As the simplest example, consider a diatomic molecule. On the assumption of simple harmonic oscillator, the vibrational energy of a diatomic molecule in an electronic state m can be expressed as follows:

$$E_{vib} = (v + \tfrac{1}{2})h\nu'_m, \tag{1.2}$$

where h is the Planck constant, ν'_m is the fundamental vibrational frequency for the electronic state, given by $(1/2\pi)(k_m/\mu)^{1/2}$, k_m and μ being the force constant for the electronic state and the reduced mass of the two atoms, respectively, and v is the vibrational quantum number. The vibrational quantum number v can take integral values 0, 1, 2, . . . , and the energy increases with increasing v. According to this expression, all the energy separations of successive vibrational energy levels must be equal. Actually, however, because of anharmonicity of the vibration, the energy separation becomes progressively smaller as the vibrational quantum number increases.

Usually, the energy separations of successive vibrational levels of relatively low quantum numbers are of the order of 1–10 kcal/mole.

The rotational energy is approximately expressed as

$$E_{\text{rot}} = J(J + 1)hB_v\,, \tag{1.3}$$

where J is the rotational quantum number, and B_v is the rotational constant for the vth vibrational state, given by $h/(8\pi^2 I_v)$, I_v representing the moment of inertia of the molecule in the vibrational state with respect to the rotational axis. The rotational quantum number J can take integral values 0, 1, 2, . . . , and the rotational energy increases quadratically with increasing J. Usually, the energy separations between two neighboring rotational levels are of the order of 10^{-3}–10^{-2} kcal/mole.

The energy separations of electronic levels are much greater: they are mostly of the order of 100 kcal/mole.

In a thermal equilibrium the number of molecules in each of the energy states is proportional to the Boltzmann factor $e^{-E/kT}$, in which k is the Boltzmann constant and T is the absolute temperature. At the ordinary temperature, at which kT corresponds to an energy of about 0.6 kcal/mole, most of molecules are in a large number of rotational states associated with the lowest vibrational ($v = 0$) state belonging to the lowest electronic state (the ground electronic state). The number of molecules in higher vibrational states falls off very rapidly, and in general, higher electronic states (excited electronic states) are never thermally populated.

1.1.2. Molecular Spectra

When continuous electromagnetic radiation passes through a transparent material, a portion of the radiation may be absorbed and the residual radiation, when passed through a prism, may yield a spectrum with gaps in it. Such a spectrum is called an absorption spectrum. During the absorption process the atoms or molecules of the material pass from a state of low energy to one of higher energy. When atoms or molecules in an excited state revert to a state of lower energy, losing energy, they may emit radiation and give rise to an emission spectrum.

In both absorption and emission processes, the relation between the energy change in the atom or molecule and the frequency of the radiation absorbed or emitted is given by the well-known Bohr condition:

$$|\Delta E| = |E_f - E_i| = h\nu', \tag{1.4}$$

where ν' is the frequency of the radiation, and E_i and E_f are the energies of the atom or molecule in the initial and final states, respectively. The absolute value of the energy difference ΔE is called the transition energy. For absorption the value of ΔE is positive; for emission it is negative. The frequency ν' is related to the wavelength λ by $\nu' = c/\lambda$, in which c is the velocity of light (2.998×10^{10} cm/sec in vacuum). Therefore, the smaller the transition energy, the longer is the wavelength of the radiation absorbed or emitted.

Absorption spectra of molecules, that is, molecular absorption spectra, are classified into three types: rotation, vibration, and electronic.

The rotation spectrum of a molecule is associated with changes which occur in the rotational states of the molecule (selection rule: $\Delta J = \pm 1$) without simultaneous changes in the vibrational and electronic states. Since the energy levels for the rotation of the molecule are relatively close to one another, the frequency of the absorbed radiation is low, and the wavelength long. Actually, the rotation spectra of light molecules occur in the far infrared region, at wavelengths beyond 20 μ, and those of heavy molecules are situated in the microradiowave region, at wavelengths of the order of 0.1–10 cm.

The vibration (or vibration–rotation) spectrum of a molecule is associated with changes which occur in the vibrational states of the molecule (selection rule: $\Delta v = \pm 1$) without simultaneous changes in the electronic state. The changes in the vibrational states are generally accompanied with changes in rotational states, so that the vibration spectrum has generally rotational fine structure. The most important vibration spectra occur in the near and middle infrared region from about 1 to 30 μ. Absorptions associated with transitions between energy levels of vibrational modes with relatively small force constants, such as torsional vibrations about single bonds, occur at longer wavelengths, e.g., about 60 μ (far infrared region).

Finally, electronic spectra arise from transitions between electronic states. The electronic spectra occur in the ultraviolet (at wavelengths below 400 mμ) and visible (wavelengths 400–800 mμ) regions, and sometimes extend into the near infrared region. Electronic transitions are accompanied with simultaneous changes in the vibrational and rotational states. Therefore, a transition between two electronic states does not result in only a single spectral line, but results in a large number of lines not widely spaced from one another. In fact, electronic spectra of gaseous molecules may assume a very complicated structure arising from the superposition of vibrational and rotational changes on the electronic transition. In electronic spectra of liquids, solutions, and solids, the rotational structure is unresolvable, and even the vibrational structure is not always resolvable. The electronic spectra thus consist of relatively broad bands.

1.2. Electronic Absorption Spectra of Organic Compounds

Electronic absorption spectra of most saturated organic compounds occur in the far ultraviolet region, that is, at wavelengths shorter than 200 mμ. Absorptions occurring in the near ultraviolet (wavelengths 200–400 mμ) and visible (wavelengths 400–800 mμ) region are mostly due to unsaturated compounds.

A transition from a lower electronic state of a molecule to a higher electronic state can be regarded approximately as a promotion of one or more electrons from an occupied molecular orbital (MO) to a higher energy unoccupied molecular orbital. Figure 1.1 shows schematically the energy

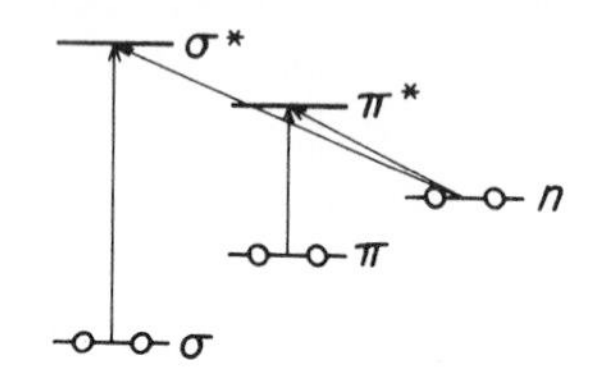

FIG. 1.1. Various types of electronic transitions in molecules.

levels of molecular orbitals of different types. In general, bonding π orbitals have higher energies than bonding σ orbitals, and antibonding π orbitals (π* orbitals) have lower energies than antibonding σ orbitals (σ* orbitals). The absorptions due to electronic transitions from bonding σ orbitals to antibonding σ orbitals (σ–σ* transitions) usually occur in the far ultraviolet region, and the absorptions due to electronic transitions from bonding π orbitals to antibonding π orbitals (π–π* transitions) usually occur at longer wavelengths, mostly in the near ultraviolet and visible region. In compounds having nonbonding lone-pair electrons, the nonbonding orbital (n orbital) may have a higher energy than bonding π orbitals. The absorption bands arising from electronic transitions from nonbonding orbitals to antibonding π orbitals (n–π* transitions) usually occur in the near ultraviolet and visible region. Some of absorption bands due to electronic transitions from nonbonding orbitals to antibonding σ orbitals (n–σ* transitions) occur at about 200 mμ or longer wavelengths.

The σ orbitals are to a first approximation considered to be localized on the respective bonds, and hence the σ–σ* transitions are not characteristic of the overall structure of the molecule, but are rather properties of the respective bonds. On the other hand, the π orbitals are delocalized over the whole conjugated system of the molecule, and hence electronic transitions involving π orbitals, that is, π–π* and n–π* transitions, are characteristic of the whole conjugated system. In addition, while the σ orbitals are not affected by twist of the bonds, the π orbitals, and hence the transitions involving π orbitals, are sensitive to changes in the geometry of the conjugated system, such as twist of bonds. For these reasons, in this book the discussion is confined to the absorption bands due to transitions involving π orbitals in unsaturated, especially conjugated compounds.

1.3. Wavelength, Wave Number, Energy, and Their Conversion Factors

The absorption spectrum is completely described by giving the absorption intensity as a function of the wavelength, and is presented usually as an absorption curve in a graph with the absorption intensity as the ordinate and the wavelength as the abscissa. In electronic absorption spectroscopy the wavelength λ is usually expressed in the millimicron unit (1 mμ = 10^{-7} cm) or the angstrom unit (1 Å = 10^{-8} cm). The millimicron unit seems to be preferable because in most cases the one less significant figure in this unit as compared with the angstrom unit gives a better indication of the true reliability of the experimental data.

In theoretical discussions the frequency $\nu' = c/\lambda$ is much more important than the wavelength. The frequency is the number of waves that pass a particular point per second. However, the frequency unit as such is not often used today. Instead, the reciprocal of the wavelength, that is, the wave number $\nu = 1/\lambda = \nu'/c$, is usually used, which is proportional to the frequency and is measured in cm^{-1} (number of waves per centimeter). In electronic absorption spectroscopy sometimes the unit kilokayser (1 kk = 10^3 cm^{-1}) is used.

The frequency is a more fundamental property of the radiation than the wave number and the wavelength, because the frequency does not depend on the medium through which the radiation transverses, while the wave number and the wavelength depend on the medium. This is because the velocity of the radiation, c, depends on the medium, being given by $c = c_{vac}/n$, where c_{vac} is the velocity of light in vacuum, a universal constant independent of the frequency of the light, and n is the refractive index of the medium, relative to the vacuum, at the frequency of the radiation. However, the wavelengths and the wave numbers in spectra in the region above about 200 mμ are usually given as measured in air, and the refractive index of air under the ordinary condition is as small as about 1.00033 at 200 mμ, and gradually decreases further with the increasing wavelength, to about 1.00027 at 1000 mμ. Therefore, in the visible and near ultraviolet region, with which we are mainly concerned, the differences between the values of wavelength and wave number in air and those in vacuum can usually be safely neglected.

Since the frequency is related to the energy E by the fundamental Planck–Einstein relation $E = h\nu'$, we have another set of units for defining the radiation. The units of energy which are most commonly used are the electron volt (ev) per molecule and the kilocalorie per mole (kcal/mole). Since the wave number is proportional to the energy, it is often taken as a measure of the energy.

The wavelength λ, the wave number ν, and the energy E in units of electron volt per molecule and kilocalorie per mole are converted to one another by the following equations*:

$$\begin{aligned}
\lambda\,(\mathrm{m}\mu) &= 10^7/\nu\,(\mathrm{cm}^{-1}) = 28{,}591.2/E\,(\mathrm{kcal/mole}) \\
&= 1239.81/E\,(\mathrm{ev}) \\
\nu\,(\mathrm{cm}^{-1}) &= 0.349758 \times 10^3 \times E\,(\mathrm{kcal/mole}) \\
&= 8.06575 \times 10^3 \times E\,(\mathrm{ev}) = 10^7/\lambda\,(\mathrm{m}\mu) \\
E\,(\mathrm{kcal/mole}) &= 23.0609 \times E\,(\mathrm{ev}) = 2.85912 \times 10^{-3} \times \nu\,(\mathrm{cm}^{-1}) \\
&= 28{,}591.2/\lambda\,(\mathrm{m}\mu) \\
E\,(\mathrm{ev}) &= 0.123981 \times 10^{-3} \times \nu\,(\mathrm{cm}^{-1}) \\
&= 4.33634 \times 10^{-2} \times E\,(\mathrm{kcal/mole}) = 1239.81/\lambda\,(\mathrm{m}\mu)
\end{aligned}$$

For example, the wavelength of 200 mμ corresponds to the wave number of 50,000 cm^{-1} or 50 kk and the energy of 6.199 ev/molecule or 143.0 kcal/mole, and the wavelength of 400 mμ corresponds to the wave number of 25,000 cm^{-1} or 25 kk and the energy of 3.100 ev/molecule or 71.5 kcal/mole.

1.4. Practical Measures of Absorption Intensity

1.4.1. Laws of Light Absorption and the Molar Extinction Coefficient

A practical measure of the intensity of light absorption is defined on the basis of two principal laws of light absorption: Lambert's law and Beer's law. Lambert's law states that the proportion of light absorbed by a

* The conversion factors in these equations have been calculated by use of the following values of fundamental constants, which were adopted in 1963 by IUPAC (International Union of Pure and Applied Chemistry) and IUPAP (International Union of Pure and Applied Physics):

$$\begin{aligned}
\text{Planck constant } h &= 6.6256 \times 10^{-27}\ \text{erg sec}, \\
\text{Velocity of light in vacuum } c_{\mathrm{vac}} &= 2.997925 \times 10^{10}\ \text{cm sec}^{-1}, \\
\text{Avogadro number } N &= 6.02252 \times 10^{23}\ \text{mole}^{-1}, \\
\text{Faraday constant } F &= 96{,}487.0\ \text{coulomb equiv.}^{-1} \\
&= 23{,}060.9\ \text{cal. volt}^{-1}\ \text{equiv.}^{-1}, \\
\text{Charge of electron } e &= F/N = 1.60210 \times 10^{-19}\ \text{coulomb} \\
&= 4.80298 \times 10^{-10}\ \text{esu}, \\
1\ \text{cal} = 4.1840\ \text{joule},\ 1\ \text{ev} &= 1.60210 \times 10^{-19}\ \text{joule}.
\end{aligned}$$

transparent medium is independent of the intensity of the incident light and that each successive layer of the medium absorbs an equal fraction of the light passing through it. This law is given the following expression:

$$-dI_x/dx = \alpha I_x\,, \tag{1.5}$$

where I_x is the intensity of the light at a point in the medium, x is the path length of the light in the medium up to that point, and α is a constant characteristic of the medium and of the frequency of the light. On integration of this equation from $x = 0$ to $x = L$ (the thickness of the medium layer), the following expression is obtained:

$$\ln (I_0/I) = \alpha L, \quad \text{or} \quad \log (I_0/I) = \alpha L/\ln 10 = KL, \tag{1.6}$$

where I_0 is the intensity of the incident light and I is the intensity of the transmitted light. K, which is equal to $\alpha/\ln 10 \doteqdot 0.4343\alpha$, is called the Bunsen–Roscoe extinction coefficient, and has a meaning of the reciprocal of the thickness of the medium required to weaken the light to one-tenth of the incident intensity. The quantity that is actually measured with the usual spectrophotometer is $\log (I_0/I)$. This quantity is called the absorbance or the optical density. I/I_0 is called the transmittance, and is denoted by T. Of course, the absorbance can be expressed as $-\log T$.

The second important law of light absorption, Beer's law, states that the amount of light absorbed is proportional to the number of molecules of the absorbing substance in the light path. This amounts to saying that the absorption coefficient (α or K) of a medium is proportional to the number of molecules of the absorbing substance in a unit volume. If the absorbing substance is dissolved in a transparent solvent, the absorption coefficient of the solution will be proportional to its concentration, C. Therefore, we can write $K = kC$, where k is a constant called the absorptivity, which is characteristic of the absorbing substance and independent of its concentration.

When the concentration C is expressed in units of moles per liter and the cell length (that is, the thickness of the absorbing layer) L in centimeters, the absorptivity k is called the molar absorptivity or the molar extinction coefficient, and usually is denoted by symbol ϵ. That is, the molar extinction coefficient ϵ is expressed as follows:

$$\epsilon = \log (I_0/I)/L(\text{cm})C(\text{moles/liter}). \tag{1.7}$$

Usually, the molar extinction coefficient is given without units, although it has units of liter $\text{mole}^{-1}\,\text{cm}^{-1}$ and a dimension of area per mole. The highest values of ϵ observed in electronic absorption spectra are of the order of 10^5.

The molar extinction coefficient is, as indicated, a fundamental molecular property independent of concentration and of cell length, as long as the aforementioned two laws of light absorption hold good. Therefore, it is most desirable to use this quantity as the measure of the absorption intensity, as far as it is possible. However, when it is not possible, that is, when the molecular weight of the absorbing substance under examination is unknown, we are compelled to use the absorbance for a specified concentration and cell length as a measure of the absorption intensity.

Lambert's law holds good for any homogeneous substances. There are, however, many exceptions to Beer's law. Deviation from Beer's law can occur owing either to real or to apparent causes. Real deviations from Beer's law are expected when the concentration of the absorbing molecules is high enough for these molecules to exert perturbing forces on one another, and when the absorbing substances dissociate or associate in solutions, the degree of dissociation or association being a function of the concentration. For example, many acids, bases, and salts in solution do not obey Beer's law, because they are more completely ionized with increasing dilution, and because in most cases the light absorption of the ions markedly differs from that of the parent nonionized molecule. Aqueous solutions of dyes which are prone to dimerize or polymerize do not also obey Beer's law. However, most nonionizing, and especially, nonpolar organic compounds obey Beer's law, at least approximately, over a considerably wide concentration range.

Since the extinction coefficient depends on wavelength, Beer's law can be strictly true only for monochromatic light. The exit beam from any monochrometer is not purely monochromatic, but always comprises lights of wavelengths extended over a range, which is determined by the resolving power and dispersion of the spectrometer and the slit width used. The nominal wavelength of the beam corresponds to the center of the wavelength range. Accordingly, the observed absorbance at a given wavelength may vary, depending on the wavelength range of the so-called monochromatic beam, and hence deviation from Beer's law may occur, because usually it is necessary to use the wider slit width for the longer cell length. The intensity of the incident light is distributed in the wavelength range of the so-called monochromatic beam in such a way that the intensity at the nominal wavelength is highest. Therefore, as far as the spectrum is measured under the pertinent conditions, the effect of the nonmonochromaticity of the incident light is not strong, and usually can be safely neglected. However, this effect cannot be neglected for sharp absorption bands, especially, absorption bands having distinct vibrational fine structure.

1.4.2. The Integrated Intensity and the Oscillator Strength

The intensity of an absorption band is usually described in terms of the maximum value of the molar extinction coefficient for that band, ϵ_{max}. However, the maximum molar extinction coefficient is only a rough measure of the band intensity. It depends, as mentioned already, more or less markedly on the resolving power of the spectrometer and the slit width used, and it is further affected by change of the phase or of the solvent. On the other hand, a quantity called the integrated intensity or the integrated molar extinction coefficient, in most cases, remains constant at least approximately, and thus affords a better measure of the band intensity.

The integrated intensity, A, is defined as follows:

$$A = \int_{\nu_1}^{\nu_2} \epsilon_\nu \, d\nu, \tag{1.8}$$

where ϵ_ν is the molar extinction coefficient at wave number ν, and ν_1 and ν_2 are, respectively, the lower and higher wave number limits of the absorption band under consideration. The integrated intensity is equivalent to the area under the band envelope in the spectrum with the abscissa linear in the wave number and the ordinate linear in the molar extinction coefficient, and has units of liter $mole^{-1}$ cm^{-2} or 10^3 cm $mole^{-1}$.

Frequently, the intensity of an absorption band is expressed in terms of the oscillator strength, f, which is related to the integrated intensity by the following equation:

$$f = (10^3 \ln 10)(mc^2/\pi e^2 N) \int \epsilon_\nu \, d\nu \doteqdot 4.319 \times 10^{-9} \int \epsilon_\nu \, d\nu, \tag{1.9}$$

where m and e are the mass and charge of electron, respectively, c is the velocity of light in vacuum, and N is the Avogadro number. The oscillator strength as well as the integrated intensity can be taken as a measure of the total probability of the corresponding electronic transition. For intense absorption bands the oscillator strength is of the order of 1.

Evaluation of the integrated intensity requires that the absorption band under consideration be isolated from all other bands. Actually, however, electronic absorption spectra rarely contain isolated bands. Therefore, we must generally extrapolate the band envelope to the abscissa in a somewhat arbitrary manner. Alternatively, when the overlap of the band with other bands is relatively small, the band area between the absorption minima on both sides of the band, which are called the effective wave number limits, is often taken as an approximation to the true integrated intensity. In either case, the evaluation of the integrated intensity inevitably involves arbitrariness and ambiguity more or less. In cases of strongly overlapping bands, particularly in cases of partly submerged bands, the segregation of a band and the evaluation of its integrated intensity are difficult. In cases of smooth and symmetrical isolated bands, the integrated intensity is sometimes approximated as follows:

$$A \fallingdotseq \epsilon_{max} \cdot (\Delta\nu)_{1/2}, \tag{1.10}$$

where $(\Delta\nu)_{1/2}$ is the so-called half-band-width, that is, the width of the band in cm^{-1} where $\epsilon = \frac{1}{2}\epsilon_{max}$. In comparison of bands of similar electronic origin in spectra of closely related compounds, it is frequently assumed that ϵ_{max} is proportional to the integrated intensity.

Chapter 2

Molecular Orbital Theory

2.1. Fundamental Postulates of Quantum Theory

A satisfactory description of various states of molecules and hence a satisfactory understanding of the relation between the electronic spectrum and the molecular structure invoke quantum theory. In this place some introductory remarks on fundamental postulates of the theory are made.

Stationary states of a system consisting of nuclei and electrons are characterized by wave functions Ψ which are solutions (eigenfunctions) of the operator equation

$$\mathbf{H}\Psi = E\Psi, \tag{2.1}$$

where $\mathbf{H}$ is the quantum-mechanical Hamiltonian, that is, the operator whose eigenvalues E are the energies of the different states of the system. This equation is called the Schrödinger equation.

The quantum-mechanical Hamiltonian $\mathbf{H}$ can be easily derived from the classical Hamiltonian, which is the expression for the total energy of the system in terms of the coordinates and momenta of all the particles in the system. For stationary states the sum of the kinetic energy and the potential energy is constant and is equal to the total energy. For a single spin-free particle of mass m, moving in a potential field $V(x, y, z)$, the classical Hamiltonian is expressed as follows:

$$H = (1/2m)(p_x^2 + p_y^2 + p_z^2) + V(x, y, z), \tag{2.2}$$

where p_x, p_y, and p_z are the components of the momentum of the particle along the coordinate axes defined to the system. The first term on the right-hand side of this equation expresses the kinetic energy of the particle. The quantum-mechanical Hamiltonian can be obtained from the classical Hamiltonian by replacing the components of the momentum p_x, p_y, and p_z by the partial differential operators

$$\frac{h}{2\pi i}\frac{\partial}{\partial x}, \qquad \frac{h}{2\pi i}\frac{\partial}{\partial y}, \qquad \text{and} \qquad \frac{h}{2\pi i}\frac{\partial}{\partial z},$$

respectively, where h is the Planck constant and $i = \sqrt{-1}$. Thus we obtain the quantum-mechanical Hamiltonian for the single particle in the following form:

$$\mathbf{H} = -(h^2/8\pi^2 m)\nabla^2 + V(x, y, z), \tag{2.3}$$

where ∇^2 is the Laplacian operator:

$$\nabla^2 = \frac{\partial^2}{\partial x^2} + \frac{\partial^2}{\partial y^2} + \frac{\partial^2}{\partial z^2}. \tag{2.4}$$

The first term on the right-hand side of Eq. (2.3) is called the kinetic energy operator, usually denoted by $\mathbf{T}$, and the second term is called the potential energy operator, usually denoted by $\mathbf{V}$. On substituting expression (2.3) into (2.1) we obtain

$$[-(h^2/8\pi^2 m)\nabla^2 + \mathbf{V}]\Psi = E\Psi. \tag{2.5}$$

This equation can be rewritten in the following form:

$$\nabla^2\Psi + (8\pi^2 m/h^2)(E - \mathbf{V})\Psi = 0. \tag{2.6}$$

This is the Schrödinger equation for the single particle. The eigenfunctions Ψ are of course functions only of the coordinates of the particle.

If we forget about the electronic and nuclear spin for the moment, the Hamiltonian operator for a molecule consisting of nuclei ($I, J, \ldots$) and electrons ($i, j, \ldots$) is as follows:

$$\mathbf{H} = \sum_I \mathbf{T}_I + \sum_i \mathbf{T}_i + \sum_{IJ} \mathbf{V}_{IJ} + \sum_{ij} \mathbf{V}_{ij} + \sum_{iI} \mathbf{V}_{iI}, \tag{2.7}$$

where $\mathbf{T}_I = -(h^2/8\pi^2 M_I)\nabla_I^2$, $\mathbf{T}_i = -(h^2/8\pi^2 m)\nabla_i^2$, $\mathbf{V}_{IJ} = Z_I Z_J e^2/r_{IJ}$, $\mathbf{V}_{ij} = e^2/r_{ij}$, and $\mathbf{V}_{iI} = -Z_I e^2/r_{iI}$. M_I is the mass of nucleus I, m is the mass of an electron, Z_I is the atomic number of nucleus I, e is the electronic charge, and r_{IJ}, r_{ij}, and r_{iI} are the distances between nuclei I and J, between electrons i and j, and between electron i and nucleus I, respectively. The first two terms in expression (2.7) are the kinetic energy operators of the nuclei and of the electrons, respectively. The third term is the mutual potential energy of repulsion of the nuclei ($\sum_{IJ}$ means a sum over all pairs of nuclei). The fourth term is the repulsion energy of the electrons ($\sum_{ij}$ means a sum over all pairs of electrons). The last term is the mutual potential energy of attraction between the electrons and the nuclei ($\sum_{iI}$ means a sum over all pairs of an electron and a nucleus). On substituting expression (2.7) into (2.1), we obtain the Schrödinger equation for the molecule. The eigenfunctions, that is, the molecular state functions, are, of course, functions of the space coordinates of all the nuclei and electrons.

The product $\Psi_n^*\Psi_n$ is called the probability density associated with the state n, characterized by wavefunction Ψ_n. Ψ_n^* is the complex conjugate of Ψ_n. If Ψ_n is a real function, Ψ_n^2 is the probability density. For a single particle, $\Psi_n^*\Psi_n\, dx\, dy\, dz$ is the probability that the particle in the state n is in the small volume element $dx\, dy\, dz$ ($\equiv d\tau$) surrounding the point (x, y, z). Since the particle must be somewhere in space, the total probability is unity, so that

$$\int \Psi_n^*\Psi_n\, d\tau \equiv \langle\Psi_n \mid \Psi_n\rangle = 1. \tag{2.8}$$

This expression is called the normalization condition, and wavefunctions satisfying this condition are said to be normalized. For a system consisting of more than one particle, $\Psi_n^*\Psi_n\, d\tau$ is the probability of finding the system in the volume element $d\tau$ of the many-dimensional space which has as orthogonal coordinates the spin and space coordinates of all the particles. The state functions for such a system must also be normalized.

If two wavefunctions satisfy the relation (orthogonality condition)

$$\int \Psi_n^*\Psi_m\, d\tau \equiv \langle\Psi_n \mid \Psi_m\rangle = 0, \tag{2.9}$$

the wavefunctions are said to be mutually orthogonal. The eigenfunctions of an operator associated with different eigenvalues are orthogonal to one another. A set of functions which are normalized and mutually orthogonal is called an orthonormal set. The normalization and orthogonality conditions can be combined into the following one condition:

$$\int \Psi_n^*\Psi_m\, d\tau \equiv \langle\Psi_n \mid \Psi_m\rangle = \delta_{nm}, \tag{2.10}$$

where δ_{nm} is the Kronecker symbol, which is unity when $n = m$, and otherwise zero. If there are two or more different orthonormal eigenfunctions of the Hamiltonian associated with the same eigenvalue, they are said to be degenerate.

The molecular state functions determine the value of any physically observable property of the molecule. For each observable property there is an associated operator **O**. The expected value of that observable property in the state n is given by the integral

$$\int \Psi_n^*\mathbf{O}\Psi_n\, d\tau \equiv \langle\Psi_n| \mathbf{O} |\Psi_n\rangle. \tag{2.11}$$

The explicit form of an operator is usually obtained from the corresponding classical expression in the same way that the Hamiltonian operator was obtained from the classical Hamiltonian.

2.2. Molecular State Functions

2.2.1. Factorization of Molecular State Functions into Nuclear and Electronic Parts

The Schrödinger equation can be solved exactly for systems containing a single electron and a single nucleus, such as the hydrogen atom, but not for many-electron atoms and molecules, since the Hamiltonian for such a system contains terms simultaneously involving the coordinates of two particles as variables, and hence the equation can not be separated into equations each of which is a function of only one variable. Therefore, to make any progress, a series of approximations and assumptions is necessary.

The first approximation is to separate the electronic and nuclear motions. This can be safely done because the motion of the electrons is considered to be so much faster than the motion of the nuclei that the electrons adjust themselves instantaneously to the position of the nuclei. This separation is known as the Born–Oppenheimer approximation. The Hamiltonian operator (2.7) can be rewritten as follows:

$$\mathbf{H} = \mathbf{H}_e(q, Q) + E_R(Q) + \mathbf{H}_N(Q), \tag{2.12}$$

where

$$\mathbf{H}_e(q, Q) = \sum_i \mathbf{T}_i + \sum_{ij} \mathbf{V}_{ij} + \sum_{iI} \mathbf{V}_{iI}, \tag{2.13}$$

$$E_R(Q) = \sum_{IJ} \mathbf{V}_{IJ}, \tag{2.14}$$

and

$$\mathbf{H}_N(Q) = \sum_I \mathbf{T}_I. \tag{2.15}$$

$\mathbf{H}_e(q, Q)$, the electronic Hamiltonian, is the Hamiltonian for a fixed arrangement of the nuclei; it is a function of both the electronic coordinates (q) and the nuclear coordinates (Q). $E_R(Q)$ is just the nuclear repulsion energy, and $\mathbf{H}_N(Q)$ is just the kinetic energy operator of the nuclei; they are functions only of the nuclear coordinates. In the Born–Oppenheimer approximation, a complete molecular state function is written as a product of an electronic and a nuclear part as follows:

$$\Psi_{Ne}(q, Q) = \Psi_e(q, Q)\mathrm{X}_{Ne}(Q), \tag{2.16}$$

where $\Psi_e(q, Q)$ and $\mathrm{X}_{Ne}(Q)$ are solutions of the following wave equations:

$$\mathbf{H}_e(q, Q)\Psi_e(q, Q) = E_e(Q)\Psi_e(q, Q), \tag{2.17}$$

$$[\mathbf{H}_N(Q) + (E_e(Q) + E_R(Q))]\mathrm{X}_{Ne}(Q) = E_{Ne}\mathrm{X}_{Ne}(Q). \tag{2.18}$$

That is, $\Psi_e(q, Q)$ is the electronic wavefunction characterizing an electronic state of the molecule, and its eigenvalue $E_e(Q)$ is the electronic energy of the state. The electronic wavefunction is a function of both the electronic and nuclear coordinates. For a specified nuclear configuration the electronic wavefunction contains only the electronic coordinates as variables. The electronic energy of an electronic state, as well as the nuclear repulsion energy, is a function only of the nuclear configuration, and hence the sum of these energies, that is, the energy $(E_e(Q) + E_R(Q))$, is also a function only of the nuclear configuration. A curve relating the energy $(E_e(Q) + E_R(Q))$ to the nuclear coordinates Q is the potential energy curve for the molecule in the electronic state. The energy $(E_e(Q) + E_R(Q))$ is used as the potential energy acting on the nuclei in the nuclear wave equation (2.18). Therefore, the nuclear wavefunction $X_{Ne}(Q)$ depends on the electronic state of the molecule. Each electronic state thus has its own set of nuclear wavefunctions. Each nuclear wavefunction can be further factorized into a product of a rotational and a vibrational wavefunction.

2.2.2. One-Electron Wavefunctions (Orbitals)

The electronic Hamiltonian (2.13) can be rewritten as follows:

$$\mathbf{H}_e = \sum_i \mathbf{h}_i^N + \sum_{ij} \mathbf{V}_{ij}, \tag{2.19}$$

where

$$\mathbf{h}_i^N = \mathbf{T}_i + \sum_I \mathbf{V}_{iI}. \tag{2.20}$$

Once the nuclear configuration has been specified, the electronic Hamiltonian and its eigenfunctions, that is, the electronic wavefunctions, contain only the electronic coordinates as variables. The first term in expression (2.19) is the sum of terms each containing the coordinates of only one electron as variables, but the second term, the electronic repulsion energy term, is the sum of terms simultaneously containing the coordinates of two electrons as variables. Because of this two-electron term the electronic wave equation (2.17) can not be exactly solved. Therefore, we must make some assumptions.

In molecular orbital (MO) theory it is assumed that the electronic Hamiltonian can be expressed as a sum of one-electron operators, as follows:

$$\mathbf{H}_e = \sum_i \mathbf{h}_i, \tag{2.21}$$

where $\mathbf{h}_i$ is a one-electron operator, that is, an operator containing only the space coordinates of a single electron as variables. Then an eigenfunction of the electronic Hamiltonian can be written as a product of one-electron functions:

$$\Psi_{\mathrm{e}} = \psi_a(1)\psi_b(2) \cdots \psi_k(n), \tag{2.22}$$

where $\psi_a(1)$, for example, is a function only of the space coordinates of electron 1. Thus, the electronic wave equation (2.17) can be separated into n equations each containing the coordinates of only one electron as variables:

$$\mathbf{h}_i\psi(i) = \varepsilon\psi(i). \tag{2.23}$$

If the explicit form of the one-electron operator $\mathbf{h}$ for a specified nuclear configuration is known, these one-electron equations will be solved to give the one-electron functions ψ_a, ψ_b, . . . , ψ_k. Such one-electron functions are called orbitals. The electronic energy of the system is given as the sum of the orbital energies ε:

$$E_{\mathrm{e}} = \sum_k n_k\varepsilon_k, \tag{2.24}$$

where n_k is the number of electrons in the orbital ψ_k.

The one-electron operator $\mathbf{h}_i$ can be written as

$$\mathbf{h}_i = \mathbf{h}_i^{\mathrm{N}} + \mathbf{V}_i, \tag{2.25}$$

where $\mathbf{V}_i$ is the effective potential acting on an electron i due to all the other electrons. The $\mathbf{h}$ is called the effective one-electron Hamiltonian. The explicit form of $\mathbf{h}_i^{\mathrm{N}}$ is determined when the nuclear configuration has been specified. However, to express $\mathbf{V}_i$ in the explicit form, it is necessary to know the distribution of all the electrons other than i, that is, the form of the orbitals occupied by the electrons other than i. In calculation of atomic orbitals there is a well-established technique for obtaining the potentials $\mathbf{V}_i$ and the atomic orbitals ψ_i. This technique consists of the following iterative procedures: a trial set of orbitals is used to calculate $\mathbf{V}_i$, and from this a new set of orbitals is obtained. These new orbitals are then used to calculate a better potential, and the whole procedure is repeated until self-consistency is reached, that is, until a new set of orbitals does not differ significantly from the precursor. This technique, which was developed by Hartree and Fock, is known as the self-consistent-field (SCF) method. A similar method, which is applied to the calculation of molecular orbitals, the so-called SCF MO method, has been developed largely by Roothaan. However, in the calculation of molecular orbitals, the effective one-electron Hamiltonian can be used, to a first approximation, without being given any explicit expression. The method using such effective one-electron Hamiltonians whose explicit form is not specified is called the simple MO method.

2.2.3. Electron Spin Functions

It is well known that for a single electron there are two spin states. If we choose any arbitrary direction the component of the electron spin angular momentum in this direction can have only two values $m_s(h/2\pi)$, where m_s is the spin quantum number and can only assume the values $\pm\frac{1}{2}$. The normalized wavefunction of the spin state with $m_s = +\frac{1}{2}$ is usually denoted by α, and that of the spin state with $m_s = -\frac{1}{2}$ by β. If the operators for the spin angular momentum and its component in the chosen direction (z direction) are denoted by $\mathbf{S}$ and $\mathbf{S}_z$, respectively, the following relations must hold:

$$\mathbf{S}^2\alpha = \tfrac{1}{2}(\tfrac{1}{2}+1)(h/2\pi)^2\alpha; \qquad \mathbf{S}_z\alpha = +\tfrac{1}{2}(h/2\pi)\alpha;$$
$$\mathbf{S}^2\beta = \tfrac{1}{2}(\tfrac{1}{2}+1)(h/2\pi)^2\beta; \qquad \mathbf{S}_z\beta = -\tfrac{1}{2}(h/2\pi)\beta. \tag{2.26}$$

Thus, while the two electron spin functions α and β are eigenfunctions of operator $\mathbf{S}^2$ associated with the same eigenvalue, they are eigenfunctions of operator $\mathbf{S}_z$ associated with different eigenvalues. Therefore, they are mutually orthogonal:

$$\int \alpha^*\beta \, d\sigma \equiv \langle \alpha \mid \beta \rangle = 0, \tag{2.27}$$

where σ represents a coordinate called the electron spin coordinate. Two electrons having the same spin function α or β are said to have parallel spins; two electrons having different spin functions α and β are said to have opposed, or antiparallel, spins.

If there is more than one electron in the atom or molecule, we can define the operators for the total electron spin angular momentum and its component in the z direction as follows:

$$\mathbf{S}^2 = \mathbf{S}_1{}^2 + \mathbf{S}_2{}^2 + \cdots; \tag{2.28}$$

$$\mathbf{S}_z = \mathbf{S}_{z1} + \mathbf{S}_{z2} + \cdots, \tag{2.29}$$

where the subscripts 1, 2, etc., identify the different electrons. These operators have eigenfunctions F which satisfy the following equations:

$$\mathbf{S}^2F = s(s+1)(h/2\pi)^2F; \tag{2.30}$$

$$\mathbf{S}_zF = m_s(h/2\pi)F, \tag{2.31}$$

where s is zero, integral, or half integral, and for each value of s there are $2s+1$ values of m_s given by

$$m_s = s, s-1, \ldots, -(s-1), -s. \tag{2.32}$$

$2s + 1$ is called the electron spin multiplicity. States with $s = 0, \frac{1}{2}$, and 1 have the spin multiplicities of 1, 2, and 3, respectively, and these states are called singlet, doublet, and triplet states, respectively. The value of m_s is equal to the sum of the values of m_s of the individual electrons.

For a single electron we have a doublet spin state with wavefunctions α and β. For two electrons (labeled 1 and 2), we can construct four spin functions, $\alpha(1)\alpha(2)$, $\alpha(1)\beta(2)$, $\beta(1)\alpha(2)$, and $\beta(1)\beta(2)$. However, $\alpha(1)\beta(2)$ and $\beta(1)\alpha(2)$ are not eigenfunctions of $\mathbf{S}^2$, that is, do not satisfy Eq. (2.30). $\alpha(1)\beta(2)$ means that electron 1 has spin α and electron 2 spin β, and $\beta(1)\alpha(2)$ means that electron 1 has spin β and electron 2 spin α. Since electrons are essentially indistinguishable, that is, we cannot tell which electron is which, these functions cannot be true spin state functions, but only the symmetry combinations of these can be true wavefunctions. Thus, we have the following correct normalized spin state functions, which are all eigenfunctions of both the operators $\mathbf{S}^2$ and $\mathbf{S}_z$:

	s	m_s		
$F_{1,1} = \alpha(1)\alpha(2)$	1	1		
$F_{1,0} = \sqrt{\tfrac{1}{2}}\,\{\alpha(1)\beta(2) + \beta(1)\alpha(2)\}$	1	0	a triplet (T)	(2.33)
$F_{1,-1} = \beta(1)\beta(2)$	1	-1		
$F_{0,0} = \sqrt{\tfrac{1}{2}}\,\{\alpha(1)\beta(2) - \beta(1)\alpha(2)\}$	0	0	a singlet (S)	(2.34)

These spin state functions are of course mutually orthogonal. If we interchange the labels of the two electrons, we find that the singlet spin function (S) is multiplied by -1 and the triplet spin functions (T) remain unchanged. Thus, we say that the singlet spin function is antisymmetrical with respect to interchange of the electrons and the triplet spin functions are symmetrical.

The orbital motion of electrons in a central field of an atom has a quantized angular momentum of a magnitude of $M = [l(l+1)]^{1/2}(h/2\pi)$, where l is either zero or an integral. The component of this momentum in a chosen direction (z direction) is also quantized and can have values of $M_z = m_l(h/2\pi)$, where

$$m_l = l, l-1, \ldots, 0, \ldots, -(l-1), -l. \tag{2.35}$$

That is, if the operators for the angular momentum and its component in the z direction are denoted by $\mathbf{L}$ and $\mathbf{L}_z$, respectively, the correct electronic space functions Ψ_e are to be eigenfunctions of $\mathbf{L}^2$ and $\mathbf{L}_z$, satisfying the following equations:

$$\mathbf{L}^2\Psi_e = l(l+1)(h/2\pi)^2\Psi_e\,; \tag{2.36}$$

$$\mathbf{L}_z\Psi_e = m_l(h/2\pi)\Psi_e\,. \tag{2.37}$$

The relation (2.37) also holds good for linear molecules, which possess an axial symmetry along the molecular axis (z axis).

Since the electron is an electrically charged body, the orbital angular momentum is accompanied with a magnetic moment in the z direction, the magnitude of which is given by $\mu = (e/2mc)M_z = (e/2mc)m_l(h/2\pi)$. The spin angular momentum of electrons is also accompanied with a magnetic moment in the z direction, the magnitude of which is given by $\mu_s = (e/mc) \times m_s(h/2\pi)$. The interaction of this magnetic moment with the magnetic field produced by the orbital motion of the electrons is responsible for the fine structure in the spectra of alkali atoms. This interaction is called the spin–orbit coupling. However, a study of atomic spectra shows that there is very little spin–orbit coupling except for heavy atoms. If the spin–orbit coupling can be neglected, we can incorporate the spins of the electrons by multiplying the electronic space function Ψ, which is a function of only the space coordinates of the electrons, by an appropriate spin function F, which is of course a function of only the spin coordinates of the electrons. Usually, an electronic state is expressed by writing the spin multiplicity as a presuperfix to the symbol for the electronic space function: for a singlet state, ${}^1\Psi$; for a doublet state, ${}^2\Psi$; for a triplet state, ${}^3\Psi$; etc.

For molecules consisting of only light atoms, the spin–orbit coupling is negligible, and the spin numbers s and m_s are good quantum numbers. For such molecules the electronic state function of a given electronic state can be written as the product of an electronic space function and an electronic spin function. However, the spin–orbit coupling becomes much more important if heavy atoms are present. In such a case the spin quantization breaks down, and the separation of the total electronic function into a space part and a spin part is no longer a good approximation.

2.2.4. Electron Configurations and Slater Determinants

As mentioned above, if the spin–orbit coupling is neglected, an electronic wavefunction can be expressed as the product of an electronic space function and an electronic spin function. When we have found a set of one-electron space functions (i.e., orbitals) ψ_k for the atom or molecule under consideration, with each of these orbitals we can associate a one-electron spin function, α or β, and obtain the product functions $\psi_k\alpha$, $\psi_k\beta$, etc. These product functions are called spin-orbitals. For brevity the spin-orbitals $\psi_k\alpha$ and $\psi_k\beta$ will be written as ψ_k^α and ψ_k^β, respectively. Sometimes, if no confusion with simple orbitals can arise, they are also written as ψ_k and $\bar{\psi}_k$, respectively.

According to the Pauli exclusion principle, each spin-orbital can be occupied by only a single electron. If there are n electrons in our system, each is allotted to a spin-orbital, and a product function is built up as follows:

$$\psi_a^{\alpha}(1)\psi_a^{\beta}(2)\psi_b^{\alpha}(3)\cdots\psi_k^{\alpha}(i)\cdots\psi_m^{\beta}(n) \tag{2.38}$$

However, precisely the same point arises as arose when discussing the spin functions. That is, electrons are all identical: they carry essentially no labels with which we can identify them. Therefore, any one of the $n!$ functions which are obtained by permuting the n electrons among the n available spin-orbitals is as much acceptable as the function (2.38), and the most general solution of the electronic wave equation must be some combination of all these $n!$ functions. The correct combination is determined by the fact that an acceptable wavefunction for a system containing two or more electrons must be antisymmetrical with respect to interchange of any two electrons, that is, must change sign on interchanging the labels of any two electrons. In general, it is shown by experiment that for all particles with an odd spin (the spin quantum number m_s is half integral), such as electrons, protons, and neutrons, only antisymmetrical wavefunctions occur; for all particles with an even spin (m_s is zero or integral), such as α particles and photons, only symmetrical wavefunctions are found.

For two electrons having the same orbital function, say ψ_a, the combination

$$\sqrt{\tfrac{1}{2}}\{\psi_a^{\alpha}(1)\psi_a^{\beta}(2) - \psi_a^{\alpha}(2)\psi_a^{\beta}(1)\} \tag{2.39}$$

is the only acceptable function. This function may be written as a determinant

$$\sqrt{\tfrac{1}{2}}\begin{vmatrix} \psi_a^{\alpha}(1) & \psi_a^{\beta}(1) \\ \psi_a^{\alpha}(2) & \psi_a^{\beta}(2) \end{vmatrix}. \tag{2.40}$$

In general, a determinant of the form

$$(n!)^{-1/2}\begin{vmatrix} \psi_a^{\alpha}(1) & \psi_a^{\beta}(1) & \cdots & \psi_m^{\beta}(1) \\ \psi_a^{\alpha}(2) & \psi_a^{\beta}(2) & \cdots & \psi_m^{\beta}(2) \\ \cdots & \cdots & \cdots & \cdots \\ \psi_a^{\alpha}(n) & \psi_a^{\beta}(n) & \cdots & \psi_m^{\beta}(n) \end{vmatrix} \tag{2.41}$$

satisfies the condition that an acceptable wavefunction must be antisymmetrical with respect to interchange of any two electrons, changing sign on interchanging two of its columns or rows. Such a normalized determinant is called a Slater determinant, and is usually written simply as

$$|\psi_a^{\alpha}\psi_a^{\beta}\cdots\psi_m^{\beta}| \tag{2.42}$$

or

$$(n!)^{-1/2}[\psi_a^{\alpha}\psi_a^{\beta}\cdots\psi_m^{\beta}]. \tag{2.43}$$

Here the symbol | | is used to represent a normalized determinant and the symbol ⌈ ⌋ an unnormalized one. If two electrons occupy the same spin-orbital, the Slater determinant will vanish since two of its columns are identical. This means that such a situation cannot exist, in accordance with the Pauli exclusion principle.

The Slater determinant (2.41) is equivalent to the following expression:

$$(n!)^{-1/2} \sum_{\mathrm{P}} (-1)^{\mathrm{P}} \mathbf{P}\{\psi_a^{\alpha}(1)\psi_a^{\beta}(2) \cdots \psi_m^{\beta}(n)\}, \tag{2.44}$$

where **P** is the operator that interchanges electron labels, i.e., the permutation operator, and $(-1)^{\mathrm{P}}$ means that the terms arising from interchanges of odd times are signed by minus, and that the terms arising from interchanges of even times are signed by plus.

As mentioned already, if two or more mutually orthogonal normalized eigenfunctions of the Hamiltonian have the same eigenvalue they are said to be degenerate. If, for example, orbitals ψ_a and ψ_b are degenerate eigenfunctions of the same one-electron Hamiltonian **h**, having the same orbital energy ε, any linear combination of these orbitals is an equally good eigenfunction of the Hamiltonian, as illustrated below:

$$\mathbf{h}(c_a\psi_a + c_b\psi_b) = c_a\varepsilon\psi_a + c_b\varepsilon\psi_b = \varepsilon(c_a\psi_a + c_b\psi_b), \tag{2.45}$$

where the c's are constants. Thus, when orbitals occur in a degenerate set, no unique forms can be assigned to them: any set of normalized linear combinations of one set of degenerate orbitals is also an equally satisfactory set, provided that these linear combinations are mutually orthogonal.

The Pauli exclusion principle amounts to saying that each orbital can contain maximally two electrons which have different spin functions. Any description of an electronic state of an atom or molecule requires that we know which orbitals are occupied, and by how many electrons. Such information provides us with what is usually called an electron configuration. That is, an electronic wavefunction approximated as a product of orbitals, or in other words, a description of an electronic state of an atom or molecule in terms of a product of orbitals, is an electron configuration. The lowest energy electron configuration (ground electron configuration) is constructed by following the so-called Aufbau or building-up principle: the electrons are fed into the orbitals so that each of the orbitals of lower energy is filled with two electrons having antiparallel spins before any electron is fed into an orbital of higher energy. In the case of degenerate orbitals Hund's rule applies: electrons with the same spin are fed one into one of the degenerate orbitals before the second electron is fed into any one of the degenerate orbitals.

A set of mutually orthogonal degenerate orbitals is said to constitute a shell. A nondegenerate orbital constitutes a shell by itself. An r-fold degenerate shell, that is, a shell consisting of r degenerate orbitals, can be occupied by $2r$ electrons. An electron configuration having only completely occupied shells is called a closed-shell configuration, and an electron configuration having one or more incompletely occupied shells is called an open-shell configuration.

A closed-shell configuration is of course singlet, all the electrons being paired with opposed spins, and can be described by a single Slater determinant. For ordinary molecules, having an even number of electrons, the ground electron configuration is a closed-shell configuration. The ground electron configuration in which the m lowest-energy orbitals from ψ_a through ψ_m are occupied by $2m$ electrons is described by the Slater determinant

$$^1\Psi_0 = V_0 = |\psi_a{}^\alpha\psi_a{}^\beta \cdots \psi_k{}^\alpha\psi_k{}^\beta \cdots \psi_m{}^\alpha\psi_m{}^\beta| \,. \tag{2.46}$$

Such an antisymmetrized wavefunction of the ground electron configuration will be denoted by a symbol $^1\Psi_0$ or V_0, $^1\Psi$ and V being used as symbols for singlet configurations.

For an open-shell configuration more than one space or spin combination may be possible. For example, an electron configuration having two electrons in different nondegenerate orbitals can be either singlet or triplet. The singlet and triplet configurations arising from the ground configuration by promotion of one electron from an occupied orbital ψ_k to an unoccupied orbital ψ_r are denoted by $^1\Psi_1^{kr}$ or V_1^{kr} and by $^3\Psi_1^{kr}$ or T_1^{kr}, respectively, where V and T are symbols for singlet and triplet configurations, respectively, and the subscript 1 is used to indicate a singly excited or mono-excited configuration. The antisymmetrized wavefunctions of these singly excited configurations are

$$V_1^{kr} = \sqrt{\tfrac{1}{2}}(|\psi_a{}^\alpha\psi_a{}^\beta \cdots \psi_k{}^\alpha \cdots \psi_m{}^\alpha\psi_m{}^\beta\psi_r{}^\beta| - |\psi_a{}^\alpha\psi_a{}^\beta \cdots \psi_k{}^\beta \cdots \psi_m{}^\alpha\psi_m{}^\beta\psi_r{}^\alpha|), \tag{2.47}$$

$$T_1^{kr} = \begin{cases} |\psi_a{}^\alpha\psi_a{}^\beta \cdots \psi_k{}^\alpha \cdots \psi_m{}^\alpha\psi_m{}^\beta\psi_r{}^\alpha| \\ \sqrt{\tfrac{1}{2}}(|\psi_a{}^\alpha\psi_a{}^\beta \cdots \psi_k{}^\alpha \cdots \psi_m{}^\alpha\psi_m{}^\beta\psi_r{}^\beta| + |\psi_a{}^\alpha\psi_a{}^\beta \cdots \psi_k{}^\beta \cdots \psi_m{}^\alpha\psi_m{}^\beta\psi_r{}^\alpha|) \\ |\psi_a{}^\alpha\psi_a{}^\beta \cdots \psi_k{}^\beta \cdots \psi_m{}^\alpha\psi_m{}^\beta\psi_r{}^\beta| \,. \end{cases} \tag{2.48}$$

The values of the total spin quantum number m_s for the three functions of the triplet configuration are evidently 1, 0, and -1, respectively. The energies associated with these three functions are to a first approximation identical. When the spin–orbit coupling is included, the three energies are slightly separated.

If the Slater determinant for the electron configuration having $2m - 2$ electrons in $m - 1$ orbitals from ψ_a through ψ_m except for ψ_k is denoted by

$V_{(k)}$, then, on the basis of the assumption that these electrons can be separated from the electrons occupying ψ_k and ψ_r, the functions given in (2.46)–(2.48) can be rewritten as follows:

$$V_0 = V_{(k)} \cdot \psi_k(1)\psi_k(2)S, \tag{2.46'}$$

$$V_1^{kr} = V_{(k)} \cdot \sqrt{\tfrac{1}{2}}\{\psi_k(1)\psi_r(2) + \psi_k(2)\psi_r(1)\}S, \tag{2.47'}$$

$$T_1^{kr} = V_{(k)} \cdot \sqrt{\tfrac{1}{2}}\{\psi_k(1)\psi_r(2) - \psi_k(2)\psi_r(1)\}T. \tag{2.48'}$$

In these expressions, S represents the singlet spin-state function given in (2.34), and T represents the triplet spin-state function given in (2.33). Evidently, in functions for singlet configurations the space part is symmetrical and the spin part antisymmetrical, and on the other hand, in functions for triplet configurations, the space part is antisymmetrical and the spin part symmetrical.

2.3. The Method of Linear Combinations and the Variation Principle

2.3.1. The Variation Principle

To determine approximately electronic wavefunctions and their energies the variation method has been found to be the most effective technique. Now we return to the Schrödinger equation

$$\mathbf{H}\Psi_m{}^0 = E_m{}^0\Psi_m{}^0, \tag{2.49}$$

where $\mathbf{H}$ is an electronic Hamiltonian operator, the exact form of which will be of no interest in this discussion, and $\Psi_m{}^0$ and $E_m{}^0$ are the exact eigenfunctions and eigenvalues, respectively. If we multiply both sides of this equation by $\Psi_m{}^{0*}$ and integrate over all the coordinates involved, we obtain the result

$$\int \Psi_m^{0*}\mathbf{H}\Psi_m{}^0\, d\tau = E_m{}^0 \int \Psi_m^{0*}\Psi_m{}^0\, d\tau, \tag{2.50}$$

or

$$E_m{}^0 = \int \Psi_m^{0*}\mathbf{H}\Psi_m{}^0\, d\tau \Big/ \int \Psi_m^{0*}\Psi_m{}^0\, d\tau, \tag{2.51}$$

where $\int \cdots d\tau$ implies, as usual, an integration over all the coordinates involved in the integrand. When $\Psi_m{}^0$ are normalized, expression (2.51) can be reduced to the following simple form:

$$E_m{}^0 = \int \Psi_m^{0*}\mathbf{H}\Psi_m{}^0\, d\tau \equiv \langle \Psi_m{}^0 | \mathbf{H} | \Psi_m{}^0 \rangle. \tag{2.52}$$

For an approximate solution Ψ_m, the corresponding wave equation,

$$\mathbf{H}\Psi_m = E_m \Psi_m, \tag{2.53}$$

does not hold accurately at each point in space: the expression $\mathbf{H}\Psi_m/\Psi_m$ will be a function of the coordinates of the system, and, as such, it will vary from place to place and cannot be equated to a constant E_m. Nevertheless, we can consider the quantity E_m, with dimensions of energy, which is related to Ψ_m by the equation

$$E_m = \int \Psi_m{}^* \mathbf{H} \Psi_m \, d\tau \Big/ \int \Psi_m{}^* \Psi_m \, d\tau. \tag{2.54}$$

E_m is called the average energy associated with the approximate function Ψ_m. If we choose Ψ_m so that it is normalized, E_m can be expressed as

$$E_m = \int \Psi_m{}^* \mathbf{H} \Psi_m \, d\tau \equiv \langle \Psi_m | \, \mathbf{H} \, | \Psi_m \rangle. \tag{2.55}$$

Since Ψ_m is an approximate wavefunction for the system, E_m is only an approximate energy of the system. The variation principle states that the approximate energy E calculated by Eq. (2.54) or (2.55) for any approximate wavefunction Ψ is always higher than the energy $E_0{}^0$ associated with the lowest-energy correct wavefunction $\Psi_0{}^0$. Therefore, minimizing the energy of an approximate function is a method for obtaining the best approximation to the correct wavefunction of the system. This method is called the variation method.

The variation principle can be easily proved. A fundamental theorem of mathematics states that it is always possible to expand any arbitrary function f as a series of a set of functions g of the same variables as long as the set of functions g is what is known as a complete set of orthogonal functions. The set of all the functions Ψ^0 which are exact solutions of the wave equation (2.49) is a complete set of orthogonal functions. Therefore, it is possible to express an arbitrary approximate function Ψ_m as a linear combination of the correct eigenfunctions Ψ^0:

$$\Psi_m = \sum_k c_{mk} \Psi_k{}^0, \tag{2.56}$$

with the normalization condition,

$$\sum_k c_{mk}^* c_{mk} = 1. \tag{2.57}$$

The average energy associated with Ψ_m is given by

$$E_m = \int \Psi_m{}^* \, \mathbf{H} \, \Psi_m \, d\tau = \sum_k \sum_l c_{mk}^* c_{ml} \int \Psi_k^{0*} \, \mathbf{H} \, \Psi_l{}^0 \, d\tau = \sum_k c_{mk}^* c_{mk} E_k{}^0. \tag{2.58}$$

Subtracting the lowest energy $E_0{}^0$ and remembering the normalization condition (2.57), we can write

$$E_m - E_0{}^0 = \sum_k c_{mk}^* c_{mk} (E_k{}^0 - E_0{}^0). \tag{2.59}$$

Since $c_{mk}^* c_{mk} \geq 0$ and $E_k^0 \geq E_0^0$, the quantity $E_m - E_0^0$ must be either positive or zero. This means that the approximate energy E_m must be greater than the exact lowest energy E_0^0, except when Ψ_m happens to be identical with Ψ_0^0; in this case, of course, E_m is equal to E_0^0.

2.3.2. The Method of Linear Combinations

A standard technique for obtaining approximate wavefunctions is to write trial functions Ψ as linear combinations of a known set of normalized functions ϕ of a finite number, say n, as

$$\Psi = \sum_{i=1}^{n} c_i \phi_i, \tag{2.60}$$

and then to determine the best values for the coefficients c_i by applying the variation principle. This particular form of the variation method is sometimes called the Rayleigh–Ritz, or Ritz, method.

In the case of molecules it is most convenient to express the one-electron wavefunctions, that is, the molecular orbitals (MO's), ψ, as linear combinations of the atomic orbitals (AO's) χ of the constituent atoms:

$$\psi = \sum_{i} c_i \chi_i, \tag{2.61}$$

where χ_i are the atomic orbitals and c_i are coefficients to be determined by use of the variation principle. This approximation is called the LCAO (linear-combinations-of-atomic-orbitals) approximation to the molecular orbitals, and the method of determining the molecular orbitals as linear combinations of atomic orbitals by use of the variation principle is called the LCAO MO method. The LCAO MO method is thus a special form of the Rayleigh–Ritz method. This method will be detailed in later chapters. In this place the principle of the method of linear combinations (i.e., the Rayleigh–Ritz method) is described in its general form.

Substituting expression (2.60) into (2.54), we obtain the following expression for the average energy associated with the trial function:

$$E = \sum_{i}\sum_{j} c_i^* c_j H_{ij} \Big/ \sum_{i}\sum_{j} c_i^* c_j S_{ij}, \tag{2.62}$$

where

$$H_{ij} = \int \phi_i^* \mathbf{H} \phi_j \, d\tau \equiv \langle \phi_i | \, \mathbf{H} \, | \phi_j \rangle, \tag{2.63}$$

and

$$S_{ij} = \int \phi_i^* \phi_j \, d\tau \equiv \langle \phi_i \mid \phi_j \rangle. \tag{2.64}$$

H_{ij} is called the matrix element of the Hamiltonian $\mathbf{H}$, or the energy matrix element, between the functions ϕ_i and ϕ_j, since we can construct a square table (or matrix) of these integrals in the following way:

	1	2	3	$\cdots$
1	H_{11}	H_{12}	H_{13}	$\cdots$
2	H_{21}	H_{22}	H_{23}	$\cdots$
3	H_{31}	H_{32}	H_{33}	$\cdots$
$\cdot$	$\cdot$	$\cdot$	$\cdot$	
$\cdot$	$\cdot$	$\cdot$	$\cdot$	$\cdots$
$\cdot$	$\cdot$	$\cdot$	$\cdot$	

Since the Hamiltonian is a Hermitian operator, if the functions ϕ are complex, then, by definition, $H_{ij} = H_{ji}^*$. Since $H_{ii} = H_{ii}^*$, the diagonal elements of the matrix (energy matrix) are always real. If the functions ϕ are real, all the matrix elements are real, and $H_{ij} = H_{ji}$. In the same way we could call S_{ij} the matrix element of unity, but it is more usual to refer to it as the overlap integral between the functions ϕ_i and ϕ_j. If the functions ϕ are complex, $S_{ij} = S_{ji}^*$; if the functions are real, $S_{ij} = S_{ji}$. If the functions ϕ are separately normalized, $S_{ii} = 1$, and $|S_{ij}|$ $(i \neq j) < 1$. In the ensuing discussions it is assumed, for simplicity, that the functions ϕ are real. Furthermore, it is assumed that the coefficients c's are real. Then, of course, $c_i{}^* = c_i$.

The set of coefficients c_i that yields the lowest value for E can be found by successively partially differentiating Eq. (2.62) with respect to each c_i and setting the partial differential coefficients $\partial E/\partial c_i$ equal to zero. Equation (2.62) is rearranged to give

$$E \sum_i \sum_j c_i c_j S_{ij} = \sum_i \sum_j c_i c_j H_{ij}. \tag{2.65}$$

When this equation is differentiated with respect to a particular coefficient, say c_k, and $\partial E/\partial c_k$ is set equal to zero, the following equation is obtained:

$$E\left(\sum_i c_i S_{ik} + \sum_j c_j S_{kj}\right) = \sum_i c_i H_{ik} + \sum_j c_j H_{kj}, \tag{2.66}$$

or

$$\sum_i c_i \{(H_{ik} + H_{ki}) - E(S_{ik} + S_{ki})\} = 0. \tag{2.67}$$

Since $H_{ik} = H_{ki}$ and $S_{ik} = S_{ki}$, Eq. (2.67) reduces to

$$\sum_i c_i(H_{ki} - ES_{ki}) = 0. \tag{2.68}$$

We have an equation of this type for each value of k: we have as many equations of this type as there are unknown coefficients c's, that is, as there are the expansion functions ϕ's. These equations are called the secular equations.*

The set of the secular equations has nontrivial solutions (that is, solutions other than all the c's equal to zero) only for certain values of the energy E. These values of E are given by the condition that the determinant of the quantities that multiply the c's vanishes:

$$D \equiv \begin{vmatrix} H_{11} - ES_{11} & H_{12} - ES_{12} & \cdots & H_{1n} - ES_{1n} \\ H_{21} - ES_{21} & H_{22} - ES_{22} & \cdots & H_{2n} - ES_{2n} \\ \cdots & \cdots & \cdots & \cdots \\ H_{n1} - ES_{n1} & H_{n2} - ES_{n2} & \cdots & H_{nn} - ES_{nn} \end{vmatrix} = 0. \quad (2.69)$$

This determinant, D, is known as the secular determinant. Equation (2.69), the secular determinantal equation, is sometimes written in the short notation as

$$D = |H_{ki} - ES_{ki}| = 0. \quad (2.70)$$

If there are n functions in the expansion set, the secular determinant has n rows and n columns, that is, it is an nth order or n-dimensional determinant, and it can be expanded into an nth order polynomial in E in the manner described in the footnote.† This means that if appropriate values can be

* These secular equations can also be reached through the following shortcut. It is assumed that the following equation could be written:

$$\mathbf{H} \sum_i c_i \phi_i = E \sum_i c_i \phi_i .$$

Multiplying this equation on the left by ϕ_k^* and integrating, we obtain

$$\sum_i c_i H_{ki} = E \sum_i c_i S_{ki} .$$

This equation can be rearranged to give Eq. (2.68).

† When the element of ith row and jth column of an nth order determinant D is denoted by a_{ij}, the $(n - 1)$th order determinant with the ith row and jth column removed from the original determinant D is called the minor or subdeterminant of a_{ij} in D, and is denoted by $\bar{D}(_j^i)$. The minor signed by plus when $i + j$ is even and by minus when $i + j$ is odd is called the cofactor of a_{ij}, and is denoted by A_{ij} : $A_{ij} = (-1)^{i+j} \bar{D}(_j^i)$. Then, the nth order determinant D can be expressed as the sum of the products of n elements in any row (or column) and the corresponding cofactors:

$$D = \sum_j a_{ij} A_{ij} , \quad (2.71)$$

or

$$D = \sum_i a_{ij} A_{ij} . \quad (2.72)$$

To expand D according to Eq. (2.71) is said to expand D with respect to the ith row, and to expand D according to (2.72) is said to expand D with respect to the jth column. By repeating such a procedure D can be broken down to an nth order polynomial.

substituted for all the H_{ki}'s and S_{ki}'s, Eq. (2.69) gives a simple equation of the nth order in E. By solving this equation we can obtain n values of E. By substituting one of the n solutions E, say E_r, back into the n secular equations (2.68) and solving the resulting n simultaneous homogeneous linear equations in n unknowns, we can determine the ratios of all the n coefficients c_{ri} to one of them, for example, c_{rk}. When the ratio of c_{ri} to c_{rk} is denoted by k_{ri}, the coefficient c_{ri} can be expressed as $N_r k_{ri}$. Here N_r is a constant which is chosen so that the resulting function $\Psi_r = \sum_i c_{ri}\phi_i$ is normalized.* That is, from the normalization condition,

$$\int \Psi_r^2 \, d\tau = \sum_i \sum_j c_{ri} c_{rj} S_{ij} = N_r^2 \sum_i \sum_j k_{ri} k_{rj} S_{ij} = 1, \tag{2.73}$$

the constant N_r is given by

$$N_r = \left(\sum_i \sum_j k_{ri} k_{rj} S_{ij} \right)^{-1/2}. \tag{2.74}$$

The N_r is the so-called normalization factor. When $S_{ij} = \delta_{ij}$, where δ_{ij} is the Kronecker symbol, Eq. (2.74) reduces to the following simple form:

$$N_r = \left(\sum_i k_{ri}^2 \right)^{-1/2}. \tag{2.75}$$

This completes the determination of the wavefunction Ψ_r, which corresponds to the energy E_r. By successively repeating the same procedure for other solutions E, all the n wavefunctions corresponding to the n energy values can be determined.

The procedure which has just been outlined is one which minimizes the energy of the lowest energy wavefunction to obtain the best approximation to the lowest energy correct wavefunction. This means that the wavefunction corresponding to the lowest of the n values of E is the best approximate lowest energy wavefunction which can be obtained from the given expansion

* The ratios between the coefficients c_{ri} can be calculated systematically and straightforwardly by use of the method of cofactors. The coefficients $c_{r1}, c_{r2}, \ldots, c_{rn}$ are proportional to the cofactors of the corresponding elements in any row (or column) of the secular determinant D (2.69) with E_r substituted for E. Accordingly, the ratios k_{ri} can be expressed, for example, as follows:

$$k_{r1} = A_{i1}/A_{ik}, \; k_{r2} = A_{i2}/A_{ik}, \ldots, k_{rn} = A_{in}/A_{ik},$$

where $A_{i1}, A_{i2}, \ldots, A_{in}$ are the cofactors along the ith row of the secular determinant with E_r substituted for E.

set. Moreover, this procedure is one which gives the higher energy approximate wavefunctions which are orthogonal to all the approximate wavefunctions of lower energy. Thus, by this procedure, we can obtain the good approximate wavefunctions not only for the lowest energy level but also for the higher energy levels

A measure of the extent to which a particular function ϕ_r of an expansion set deviates from an approximate wavefunction Ψ which is formed from the expansion set is provided by the magnitude of the off-diagonal matrix elements H_{ri}. The smaller H_{ri}'s are, the better is the function ϕ_r. This is evident from the fact that the matrix of $\mathbf{H}$ for the correct eigenfunctions of $\mathbf{H}$ has all its off-diagonal elements zero, since

$$\int \Psi_s^* \mathbf{H} \Psi_r \, d\tau = \int \Psi_s^* E_r \Psi_r \, d\tau = E_r \int \Psi_s^* \Psi_r \, d\tau = E_r \delta_{rs} . \tag{2.76}$$

Therefore, we can say that the best approximate wavefunctions are obtained from a set of functions by diagonalizing the Hamiltonian matrix.

If ϕ_r is a good approximation to the wavefunction Ψ_r, that is, if all the off-diagonal matrix elements H_{ri}'s are sufficiently small, the function Ψ_r and its energy can be approximately determined by the perturbation method, without solving the secular equations exactly. The principle of this method will be described in the next section.

2.4. The Perturbation Method

2.4.1. The Perturbation Method for Nondegenerate Systems

Effects of introduction of substituents and those of interaction with other molecules on the electronic state functions and molecular orbitals of a molecule can frequently be expressed in analytical forms by means of perturbation theory. In this section the principle of this theory is briefly described. Examples of its application will be given later.

Suppose that some problem whose Hamiltonian operator is $\mathbf{H}^0$ has already been solved, with a set of eigenvalues $E_1^0, E_2^0, \ldots, E_r^0, \ldots$, and the corresponding eigenfunctions $\Phi_1^0, \Phi_2^0, \ldots, \Phi_r^0, \ldots$ satisfying the equation

$$\mathbf{H}^0 \Phi_r^0 = E_r^0 \Phi_r^0, \tag{2.77}$$

and suppose that we wish to solve a similar problem whose Hamiltonian operator $\mathbf{H}$ is only slightly different from $\mathbf{H}^0$. The perturbation method is an approximate method of expressing the eigenvalues F and eigenfunctions Ψ of $\mathbf{H}$ for the new problem (called the perturbed problem) in terms of the eigenvalues E^0 and eigenfunctions Φ^0 of $\mathbf{H}^0$ for the original problem (called the unperturbed problem), instead of solving exactly the equation

$$\mathbf{H}\Psi_r = F_r\Psi_r\,. \tag{2.78}$$

The Hamiltonian for the new problem is written as

$$\mathbf{H} = \mathbf{H}^0 + \lambda\mathbf{H}', \tag{2.79}$$

where λ is some parameter, and the term $\lambda\mathbf{H}'$, which is called a perturbation operator, is small by assumption in comparison with $\mathbf{H}^0$. Then Eq. (2.78), which we wish to solve, becomes

$$(\mathbf{H}^0 + \lambda\mathbf{H}')\Psi_r = F_r\Psi_r\,. \tag{2.80}$$

If λ is placed equal to zero, this equation reduces to Eq. (2.77). Therefore, it is natural to assume that for small values of λ the solutions of Eq. (2.80) will lie close to those of (2.77). For the present it is assumed that no two of the E^0's are equal, that is, there are no degenerate energy levels in the unperturbed problem. Then, to each level in the unperturbed problem there will correspond a level in the perturbed problem. Now suppose that Ψ_r and F_r are the eigenfunction and eigenvalue of $\mathbf{H}$ which approach Φ_r^0 and E_r^0 as λ approaches zero. Since Ψ_r and F_r will be functions of λ, we may expand them in the form of a power series as

$$\Psi_r = N_r(\Phi_r^0 + \lambda\Psi_r^{(1)} + \lambda^2\Psi_r^{(2)} + \cdots), \tag{2.81}$$

$$F_r = E_r^0 + \lambda E_r^{(1)} + \lambda^2 E_r^{(2)} + \cdots, \tag{2.82}$$

where $\Psi_r^{(1)}, \Psi_r^{(2)}, \ldots$; $E_r^{(1)}, E_r^{(2)}, \ldots$ are independent of λ, and N_r is the normalization factor. Since the functions Φ^0's form a complete set of orthonormal functions, the new functions Ψ's can be expanded in terms of the set of Φ^0's. This amounts to saying that $\Psi_r^{(1)}, \Psi_r^{(2)}, \ldots$ can be expanded as

$$\Psi_r^{(1)} = \sum_{j\neq r} a_{rj}\Phi_j^0, \tag{2.83}$$

$$\Psi_r^{(2)} = \sum_{k\neq r} b_{rk}\Phi_k^0, \tag{2.84}$$

where the a's and the b's are constants to be determined.

Substituting Eqs. (2.81) and (2.82) into (2.80) and equating coefficients of like powers of λ on the two sides of the resulting equation gives the series of equations

$$\mathbf{H}^0\Phi_r^0 = E_r^0\Phi_r^0, \tag{2.85}$$

$$(\mathbf{H}^0 - E_r^0)\Psi_r^{(1)} = E_r^{(1)}\Phi_r^0 - \mathbf{H}'\Phi_r^0, \tag{2.86}$$

$$(\mathbf{H}^0 - E_r^0)\Psi_r^{(2)} = E_r^{(2)}\Phi_r^0 + E_r^{(1)}\Psi_r^{(1)} - \mathbf{H}'\Psi_r^{(1)}. \tag{2.87}$$

The first of these, by assumption, is already solved. The function $\Phi_r{}^0$ and eigenvalue $E_r{}^0$ are called the zero-order approximations to Ψ_r and F_r.

The function $\mathbf{H}'\Phi_r{}^0$ can be expanded into the series

$$\mathbf{H}'\Phi_r{}^0 = \sum_j H'_{jr}\Phi_j{}^0, \tag{2.88}$$

where

$$H'_{jr} = \int \Phi_j^{0*}\mathbf{H}'\Phi_r{}^0\, d\tau. \tag{2.89}$$

Substituting Eqs. (2.83) and (2.88) into (2.86) and using (2.85), we obtain

$$\sum_{j\neq r} a_{rj}(E_j{}^0 - E_r{}^0)\Phi_j{}^0 = (E_r^{(1)} - H'_{rr})\Phi_r{}^0 - \sum_{j\neq r} H'_{jr}\Phi_j{}^0. \tag{2.90}$$

The coefficient of each $\Phi_j{}^0$ must be equal on both sides of the equation. Therefore, we obtain

$$E_r^{(1)} - H'_{rr} = 0, \tag{2.91}$$

$$a_{rj}(E_j{}^0 - E_r{}^0) = -H'_{jr}, \tag{2.92}$$

or

$$a_{rj} = H'_{jr}/(E_r{}^0 - E_j{}^0), \qquad \text{where } j \neq r. \tag{2.93}$$

Thus, the first-order perturbation energy $E_r^{(1)}$ and the first-order perturbation function $\Psi_r^{(1)}$ can be expressed as

$$E_r^{(1)} = H'_{rr}\,; \tag{2.94}$$

$$\Psi_r^{(1)} = \sum_{j\neq r} [H'_{jr}/(E_r{}^0 - E_j{}^0)]\Phi_j{}^0. \tag{2.95}$$

$E_r^{(2)}$ and $\Psi_r^{(2)}$ can be obtained by a similar process from Eq. (2.87). The function $\mathbf{H}'\Psi_r^{(1)}$ can be expanded as follows by use of Eqs. (2.95) and (2.88):

$$\mathbf{H}'\Psi_r^{(1)} = \sum_k \sum_{j\neq r} [H'_{kj}H'_{jr}/(E_r{}^0 - E_j{}^0)]\Phi_k{}^0. \tag{2.96}$$

Substituting Eqs. (2.95), (2.96), and (2.84) into (2.87), and equating the coefficients of $\Phi_r{}^0$ on the two sides of the resulting equation, we obtain

$$E_r^{(2)} = \sum_{j\neq r} H'_{rj}H'_{jr}/(E_r{}^0 - E_j{}^0). \tag{2.97}$$

From equating the coefficients of $\Phi_k{}^0$ ($k \neq r$), the expression for b_{rk}, and hence the expression for $\Psi_r^{(2)}$, can be obtained. The perturbed energy, correct to the second order in λ, and the perturbed function, correct to the first order in λ, can therefore be written as follows:

$$F_r \fallingdotseq E_r^0 + \lambda H'_{rr} + \lambda^2 \sum_{j \neq r} H'_{rj} H'_{jr}/(E_r^0 - E_j^0) \equiv E_r^{\mathrm{p}}; \tag{2.98}$$

$$\Psi_r \fallingdotseq N_r \left(\Phi_r^0 + \lambda \sum_{j \neq r} [H'_{jr}/(E_r^0 - E_j^0)] \Phi_j^0 \right) \equiv \Phi_r^{\mathrm{p}}. \tag{2.99}$$

When $|\lambda H'_{jr}|$ is sufficiently smaller than $|E_r^0 - E_j^0|$ for all the j's except $j = r$, the terms of the higher orders in λ are negligibly small, and hence these expressions are considerably good approximations to F_r and Ψ_r. Of course, in this case the normalization factor N_r is close to unity.

In most applications it is convenient to denote the perturbation operator $\lambda \mathbf{H}'$ simply by $\mathbf{H}'$, the parameter λ being absorbed into $\mathbf{H}'$, or in other words, to place λ equal to unity in the above equations. Furthermore, if the functions Φ^0 are real functions, it follows that $H'_{rj} = H'_{jr}$. Then, Eqs. (2.98) and (2.99) can be rewritten as follows:

$$F_r \fallingdotseq E_r^0 + H'_{rr} + \sum_{j \neq r} (H'_{jr})^2/(E_r^0 - E_j^0) \equiv E_r^{\mathrm{p}}; \tag{2.100}$$

$$\Psi_r \fallingdotseq N_r \left(\Phi_r^0 + \sum_{j \neq r} [H'_{jr}/(E_r^0 - E_j^0)] \Phi_j^0 \right) \equiv \Phi_r^{\mathrm{p}}. \tag{2.101}$$

These are the most common expressions of perturbation theory. It is to be noted in particular that, since $(H'_{jr})^2$ is always positive, the functions Φ_j^0 of energy higher than Φ_r^0 contribute to the perturbed function Ψ_r with an effect of lowering its energy, and the functions of lower energy, with an effect of raising its energy.

2.4.2. The Perturbation Method for Degenerate Systems

If two or more orthogonal eigenfunctions of the unperturbed problem have the same eigenvalue, evidently Eqs. (2.98)–(2.101) cannot be used and the perturbation problem needs another formulation.

Let us assume that the unperturbed Hamiltonian $\mathbf{H}^0$ has m orthonormal eigenfunctions $\Phi_1^0, \Phi_2^0, \ldots, \Phi_m^0$ all with the same eigenvalue: $E_1^0 = E_2^0 = \cdots = E_m^0$. That is, these functions are m-fold degenerate functions. As mentioned already, any linear combination of degenerate eigenfunctions is also an eigenfunction having the same eigenvalue:

$$\mathbf{H}^0 \left(\sum_{j=1}^{m} c_j \Phi_j^0 \right) = \sum_{j=1}^{m} c_j \mathbf{H}^0 \Phi_j^0 = E_1^0 \left(\sum_{j=1}^{m} c_j \Phi_j^0 \right). \tag{2.102}$$

Now suppose that we wish to find the solution of the equation

$$(\mathbf{H}^0 + \lambda \mathbf{H}') \Psi_k = F_k \Psi_k, \tag{2.103}$$

where the eigenvalue F_k approaches the m-fold degenerate eigenvalue $E_1{}^0$ as λ approaches zero. As λ approaches zero, Ψ_k must approach some eigenfunction of $\mathbf{H}^0$ whose eigenvalue equals $E_1{}^0$, that is, some normalized linear combination

$$\Psi_k{}^0 = \sum_{j=1}^{m} c_{kj}\Phi_j{}^0, \tag{2.104}$$

where the c's are constants to be determined. This linear combination, $\Psi_k{}^0$, is called the zero-order approximation to Ψ_k. The expansion of F_k and Ψ_k in powers of λ must therefore be of the form

$$F_k = E_1{}^0 + \lambda E_k^{(1)} + \lambda^2 E_k^{(2)} + \cdots, \tag{2.105}$$

$$\Psi_k = N_k(\Psi_k{}^0 + \lambda\Psi_k^{(1)} + \lambda^2\Psi_k^{(2)} + \cdots), \tag{2.106}$$

where N_k is the normalization factor, and $\Psi_k^{(1)}$ and $\Psi_k^{(2)}$ are expressed as follows:

$$\Psi_k^{(1)} = \sum_{j>m} a_{kj}\Phi_j{}^0, \tag{2.107}$$

$$\Psi_k^{(2)} = \sum_{j>m} b_{kj}\Phi_j{}^0, \tag{2.108}$$

where the a's and the b's are constants to be determined.

Substituting Eqs. (2.104)–(2.106) into (2.103) and equating coefficients of like powers of λ, we obtain

$$\mathbf{H}^0\sum_{j=1}^{m} c_{kj}\Phi_j{}^0 = E_1{}^0\sum_{j=1}^{m} c_{kj}\Phi_j{}^0, \tag{2.109}$$

or

$$\mathbf{H}^0\Psi_k{}^0 = E_1{}^0\Psi_k{}^0, \tag{2.109$'$}$$

$$(\mathbf{H}^0 - E_1{}^0)\Psi_k^{(1)} = \sum_{j=1}^{m} c_{kj}(E_k^{(1)} - \mathbf{H}')\Phi_j{}^0, \tag{2.110}$$

$$(\mathbf{H}^0 - E_1{}^0)\Psi_k^{(2)} = E_k^{(2)}\Psi_k{}^0 + E_k^{(1)}\Psi_k^{(1)} - \mathbf{H}'\Psi_k^{(1)}. \tag{2.111}$$

Equation (2.109), or (2.109′), is already satisfied [see (2.102) and (2.104)].

As before, $\mathbf{H}'\Phi_j{}^0$ is expanded into the series

$$\mathbf{H}'\Phi_j{}^0 = \sum_i H'_{ij}\Phi_i{}^0. \tag{2.112}$$

Then, it follows that

$$\sum_{j=1}^{m} c_{kj}\mathbf{H}'\Phi_j{}^0 = \sum_{j=1}^{m}\sum_i c_{kj}H'_{ij}\Phi_i{}^0 = \sum_j\sum_{i=1}^{m} c_{ki}H'_{ji}\Phi_j{}^0. \tag{2.113}$$

Substituting Eqs. (2.107) and (2.113) into (2.110) gives

$$\sum_{j>m} (E_j^{\,0} - E_1^{\,0}) a_{kj} \Phi_j^{\,0} = \sum_{j=1}^{m} E_k^{(1)} c_{kj} \Phi_j^{\,0} - \sum_j \left(\sum_{i=1}^{m} c_{ki} H'_{ji} \right) \Phi_j^{\,0}. \tag{2.114}$$

Equating the coefficients of $\Phi_j^{\,0}$ on both sides of this equation gives the following equations:

$$(E_j^{\,0} - E_1^{\,0}) a_{kj} = - \sum_{i=1}^{m} c_{ki} H'_{ji}, \quad \text{for } j > m; \tag{2.115}$$

$$\sum_{i=1}^{m} c_{ki} H'_{ji} - E_k^{(1)} c_{kj} = 0, \quad \text{for } j \leq m. \tag{2.116}$$

Substituting the values 1, 2, . . . , m for j in Eq. (2.116), we obtain a set of m simultaneous equations for the c_{kj}'s. This set of equations has nontrivial solutions only if the determinant of the coefficients of the c's vanishes, that is, only if the following determinantal secular equation is satisfied:

$$\begin{vmatrix} H'_{11} - E_k^{(1)} & H'_{12} & \cdots & H'_{1m} \\ H'_{21} & H'_{22} - E_k^{(1)} & \cdots & H'_{2m} \\ \cdots & \cdots & \cdots & \cdots \\ H'_{m1} & H'_{m2} & \cdots & H'_{mm} - E_k^{(1)} \end{vmatrix} = 0. \tag{2.117}$$

Since the H'_{ij}'s are known constants, this equation is an equation of the mth degree in $E_k^{(1)}$, and therefore has m roots. Let these roots be $E_k^{(1)}$'s in which $k = 1, 2, \ldots, m$. Unless some of these roots happen to be equal, there are m perturbed functions with different eigenvalues which approach the m-fold degenerate eigenvalue $E_1^{\,0}$ of the unperturbed problem as the perturbation approaches zero. The coefficients of $\Phi_j^{\,0}$'s in the zero-order perturbed function $\Psi_k^{\,0}$ [compare Eq. (2.104)], that is, c_{kj}'s, can be found by the usual procedure (see Subsection 2.3.2).

The coefficients of $\Phi_j^{\,0}$'s in the first-order perturbation function $\Psi_k^{(1)}$ [compare Eq. (2.107)] can be derived from Eq. (2.115), as follows:

$$a_{kj} = \sum_{i=1}^{m} c_{ki} H'_{ji} / (E_1^{\,0} - E_j^{\,0}). \tag{2.118}$$

Hence the first-order perturbation function can be expressed as

$$\Psi_k^{(1)} = \sum_{j>m} \left[\sum_{i=1}^{m} c_{ki} H'_{ji} / (E_1^{\,0} - E_j^{\,0}) \right] \Phi_j^{\,0}. \tag{2.119}$$

Thus, if Ψ_k is the eigenfunction whose zero-order approximation is $\Psi_k^{\,0}$, Ψ_k can be expressed to the first-order approximation as follows:

$$\Psi'_k \doteqdot N_k\left(\sum_{i=1}^{m} c_{ki}\Phi_i^0 + \lambda \sum_{j>m}\left[\sum_{i=1}^{m} c_{ki}H'_{ji}/(E_1^0 - E_j^0)\right]\Phi_j^0\right). \tag{2.120}$$

The function $\mathbf{H}'\Phi_j^0$ can be expanded as follows:

$$\mathbf{H}'\Phi_j^0 = \sum_i H'_{ij}\Phi_i^0 = \sum_{k=1}^{m} H'_{kj}\Psi'^0_k + \sum_{n>m} H'_{nj}\Phi_n^0, \tag{2.121}$$

where

$$H'_{kj} = \int \Psi'^{0*}_k \mathbf{H}'\Phi_j^0 \, d\tau = \sum_{i=1}^{m} c_{ki}^* H'_{ij}. \tag{2.122}$$

Therefore,

$$\begin{aligned}\mathbf{H}'\Psi'^{(1)}_k &= \sum_{j>m}\left[\sum_{i=1}^{m} c_{ki}H'_{ji}/(E_1^0 - E_j^0)\right]\mathbf{H}'\Phi_j^0 \\ &= \sum_{j>m}\left[\sum_{i=1}^{m} c_{ki}H'_{ji}/(E_1^0 - E_j^0)\right]\left(\sum_{k=1}^{m}\sum_{i=1}^{m} c_{ki}^* H'_{ij}\Psi'^0_k + \sum_{n>m} H'_{nj}\Phi_n^0\right).\end{aligned} \tag{2.123}$$

Substituting this expression into Eq. (2.111) and equating the coefficients of Ψ'^0_k on the two sides of the resulting equation, we obtain

$$E_k^{(2)} = \sum_{j>m}\sum_{i=1}^{m} c_{ki}^* c_{ki} H'_{ji}H'_{ij}/(E_1^0 - E_j^0). \tag{2.124}$$

Thus, the eigenvalue of Ψ'_k can be expressed to the second-order approximation as follows:

$$F_k \doteqdot E_1^0 + \lambda E_k^{(1)} + \lambda^2 \sum_{j>m}\sum_{i=1}^{m} c_{ki}^* c_{ki} H'_{ji}H'_{ij}/(E_1^0 - E_j^0). \tag{2.125}$$

If Φ_j^0's are real functions, this expression can be rewritten in the following form:

$$F_k \doteqdot E_1^0 + \lambda E_k^{(1)} + \lambda^2 \sum_{j>m}\sum_{i=1}^{m} c_{ki}^2 (H'_{ij})^2/(E_1^0 - E_j^0). \tag{2.126}$$

Substituting expressions (2.108), (2.119), and (2.123) into Eq. (2.111) and equating the coefficients of Φ_j^0 $(j > m)$ on both sides of the resulting equation, we can determine the coefficient of Φ_j^0 in the second-order perturbation function $\Psi'^{(2)}_k$ [see Eq. (2.108)], that is, b_{kj}.

If one or more unperturbed functions Φ_j^0 have eigenvalues which are close to that of Φ_r^0, that is, if one or more unperturbed functions Φ_j^0 are quasi-degenerate with Φ_r^0, the formulas (2.98)–(2.101) cannot be good approximations to the perturbed function Ψ'_r and its eigenvalue F_r. Here again, a secular equation must be set up based on the set of the quasi-degenerate functions.

Chapter 3

Group Theory

3.1. Symmetry Operations and Symmetry Groups

If a molecule has symmetry its wavefunctions have some symmetry properties limited by the symmetry of the molecule. The symmetry properties of molecules and molecular wavefunctions are most conveniently described in terms of a special branch of mathematics known as group theory. Group theory is very useful for determination of electronic wavefunctions of molecules and is essential for understanding the so-called symmetry selection rule of electronic transition, which will be dealt with in the next chapter. It seems convenient here to give an outline of this theory.

An operation which transforms a symmetrical figure into itself is called a symmetry operation. That is, a symmetry operation is an operation which, when applied to an object, results in a new object which is indistinguishable from the original one and hence superimposable on it. For example, let us consider the ethylene molecule. Needless to say, in the ground state this molecule has a planar structure. Labeling the coordinate axes as in Fig. 3.1, we see that there are the following symmetry operations for the structure

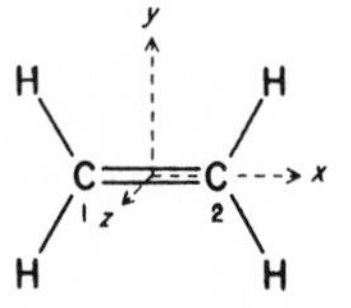

FIG. 3.1. The ethylene molecule.

of this molecule: (1) the identity operation (E or I), which leaves each point unchanged; (2) a rotation about the z axis by 180° ($C_2(z)$); (3) a rotation about the y axis by 180° ($C_2(y)$); (4) a rotation about the x axis by 180° ($C_2(x)$); (5) the inversion in the center of symmetry (i); (6) a reflection in the xy plane ($\sigma(xy)$); (7) a reflection in the zx plane ($\sigma(zx)$); and (8) a reflection in the yz plane ($\sigma(yz)$). The operation $\sigma(xy)$ is equivalent to the operation $C_2(z)$ followed by the operation i. This relation is expressed as $\sigma(xy) = iC_2(z)$. Similarly, there are following relations: $\sigma(zx) = iC_2(y)$; $\sigma(yz) = iC_2(x)$.

A set of symmetry operations is said to form a symmetry group if the following requirements are satisfied: (1) The identity operation must be in the set; (2) Any two operations in the set applied successively must be equivalent to an operation in the set; (3) The associative law of multiplication must hold: if operations in the set are denoted by A, B, C, etc., there must be the relations, $A(BC) = (AB)C$, etc.; (4) Every operation in the set must have its reciprocal such that, if the reciprocal of an operation R is denoted by R^{-1}, $RR^{-1} = R^{-1}R = E$. The commutative law of multiplication ($AB = BA$) need not hold.

It is easily verified that all these requirements are satisfied by the set of operations given above for the ethylene molecule. This set of operations forms a symmetry group designated as D_{2h}. Thus, the ethylene molecule in the equilibrium geometry is said to belong to the symmetry group D_{2h}. Since all these operations leave the center of gravity of the molecule unchanged, such a symmetry group is called a point group. The number of operations in a group is called the order (h) of the group: the order of the group D_{2h} is 8.

Each symmetry operation implies the existence of an element of symmetry. For example, the ethylene molecule can be said to possess three twofold symmetry axes, corresponding to the three C_2 operations, a center of symmetry, corresponding to the i operation, and three symmetry planes, corresponding to the three σ operations. Thus the symmetry of an object can be indicated by specifying the symmetry group to which that object belongs.

Symmetry operations and corresponding symmetry elements are denoted by symbols. Some of symbols for operations of interest to us are given below.

- E (or I) Identity operation.
- C_n Rotation about an axis of symmetry by an angle $2\pi/n$. The axis is called an n-fold symmetry (or rotational) axis. The symmetry axis with the largest value of n in a figure is called the principal axis of symmetry of the figure and is usually taken as the z coordinate axis.
- σ_h Reflection in the plane of symmetry perpendicular to the principal axis, that is, the "horizontal" symmetry (or mirror) plane.
- σ_v Reflection in the plane of symmetry containing the principal axis, that is, the "vertical" symmetry (or mirror) plane.
- σ_d Reflection in the plane of symmetry containing the principal axis and bisecting the angle between two successive twofold axes which are perpendicular to the principal axis, that is, the "diagonal" symmetry plane.
- i Inversion at a center of symmetry.

Common symbols for several point groups of interest in consideration of molecular structure are given below.

C_n The molecule belonging to this group has a C_n axis only. This axis is usually taken as the z coordinate axis.

C_{nv} The molecule belonging to this group has the symmetry elements C_n and σ_v.

C_{nh} The molecule belonging to this group has the symmetry elements C_n and σ_h. When n is even, the existence of C_n and σ_h implies the existence of a center of symmetry. Such a point group can be expressed as the combination of the corresponding C_n group and the inversion i: for example, $C_{4h} = C_4 \times i$, $C_{6h} = C_6 \times i$.

D_n The molecule belonging to this group has one n-fold principal axis and n twofold axes perpendicular to the principal axis. The molecule belonging to the symmetry group D_2 has three twofold axes, no one of which stands out over the other; in place of D_2 a special symbol V is often used.

D_{nh} The molecule belonging to this group has, in addition to the symmetry elements of the group D_n, a horizontal symmetry plane (σ_h), and hence n vertical symmetry planes (σ_v), each the plane containing one C_2 axis and the principal axis. When n is even, a center of symmetry (i) is implied; in such a case, the group can be expressed as the combination of the corresponding D_n group and the inversion as follows: $D_{2h} = D_2 \times i$, $D_{4h} = D_4 \times i$, $D_{6h} = D_6 \times i$. In place of D_{2h} the special designation V_h is often used.

D_{nd} The molecule belonging to this group has, besides the symmetry axes defining D_n, n diagonal symmetry planes (σ_d).

3.2. Representations of Symmetry Groups

The successive application of any two of the operations belonging to a point group is equivalent to some single operation in the group. For example, in the group D_2, which contains four symmetry operations, E, $C_2(z)$, $C_2(y)$, and $C_2(x)$, operation $C_2(z)$ followed by operation $C_2(y)$ is equivalent to operation $C_2(x)$. That is, we can say that the product of two operations $C_2(y)$ and $C_2(z)$ is equal to operation $C_2(x)$: $C_2(y)C_2(z) = C_2(x)$. If we work out all possible products of two operations in this way, the multiplication table for the group is obtained which shows the results of successive operations. The multiplication table for the group D_2 is given in Table 3.1, where the letters

TABLE 3.1

THE MULTIPLICATION TABLE OF POINT GROUP D_2

	E	Z	Y	X
E	E	Z	Y	X
Z	Z	E	X	Y
Y	Y	X	E	Z
X	X	Y	Z	E

E, Z, Y, and X are used for the operations E, $C_2(z)$, $C_2(y)$, and $C_2(x)$, respectively. The operation which is to be applied first is written in the top row, and the following operation in the first left-hand column. The result is found at the intersection of the row and column.

Any set of elements which multiply according to the multiplication table of a group is said to form a representation of the group. The set of elements which form a representation may be a set of numbers and may be a set of matrices. In cases of matrices, the multiplication should, of course, be carried out according to the rule for matrix multiplication.* The dimension of a matrix of a representation is called the dimension of the representation. The dimension of a representation consisting of a set of simple numbers is, of course, unity.

If the set of matrices, **a**, **b**, **c**, . . . , is a representation of a group, a set of matrices formed from these matrices by a similarity transformation,† for example, the set of matrices, $\mathbf{a}' = \boldsymbol{\beta}\mathbf{a}\boldsymbol{\beta}^{-1}$, $\mathbf{b}' = \boldsymbol{\beta}\mathbf{b}\boldsymbol{\beta}^{-1}$, $\mathbf{c}' = \boldsymbol{\beta}\mathbf{c}\boldsymbol{\beta}^{-1}, \ldots$, where $\boldsymbol{\beta}$ is a matrix and $\boldsymbol{\beta}^{-1}$ is its inverse matrix, is also a representation of the group. If it is possible to find a similarity transformation which transforms all the matrices of a representation, **a**, **b**, **c**, . . . , into the block form

$$\mathbf{a}' = \boldsymbol{\beta}\mathbf{a}\boldsymbol{\beta}^{-1} = \begin{bmatrix} \mathbf{a}_1' & 0 & \cdot \\ 0 & \mathbf{a}_2' & \cdot \\ & & \cdot \\ \cdot & \cdots & \cdot \end{bmatrix}, \tag{3.1}$$

where $\mathbf{a}_1'$ is a square matrix of the same dimension as $\mathbf{b}_1'$, $\mathbf{c}_1'$, . . . , and where there are only zeros outside the squares, the sets of matrices, $\mathbf{a}_1'$, $\mathbf{b}_1'$, $\mathbf{c}_1'$, . . . , $\mathbf{a}_2'$, $\mathbf{b}_2'$, $\mathbf{c}_2'$, . . . , etc., also form representations of the group. The representation consisting of matrices **a**, **b**, **c**, . . . is said to be reducible and to have been

* If **a** and **b** are both square matrices of dimension n, that is, matrices with n rows and n columns, the product $\mathbf{ab} = \mathbf{c}$ is also a square matrix of dimension n. If, for example, the component of matrix **a** of ith row and jth column is denoted by a_{ij}, the components of the product matrix **c** are given by the relation $c_{ij} = \sum_k a_{ik} b_{kj}$.

† If **B** is a matrix, the matrix $\mathbf{B}^{-1}$ which satisfies the following relation is called the inverse matrix of **B**: $\mathbf{BB}^{-1} = \mathbf{1}$, where **1** is a unit matrix, that is, a matrix in which all the diagonal elements are unity and all the off-diagonal elements are zero. If the matrix **B** is an orthogonal matrix, that is, a matrix satisfying the condition $\sum_k b_{ik} b_{jk} = \delta_{ij}$, where b_{ik} is the element of the matrix **B** of row i column k and δ_{ij} is the Kronecker symbol, the elements b'_{ij} of the inverse matrix $\mathbf{B}^{-1}$ are related to the components of the original matrix **B** by $b'_{ij} = b_{ji}$. That is, in this case, the inverse matrix $\mathbf{B}^{-1}$ is identical with the matrix obtained from the original matrix **B** by changing columns into rows, that is, the transposed matrix $\tilde{\mathbf{B}}$. The transformation of a matrix, say **A**, into a different matrix, **A**′, by the following multiplication is called a similarity transformation: $\mathbf{BAB}^{-1} = \mathbf{A}'$.

reduced by the similarity transformation with the matrix **β** into representations of lower dimension. If it is not possible to find a similarity transformation which will further reduce a given representation, the representation is said to be irreducible. That is, a reducible representation is a representation which can be transformed to a set of irreducible representations by a similarity transformation. Two irreducible representations which differ only by a similarity transformation are said to be equivalent.

An irreducible representation of unit dimension is said to be nondegenerate. An irreducible representation of dimension $n > 1$ is said to be n-fold degenerate. In point groups containing no rotational axis higher than twofold, all irreducible representations are nondegenerate. On the other hand, any point group containing an n-fold symmetry axis with $n > 2$ has one or more doubly degenerate irreducible representations. The sum of the squares of the dimensions of all the nonequivalent irreducible representations of a group is equal to the order h of the group.

The sum of the diagonal elements of a matrix, that is, the trace of a matrix, is called the character of the matrix. That is, if **a** is a square matrix and if its element of row i column j is denoted by a_{ij}, the character of the matrix **a** is given by $\sum_k a_{kk}$. The character of a matrix is unchanged by a similarity transformation. For nondegenerate representations, the characters are obviously identical with the elements of the representation.

3.3. Character Tables for Point Groups and Symmetry Properties of Molecules

3.3.1. Point Groups Containing No Rotational Axis Higher than Twofold

For point groups which have no n-fold rotational axis with $n > 2$, all the irreducible representations are nondegenerate, and all the characters of the irreducible representations, which are identical with the elements of the representations, are either $+1$ or -1. The number of irreducible representations is, of course, equal to the order of the group. As examples, the characters of the irreducible representations of point groups D_2 and D_{2h} are shown in Tables 3.2 and 3.3. Such tables are called character tables.

If a molecule has symmetry the form of its electronic wavefunctions is restricted by the molecular symmetry. For example, any physical measurement that is made on ethylene must be consistent with the two carbon atoms being indistinguishable. This requirement places some restriction on the form of the wavefunctions. Suppose that a π orbital ψ is to be formed as a linear combination of the $2p_z$ atomic orbitals of the two carbon atoms, χ_1 and χ_2.

TABLE 3.2

CHARACTER TABLE OF POINT GROUP D_2

D_2	E	$C_2(z)$	$C_2(y)$	$C_2(x)$	
A_1	1	1	1	1	
B_1	1	1	−1	−1	z, R_z
B_2	1	−1	1	−1	y, R_y
B_3	1	−1	−1	1	x, R_x

Since the square of the orbital represents an electron density, and since this density must be the same about each carbon atom, there are only two possible forms for the orbital: either $N(\chi_1 + \chi_2) \equiv \psi_{+1}$, or $N(\chi_1 - \chi_2) \equiv \psi_{-1}$, where N is a normalization constant. It is evident that these two forms have some symmetry properties. When one of symmetry operations in the symmetry group D_{2h} is applied to either ψ_{+1} or ψ_{-1}, the function either remains unchanged or only changes its sign. Thus, if an operation is represented by operator **R**, for any one of operations in the group D_{2h} the following relation exists:

$$\mathbf{R}\psi = \alpha\psi, \tag{3.2}$$

where α is a constant whose value is either $+1$ or -1. The value of $+1$ means that the wavefunction ψ remains unchanged under the operation R, and the value of -1 means that the wavefunction is changed only in sign. When α is $+1$ the function ψ is said to be symmetric with respect to the operation R, and when α is -1 the function is said to be antisymmetric with respect to the operation R. By taking account of the fact that each of the carbon $2p_z$ atomic orbitals is antisymmetric with respect to the reflection in the molecular

TABLE 3.3

CHARACTER TABLE OF POINT GROUP D_{2h}

D_{2h} $= D_2 \times i$	E	$C_2(z)$	$C_2(y)$	$C_2(x)$	i	$\sigma(xy)$ $= C_2(z) \times i$	$\sigma(zx)$ $= C_2(y) \times i$	$\sigma(yz)$ $= C_2(x) \times i$	
A_g	1	1	1	1	1	1	1	1	
A_u	1	1	1	1	−1	−1	−1	−1	
B_{1g}	1	1	−1	−1	1	1	−1	−1	R_z
B_{1u}	1	1	−1	−1	−1	−1	1	1	z
B_{2g}	1	−1	1	−1	1	−1	1	−1	R_y
B_{2u}	1	−1	1	−1	−1	1	−1	1	y
B_{3g}	1	−1	−1	1	1	−1	−1	1	R_x
B_{3u}	1	−1	−1	1	−1	1	1	−1	x

plane, it can be easily verified that the values of α for ψ_{+1} are $+1$ for operations E, $C_2(z)$, $\sigma(zx)$, $\sigma(yz)$, and are -1 for operations $C_2(y)$, $C_2(x)$, i, $\sigma(xy)$. This set of values of α is identical with the set of elements (or characters) of the fourth irreducible representation of the symmetry group D_{2h} (cf. Table 3.3). Thus we can say that the transformation properties of ψ_{+1} under the various symmetry operations, or in other words, the symmetry properties of ψ_{+1}, are specified by the fourth irreducible representation of D_{2h}. Similarly, it can be easily verified that the symmetry properties of ψ_{-1} are specified by the fifth irreducible representation of D_{2h}.

More generally, the connection between molecular wavefunctions and the irreducible representations of the group to which the molecule belongs can be explained as follows. Any molecular wavefunction is to be a solution of the wave equation

$$\mathbf{H}\phi_n = E_n\phi_n. \tag{3.3}$$

If $\mathbf{H}$ is the total electronic Hamiltonian, ϕ_n is an electronic state function; if $\mathbf{H}$ is a one-electron Hamiltonian, ϕ_n is a one-electron function, that is, a molecular orbital. In any case, the Hamiltonian $\mathbf{H}$ reflects the molecular symmetry insofar as it remains unchanged under any symmetry operation of the symmetry group to which the molecule belongs. As before, one of the symmetry operations of the symmetry group to which the molecule belongs is represented by operator $\mathbf{R}$. When both sides of Eq. (3.3) are subjected to $\mathbf{R}$, we obtain

$$\mathbf{RH}\phi_n = \mathbf{R}E_n\phi_n. \tag{3.4}$$

But since $\mathbf{R}$ leaves $\mathbf{H}$ unchanged, and since E_n is just a constant, we have

$$\mathbf{H}(\mathbf{R}\phi_n) = E_n(\mathbf{R}\phi_n). \tag{3.5}$$

This means that the function $\mathbf{R}\phi_n$ is also an eigenfunction of $\mathbf{H}$ which has the same eigenvalue as ϕ_n. If ϕ_n was nondegenerate then $\mathbf{R}\phi_n$ must be equal to ϕ_n, or just differ from it by a change of sign. That is, in this case, again we can write

$$\mathbf{R}\phi_n = \alpha\phi_n, \tag{3.6}$$

where α is a constant whose value is either $+1$ or -1. The set of values of α for all the symmetry operations corresponds to an irreducible representation of the symmetry group.

In point groups containing no rotational symmetry axis higher than two-fold, the successive two applications of an operation are equivalent to the identity operation, that is, leave the object unchanged. Therefore, we have

$$\mathbf{R}(\mathbf{R}\phi_n) = \mathbf{R}(\alpha\phi_n) = \alpha^2\phi_n = \phi_n. \tag{3.7}$$

It follows that $\alpha^2 = 1$. This means that the value of α must be either $+1$ or -1. This conclusion is the same as that deduced above from the nondegeneracy of the molecular wavefunction. This means that in molecules belonging to point groups containing no rotational symmetry higher than twofold all the molecular wavefunctions must be nondegenerate, except for the rare case of accidental degeneracy, where two or more eigenfunctions have the same eigenvalue but behave differently under symmetry operations.

In general, if a molecule has symmetry, any wavefunction of the molecule has symmetry properties specified by one of the irreducible representations of the point group to which the molecule belongs. An irreducible representation can be said to represent a symmetry species. When a given wavefunction behaves under the symmetry operations in the manner specified by an irreducible representation, the wavefunction is said to form a basis for that representation, and is also said to belong to the corresponding symmetry species.

For symmetry species the following notation is generally used. Species corresponding to nondegenerate representations are designated by letters A and B. Species A are symmetric, and species B are antisymmetric, with respect to the highest rotational symmetry, that is, the rotation $C_n(z)$ with the highest value of n. In the case of D_2, as well as D_{2h}, since no one of the three twofold symmetry axes stands out over the others, all the species antisymmetric with respect to any one of the three C_2 operations are denoted by B. To distinguish species which are similar with respect to the principal rotational symmetry but differ with respect to other symmetry operations, subscripts 1, 2, etc., are used. In groups containing the inversion i in the center of symmetry, species symmetric with respect to the inversion are denoted by subscript g (from the German word, *gerade*), and species antisymmetric with respect to the inversion are denoted by subscript u (from the German word, *ungerade*).

The symmetry species corresponding to the representation in which all the characters are $+1$ is said to be totally symmetric, and is designated as A, A_1, or A_{1g}, depending on the type of the symmetry group. If a molecule belongs to a point group, the Hamiltonian for the molecule, the wavefunction of its electronic ground state, and observable static properties of the molecule in the ground state, such as the charge distribution, belong to the totally symmetric species of the group. Electronic states and electron configurations of the molecule are frequently designated by the symbols of the symmetry species to which their wavefunctions belong, sometimes prefixed by the superscripts indicating the spin multiplicities. Thus, the ground state of an ordinary molecule is designated, for example, as ${}^1A_{1g}$. While the symmetry species of electronic states, as well as electron configurations, are written in capital letters as above, those of molecular orbitals are conventionally written in lower-case letters. For example, since the π orbitals of ethylene, ψ_{+1} and ψ_{-1}, belong to symmetry species B_{1u} (corresponding to the fourth irreducible representation) and B_{2g} (corresponding to the fifth irreducible representation) of the point group D_{2h}, respectively, these orbitals can be called a b_{1u} and a b_{2g} orbital, respectively.

Each of the coordinates x, y, z belongs to a symmetry species of the point group. Each of the rotations, R_x, R_y, R_z, about the x, y, z axes, respectively, also belongs to a symmetry species of the point group. For example, the x axis remains unchanged under operations such as E and $C_2(x)$, but changes its sign under operations such as $C_2(y)$ and $C_2(z)$. If the operation i is present in the group, the coordinates invariably belong to u symmetry species, and the rotations about the coordinate axes invariably belong to g symmetry species. Usually, in character tables the transformation properties of the coordinates and the rotations about the coordinate axes are indicated (see Tables 3.2 and 3.3).

3.3.2. Point Groups Containing a Rotational Axis Higher than Twofold

The successive two applications of operation C_n with $n > 2$ are not equivalent to the identity operation. Therefore, it will be understandable that the situation with point groups containing a rotational axis higher than twofold will be somewhat different from that with point groups containing no such an axis. Actually, while all the irreducible representations of point groups containing no rotational axis higher than twofold are nondegenerate, point groups containing a rotational axis higher than twofold have some doubly degenerate irreducible representations. Doubly degenerate representations, differently from nondegenerate ones, do not express simply the symmetrical or antisymmetrical behavior of a function under the symmetry operations of the point group.

Molecules belonging to point groups containing a rotational axis higher than twofold have invariably some sets of doubly degenerate wavefunctions. As mentioned already, any linear combination of members of a set of degenerate functions is also an eigenfunction with the same eigenvalue [see Subsection 2.2.4, Eq. (2.45)]. Therefore, if ϕ_n is a member of a set of degenerate functions, the relation (3.5) only requires that $\mathbf{R}\phi_n$ must also belong to that set, or must be some linear combination of members of that set. If ϕ_1 and ϕ_2 are doubly degenerate wavefunctions, the results of application of a symmetry operation R on these functions can be expressed as follows:

$$\mathbf{R}\phi_1 = a_{11}\phi_1 + a_{12}\phi_2\,; \qquad \mathbf{R}\phi_2 = a_{21}\phi_1 + a_{22}\phi_2\,, \tag{3.8}$$

where the a's are constants. This set of equations can be written in the form

$$\begin{pmatrix} \mathbf{R}\phi_1 \\ \mathbf{R}\phi_2 \end{pmatrix} = \begin{pmatrix} a_{11} & a_{12} \\ a_{21} & a_{22} \end{pmatrix} \begin{pmatrix} \phi_1 \\ \phi_2 \end{pmatrix}. \tag{3.9}$$

Thus, to any symmetry operation there corresponds a two-row square matrix

$$\mathbf{A}(R) = \begin{pmatrix} a_{11} & a_{12} \\ a_{21} & a_{22} \end{pmatrix}. \tag{3.10}$$

The set of matrices of the transformations for all the symmetry operations of the point group makes up an irreducible representation of the group. The set of doubly degenerate wavefunctions ϕ_1 and ϕ_2 is said to form a basis for that doubly degenerate irreducible representation.

The character of a degenerate representation for a symmetry operation R is the sum of the diagonal elements of the corresponding matrix $\mathbf{A}(R)$. In the above example, the character of the representation for the operation R is $a_{11} + a_{22}$.

Even degenerate functions may be symmetric or antisymmetric with respect to some of the "simple" operations, such as i, σ's, and C_2's. If, for example, both the degenerate functions ϕ_1 and ϕ_2 are symmetric with respect to an operation R, the matrix of the transformation for that operation is

$$\mathbf{A}(R) = \begin{pmatrix} 1 & 0 \\ 0 & 1 \end{pmatrix}, \tag{3.11}$$

and the corresponding character is 2.

Symmetry species corresponding to doubly degenerate irreducible representations are designated by a symbol E.

Let us take groups $\mathbf{D}_6$ and $\mathbf{D}_{6h}$ as examples of point groups containing one rotational axis higher than twofold. The character table for group $\mathbf{D}_6$ is shown in Table 3.4. Point group $\mathbf{D}_{6h}$ can be expressed as the combination of point group $\mathbf{D}_6$ and the inversion: $\mathbf{D}_{6h} = \mathbf{D}_6 \times i$. The character table for $\mathbf{D}_{6h}$ can easily be constructed from that for $\mathbf{D}_6$ in the manner illustrated by comparison of the character tables for $\mathbf{D}_2$ and $\mathbf{D}_{2h}$ (Tables 3.2 and 3.3). Thus, each representation of $\mathbf{D}_6$ splits into a pair of g and u representations in $\mathbf{D}_{6h}$. The characters of the two representations of the pair for the operations

TABLE 3.4

CHARACTER TABLE OF POINT GROUP D_6

D_6	E	C_2	$2C_3$	$2C_6$	$3C_2'$	$3C_2''$	
A_1	1	1	1	1	1	1	
A_2	1	1	1	1	−1	−1	z, R_z
B_1	1	−1	1	−1	1	−1	
B_2	1	−1	1	−1	−1	1	
E_1	2	−2	−1	1	0	0	$\begin{cases}(x, y)\\(R_x, R_y)\end{cases}$
E_2	2	2	−1	−1	0	0	

in D_6 (R's) are equal to the characters of the corresponding representation of D_6. The characters of the g representation for the operations iR's are equal to those for the corresponding R's, and the characters of the u representation for the operations iR's are equal to -1 times the characters for the corresponding R's. In D_{6h}, the z coordinate axis and the rotation R_z form bases for nondegenerate representations corresponding to symmetry species A_{2u} and A_{2g}, respectively, and the pair of the x and y coordinate axes and the pair of the rotations R_x and R_y form bases for doubly degenerate representations corresponding to symmetry species E_{1u} and E_{1g}, respectively.

A planar regular hexagon, shown in Fig. 3.2 with the coordinate system, belongs to point group D_{6h}. This figure can be considered to represent the carbon skeleton of the benzene molecule in the equilibrium geometry, in which six identical carbon nuclei are situated at the vertices 1–6 of the hexagon. That is, the benzene molecule in the equilibrium geometry belongs to D_{6h}. On the basis of this figure, the symmetry operations of D_{6h}, as well as D_6, will be explained as follows.

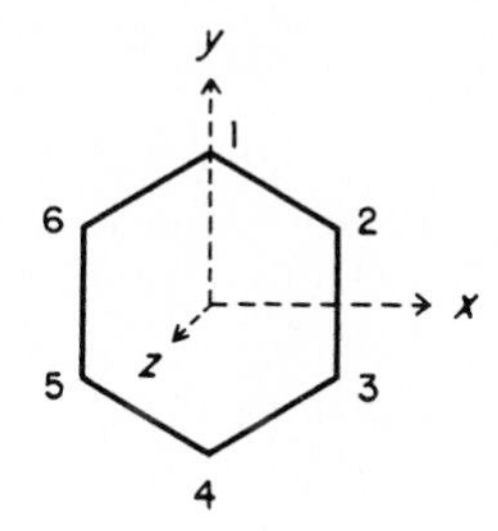

FIG. 3.2. The skeleton of the benzene molecule.

In this figure the principal rotational axis, that is, the sixfold rotational axis, is the axis (z axis) through the center of the figure and perpendicular to the plane (xy plane) of the figure. C_2 is the rotation about the z axis by 180°. C_3 is the rotation about the z axis by 120°. There are two operations of this type: the clockwise rotation (C_3^+) and the counterclockwise rotation (C_3^-). These two operations are said to form a class. Precisely speaking, two elements A and B of a group which satisfy the relation $X^{-1}AX = A$ or B, where X is any element of the group and X^{-1} is its reciprocal, are said to belong to the same class. In general, since the character of a matrix is unchanged by a similarity transformation, a given representation has the same character for operations belonging to the same class. In character tables, the operations are collected into classes, as $2C_3$, and the characters for one of operations belonging to each class are shown.

C_6 is the rotation about the z axis by 60°. Two operations belong to this class: the clockwise rotation (C_6^+) and the counterclockwise rotation (C_6^-). C_2' is the rotation through 180° around an axis passing through two opposite vertices. Three such axes are possible. Accordingly, there are three operations belonging to the class C_2': the rotations around the axes 1–4, 2–5, and 3–6, which are provisionally designated as $C_2'(1)$, $C_2'(2)$, and $C_2'(3)$, respectively. C_2'' is the rotation through 180° around an axis passing through the middle points of two opposite sides. Three such axes are possible. Accordingly, there are three operations belonging to the class C_2'': the rotation about the axis passing through the middle points of the sides 1–2 and 4–5 is provisionally designated as $C_2''(12)$, and $C_2''(23)$ and $C_2''(34)$ are analogously defined.

Operation i is, needless to say, the inversion in the center of the figure. The remaining operations in D_{6h} are the combinations of the operations in D_6 and the inversion i. For example, iC_2 is equivalent to the reflection in the plane of the figure, that is, σ_h.

The order (that is, the number of operations) of D_6 is 12, and that of D_{6h} is 24. The number of classes of operations in D_6 is 6, and that in D_{6h} is 12. In general, while the sum of the squares of the dimensions of all the irreducible representations of a group is equal to the order of the group, the number of the irreducible representations of a group is equal to the number of the classes of operations in the group.

As will be detailed in a later chapter, when the six $2p\pi$ atomic orbitals on the six carbon atoms are denoted by $\chi_1, \chi_2, \ldots, \chi_6$, the six possible molecular π orbitals ψ of benzene as linear combinations of these atomic orbitals can be expressed as shown in Table 3.5. Molecular orbitals ψ_{+1} and ψ_{+2} are degenerate, having the same energy of $\alpha + \beta$, where α and β

are negative energy quantities known as the Coulomb and resonance integrals, respectively. Molecular orbitals ψ_{-1} and ψ_{-2} form another pair of degenerate orbitals, having the same energy of $\alpha - \beta$. Molecular orbitals ψ_{+3} and ψ_{-3} are nondegenerate. The six AO's may be considered as the six unit vectors defining a six-dimensional coordinate system, and the six MO's may be considered as six independent orthogonal unit vectors in the six-dimensional space.

TABLE 3.5

THE π ORBITALS OF BENZENE

Symmetry	Molecular π orbital	Energy
B_{2g}	$\psi_{-3} = -6^{-1/2}(\chi_1 - \chi_2 + \chi_3 - \chi_4 + \chi_5 - \chi_6)$	$\alpha - 2\beta$
E_{2u}	$\psi_{-2} = 2^{-1}(\chi_2 - \chi_3 + \chi_5 - \chi_6)$ $\psi_{-1} = -12^{-1/2}(2\chi_1 - \chi_2 - \chi_3 + 2\chi_4 - \chi_5 - \chi_6)$	$\alpha - \beta$
E_{1g}	$\psi_{+1} = 12^{-1/2}(2\chi_1 + \chi_2 - \chi_3 - 2\chi_4 - \chi_5 + \chi_6)$ $\psi_{+2} = 2^{-1}(\chi_2 + \chi_3 - \chi_5 - \chi_6)$	$\alpha + \beta$
A_{2u}	$\psi_{+3} = 6^{-1/2}(\chi_1 + \chi_2 + \chi_3 + \chi_4 + \chi_5 + \chi_6)$	$\alpha + 2\beta$

In general, application of a symmetry operation to a vector may be interpreted as meaning the transformation of the vector in the fixed coordinate system, or alternatively, may be interpreted as meaning the transformation of the coordinate system, leaving the vector unchanged. Thus, **Rr** means, according to the first interpretation, the vector obtained by applying operator **R** to vector **r**, expressed in terms of the fixed coordinate system, or alternatively, according to the second interpetation, it means the vector **r** expressed in terms of the new coordinate system resulted from application of operator **R** to the original coordinate system.

The first interpretation is adopted here. Then, $\mathbf{C}_6{}^+\psi$, for example, means the MO resulted from the clockwise rotation of the original MO through 60° about the principal axis, by which the AO's χ_1, χ_2, χ_3, etc., in the expression of the original MO ψ are converted to the AO's χ_2, χ_3, χ_4, etc., respectively. It is to be noted here that the AO's have a nodal plane coinciding with the molecular plane, that is, they change sign under the reflection in the molecular plane, that is, operation σ_h. Therefore, application of operation $C_2'(1)$, for example, to an MO results in the changes of the AO's χ_1, χ_2, χ_3, etc., in the expression of the original MO to $-\chi_1$, $-\chi_6$, $-\chi_5$, etc., respectively.

The nondegenerate MO ψ_{+3} remains unchanged by application of operations E, C_2, C_3, C_6, etc., and changes its sign by application of operations

C_2', C_2'', i, iC_2 $(= \sigma_h)$, etc. Thus, the transformation properties of this MO are specified by the one-dimensional representation corresponding to the symmetry species A_{2u} of point group D_{6h}. That is, we can say that this MO belongs to the symmetry species A_{2u}, and also that this MO forms a basis for the corresponding nondegenerate representation. Analogously, it is easily seen that another nondegenerate MO, ψ_{-3}, belongs to the symmetry species B_{2g}, or in other words, it forms a basis for the nondegenerate representation corresponding to the symmetry species B_{2g}.

A single one of the orbitals of a degenerate set is not always either symmetric or antisymmetric with respect to the symmetry operations. For example, application of operation C_6^+ to one of doubly degenerate MO's ψ_{+1} and ψ_{+2} results in a linear combination of these two MO's:

$$\begin{aligned} \mathbf{C}_6^+\psi_{+1} &= (1/\sqrt{12})(2\chi_2 + \chi_3 - \chi_4 - 2\chi_5 - \chi_6 + \chi_1) \\ &= +\tfrac{1}{2}\psi_{+1} + \tfrac{3}{2}\psi_{+2}\,; \\ \mathbf{C}_6^+\psi_{+2} &= \tfrac{1}{2}(\chi_3 + \chi_4 - \chi_6 - \chi_1) \\ &= -\tfrac{3}{2}\psi_{+1} + \tfrac{1}{2}\psi_{+2}\,. \end{aligned} \tag{3.12}$$

This transformation of the pair of MO's under $\mathbf{C}_6^+$ can be expressed as

$$\begin{pmatrix} \mathbf{C}_6^+\psi_{+1} \\ \mathbf{C}_6^+\psi_{+2} \end{pmatrix} = \mathbf{A}(C_6^+)\begin{pmatrix} \psi_{+1} \\ \psi_{+2} \end{pmatrix}, \tag{3.13}$$

where $\mathbf{A}(C_6^+)$ is the matrix of the transformation:

$$\mathbf{A}(C_6^+) = \begin{pmatrix} +\tfrac{1}{2} & +\tfrac{3}{2} \\ -\tfrac{3}{2} & +\tfrac{1}{2} \end{pmatrix}. \tag{3.14}$$

The character of this matrix is +1. The matrix of the transformation of this pair of degenerate MO's for operation C_6^- is also easily verified to be the following:

$$\mathbf{A}(C_6^-) = \begin{pmatrix} +\tfrac{1}{2} & -\tfrac{3}{2} \\ +\tfrac{3}{2} & +\tfrac{1}{2} \end{pmatrix}. \tag{3.15}$$

The character of this matrix is also +1.

The matrices of transformation of the set of doubly degenerate MO's ψ_{+1} and ψ_{+2} for some of the symmetry operations of point group D_{6h} and their characters are shown in Table 3.6. It will be seen that the characters for operations belonging to a class have invariably the same value. The set of the two-dimensional square matrices of transformation of this pair

TABLE 3.6

MATRICES OF TRANSFORMATION OF THE PAIR OF DEGENERATE MO'S ψ_{+1} AND ψ_{+2} OF BENZENE FOR SOME OF THE SYMMETRY OPERATIONS OF POINT GROUP D_{6h} AND THEIR CHARACTERS

Operation	E	C_2	C_3^+	C_3^-
Matrix	$\begin{bmatrix} +1 & 0 \\ 0 & +1 \end{bmatrix}$	$\begin{bmatrix} -1 & 0 \\ 0 & -1 \end{bmatrix}$	$\begin{bmatrix} -\frac{1}{2} & +\frac{\sqrt{3}}{2} \\ -\frac{\sqrt{3}}{2} & -\frac{1}{2} \end{bmatrix}$	$\begin{bmatrix} -\frac{1}{2} & -\frac{\sqrt{3}}{2} \\ +\frac{\sqrt{3}}{2} & -\frac{1}{2} \end{bmatrix}$
Character	$+2$	-2	-1	-1
Operation	$C_2'(1)$	$C_2'(2)$	$C_2'(3)$	i
Matrix	$\begin{bmatrix} -1 & 0 \\ 0 & +1 \end{bmatrix}$	$\begin{bmatrix} +\frac{1}{2} & -\frac{\sqrt{3}}{2} \\ -\frac{\sqrt{3}}{2} & -\frac{1}{2} \end{bmatrix}$	$\begin{bmatrix} +\frac{1}{2} & +\frac{\sqrt{3}}{2} \\ +\frac{\sqrt{3}}{2} & -\frac{1}{2} \end{bmatrix}$	$\begin{bmatrix} +1 & 0 \\ 0 & +1 \end{bmatrix}$
Character	0	0	0	$+2$

of degenerate MO's for all the symmetry operations of point group D_{6h} forms a doubly degenerate irreducible representation of the group, which corresponds to symmetry species E_{1g}. That is, this pair of doubly degenerate MO's forms a basis for the representation corresponding to the symmetry species E_{1g}, or in other words, it belongs to the symmetry species E_{1g}. Analogously, it is easily verified that the pair of doubly degenerate MO's ψ_{-1} and ψ_{-2} forms a basis for the doubly degenerate irreducible representation corresponding to symmetry species E_{2u}.

As has been so far illustrated, if a molecule has symmetry its wavefunctions must have the symmetry properties specified by the irreducible representations

If the second interpretation of the meaning of application of operations is adopted, $\mathbf{C}_6^+\psi$, for example, means the MO ψ expressed in terms of the new six-dimensional coordinate system resulted from application of $\mathbf{C}_6^+$ to the original one, in which the original AO's χ_1, χ_2, χ_3, etc., are replaced by χ_6, χ_1, χ_2, etc., respectively. This transformation is equivalent to the application of the reciprocal of $\mathbf{C}_6^+$, that is, $\mathbf{C}_6^-$, to the MO ψ in the fixed coordinate system. In general, the effect of the application of an operation according to the second interpretation is the same as that of the application of its reciprocal according to the first interpretation. The reciprocals of the "simple" operations are identical with the original operations. The reciprocal of any "complicated" operation such as a rotation through $2\pi/n$ with $n > 2$ or its combination with the inversion belongs to the same class, and hence has the same character, as the original operation for any representation. Therefore, the foregoing assignment of the π MO's of benzene to symmetry species, made on the basis of the first interpretation of the meaning of application of symmetry operations, remains valid even when the second interpretation is adopted.

of the point group to which the molecule belongs. Therefore, from knowledge of the irreducible representations of the group, we can determine the form of the possible wavefunctions to a large extent without the labor of solving the wave equation, and furthermore, we can know immediately what degrees of degeneracy are possible.

3.4. Symmetry Properties of Products of Functions

3.4.1. Symmetry Properties of Products of Functions Belonging to Nondegenerate Symmetry Species

If functions ϕ_i and ϕ_j are nondegenerate and belong to nondegenerate irreducible representations of a point group, Γ_i and Γ_j, respectively, the product of the two functions, $\phi_i\phi_j$, belongs to a nondegenerate irreducible representation of the same point group, Γ_{ij}, whose character for any symmetry operation is equal to the product of the characters of the representations Γ_i and Γ_j for that operation. That is, if the characters of representations Γ_i, Γ_j, and Γ_{ij} for operation R are denoted by $X_i(R)$, $X_j(R)$, and $X_{ij}(R)$, respectively, there exists the following relation:

$$X_{ij}(R) = X_i(R)X_j(R). \tag{3.16}$$

This relation is also written as follows:

$$\Gamma_{ij} = \Gamma_i \times \Gamma_j . \tag{3.17}$$

This fact may be explained as follows. Functions ϕ_i and ϕ_j are either symmetric (character $+1$) or antisymmetric (character -1) with respect to any symmetry operation of the point group. If the two functions are both symmetric or antisymmetric with respect to a given operation, the product of the functions must be symmetric with respect to that operation. If one of the two functions is symmetric and the other is antisymmetric with respect to a given operation, the product must be antisymmetric with respect to that operation.

An electronic structure of a molecule is described, to a first approximation, in terms of a product of occupied orbitals, that is, in terms of an electron configuration. For example, the electron configuration in which two electrons are placed in a molecular orbital ψ_i is expressed as ψ_i^2, and the electron configuration in which one electron is in a molecular orbital ψ_i and another electron in a molecular orbital ψ_j is expressed as $\psi_i\psi_j$. The symmetry properties of an electron configuration are characterized by the representation to which the product of orbitals belongs, that is, by the product of the representations to which the individual orbitals belong.

The multiplication of a nondegenerate representation by itself always gives the totally symmetric representation. Therefore, a nondegenerate molecular orbital occupied by two electrons is always totally symmetric. In general, closed-shell configurations are totally symmetric.

Multiplication of two *g* or two *u* representations always gives a *g* representation, and multiplication of a *g* representation and a *u* representation always gives a *u* representation. For example, inspection of the character table for D_{2h} (Table 3.3) shows that $A_{1u} \times B_{2g} = B_{2u}$, $A_{1u} \times B_{3g} = B_{3u}$, $B_{1u} \times B_{2g} = B_{3u}$, $B_{1u} \times B_{3g} = B_{2u}$, etc.

Since the ground electron configurations of ordinary molecules are closed-shell configurations, they are totally symmetric. The excited electron configuration arising from the ground electron configuration by promoting one of two electrons occupying an orbital ψ_i to an unoccupied orbital ψ_j belongs to the symmetry species determined by the product $\psi_i\psi_j$. For example, the ground electron configuration of ethylene, in which a π orbital ψ_{+1} is occupied by two electrons, belongs to the A_g species of point group D_{2h}, and the excited electron configuration arising from the ground electron configuration by promoting one electron from the occupied π orbital ψ_{+1} to the unoccupied π orbital ψ_{-1} belongs to the B_{3u} species of the group, since, as explained in the preceding section, the π orbitals ψ_{+1} and ψ_{-1} belong to symmetry species B_{1u} and B_{2g}, respectively, and $B_{1u} \times B_{2g} = B_{3u}$.

3.4.2. Symmetry Properties of Products of Functions Belonging to Degenerate Symmetry Species

If a pair of functions ϕ_{i1} and ϕ_{i2} and a pair of functions ϕ_{j1} and ϕ_{j2} form bases for doubly degenerate representations of a point group, Γ_i and Γ_j, respectively, the set of four products $\phi_{i1}\phi_{j1}$, $\phi_{i1}\phi_{j2}$, $\phi_{i2}\phi_{j1}$, and $\phi_{i2}\phi_{j2}$ is called the direct product of the two pairs of functions, and it forms a basis for a four-dimensional representation of the group, Γ_{ij}. In general, if two sets of functions $I_1, I_2, \ldots, I_m$ and $J_1, J_2, \ldots, J_n$ form bases for an m-dimensional representation Γ_i and an n-dimensional representation Γ_j of a group, respectively, the set of $m \times n$ products IJ's is called the direct product of the sets of functions I's and J's, and it forms a basis for an $m \times n$-dimensional representation of the group, Γ_{ij}. The character of the representation of the direct product, Γ_{ij}, for any symmetry operation of the group is equal to the product of the characters of the individual representations Γ_i and Γ_j for that operation. That is, the relation (3.16) generally holds.

The representation of the direct product of two degenerate irreducible representations, Γ_{ij}, is in general a reducible representation and can be expressed in terms of irreducible representations so that the following relation holds for all the symmetry operations of the group:

$$X_{ij}(R) = \sum_k a_k X_k(R), \tag{3.18}$$

where $X_{ij}(R)$ and $X_k(R)$ are the characters for an operation R of the reducible representation Γ_{ij} and of the kth irreducible representation, Γ_k, respectively, and a_k is a constant. This relation may be expressed as follows:

$$\Gamma_{ij} = \sum_k a_k \Gamma_k. \tag{3.19}$$

The constant a_k has the meaning of the number of times the kth irreducible representation occurs in the reducible representation Γ_{ij}. This constant can be found by means of the following equation:

$$a_k = \frac{1}{h} \sum_R X_{ij}(R) X_k(R), \tag{3.20}$$

where h is the number of symmetry operations in the group, that is, the order of the group, and the summation is carried out over all the operations in the group.

When the representation of the direct product is reducible to irreducible representations, as expressed by Eq. (3.19), it is always possible to find a_k sets of linear combinations of functions of the direct product set which form bases for the kth irreducible representation.

For example, let us consider the benzene molecule. As mentioned already, this molecule in the equilibrium geometry belongs to point group D_{6h}, and has the six π orbitals shown in Table 3.5. In the ground electron configuration all the bonding π orbitals, ψ_{+1}, ψ_{+2}, and ψ_{+3}, are doubly occupied. That is, the ground electron configuration is expressed as $\psi_{+1}^2\psi_{+2}^2\psi_{+3}^2$. $\psi_{+1}^2\psi_{+2}^2$ and ψ_{+3}^2 are closed shells. Every closed shell remains unchanged under all the symmetry operations of the group, that is, it belongs to the totally symmetric representation of the group. The ground electron configuration of benzene is, of course, a closed-shell configuration, and belongs to symmetry species A_{1g} of point group D_{6h}.

As usual, the excited singlet electron configuration arising from the ground electron configuration by promoting one electron in an occupied orbital ψ_{+m} to an unoccupied orbital ψ_{-n} is denoted by V_1^{mn}, and its triplet counterpart by T_1^{mn}. Evidently, four degenerate, lowest-energy excited singlet electron configurations V_1^{11}, V_1^{12}, V_1^{21}, V_1^{22} and the corresponding four triplet electron configurations should arise from one-electron transitions from the pair of the highest occupied doubly degenerate orbitals, ψ_{+1} and ψ_{+2}, to the pair of the lowest unoccupied doubly degenerate orbitals, ψ_{-1} and ψ_{-2}. The set of the four degenerate excited singlet configurations, as well as the set of the four degenerate excited triplet configurations, behaves like the direct product of the two pairs of degenerate orbitals, and hence forms a basis for the four-dimensional representation of the direct product, $E_{1g} \times E_{2u}$. This four-dimensional representation can be reduced to $E_{1u} + B_{1u} + B_{2u}$ by use of Eq. (3.20), implying that it must be possible, from the four excited singlet configurations as well as from the four excited triplet configurations, to form

TABLE 3.7

EXCITED ELECTRONIC STATES OF BENZENE

State	Wavefunction			Symmetry	Energy, ev[a]
α	$2^{-1/2}(V_1^{12} - V_1^{21}) \equiv {}^{-}[V]_1^{12}$			$^1B_{2u}$	4.90
p	$2^{-1/2}(V_1^{11} + V_1^{22}) \equiv [V]_1^{11+22}$	or	$[V]_1^{11}$	$^1B_{1u}$	5.31
β	$2^{-1/2}(V_1^{12} + V_1^{21}) \equiv {}^{+}[V]_1^{12}$			$^1E_{1u}$	6.95
β'	$2^{-1/2}(V_1^{11} - V_1^{22}) \equiv [V]_1^{11-22}$	or	$[V]_1^{22}$		
t_p	$2^{-1/2}(T_1^{11} + T_1^{22}) \equiv [T]_1^{11+22}$	or	$[T]_1^{11}$	$^3B_{1u}$	4.01
t_β	$2^{-1/2}(T_1^{12} + T_1^{21}) \equiv {}^{+}[T]_1^{12}$			$^3E_{1u}$	4.45
$t_\beta{}'$	$2^{-1/2}(T_1^{11} - T_1^{22}) \equiv [T]_1^{11-22}$	or	$[T]^{22}$		
t_α	$2^{-1/2}(T_1^{12} - T_1^{21}) \equiv {}^{-}[T]_1^{12}$			$^3B_{2u}$	4.90

[a] The energy relative to the ground state, calculated by the Pariser–Parr method (see Subsection 11.5.4).

two nondegenerate linear combinations which will form bases for nondegenerate representations B_{1u} and B_{2u}, and a set of doubly degenerate linear combinations which will form a basis for doubly degenerate representation E_{1u}. Actually, as will be given a detailed explanation in a later chapter, an advanced MO treatment explicitly taking account of the interelectronic repulsion shows that the four degenerate singlet configurations and their triplet counterparts split into the electronic states shown in Table 3.7.

3.4.3. Requirements for Integrals To Be Nonvanishing

If a function ϕ belongs to a representation of a point group, the integral $\int \phi \, d\tau$ can be nonzero only if the representation to which the integrand belongs, Γ_{int}, is totally symmetric or can be reduced to irreducible representations, of which at least one is totally symmetric. If the integrand is the product of two functions ϕ_i and ϕ_j belonging to irreducible representations of a group, Γ_i and Γ_j, the representation Γ_{int}, which is equal to $\Gamma_i \times \Gamma_j$, is totally symmetric or contains the totally symmetric representation and hence the integral can be nonzero only if Γ_i and Γ_j are the same. Furthermore, the integral $\int \phi_i f \phi_j \, d\tau$ can, of course, be different from zero only if the representation $\Gamma_i \times \Gamma_j$ is equal to, or contains, the representation to which the function f belongs.

These theorems are of special interest to us, because they afford a basis for simplification of secular equations for determination of molecular orbitals and electronic state functions, and also a basis for the symmetry selection rule of electronic transition. Actual applications of these theorems to these subjects will be given in later chapters.

Chapter 4

Absorption Intensity and Selection Rules

4.1. Transition Probability

Light possesses the dual properties of a stream of particles, called photons, and an electromagnetic wave. The wavelength λ, as well as the wave number ν of the wave, and the momentum p of the equivalent photon are connected by the de Broglie relationship

$$p = h/\lambda = h\nu, \tag{4.1}$$

where h is the Planck constant. The energy of a photon is $hc\nu$, where c is the light velocity. An electromagnetic wave is characterized by an electric vector and a magnetic vector, which form a mutually perpendicular set of axes with the propagation direction, and whose amplitudes vary periodically. If the light consists of waves whose electric vectors all lie in parallel planes, it is called plane-polarized or linearly polarized light, and the direction of the electric vectors is called the direction of (electric) polarization of the light.* If the light consists of waves whose electric vectors lie evenly in all the possible directions, it is called natural light.

The absorption of light by a molecule occurs mainly through the interaction of the electric vector of the light with the charged particles, electrons and nuclei, in the molecule. A molecule which is initially in a state m can absorb a photon of energy $hc\nu$ if there is a state n which satisfies the following condition:

$$E_n - E_m = hc\nu, \tag{4.2}$$

where E_n and E_m are the energies of the states n and m, respectively. If this condition, the so-called Bohr condition, is satisfied, the probability of the photon being absorbed depends on the magnitude of a vector quantity called the matrix element of the electric dipole moment between the two states. This

* In old usage, the plane of polarization refers to the plane containing the magnetic vector and the propagation direction.

is defined by

$$\mathbf{R}_{mn} = e\int \Psi_m^* \left(\sum_i \mathbf{r}_i - \sum_I Z_I \mathbf{r}_I\right) \Psi_n \, d\tau = e\mathbf{M}_{mn}, \tag{4.3}$$

where Ψ_m and Ψ_n are the wavefunctions of the states m and n, respectively, $\mathbf{r}_i$ is the position vector of the ith electron, $\mathbf{r}_I$ is that of the Ith nucleus of atomic number Z_I, in the molecule, and the summation is carried out over all the particles in the molecule. The quantity $(\sum_i \mathbf{r}_i - \sum_I Z_I \mathbf{r}_I)$ is called the electric dipole moment operator for the molecule. The quantity $\mathbf{R}_{mn}$ can be considered as a measure of a charge migration during the transition between the two states m and n. If $n = m$, then $\mathbf{R}_{mm}$ is just the dipole moment of the state m. $\mathbf{R}_{mn}$ or $\mathbf{M}_{mn}$ is called the (electric) transition moment of the transition between the states m and n. In this book the convention will be adopted that the transition moment refers to $\mathbf{M}_{mn}$.

The transition moment $\mathbf{M}$ is a vector, that is, it has magnitude and direction. The magnitude of the transition moment is sometimes called the transition moment length, and is usually given in units of angstrom. The square of the transition moment length, M^2, is often called the dipole strength of the transition, and is denoted by a symbol D. The direction of the transition moment is called the direction of (electric) polarization of the transition. If the Cartesian coordinate system is assigned to the molecule, the transition moment can be resolved into three components along the three Cartesian coordinate axes:

$$D_{mn} = M_{mn}^2 = M_{mn(x)}^2 + M_{mn(y)}^2 + M_{mn(z)}^2, \tag{4.4}$$

where, for example,

$$M_{mn(x)} = \int \Psi_m^* \left(\sum_i x_i - \sum_I Z_I x_I\right) \Psi_n \, d\tau, \tag{4.5}$$

in which x_i and x_I are the x coordinates of the ith electron and the Ith nucleus, respectively. If the molecule has symmetry axes, and if a transition has its transition moment in the direction of the long symmetry axis of the molecule, the transition, as well as the corresponding absorption band, is said to be longitudinally polarized; if a transition has its transition moment in the direction perpendicular to the long axis, the transition, as well as the corresponding absorption band, is said to be transversely polarized. Determination of the direction of polarization of an absorption band is of great help in assigning the underlying transition.

If the incident light is linearly polarized, the probability of a photon of wave number ν_{mn} being absorbed by the molecule depends not on the magnitude of the transition moment $\mathbf{M}_{mn}$ but on its component in the direction of polarization of the light. This means that the light is absorbed most effectively when the direction of polarization of the incident light and that of the transition coincide. Since molecules in vapor and liquid phases normally are randomly oriented with respect to the incident light, there are many possible relations between the direction of polarization of the light and that of the transition, and the absorption probability depends on the mean magnitude of the squares of the components of the randomly oriented transition moments in the direction of polarization of the light.

By use of a single crystal or of a monomolecular film of oriented molecules it is possible to achieve a fixed orientation of the molecules with respect to the direction of polarization of the incident light and hence it is possible, at least in principle, to determine the direction of polarization of a given absorption band. However, interpretation of data obtained by using crystalline samples is complicated by the fact that the molecules so oriented are not free, but are in the crystal field, which, as will be explained in a later chapter, appreciably affects the spectra. In some cases, a solid solution in an inert hydrocarbon crystal of the compound under examination can be used as a sample of oriented molecules of that compound. With highly polar compounds, by placing a solution of the compound in a strong electric field, it is possible to orient a larger number of molecules in a direction than in other directions. Examination of the effects of substituents on spectra is often of some help in determining the direction of polarization of absorption bands of the parent compounds. In general, extension of conjugation by substitution of a mesomeric group in a given direction affects primarily a band polarized in that direction.

Under the condition that the incident light is natural and the molecules are randomly oriented, the probability of a photon of wave number ν_{mn} being absorbed by a molecule in unit time is given by $B_{mn} \cdot \rho(\nu_{mn})$, where $\rho(\nu_{mn})$ is the radiation density at the wave number ν_{mn}, and B_{mn} is the Einstein coefficient of absorption, which is related to the transition moment by the expression

$$B_{mn} = \frac{8\pi^3 e^2}{3h^2 c} M_{mn}^2 . \tag{4.6}$$

The oscillator strength of the transition $m \rightarrow n$, f_{mn}, is a dimensionless quantity which was defined originally as the ratio of the quantum mechanical and classical contributions of the transition $m \rightarrow n$ to the polarizability of the molecule where the classical value was calculated on the assumption of a single electron oscillating harmonically with the frequency $c\nu_{mn}$. The oscillator strength is related to the Einstein coefficient of absorption by the expression

$$f_{mn} = \frac{mhc^2\nu_{mn}}{\pi e^2} B_{mn} . \tag{4.7}$$

Accordingly, the oscillator strength is related to the transition moment of that transition by the expression

$$f_{mn} = \frac{8\pi^2 mc}{3h} \nu_{mn} M_{mn}^2 . \tag{4.8}$$

When M_{mn} is given in units of angstrom, this expression reduces to

$$f_{mn} \fallingdotseq 1.085 \times 10^{-5} \nu_{mn} M_{mn}^2 . \tag{4.9}$$

As mentioned already (see Subsection 1.4.2.), the oscillator strength of an electronic transition is related to the integrated intensity of the corresponding absorption band by the expression

$$f \doteqdot 4.319 \times 10^{-9} \int \epsilon_\nu \, d\nu. \tag{4.10}$$

The value of the oscillator strength calculated from the theoretically determined transition moment length by use of the expression (4.9) will be referred to as the theoretical f value, and that calculated from the integrated intensity of the experimentally determined absorption band by use of the expression (4.10) will be referred to as the experimental f value. For strongly allowed transitions ($\epsilon_{max} \simeq 10^5$), the value of the oscillator strength is of the order of 1. For a transition at about 25,000 cm^{-1} (400 mμ), a value of f of about 1 corresponds to a value of M of about 2 Å.*

When a molecule in the excited state n is illuminated by light of wave number ν_{mn}, it may be induced to revert to the lower state m, emitting a photon of this same wave number. The probability of this induced emission to occur is the same as the corresponding probability of absorption, $B_{mn} \cdot \rho(\nu_{mn})$. The molecule may spontaneously emit a photon even in the absence of a radiation field. The probability of the spontaneous radiative transition from the excited state n to the lower state m is given by

$$A_{nm} = 8\pi hc\nu_{mn}^3 B_{mn} = \frac{64\pi^4 e^2}{3h} \nu_{mn}^3 M_{mn}^2 = \frac{8\pi^2 e^2}{mc} \nu_{mn}^2 f_{mn} \fallingdotseq 0.667 \, \nu_{mn}^2 f_{mn} . \tag{4.11}$$

The quantity A_{nm} is called the Einstein coefficient of spontaneous emission.

* Even if the matrix element of the electric dipole moment is zero for a transition, if the matrix element of the magnetic dipole moment or that of the electric quadrupole moment is different from zero, the transition can occur. However, such a transition has quite a small probability. Calculation shows that the probabilities of magnetic dipole transitions and those of electric quadrupole transitions are only about 10^{-5} and 10^{-8} of those of electric dipole transitions, respectively.

If at some instant there are $N_n(0)$ molecules in the excited state n, and if the transition from n to m is the only spontaneous radiative transition possible in the state n, the number of molecules in the state n at a later time t is given by the exponential function

$$N_n(t) = N_n(0) \exp(-A_{nm}t). \tag{4.12}$$

If several transitions are possible, expression (4.12) has to be replaced by

$$N_n(t) = N_n(0) \exp(-\sum_m A_{nm}t). \tag{4.13}$$

After a time $\tau_n = 1/A_{nm}$ or $1/\sum_m A_{nm}$ the number of molecules left in the state n is $1/e$ of the initial number $N_n(0)$, where e represents the base of the natural logarithms, 2.71828 . . . . This time τ_n is called the radiative lifetime of the state n. Evidently τ_n can be expressed as

$$\tau_n = (mc/8\pi^2e^2)\left(1/\sum_m \nu_{mn}^2 f_{mn}\right) \doteqdot 1.50/\sum_m \nu_{mn}^2 f_{mn} \qquad \text{(in sec.).} \tag{4.14}$$

For excited states of strongly allowed transitions ($f \simeq 1$) τ is of the order of 10^{-9}–10^{-8} sec. If no allowed transitions can occur from any lower state to the excited state n, the lifetime of the state n is much larger: in most cases the lifetime of the lowest triplet excited state of a molecule is of the order of 10^{-3} sec. or longer.

4.2. Total Probability of an Electronic Transition and Its Distribution among Vibrational Components

According to the Born–Oppenheimer approximation (see Subsection 2.2.1), a complete state function of a molecule can be factorized into a product of an electronic and a nuclear wavefunction as expression (2.16). The nuclear function can be further factorized into a product of a rotational and a vibrational function. As mentioned already, the energy separation of different electronic states is of the order of 10,000 cm^{-1}; the separation of different vibrational states associated with the same electronic state is of the order of 1000 cm^{-1}; the separation of different rotational states belonging to the same vibronic state is much smaller, being of the order 10–100 cm^{-1}. By the way, the word "vibronic" has been coined to designate an electronic-vibrational state: a vibrational state belonging to an electronic state is called a vibronic state, and the transition from a vibrational state belonging to an electronic state to a vibrational state belonging to a different electronic state is called a vibronic transition.

In an equilibrium system, in the absence of a radiation field, the number of molecules in each of the stationary states is proportional to the Boltzmann factor, $\exp(-E/kT)$, where E is the energy of the state relative to the lowest energy state, k is the Boltzmann constant, and T is the absolute temperature. The quantity kT at room temperature corresponds to a wave number of about 200 cm^{-1}. Therefore, at ordinary temperature, excited electronic states are never thermally populated, and almost all the molecules will be in a large number of rotational levels of the lowest vibrational state belonging to the ground electronic state. Accordingly, the main part of an electronic absorption band will be due to transitions from the ground vibronic state to various vibrational states belonging to the excited electronic state. The population of a large number of rotational states belonging to the ground vibronic state and the presence of a large number of rotational states belonging to each vibrational state of the excited electronic state will just broaden out the vibrational components of the band.

For the purpose of simplification, we consider a diatomic molecule, which has only one vibrational mode, that is, the stretching vibration. Let us denote electronic and vibrational wavefunctions by symbols $\Psi(q, Q)$ and $X(Q)$, respectively, where q and Q are general symbols for electronic and nuclear coordinates, respectively. In the case of diatomic molecules, Q can be taken to simply represent the internuclear distance. The transition moment of the vibronic transition from the lowest vibrational state (the vibrational quantum number $v^{\mathrm{G}} = 0$) of the ground electronic state, $\Psi_{\mathrm{G}}(q, Q)X_{\mathrm{G}0}(Q)$, to the kth vibrational state (the vibrational quantum number $v^{\mathrm{E}} = k$) of an excited electronic state, $\Psi_{\mathrm{E}}(q, Q)X_{\mathrm{E}k}(Q)$, is given by the integral [see Eq. (4.3)]

$$\mathbf{M}_{\mathrm{G}0,\mathrm{E}k} = \int \Psi_{\mathrm{G}}^*(q, Q)X_{\mathrm{G}0}^*(Q)\left(\sum_i \mathbf{r}_i - \sum_I Z_I\mathbf{r}_I\right)\Psi_{\mathrm{E}}(q, Q)X_{\mathrm{E}k}(Q)\, dq\, dQ. \tag{4.15}$$

This integral may be split into two parts by separating the electronic and nuclear coordinates:

$$\mathbf{M}_{\mathrm{G}0,\mathrm{E}k} = \int X_{\mathrm{G}0}^*(Q)\left[\int \Psi_{\mathrm{G}}^*(q, Q)\left(\sum_i \mathbf{r}_i\right)\Psi_{\mathrm{E}}(q, Q)\, dq\right]X_{\mathrm{E}k}(Q)\, dQ$$
$$- \int X_{\mathrm{G}0}^*(Q)\left(\sum_I Z_I\mathbf{r}_I\right)\left[\int \Psi_{\mathrm{G}}^*(q, Q)\Psi_{\mathrm{E}}(q, Q)\, dq\right]X_{\mathrm{E}k}(Q)\, dQ. \tag{4.16}$$

Since electronic functions for the same value of Q are mutually orthogonal, the second term vanishes. Therefore, if an electronic transition moment is defined by

$$\mathbf{M}_{\mathrm{G},\mathrm{E}}(Q) = \int \Psi_{\mathrm{G}}^*(q, Q)\left(\sum_i \mathbf{r}_i\right)\Psi_{\mathrm{E}}(q, Q)\, dq, \tag{4.17}$$

the expression (4.16) is reduced to the following form:

$$\mathbf{M}_{\mathrm{G}0,\mathrm{E}k} = \int X_{\mathrm{G}0}^*(Q)\mathbf{M}_{\mathrm{G},\mathrm{E}}(Q)\, X_{\mathrm{E}k}(Q)\, dQ. \tag{4.18}$$

Since the electronic wavefunctions $\Psi^{\cdot}$ are functions of the nuclear coordinates Q, the electronic transition moment $\mathbf{M}_{G,E}(Q)$ is also a function of Q. If it is assumed that the variation of $\mathbf{M}_{G,E}(Q)$ with Q is slow, and that $\mathbf{M}_{G,E}(Q)$ may be replaced by some average value $\bar{\mathbf{M}}_{G0,E}$, then we have

$$\mathbf{M}_{G0,Ek} = \bar{\mathbf{M}}_{G0,E} S_{G0,Ek}, \tag{4.19}$$

where $S_{G0,Ek}$ is the overlap integral

$$S_{G0,Ek} = \int X_{G0}^*(Q) X_{Ek}(Q) \, dQ. \tag{4.20}$$

Vibrational wavefunctions belonging to different electronic states are not orthogonal, since they are eigenfunctions of different Hamiltonian operators [cf. Eq. (2.18)]. The probability of the vibronic transition from the G0 state to the Ek state is proportional to the square of the modulus of $\mathbf{M}_{G0,Ek}$. Thus the oscillator strength for this transition is given by

$$f_{G0,Ek} \doteqdot 1.085 \times 10^{-5} \nu_{G0,Ek} S_{G0,Ek}^2 \bar{M}_{G0,E}^2 . \tag{4.21}$$

This means that, if it is assumed that the wave number ν does not much vary over an electronic absorption band, the relative intensities of vibrational subbands of the electronic band will be proportional to the square of the corresponding vibrational overlap integrals. If $S_{G0,Ek}^2$ has its maximum value when $k = m$, then the transition from the G0 state to the Em state is the most probable of the possible vibronic transitions from the ground vibronic state to various vibrational states belonging to the excited electronic state considered. As will be shown later, the Em state is the state which has one of the classical turning points of the vibration nearly vertically above the minimum of the potential energy curve for the ground electronic state.

The overall probability of an electronic transition is the sum of contributions from the individual vibrational components. If it is assumed that all the molecules are initially in the G0 state, and hence the electronic transition can occur only from this state, the oscillator strength of the electronic transition G → E can be expressed as follows:

$$\begin{aligned} f_{G,E} &= \sum_k f_{G0,Ek} \doteqdot 1.085 \times 10^{-5} \sum_k \nu_{G0,Ek} M_{G0,Ek}^2 \\ &\doteqdot 1.085 \times 10^{-5} \bar{M}_{G0,E}^2 \sum_k \nu_{G0,Ek} S_{G0,Ek}^2 . \end{aligned} \tag{4.22}$$

All the vibrational wavefunctions for a vibrational mode associated with an electronic state form a complete set of orthogonal functions, since they are eigenfunctions of the same Hamiltonian operator. Thus the infinite set of vibrational functions belonging to the ground electronic state, X_{Gi}'s, where i is the running index identifying different vibrational functions, is a complete set, and the infinite set of vibrational functions belonging to the excited electronic state E, X_{Ek}'s, is a different complete set. As mentioned previously, it is always possible to expand any given function in terms of functions of the same variables forming a complete set. Therefore, each X_{Ek} can be expanded in terms of X_{Gi}'s:

$$X_{Ek} = \sum_i c_{ki} X_{Gi} . \tag{4.23}$$

The transformation of one complete set of orthogonal functions into another must obey the following conditions:

$$\sum_i c_{ki}^* c_{li} = \delta_{kl} ; \qquad \sum_k c_{ki}^* c_{kj} = \delta_{ij} . \tag{4.24}$$

Hence we have

$$\begin{aligned} \sum_k S_{G0,Ek}^2 &= \sum_k \left[\int X_{G0}^* \left(\sum_i c_{ki} X_{Gi} \right) dQ \cdot \int \left(\sum_i c_{ki}^* X_{Gi}^* \right) X_{G0} dQ \right] \\ &= \sum_k c_{k0}^* c_{k0} S_{G0,G0}^2 = 1. \end{aligned} \tag{4.25}$$

The average value of $\nu_{G0,Ek}$'s can be defined as

$$\bar{\nu}_{G0,E} = \sum_k \nu_{G0,Ek} S_{G0,Ek}^2 / \sum_k S_{G0,Ek}^2 = \sum_k \nu_{G0,Ek} S_{G0,Ek}^2 . \tag{4.26}$$

On substituting this expression into (4.22), the total oscillator strength of the electronic transition can be written as

$$f_{G,E} \doteqdot 1.085 \times 10^{-5} \bar{\nu}_{G0,E} \bar{M}_{G0,E}^2 . \tag{4.27}$$

From Eqs. (4.19) and (4.25) we obtain

$$\sum_k M_{G0,Ek}^2 = \bar{M}_{G0,E}^2 \sum_k S_{G0,Ek}^2 = \bar{M}_{G0,E}^2 . \tag{4.28}$$

On the other hand, substituting expression (4.23) into (4.18), we have

$$\mathbf{M}_{G0,Ek} = \sum_i c_{ki} \int X_{G0}^*(Q) \mathbf{M}_{G,E}(Q) X_{Gi}(Q) \, dQ. \tag{4.29}$$

Since X_{Gi}'s are mutually orthogonal, if $\mathbf{M}_{G,E}(Q)$ is assumed not to be a rapidly varying function of Q, the integrals in expression (4.29) can be approximated as being zero except the term where $i = 0$:

$$\mathbf{M}_{G0,Ek} \doteqdot c_{k0} \int X_{G0}^*(Q) \mathbf{M}_{G,E}(Q) X_{G0}(Q) \, dQ. \tag{4.30}$$

Accordingly, we have

$$\sum_k M^2_{\mathrm{G0,E}k} = \sum_k c^*_{k0}c_{k0}\left|\int \mathrm{X}^*_{\mathrm{G0}}(Q)\mathbf{M}_{\mathrm{G,E}}(Q)\mathrm{X}_{\mathrm{G0}}(Q)\,dQ\right|^2$$
$$= \left|\int \mathrm{X}^*_{\mathrm{G0}}(Q)\mathbf{M}_{\mathrm{G,E}}(Q)\mathrm{X}_{\mathrm{G0}}(Q)\,dQ\right|^2. \tag{4.31}$$

From Eqs. (4.28) and (4.31) we obtain

$$\bar{\mathbf{M}}_{\mathrm{G0,E}} \doteqdot \int \mathrm{X}^*_{\mathrm{G0}}(Q)\mathbf{M}_{\mathrm{G,E}}(Q)\mathrm{X}_{\mathrm{G0}}(Q)\,dQ. \tag{4.32}$$

This integral has a meaning of the average of $\mathbf{M}_{\mathrm{G,E}}(Q)$ in the lowest vibrational state of the ground electronic state.

In general, a vibrational motion whose vibrational quantum number is sufficiently low can be well approximated as a simple harmonic vibration, and the range of variation of the nuclear coordinates covered by such a vibration is relatively small. Especially, the lowest vibrational function ($v = 0$) has only one broad maximum of its probability density approximately above the minimum of the potential energy curve for the electronic state. Therefore, the integral in expression (4.32) can reasonably be approximated as $\mathbf{M}_{\mathrm{G,E}}(Q^{\mathrm{G}}_{\mathrm{eq}})$, which means the electronic transition moment evaluated at the nuclear configuration corresponding to the minimum of the potential energy curve for the ground electronic state, that is, at the equilibrium nuclear configuration in the ground electronic state; we can then write

$$\sum_k M^2_{\mathrm{G0,E}k} \doteqdot M^2_{\mathrm{G,E}}(Q^{\mathrm{G}}_{\mathrm{eq}}). \tag{4.33}$$

This relation can also be derived in the following manner. Suppose that ϕ_i's form a complete set of orthonormal functions, and that $\mathbf{O}$ is an operator. Then, it is easily verified that $\mathbf{O}\phi_i$ can be expanded as follows:

$$\mathbf{O}\phi_i = \sum_j O_{ji}\phi_j\,. \tag{4.34}$$

Therefore, we have

$$(O^2)_{ii} \equiv \int \phi_i{}^*\mathbf{O}^2\phi_i\,d\tau = \int \phi_i{}^*\mathbf{O}\mathbf{O}\phi_i\,d\tau = \int \phi_i{}^*\mathbf{O}\Big(\sum_j O_{ji}\phi_j\Big)\,d\tau$$
$$= \sum_j O_{ij}O_{ji} \equiv \sum_j\Big(\int \phi_i{}^*\mathbf{O}\phi_j\,d\tau\Big)\Big(\int \phi_j{}^*\mathbf{O}\phi_i\,d\tau\Big). \tag{4.35}$$

From Eqs. (4.29) and (4.24) we obtain

$$\begin{aligned}\sum_k M^2_{\mathrm{G0,E}k} &= \sum_k \sum_i \sum_j c^*_{ki} c_{kj} \left(\int \mathrm{X}^*_{\mathrm{G0}} \mathbf{M}_{\mathrm{G,E}} \mathrm{X}_{\mathrm{G}j} \, dQ \right) \left(\int \mathrm{X}^*_{\mathrm{G}i} \mathbf{M}_{\mathrm{G,E}} \mathrm{X}_{\mathrm{G0}} \, dQ \right) \\ &= \sum_i \sum_j \delta_{ij} \left(\int \mathrm{X}^*_{\mathrm{G0}} \mathbf{M}_{\mathrm{G,E}} \mathrm{X}_{\mathrm{G}j} \, dQ \right) \left(\int \mathrm{X}^*_{\mathrm{G}i} \mathbf{M}_{\mathrm{G,E}} \mathrm{X}_{\mathrm{G0}} \, dQ \right) \\ &= \sum_j \left(\int \mathrm{X}^*_{\mathrm{G0}} \mathbf{M}_{\mathrm{G,E}} \mathrm{X}_{\mathrm{G}j} \, dQ \right) \left(\int \mathrm{X}^*_{\mathrm{G}j} \mathbf{M}_{\mathrm{G,E}} \mathrm{X}_{\mathrm{G0}} \, dQ \right). \end{aligned} \tag{4.36}$$

By use of relation (4.35) this equation is reduced to

$$\sum_k M^2_{\mathrm{G0,E}k} = \int \mathrm{X}^*_{\mathrm{G0}} \mathbf{M}^2_{\mathrm{G,E}} \mathrm{X}_{\mathrm{G0}} \, dQ. \tag{4.37}$$

This integral is the average of the square modulus of $\mathbf{M}_{\mathrm{G,E}}(Q)$ in the ground vibrational state, and can reasonably be replaced by $M^2_{\mathrm{G,E}}(Q^{\mathrm{G}}_{\mathrm{eq}})$.

Accordingly, the total oscillator strength of the electronic transition $\mathrm{G} \to \mathrm{E}$ can be written as

$$f_{\mathrm{G,E}} \doteqdot 1.085 \times 10^{-5} \bar{\nu}_{\mathrm{G0,E}} M^2_{\mathrm{G,E}}(Q^{\mathrm{G}}_{\mathrm{eq}}). \tag{4.38}$$

The value of $\bar{\nu}_{\mathrm{G0,E}}$ can be approximated as being equal to $\nu_{\mathrm{G0,E}m}$, that is, the transition energy in units of cm^{-1} of the most probable transition of the transitions from the ground vibronic state (G0) to various vibrational states associated with the excited electronic state E. Practically, the wave number at the absorption maximum of the band, ν_{max}, can be used as an approximate value of $\bar{\nu}_{\mathrm{G0,E}}$.

It has so far been assumed that in the equilibrium system all the molecules are in the ground vibronic state and that all the possible electronic transitions start from this state. However, if the energy separation of vibrational levels is comparatively small, some higher vibrational states of the ground electronic state will be populated by some small portion of the molecules, and hence electronic transitions from these excited vibrational states will occur. The probability of the vibronic transition from the Gi state to the Ek state is given by $P_{\mathrm{G}i} f_{\mathrm{G}i,\mathrm{E}k}$, in which $P_{\mathrm{G}i}$ is the population ratio of the Gi state, that is, the ratio of the number of molecules in the Gi state to the total number of molecules in the thermal equilibrium. The total probability of the transitions from the Gi state to various vibrational states of the excited electronic state E may be given by

$$\begin{aligned} P_{\mathrm{G}i} \sum_k f_{\mathrm{G}i,\mathrm{E}k} &\doteqdot P_{\mathrm{G}i} \times 1.085 \times 10^{-5} \bar{\nu}_{\mathrm{G}i,\mathrm{E}} \bar{M}^2_{\mathrm{G}i,\mathrm{E}} \\ &\doteqdot P_{\mathrm{G}i} \times 1.085 \times 10^{-5} \bar{\nu}_{\mathrm{G}i,\mathrm{E}} M^2_{\mathrm{G,E}}(Q^{\mathrm{G}}_{\mathrm{eq}}). \end{aligned} \tag{4.39}$$

Since the population ratios of excited vibrational states of the ground electronic state rapidly decrease with lowering of temperature, the intensity of vibrational subbands due to the transitions from these excited vibrational states will also rapidly decrease with lowering of temperature. Bands whose intensity is highly temperature-dependent are called hot bands.

The total oscillator strength of the electronic transition G → E can be expressed as

$$f_{\mathrm{G,E}} \doteqdot 1.085 \times 10^{-5} \sum_i P_{\mathrm{G}i} \bar{\nu}_{\mathrm{G}i,\mathrm{E}} M^2_{\mathrm{G,E}}(Q^{\mathrm{G}}_{\mathrm{eq}})$$

$$= 1.085 \times 10^{-5} \bar{\nu}_{\mathrm{G,E}} M^2_{\mathrm{G,E}}(Q^{\mathrm{G}}_{\mathrm{eq}}), \tag{4.40}$$

where

$$\bar{\nu}_{\mathrm{G,E}} = \sum_i P_{\mathrm{G}i} \bar{\nu}_{\mathrm{G}i,\mathrm{E}} \,. \tag{4.41}$$

Since the population ratios of excited vibrational states, $P_{\mathrm{G}i}$'s for $i \neq 0$, are normally much smaller than the population ratio of the lowest vibrational state, P_{G0}, the average wave number over the electronic absorption band, $\bar{\nu}_{\mathrm{G,E}}$, can be taken to be nearly equal to $\bar{\nu}_{\mathrm{G0,E}}$. When this approximation is made, expression (4.40) becomes identical with expression (4.38).

Expression (4.38) or (4.40) is not valid for the cases where the transition moment evaluated just for the equilibrium configuration in the ground electronic state, $M_{\mathrm{G,E}}(Q^{\mathrm{G}}_{\mathrm{eq}})$, is zero. In these cases we must be more careful in calculating the average value of the transition moment in the vibrational states of the ground electronic states, $\bar{M}_{\mathrm{G}i,\mathrm{E}}$'s.

When $M_{\mathrm{G,E}}(Q^{\mathrm{G}}_{\mathrm{eq}})$ is nonzero, the electronic transition G → E is said to be allowed in the lowest approximation; when $M_{\mathrm{G,E}}(Q^{\mathrm{G}}_{\mathrm{eq}})$ is zero, the electronic transition G → E is said to be forbidden in the lowest approximation. There are simple rules which state some conditions under which the electronic transition moment for the equilibrium configuration, for example, $M_{\mathrm{G,E}}(Q^{\mathrm{G}}_{\mathrm{eq}})$, is zero. These rules are called selection rules for electronic transition.

4.3. Selection Rules

4.3.1. The Spin Selection Rule

Transitions between states of different spin multiplicities are invariably forbidden, since electrons cannot change their spin except for spin–orbit and spin–spin interactions. This rule is called the spin selection rule.

The transition moment of the electronic transition from an electronic state, say m, to an electronic state, say n, can be expressed as follows:

$$\mathbf{M}_{m,n}(Q^{m}_{\mathrm{eq}}) = \int \Psi_m{}^*(Q^{m}_{\mathrm{eq}}) \sum_i \mathbf{r}_i \Psi_n(Q^{m}_{\mathrm{eq}})\, d\tau, \tag{4.42}$$

where Q_{eq}^m represents the equilibrium nuclear configuration in the initial electronic state m, and $\Psi_m(Q_{\text{eq}}^m)$ and $\Psi_n(Q_{\text{eq}}^m)$ are the electronic wavefunctions of the m and n states, respectively, at the equilibrium configuration in the m state. If it is assumed that there is no spin–orbit interaction, we can separate an electronic wavefunction into a space function, which is a function containing only the space coordinates (q) of electrons as variables, and a spin function, which is a function only of the spin coordinates (σ) of electrons:

$$\begin{aligned} \Psi_m(Q_{\text{eq}}^m) &= \Psi_{m(\text{space})}(Q_{\text{eq}}^m)\Psi_{m(\text{spin})}\,; \\ \Psi_n(Q_{\text{eq}}^m) &= \Psi_{n(\text{space})}(Q_{\text{eq}}^m)\Psi_{n(\text{spin})}\,. \end{aligned} \tag{4.43}$$

Since $\sum_i \mathbf{r}_i$ does not contain the spin coordinates of electrons, the intensity-determining integral $\mathbf{M}$ can be separated into a space part and a spin part:

$$\mathbf{M}_{m,n}(Q_{\text{eq}}^m) = \int \Psi^*_{m(\text{space})}(Q_{\text{eq}}^m) \sum_i \mathbf{r}_i \Psi_{n(\text{space})}(Q_{\text{eq}}^m)\, dq \int \Psi^*_{m(\text{spin})}\Psi_{n(\text{spin})}\, d\sigma. \tag{4.44}$$

Spin functions are, of course, orthonormal. Therefore, if $\Psi_{m(\text{spin})}$ and $\Psi_{n(\text{spin})}$ are different from each other, the second integral in expression (4.44), that is, the overlap integral of the two spin functions, is zero, and hence the transition moment $\mathbf{M}_{m,n}(Q_{\text{eq}}^m)$ vanishes. Thus, transitions between states of different spin multiplicities, for example, transitions from a singlet state to triplet states (S–T transitions), are forbidden. On the other hand, if the two spin functions are identical, the second integral in expression (4.44) is equal to unity, and hence the expression reduces to its space part

$$\mathbf{M}_{m,n}(Q_{\text{eq}}^m) = \int \Psi^*_{m(\text{space})}(Q_{\text{eq}}^m) \sum_i \mathbf{r}_i \Psi_{n(\text{space})}(Q_{\text{eq}}^m)\, dq. \tag{4.45}$$

This means that transitions between states of the same spin multiplicity, for example, transitions from a singlet state to different singlet states (S–S transitions), are allowed if the space part of the transition moment is not zero.

It is to be noted that the spin selection rule is based on the assumption that there are no spin–orbit and spin–spin interactions. Actually, electrons can change their spin very slowly owing to spin–orbit and spin–spin interactions, and hence transitions between states of different spin multiplicities can occur with very small probabilities.

The breakdown of the spin selection rule by the spin–orbit interactions can be explained as follows. The spin–orbit interactions are taken into account by replacing the electronic Hamiltonian in the first approximation, $\mathbf{H}$, by

$\mathbf{H} + \mathbf{H}'$, where the perturbation $\mathbf{H}'$ represents the spin–orbit coupling. Since $\mathbf{H}'$ involves both the electron spin and space coordinates, the electron spin operators $\mathbf{S}^2$ and $\mathbf{S}_z$ are no longer commutable with the new Hamiltonian $\mathbf{H} + \mathbf{H}'$, and eigenfunctions of this Hamiltonian cannot be eigenfunctions of $\mathbf{S}^2$ and $\mathbf{S}_z$. That is, in this case, spin quantization breaks down. The matrix elements of $\mathbf{H}'$ between eigenfunctions Ψ of the original Hamiltonian $\mathbf{H}$ can be nonzero even when those functions have different spin multiplicities. This means that the spin–orbit coupling results in the mixing of states of different spin multiplicities.

Now consider the transition from a singlet state V to a triplet state T. Suppose that the triplet state T mixes with a singlet state V' owing to the spin–orbit coupling as

$$T^{\mathrm{p}} = N_T(T + \lambda_T V'), \tag{4.46}$$

where λ_T is a constant whose absolute value is much smaller than unity and N_T is the normalization constant, whose value is near to unity. While the transition from V to T is forbidden, the transition moment of the transition from V to V', $\mathbf{M}_V$, may be nonzero. Therefore, the transition moment of the transition from the singlet state V to the perturbed triplet state T^{p} is given by $N_T\lambda_T\mathbf{M}_V$. If $\mathbf{M}_V$ is different from zero, the transition $V \rightarrow T^{\mathrm{p}}$ can occur. Then, we say that the formally forbidden singlet–triplet transition $V \rightarrow T$ has become allowed by borrowing the intensity from the allowed singlet–singlet transition $V \rightarrow V'$.

The singlet state V may mix with a triplet state T' as a result of the spin–orbit coupling:

$$V^{\mathrm{p}} = N_V(V + \lambda_V T'). \tag{4.47}$$

The transition moment of the transition from T' to T, $\mathbf{M}_T$, may be nonzero. Therefore, the transition moment of the transition from the perturbed singlet state V^{p} to the perturbed triplet state T^{p} is given by $N_V N_T(\lambda_T\mathbf{M}_V + \lambda_V\mathbf{M}_T)$, receiving contributions from both the $V \rightarrow V'$ and $T' \rightarrow T$ transitions. The probability of this perturbed S–T transition evidently depends on the extent of the mixing of the singlet and triplet states, which in turn is proportional to the matrix element of the spin–orbit coupling $\mathbf{H}'$ between the states to be mixed and is inversely proportional to the energy separation of these states [see Section 2.4].

The spin–orbit coupling can be qualitatively understood from the following classical model: an electron moving round a nucleus gives rise to a magnetic field, and this field interacts with the magnetic moment associated with the spinning electron. The strength of the spin–orbit coupling depends strongly on the strength of the electric field due to the atomic nuclei working upon the electrons. For most organic molecules, which contain only light nuclei, the spin–orbit coupling is very small, and hence the singlet–triplet transitions have only a very small probability. The oscillator strength of these transitions is only of the order of 10^{-6} or lower.

The intensity of n–π^* singlet–triplet transitions in which the n orbital has some contribution of the 2s atomic orbital is in general considerably greater than that of π–π^* singlet–triplet transitions: the molar extinction coefficient of the absorption maximum of n–π^* S–T bands is of the order of 0.01 or 0.1. The effect of the spin–orbit coupling is considered to be greater in the n–π^* transitions since electrons in the 2s atomic orbital can be nearer to the atomic nucleus than electrons in the π orbitals are.[1]

The spin–orbit coupling becomes much more important in the neighborhood of the nuclei of heavier atoms. It is known that a heavy atom such as iodine, either introduced directly into a molecule as a substituent or contained in the solvent, enhances considerably the intensity of the singlet–triplet transition. For example, McClure[2] showed that the S–T bands of aromatic molecules such as naphthalene could be observed after substitution by bromine or iodine, and Kasha[3] showed that the S–T band of naphthalene has an enhanced intensity when measured in solvents containing dissolved ethyl iodide.

The breakdown of spin quantization can also occur as a result of the spin–spin coupling. If a molecule in a doublet spin state ($s = \frac{1}{2}$) interacts with a molecule in a singlet spin state ($s = 0$), the combined spin states of the two molecules together will be a doublet. If the spin–spin coupling occurs between a molecule in a doublet state and a molecule in a triplet state ($s = 1$), the combined spin states will be a quartet ($s = 1 + \frac{1}{2} = \frac{3}{2}$) or a doublet ($s = 1 + \frac{1}{2} - 1 = \frac{1}{2}$). Therefore, a singlet–triplet transition in a molecule will become allowed on perturbation by a molecule in a doublet state or any other paramagnetic state ($s \neq 0$), since in the complex the perturbed singlet state and one component of the perturbed triplet state have the same spin multiplicity. The extent of the spin–spin coupling strongly depends on the overlap between orbitals of the two molecules, and usually the spin–spin coupling is not complete. However, the spin quantization in a molecule is broken down by the perturbation by a paramagnetic substance, and transitions between molecular states of different spin multiplicities become partially allowed.

[1] M. Kasha, *Discussions Faraday Soc.* **9,** 14 (1950).

[2] D. S. McClure, *J. Chem. Phys.* **17,** 905 (1949); **19,** 670 (1951); **20,** 682 (1952). See also D. S. McClure, N. W. Blake, and P. L. Hanst, *J. Chem. Phys.* **22,** 255 (1954); M. Kasha and S. P. McGlynn, *Ann. Rev. Phys. Chem.* **7,** 403 (1956).

[3] M. Kasha, *J. Chem. Phys.* **20,** 71 (1952).

Actually, Evans[4] showed that the first singlet–triplet band (at about 340 mμ) of benzene was intensified by absorbed oxygen: he reported that the oscillator strength of this band measured with the liquid saturated with oxygen at the partial pressure of the atmosphere was about 4×10^{-6}. On the other hand, Craig and co-workers[5] reported that this band could not be detected even in a 22.5-meter path of the pure liquid free from oxygen, indicating that the oscillator strength must be less than 7×10^{-12}. Evans also succeeded in observing the first singlet–triplet absorption bands of many other organic compounds by saturating the solutions with oxygen or nitric oxide, sometimes under pressure.

4.3.2. The Symmetry Selection Rule

The transition moment of an electronic transition, say $m \rightarrow n$, can be resolved into three components as follows:

$$M^2_{m,n} = M^2_{m,n(x)} + M^2_{m,n(y)} + M^2_{m,n(z)}, \tag{4.48}$$

where

$$M_{m,n(x)} = \int \Psi_m{}^* \sum_i x_i \Psi_n \, d\tau; \qquad M_{m,n(y)} = \int \Psi_m{}^* \sum_i y_i \Psi_n \, d\tau;$$
$$M_{m,n(z)} = \int \Psi_m{}^* \sum_i z_i \Psi_n \, d\tau. \tag{4.49}$$

If Ψ's are Slater determinantal electron configuration functions, the transition moment $\mathbf{M}_{m,n}$ can be reduced to an integral whose integrand is a function of the space coordinates of a single electron. For example, if the initial and final states of the transition are expressed by V_0 [given by expressions (2.46) or (2.46′)] and V_1^{kr} [given by expressions (2.47) or (2.47′)], respectively, it can be easily verified that the transition moment of this transition is expressed as follows:

$$\mathbf{M}_{m,n} = \int V_0{}^* \sum_i \mathbf{r}_i V_1^{kr} \, d\tau = \frac{1}{\sqrt{2}}\left(\int \psi_k{}^*(1)\mathbf{r}_1\psi_r(1)\, d\tau_1 + \int \psi_k{}^*(2)\mathbf{r}_2\psi_r(2)\, d\tau_2\right)$$
$$= \sqrt{2}\int \psi_k{}^*\mathbf{r}\psi_r \, d\tau \equiv \mathbf{M}_{k,r}. \tag{4.50}$$

In general, the transition moment of a singlet–singlet one-electron transition from a doubly occupied orbital k to an unoccupied orbital r is given by expression (4.50). The factor $\sqrt{2}$ can be considered to be a constant introduced

[4] D. F. Evans, *Nature*, **178,** 534 (1956); *J. Chem. Soc.* **1957,** 1351, 3885; **1959,** 2753; **1960,** 1735; **1961,** 1987, 2566.

[5] D. P. Craig, J. M. Hollas, and G. W. King, *J. Chem. Phys.* **29,** 974 (1958).

to allow for the promotion of either of the two electrons in the occupied orbital. The components of the transition moment $\mathbf{M}_{k,r}$ are of course given by

$$M_{k,r(x)} = \sqrt{2}\int \psi_k{}^* x \psi_r \, d\tau; \qquad M_{k,r(y)} = \sqrt{2}\int \psi_k{}^* y \psi_r \, d\tau;$$
$$M_{k,r(z)} = \sqrt{2}\int \psi_k{}^* z \psi_r \, d\tau. \tag{4.51}$$

The principles formulated in Chapter 3, especially in Subsection 3.4.3, can be applied to decide if the transition moment vanishes or not. If a molecule belongs to a point group, its electronic state functions as well as molecular orbitals for the equilibrium nuclear configuration in the ground electronic state and the coordinates x, y, z belong to irreducible representations of the point group, and hence the integrand of each component integral of the transition moment for the equilibrium configuration, $\mathbf{M}(Q_{\mathrm{eq}}^{\mathrm{G}})$, belongs to a representation of the point group. An integral can be different from zero only if the representation to which its integrand belongs is totally symmetric or contains the totally symmetric representation. Thus, if the point group to which the molecule belongs contains only nondegenerate irreducible representations, the selection rule is as follows: the component of the transition moment of a one-electron transition in the direction of one of the coordinate axes can be nonzero only when the direct product of the representations to which the initial and final electronic state functions (or the starting and terminating orbitals) of the transition belong is the same as the representation to which that coordinate axis belongs. Since the ground state of a molecule is in general totally symmetric, this selection rule amounts to saying that the component of the transition moment of a transition from the ground state in the direction of a coordinate axis can be nonzero only when the excited state belongs to the same representation as that coordinate axis does.

For example, in point group D_{2h} the x, y, and z coordinates belong to representations B_{3u}, B_{2u}, and B_{1u}, respectively, and the ground state of a molecule belonging to this point group usually belongs to the totally symmetric representation, A_g (see Table 3.3). Therefore, transitions from the ground state to excited states of symmetry species B_{3u}, B_{2u}, and B_{1u} are allowed, and are polarized along the x, y, and z axes, respectively. As mentioned already (see Subsections 3.3.1 and 3.4.1), the ethylene molecule in the equilibrium geometry belongs to point group D_{2h}, and its bonding and antibonding π orbitals, ψ_{+1} and ψ_{-1}, belong to symmetry species B_{1u} and B_{2g}, respectively. Therefore, the excited electron configuration $\psi_{+1}\psi_{-1}$ belongs to symmetry species B_{3u}, and hence the singlet–singlet transition from the ground electron configuration to this excited electron configuration, or the singlet–singlet one-electron transition from the bonding π orbital to the antibonding π orbital, is allowed and is polarized in the direction of the x axis, that is, in the direction of the carbon—carbon bond (see Fig. 3.1).

If the direct product of the representations to which the initial and final state functions (or the starting and terminating orbitals) belong is different from all the representations to which the coordinate axes belong, the transition moment of that transition is of course zero. Such a transition is said to be symmetry-forbidden or forbidden by symmetry.

When the point group contains degenerate irreducible representations, the symmetry selection rule is as follows: the component of the transition moment in the direction of a coordinate axis can be nonzero only when the direct product of the representations to which the initial and final wavefunctions belong contains at least once the representation to which that coordinate axis belongs.

For example, in point group D_{6h} the pair of coordinates x and y forms a basis for the doubly degenerate representation E_{1u} and the z coordinate belongs to the representation A_{2u}. Therefore, in a molecule belonging to this point group, transitions from the ground state to any excited states belonging to symmetry species other than E_{1u} and A_{2u} are forbidden by symmetry. As mentioned already (see Subsections 3.3.2 and 3.4.2), the benzene molecule in the equilibrium nuclear configuration belongs to point group D_{6h}, and the lowest-energy excited electron configuration $a_{1u}^2e_{1g}^3e_{2u}$ gives rise to electronic states of symmetry E_{1u}, B_{1u}, and B_{2u}, which can be either singlet or triplet (see Tables 3.5 and 3.7). The ground electron configuration, $a_{1u}^2e_{1g}^4$, belongs to the totally symmetric representation, A_{1g}, and is of course a singlet. In the spectrum of benzene an intense band appears near 180 mμ ($\epsilon_{\max} \approx 50{,}000$; $f \approx 0.9$), which can only come from a symmetry-allowed transition and must be due to the ${}^1A_{1g} \rightarrow {}^1E_{1u}$ transition. This band has a shoulder on it near 200 mμ ($\epsilon \approx 7000$; $f \approx 0.1$) which must be due to another electronic transition. Goeppert-Mayer and Sklar[6] assigned this transition as the symmetry-forbidden ${}^1A_{1g} \rightarrow {}^1B_{1u}$ transition, and this assignment is almost generally accepted, although a different assignment was proposed by Craig,[7] and the assignment of this transition is not yet definitely settled. The weak band near 260 mμ ($\epsilon_{\max} \approx 200$; $f \approx 0.001$) is interpreted as being due to the symmetry-forbidden ${}^1A_{1g} \rightarrow {}^1B_{2u}$ transition.

[6] M. Goeppert-Mayer and A. L. Sklar, *J. Chem. Phys.* **6**, 645 (1938).

[7] D. P. Craig, *Proc. Roy. Soc.* (*London*) **A200**, 474 (1950).

If the molecule has a center of symmetry, all the coordinate axes belong to u representations. Therefore, transitions between states (or orbitals) belonging to the same g or u parity are invariably symmetry-forbidden. This selection rule, which is only a corollary of the general symmetry selection rule, is sometimes called the parity selection rule.

In general, the higher the symmetry of a molecule, the larger is the number of possible symmetry species, and hence the smaller will be the probability of finding excited states whose representation agrees with one of the representations to which the coordinate axes belong. Therefore, the higher the symmetry of the molecule, the greater the probability of a given electronic transition being symmetry-forbidden.

Of course, even when a transition is symmetry-allowed, if the overlap of the initial and final wavefunctions is negligibly small, that transition is substantially forbidden.

A symmetry-forbidden band is one for which the electronic transition moment evaluated at the equilibrium nuclear configuration in the initial electronic state, $\mathbf{M}(Q_{\text{eq}})$, is zero. However, the intensity of an electronic absorption band really depends on the average of the electronic transition moment over all the nuclear configurations taken up by the vibrating molecule, $\overline{\mathbf{M}}$, and this average is not necessarily zero when $\mathbf{M}(Q_{\text{eq}})$ is zero. When the symmetry of a molecule is periodically changed by some vibration which is not totally symmetric, the symmetry of its electronic wavefunctions is also periodically changed since, according to the Born–Oppenheimer theorem, the electrons adapt themselves instantaneously to the motion of the nuclei, and hence a symmetry-forbidden transition may become allowed. The intensity of a transition which is symmetry-forbidden but has become allowed by coupling with the vibration will, of course, be much less than that of an ordinarily allowed transition.

The total symmetry of a vibronic state is specified by the direct product of the representations to which the electronic and vibrational wavefunctions belong. Thus, when the coupling of the electronic and vibrational motions is taken into account, the symmetry selection rule is modified into the following form: the component of the transition moment of a given vibronic transition in the direction of a given coordinate axis can be different from zero only if the direct product of the representations to which the initial and final vibronic state functions belong is the same as, or contains at least once, the representation to which that coordinate axis belongs.

As an example, let us consider the weak 260-mμ benzene band. As mentioned above, this band is due to the symmetry-forbidden ${}^1A_{1g} \to {}^1B_{2u}$ transition. The total symmetry of the ground vibronic state, that is, the lowest vibrational state belonging to the ground electronic state, is of course A_{1g}. The total symmetry of the lowest vibrational state associated with the excited electronic state $^1B_{2u}$ is B_{2u}. Therefore, the transition from the ground vibronic state to the lowest vibrational state associated with the electronic state $^1B_{2u}$, that is, the 0–0 vibrational component of the $^1A_{1g} \rightarrow {}^1B_{2u}$ electronic transition is symmetry-forbidden. However, if the $^1B_{2u}$ excited electronic state is combined with a vibrational state in which a vibrational mode of symmetry E_{2g} is excited by its one quantum, the total symmetry of the vibronic state is E_{1u} ($= B_{2u} \times E_{2g}$), and hence the vibronic transition from the ground vibronic state to this excited vibronic state is expected to be allowed and to be polarized in the xy plane (that is, the molecular plane). Furthermore, vibronic transitions consisting of the allowed transition and totally symmetric vibrations superimposing on it are also allowed, since the symmetry of the integrand of the intensity-determining integral remains unchanged by the superimposition of the totally symmetric vibrations.

Actually, the 260-mμ benzene band has a marked vibrational fine structure, and the wave numbers of the absorption peaks are found to be expressed as

$$\nu_{\text{max}} = \nu_{0\text{-}0}(\text{absent}) + \nu_0 + n_1\nu_1 \ (+ \ n_2\nu_2), \tag{4.52}$$

where n_1 and n_2 are positive integers, 0, 1, 2, etc.[8] The 0–0 component does not appear, and the wave number of this component, $\nu_{0\text{-}0}$, has been presumed to be 37,840 cm^{-1} in the spectrum measured with the ethanol solution at 95°K. ν_0, ν_1, and ν_2 are one quantum of an E_{2g} deformation vibration mainly of the carbon skeleton, one quantum of the A_{1g} pulsation vibration of the carbon skeleton, and one quantum of the A_{1g} pulsation vibration of the C—H bonds, respectively, in the $^1B_{2u}$ excited electronic state, and their values are 540, 930, and 2520 cm^{-1}, respectively. The coupling of the 2520-cm^{-1} A_{1g} vibration (ν_2) with the π–π^* electronic transition is so weak that the absorption peaks due to this coupling appear only faintly. Thus, the series of vibrational subbands starting from the 0–1 band at 37,840 + 540 cm^{-1}, and having the spacing of 930 cm^{-1}, appears prominently. This series is called the M series. By the way, a series of bands whose separation is almost constant or changes rather slowly is called a progression. The absence of the 0–0 band and the appearance of a more or less well-developed vibrational fine structure are characteristics of electronic transitions which are symmetry-forbidden but are made allowed by coupling with a vibration of appropriate symmetry.

[8] H. C. Wolf, *in* "Solid State Physics" (F. Seitz and D. Turnbull, eds.), Vol. 9, p. 1. Academic Press, New York, 1959.

The 200-mμ band of benzene is probably due to the symmetry-forbidden $^1A_{1g} \rightarrow {}^1B_{1u}$ transition. This transition is also made allowed by coupling with a single excitation of an E_{2g} vibration, since $B_{1u} \times E_{2g} = E_{1u}$.

The total probability of a symmetry-forbidden electronic transition which has been made allowed by coupling with excitation of one quantum of a vibration of appropriate symmetry is considered to depend on the mean square of the transition moment of the vibronic transition from the ground vibronic state (G0) to the first excited state of that vibration associated with the excited electronic state (n), $\overline{M^2_{\mathrm{G0},n1}}(Q)$, and this total probability is considered to be distributed among the vibrational components arising from superimposition of multiple excitations of some totally symmetric vibrations, in proportion to the square of the corresponding vibrational overlap integral. The total vibrational function of a polyatomic molecule can be factored into a product of functions representing normal vibrations, each of which is a function of only one of the so-called normal coordinates. The normal coordinate of the vibration which makes the symmetry-forbidden electronic transition allowed is denoted by Q_j. Then, to evaluate the mean square of the transition moment, $\overline{M^2_{\mathrm{G0},n1}}$, it is convenient to average over the normal coordinate Q_j, instead of averaging over the nuclear coordinates. The value of Q_j is taken to be zero at the equilibrium and to change periodically with the frequency of that vibration.

When the nuclei vibrate, the electronic Hamiltonian $\mathbf{H}(Q_j)$ may be expressed as $\mathbf{H}(0) + \mathbf{H}'(Q_j)$, where $\mathbf{H}(0)$ is the Hamiltonian for the equilibrium nuclear configuration and $\mathbf{H}'(Q_j)$ is the perturbation which is given by $(\partial \mathbf{H}/\partial Q_j)Q_j$. Then the perturbed electronic wavefunctions, $\Psi(Q_j)$, can be expressed in terms of the electronic wavefunctions for the equilibrium nuclear configuration, $\Psi(0)$, as follows:

$$\Psi_n(Q_j) = N_n(Q_j)\left[\Psi_n(0) + \sum_{r \neq n} c_{nr}(Q_j)\Psi_r(0)\right]. \tag{4.53}$$

The coefficients c_{nr} can be obtained by perturbation theory (see Section 2.4). That is, c_{nr} is proportional to the matrix element of the perturbation between $\Psi_r(0)$ and $\Psi_n(0)$, H'_{rn}, and is inversely proportional to the energy difference between the two states, $E_n - E_r$. Since the perturbation $\mathbf{H}'$ has the symmetry of the vibration considered, the off-diagonal matrix elements H'_{rn} can be nonzero only when the representation to which $\Psi_r(0)$ belongs is equal to the direct product of the representations to which the vibration and $\Psi_n(0)$ belong. Thus the perturbation by a vibration can mix a state only with states of appropriate symmetry.

For example, the ${}^1B_{2u}$ and ${}^1B_{1u}$ states of benzene can be mixed with the ${}^1E_{1u}$ state under the influence of an E_{2g} vibration. Since the ${}^1A_{1g} \rightarrow {}^1E_{1u}$ transition is allowed, the admixture of the ${}^1E_{1u}$ state in the perturbed ${}^1B_{2u}$ and ${}^1B_{1u}$ states makes the transitions from the ground state to these perturbed states allowed, with the transition moment being proportional to the mixing coefficient. It is usually said that the forbidden transitions have borrowed intensity from the allowed transition through coupling with the vibration, or that the vibration has transferred intensity from the allowed transition to the forbidden transitions. It will be understood that in the spectrum of benzene the intensity gained in this way by the ${}^1A_{1g} \rightarrow {}^1B_{1u}$ transition (200-mμ band) is greater than that of the ${}^1A_{1g} \rightarrow {}^1B_{2u}$ transition (260-mμ band), because the former transition is closer to the source of its intensity, that is, the ${}^1A_{1g} \rightarrow {}^1E_{1u}$ transition (180-mμ band). The basic theory of the vibrationally induced intensity of symmetry-forbidden transitions was developed by Herzberg and Teller,[9] and was extended by Craig,[10] Murrell and Pople,[11] and Liehr.[12]

The symmetry selection rule is also broken down by perturbation of electronic states of a molecule by intermolecular interactions. For example, while the 0–0 component of the 260-mμ benzene band is completely absent from the spectrum of the vapor, it appears very faintly in the spectrum of the liquid, and especially it appears distinctly, though still weakly, in the spectrum of the crystal.[8] The breakdown of the symmetry selection rule by the crystal field is interpreted as being due to resonance interaction between different excited states of different molecules in the crystal. That is, a forbidden transition is considered to borrow intensity from an allowed transition as a result of the crystal-induced mixing of the different excited states. The effect of the crystal-induced mixing is particularly strong when the forbidden or weakly allowed transition is separated by a small amount of energy from a stronger transition.

4.3.3. The Local-Symmetry Selection Rule

Aliphatic aldehydes and ketones show a weak absorption band with fine structure in the region of 270–300 mμ ($\log \epsilon_{\max} \simeq 1, f \simeq 10^{-4}$). This band

[9] G. Herzberg and E. Teller, *Z. physik. Chem. (Leipzig)* **B21**, 410 (1933).

[10] D. P. Craig, *J. Chem. Soc.* **1950**, 59.

[11] J. N. Murrell and J. A. Pople, *Proc. Phys. Soc. (London)* **A69**, 245 (1956).

[12] A. D. Liehr, *Z. Naturforsch.* **13a**, 311, 429, 596 (1958), **16a**, 641 (1961); *Can. J. Phys.* **36**, 1588 (1958); *Rev. Modern Phys.* **32**, 436 (1960); *Ann. Rev. Phys. Chem.* **13**, 41 (1962); "Advances in Chemical Physics" (I. Prigogine, ed.), Vol. 5, p. 241. Wiley (Interscience), New York, 1963.

is due to a singlet–singlet n–π^* transition. In general, nonconjugated and conjugated compounds containing unsaturated groups such as C=O, C=S, C=N, N=N, and N=O show a band due to a singlet–singlet n–π^* transition at comparatively long wavelengths. Such a band, an n–π^* band, is in general of low intensity and, in most cases, has prominent vibrational fine structure.

The singlet–singlet n–π^* transition in carbonyl compounds is a forbidden transition. Let us take the formaldehyde molecule as an example. This molecule has a planar geometry in the ground state, which belongs to point group C_{2v}. As illustrated in Fig. 4.1, the molecular plane is taken as the xy plane, and the direction of the carbonyl bond axis is taken as the direction of the x axis. Then, the bonding π orbital (π) and antibonding

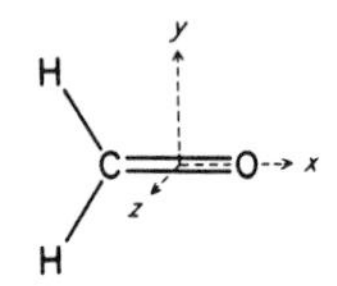

FIG. 4.1. The formaldehyde molecule.

π orbital (π^*) of the carbonyl group are considered to be formed from the $2p_z$ atomic orbitals of the carbon and oxygen atoms. The bonding σ orbital of the carbonyl bond is probably formed from an sp^2 hybrid orbital of the carbon atom and one of two sp hybrid orbitals which are formed from the $2s$ and $2p_x$ orbitals of the oxygen atom. The remaining one sp hybrid orbital and the $2p_y$ orbital of the oxygen atom are so-called nonbonding lone-pair orbitals (n orbitals), and each contains two electrons (lone-pair electrons). The absorption band under consideration is probably due to the singlet–singlet one-electron transition from the nonbonding $2p_y$ atomic orbital (ψ_n) of the oxygen atom to the antibonding π orbital (ψ_{π^*}) of the carbonyl group. Since the n orbital ψ_n is localized on the oxygen atom, in evaluating the transition moment of this transition only that part of the antibonding π orbital ψ_{π^*} is important which arises from the $2p_z$ (or $2p_\pi$) atomic orbital (χ_{pz}) of the oxygen atom. Thus, the transition moment of this transition can be approximately expressed as follows:

$$\mathbf{M}_{n-\pi^*} = \sqrt{2}\int \psi_n{}^*\mathbf{r}\psi_{\pi^*}\, d\tau \doteqdot \sqrt{2}c_{\pi^*,pz}\int \psi_n{}^*\mathbf{r}\chi_{pz}\, d\tau, \qquad (4.54)$$

where $c_{\pi^*,pz}$ is the coefficient of the $2p_z$ atomic orbital of the oxygen atom in the antibonding π orbital. Therefore, to decide if this transition moment vanishes or not, not the symmetry of the molecular orbitals involved but only the symmetry of the atomic orbitals of the oxygen atom, ψ_n and χ_{pz}, needs to be considered. When the n orbital ψ_n is the pure $2p_y$ atomic orbital (χ_{py}) as in the case of formaldehyde and other carbonyl compounds, the integral $\int \psi_n{}^*\mathbf{r}\chi_{pz}\,d\tau$ ($= \int \chi_{py}^*\mathbf{r}\chi_{pz}\,d\tau$) is evidently zero, and hence this n–π^* transition is forbidden. Such a transition is said to be local-symmetry forbidden. This conclusion, of course, is also valid for n–π^* transitions in other systems containing another heteroatom in place of the oxygen atom if the n orbital involved in the transition is the pure p atomic orbital of the heteroatom.

If the n orbital is not purely p but has some contribution of an s atomic orbital (χ_s), the nonzero transition moment in the z direction arises from that contribution of the s atomic orbital. Thus in this case the transition moment of the n–π^* transition can be expressed as follows:

$$\mathbf{M}_{n-\pi^*} = \sqrt{2}\int \psi_n{}^*\mathbf{r}\psi_{\pi^*}\,d\tau \doteqdot \sqrt{2}c_{n,s}^*c_{\pi^*,pz}\int \chi_s{}^*z\chi_{pz}\,d\tau, \qquad (4.55)$$

where $c_{n,s}$ is the coefficient of χ_s in ψ_n, and, as mentioned already, $c_{\pi^*,pz}$ is the coefficient of χ_{pz} in ψ_{π^*}. The integral $\int \chi_s{}^*z\chi_{pz}\,d\tau$ is evidently nonzero, since the s atomic orbital is centrosymmetric and the p_z atomic orbital has the same symmetry as the z coordinate. Therefore, in this case, the n–π^* transition is allowed and polarized in the direction perpendicular to the molecular plane, and its intensity is expected to depend on the magnitude of contribution of the s atomic orbital of the heteroatom to the n orbital and on the magnitude of contribution of the $p\pi$ atomic orbital of the heteroatom to the antibonding π orbital. Actually, the n–π^* bands of compounds containing a group such as C=N or N=N have relatively high intensities, probably because of contributions of the $2s$ orbital of the nitrogen atom to the n orbitals involved in the transitions. For example, the value of $\epsilon_{\max}$ of the singlet–singlet n–π^* band of pyridine appearing at about 270 mμ is as large as 450. In this compound the n orbital is probably an sp^2 hybrid orbital of the nitrogen atom. In addition, the n–π^* bands of many aza-aromatic compounds have been found to be polarized in the direction perpendicular to the molecular plane.[13]

The singlet–singlet transition from the nonbonding sp hybrid orbital to the antibonding π orbital in the carbonyl group is expected to be allowed because of the contribution of the $2s$ orbital of the oxygen atom to the n orbital. This sp hybrid n orbital ($n(sp)$) must be lower in energy than the $2p_y$ n orbital ($n(p)$), so that the $n(sp)$–π^* transition must occur at shorter wavelength than the $n(p)$–π^* transition. A band near 185 mμ of formaldehyde, as well as a band near 187 mμ ($\epsilon_{\max}$ = ca. 900) of acetone, has been attributed to the $n(sp)$–π^* transition.

[13] S. F. Mason, *J. Chem. Soc.* **1959**, 1263, 1269.

4.3.4. Prohibition of Many-Electron Transitions

When the initial and final state functions expressed as products of one-electron functions or as Slater determinants differ in more than one of the one-electron functions, the electronic transition moment evidently vanishes because of the orthogonality of the one-electron functions. Therefore, many-electron transitions are invariably forbidden.

This selection rule is based on the so-called one-electron approximation. When this approximation is removed in more refined theories, this selection rule is, of course, partially broken down.

Chapter 5

Shape of Absorption Bands and Geometry of Excited Electronic States

5.1. Shape of Absorption Bands

5.1.1. The Franck–Condon Principle

As mentioned already, an electronic transition between two electronic states is accompanied with simultaneous changes in the vibrational and rotational states, so that the absorption due to the electronic transition consists, in principle, of a large number of lines. However, in fact, the rotational fine structure is usually unresolvable, and it affords only a bandwidth to each vibrational subband. Therefore, the shape of an absorption band due to a single electronic transition may be considered to be determined by the spacing of the vibrational subbands and by the distribution of the total intensity of the electronic transition among the vibrational subbands.

The intensity distribution among the vibrational subbands is determined by the Franck–Condon principle. Franck's main idea,[1] which was later given a wave-mechanical basis by Condon,[2] is that the electronic transition in a molecule takes place so rapidly, in comparison to the vibrational motion of the nuclei, that immediately afterwards, the atomic nuclei of the molecule still have very nearly the same configuration and velocity as before the transition. Thus, the principle states that the most probable vibrational component of an electronic transition is one which involves no change in the nuclear configuration, that is, the vertical transition, which is represented by a vertical line in the potential energy diagram for the molecule.

The principle is most simply illustrated by examining a diatomic molecule. The potential energy V of a diatomic molecule is a function of the

[1] J. Franck, *Trans. Faraday Soc.* **21**, 536 (1925).

[2] E. U. Condon, *Phys. Rev.* **32**, 858 (1928).

internuclear distance R, and this functional relationship is approximated by the so-called Morse curve. Figure 5.1 is a potential energy diagram for a diatomic molecule in which the potential energy curves for the ground electronic state and one of excited electronic states and several lowest vibrational levels associated with these two electronic states are schematically shown.

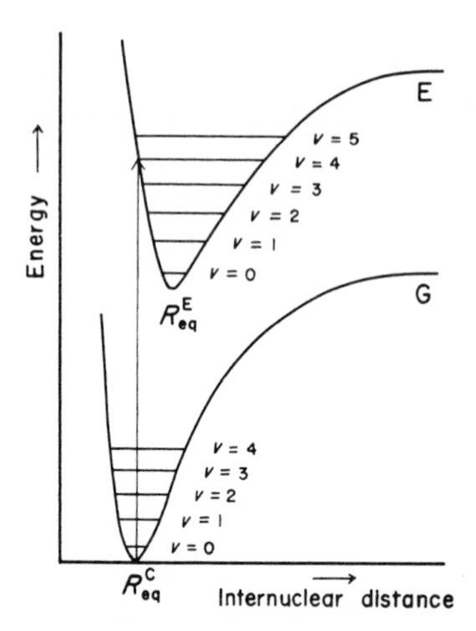

FIG. 5.1. A potential energy diagram of a diatomic molecule explaining the intensity distribution in absorption according to the Franck–Condon principle. Curve G, the potential energy curve of the ground electronic state; curve E, the potential energy curve of an excited electronic state. The arrow pointing vertically upward indicates the most probable vibronic transition.

The internuclear distance at the minimum of a potential energy curve is the equilibrium internuclear distance, R_{eq}, for the corresponding electronic state. The vibrational quantum number is denoted by v, and the values of v for vibrational levels associated with each electronic state are taken as 0, 1, 2, . . . , in the order of increasing energy. Furthermore, quantities associated with the ground and excited electronic states are indicated by superscripts G and E, respectively.

The horizontal line representing a vibrational level gives the total energy of the vibrational state, and the vertical distance of the line from the potential energy curve at a given value of R gives the kinetic energy of the vibration at the particular R. The end points of each vibrational level, that is, the points of intersection of the line with the potential energy curve, correspond to the turning points, R_t's, of the vibration in classical theory. At these points, the kinetic energy and velocity of the vibrational motion are zero, and the total energy is entirely the potential energy, while the kinetic energy and velocity of the vibrational motion are maximum near R_{eq}. Thus, the molecule in a vibrational state stays preferentially near the turning points of the vibration, and the intermediate positions are passed through very rapidly. Accordingly, an electronic transition from a vibronic state will take place most probably from the neighborhood of the turning points of the vibration.

At ordinary temperature the vast majority of molecules are in the lowest vibrational state associated with the ground electronic state, that is, the vibronic state with $v^G = 0$. If the so-called zero-point vibration in this state is disregarded, the large majority of molecules may be considered to be initially at the minimum of the potential energy curve for the ground electronic state, and the main part of an electronic transition may be considered to start from this point. According to the Franck–Condon principle, the most probable of vibronic transitions from this ground vibronic state to various vibrational states associated with an excited electronic state is the transition to the vibrational state which has one of the turning points at the same or almost the same internuclear distance as that corresponding to the starting point of the transition, that is, just or approximately vertically above the minimum of the potential energy curve for the ground electronic state. In such a vertical transition, the nuclear configuration and the kinetic energy of the vibrational motion remain almost unchanged during the transition, thus satisfying the requirement of the Franck–Condon principle. It is to be noted that the energy of the most probable vibronic transition is just the difference between the electronic energies of the excited and ground electronic states at the equilibrium nuclear configuration in the ground electronic state.

The vibrational quantum number of the excited vibronic state at which the most probable transition from the ground vibronic state terminates will be referred to as $v^E(\max)$. If the ground and excited electronic states have exactly the same equilibrium nuclear configuration, that is, if $R_{eq}^E = R_{eq}^G$, then evidently $v^E(\max) = 0$. In other words, in this case, the most probable transition is the 0–0 transition. The transition to a vibronic state with $v^E \geq 1$ would be possible only when at the moment of the transition either the nuclear configuration or the velocity of the vibration, or both, alters to an appreciable extent, so that it will take place only with a smaller probability. The transition to the higher vibrational state needs the larger alteration of the nuclear configuration or of the kinetic energy of the vibration. Therefore, the probability of the transition will rapidly decrease with the increasing value of v^E of the vibronic state at which the transition terminates.

If the equilibrium configuration of the excited and ground electronic states are different from each other, that is, if $R_{eq}^{E} \neq R_{eq}^{G}$, then the value of $v^{E}(\max)$ will be higher: in this case the 0–0 transition will be no longer the most probable. Usually R_{eq}^{E} is greater than R_{eq}^{G}. The greater the difference between R_{eq}^{E} and R_{eq}^{G}, the higher will be the value of $v^{E}(\max)$. Transitions to vibrational states successively higher and lower than the state with $v^{E}(\max)$ will have successively smaller probabilities.

The steeper the potential energy curve for the excited electronic state in the neighborhood of R_{eq}^{G}, the larger will be the number of vibrational levels associated with the excited electronic state which have the turning points in the neighborhood of R_{eq}^{G}, and hence, the slower will be the decrease in the intensity of vibrational subbands joining onto the most intense one. This means that the total intensity of the electronic transition is distributed among the larger variety of vibrational components, resulting in the longer progression. The potential energy curve rises usually with increasing steepness, especially in the region of R smaller than R_{eq}. Therefore, summarizing, we can conclude as follows: the further the minima of the potential energy curves for the two electronic states are separated from each other, and the steeper the upper potential energy curve is, the higher will be the value of $v^{E}(\max)$, and the greater will be the number of the vibrational components among which the total probability of the electronic transition is distributed.

5.1.2. Wave-Mechanical Formulation of the Franck–Condon Principle

According to wave mechanics, a vibrational state is represented by a wavefunction, X_v. For a diatomic molecule the X_v is a function only of the internuclear distance R, and the probability of a molecule in the vth vibrational state assuming a given value of R is given by the square of the vibrational function at the value of R, that is, $|X_v|^2$ or $X_v^* X_v$, which is called the probability density.

The number of nodes of each vibrational function, that is, the number of times the function goes through zero, equals the vibrational quantum number v. For the lowest vibrational state ($v = 0$), there is only one broad maximum of the vibrational function approximately above the minimum of the potential energy curve. For a higher vibrational state ($v \geq 1$), there is a broad maximum or minimum of the function near each of the classical turning point, but lying somewhat more toward the center. In addition, for $v > 1$, there are $v - 1$ smaller and narrower maxima and minima between these two terminal broad maxima or minima. The situation is illustrated in Fig. 5.2.

We have seen in Chapter 4, especially in Section 4.2, that the total intensity of an electronic transition is distributed among its vibrational components in proportion to the square of the overlap integral between the vibrational

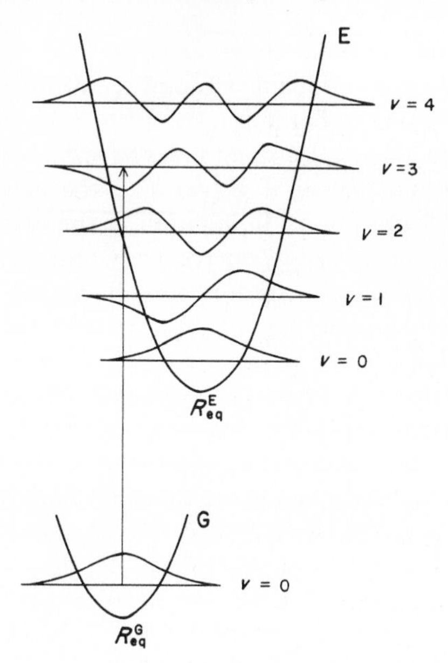

FIG. 5.2. A potential energy diagram with vibrational wavefunctions explaining wave-mechanically the intensity distribution in absorption according to the Franck–Condon principle. Curve G, the potential energy curve of the ground electronic state; curve E, the potential energy curve of an excited electronic state. The arrow pointing vertically upward indicates the most probable vibronic transition.

functions of the initial and final vibronic states, as well as to the population ratio of the initial vibronic state; that is, if it is reasonably assumed that the population ratio of the ground vibronic state ($v^{G} = 0$) is nearly unity, the total probability of the electronic transition from the ground electronic state to an excited electronic state is distributed among its vibrational components in proportion to the square of the overlap integral between the vibrational function for the state with $v^{G} = 0$ and that for the v^{E} state. This is the wave-mechanical formulation of the Franck–Condon principle. Contributions of the intermediate maxima and minima of the vibrational function of $v^{E} > 1$ to the overlap integral with the vibrational function of

$v^{G} = 0$ usually cancel one another at least to a rough approximation, or are negligibly small. Therefore, in a series of vibronic transitions starting from the ground vibronic state, the overlap integral has a maximum absolute value for the transition to the state whose vibrational function has one of its broad terminal maxima or minima vertically above the maximum of the vibrational function of the ground vibronic state, that is, vertically above the minimum of the potential energy curve for the ground electronic state. This means that the most probable vibronic transition is the so-called vertical transition, as illustrated in Fig. 5.2.

If the excited and ground electronic states have exactly the same equilibrium nuclear configuration, that is, if $R_{eq}^{E} = R_{eq}^{G}$, the most probable transition is evidently the 0–0 transition. In addition, if the two electronic states have exactly the same potential energy curves, the vibrational functions associated with the two electronic states are identical, since they are eigenfunctions of the same Hamiltonian [see Eq. (2.18)]. In this extreme case, since the vibrational functions are orthogonal to one another, the overlap integral between the vibrational function of the ground vibronic state ($v^{G} = 0$) and that of any excited vibrational state ($v^{E} \geq 1$) belonging to the excited electronic state is zero, and hence only the 0–0 transition will occur. This extreme is never realized, and hence actually the overlap integrals of the vibrational function of the state of $v^{G} = 0$ with the vibrational functions of states of $v^{E} \geq 1$ will not completely vanish, but they will be still small unless the potential energy curves for the two electronic states are much different from each other.

In a case where the two electronic states have considerably different equilibrium nuclear configurations, the 0–0 transition is no longer the most probable, and the transition to a higher vibrational state becomes the most probable. The greater the difference between R_{eq}^{E} and R_{eq}^{G}, the higher will be the value of $v^{E}(\max)$. The extension of the vibrational progression on both sides of the most intense component is determined primarily by the number of the vibrational levels of the excited electronic state whose vibrational function has one of the terminal maxima or minima in the range of R in which the vibrational function of the ground vibronic state has an appreciable value. Of course, the greater the difference between the equilibrium nuclear configurations of the two electronic states, and the steeper the potential energy curve for the excited electronic state, the greater will be the extension of the progression.

Thus we see that essentially the same conclusions are drawn from the wave-mechanical form of the Franck–Condon principle as from its classical form.

5.1.3. Shape of Electronic Absorption Bands of Polyatomic Molecules

In a molecule consisting of n atoms, there are $3n - 6$ (or, in the case of a linear molecule, $3n - 5$) normal coordinates, which specify the relative positions of the atoms with respect to one another, and there are $3n - 6$ (or $3n - 5$) corresponding normal vibrations. These coordinates are denoted by symbol Q_i with subscripts specifying the $3n - 6$ (or $3n - 5$) different ones. The total vibrational function can be expressed as the product of individual normal-vibrational functions, each of which is a function of only one of the normal coordinates:

$$\mathrm{X}_v = \prod_i \mathrm{X}_{v(i)}(Q_i), \tag{5.1}$$

where $v(i)$ represents the vibrational quantum number of the ith normal vibration.

In the case of a diatomic molecule, there is only one normal coordinate, which is identical with the internuclear distance R, and the potential energy of an electronic state is a function of R. In an n-atomic molecule, the potential energy of an electronic state is a function of all the $3n - 6$ (or $3n - 5$) normal coordinates.

The transition moment of the vibronic transition from a vibrational state of the ground electronic state characterized by a set of vibrational quantum numbers $v^{\mathrm{G}}(i)$'s to a vibrational state of a given excited electronic state characterized by a set of vibrational quantum numbers $v^{\mathrm{E}}(i)$'s can be approximately expressed as follows:

$$\mathbf{M}_{v\mathrm{G}\to v\mathrm{E}} = \bar{\mathbf{M}}_{\mathrm{G,E}} \int \mathrm{X}_{v\mathrm{G}} \mathrm{X}_{v\mathrm{E}}\, dQ = \bar{\mathbf{M}}_{\mathrm{G,E}} \prod_i \int \mathrm{X}_{v\mathrm{G}(i)} \mathrm{X}_{v\mathrm{E}(i)}\, dQ_i\,, \tag{5.2}$$

where $\bar{\mathbf{M}}_{\mathrm{G,E}}$ is the average electronic transition moment of the G–E electronic transition. This means that the total probability of the electronic transition, which is proportional to $\bar{M}^2_{\mathrm{G,E}}$, is distributed among different vibrational components in proportion to the product of the squares of $3n - 6$ (or $3n - 5$) vibrational overlap integrals, each of which involves a different one of the normal coordinates.

At ordinary temperature a large majority of molecules are in the lowest vibronic state, the state for which all the $v^{\mathrm{G}}(i)$'s are zero. On the other hand, there are many vibronic states at which transitions from the ground vibronic state can terminate, specified by a large variety of combinations of the vibrational quantum numbers $v^{\mathrm{E}}(i)$'s. Therefore, many vibronic transitions of various types may occur. Actually, in spectra of polyatomic compounds, superposition of many vibronic transitions frequently results in a substantially structureless absorption band.

Even in spectra of polyatomic compounds, sometimes a series of vibrational subbands which is the progression due to multiple excitation of one of the normal vibrations can be resolved. The arguments made for the case of diatomic molecules can be extended to the case of polyatomic molecules. If the potential energy curves with respect to a normal coordinate for the two electronic states have the same shape and the same equilibrium value of the coordinate, all vibronic transitions involving a change in quantum number of the corresponding normal vibration are forbidden, and hence, as far as this normal vibration is concerned, substantially only the 0–0 transition can occur. If the potential energy curve with respect to a particular normal coordinate changes appreciably its shape and the position of the minimum upon the electronic transition, the progression due to multiple excitation of the corresponding normal vibration will appear. The greater the differences in the shape and in the position of the minimum between the potential energy curves with respect to the normal coordinate for the two electronic states, the greater will be the extension of the corresponding progression. Furthermore, the steeper the potential energy curve for the excited electronic state, the greater will be the spacing of the progression, and hence, the greater will be the possibility of the resolution of the progression. Even when the progression is not completely resolvable, being blurred by superposition of excitation of other normal vibrations, the overall bandwidth of the electronic absorption band is considered to be mainly determined by the progression having the greatest extension.

Up to this point, it has been assumed that the total state function can be factored into the electronic and vibrational parts, and the electronic–vibrational coupling has not been explicitly considered. As mentioned in Subsection 4.3.2, this coupling is important for symmetry-forbidden electronic transitions, since it may make them allowed. For symmetry-allowed electronic transitions this coupling has only subsidiary importance. This coupling is usually not so strong that a symmetry-allowed electronic transition is made completely forbidden by coupling with nontotally-symmetric vibrations; that is, excitation of nontotally-symmetric vibrations can occur to some extent accompanying an allowed electronic transition. However, excitation of a totally-symmetric vibration can occur most strongly accompanying a symmetry-allowed electronic transition, since changes in vibational quanta of the totally-symmetric vibration do not change the symmetry of the integrand of the intensity-determining integral; that is, in an absorption band due to a symmetry-allowed electronic transition, the progression due to excitation of the totally-symmetric vibration can occur most prominently. Of course, for such a totally-symmetric vibration to give rise to a long progression in the electronic absorption band, there must be a considerable difference in the equilibrium value of the corresponding normal coordinate between the two electronic states.

5.2. Influence of Temperature and Environment upon the Band Shape

5.2.1. Influence of Temperature

In the thermal equilibrium the number of molecules in a stationary state is proportional to the Boltzmann factor. At ordinary temperature, the population ratio of the lowest vibronic state is nearly unity, and hence almost all electronic transitions are considered to start from this state. However, if the spacing of vibrational levels with respect to a normal vibrational mode in the ground electronic state is comparatively small, the next lowest and some higher vibrational states are populated by some portion of the molecules, and hence electronic transitions from these excited vibrational states can occur with the probability proportional to the population ratio. The bands due to such transitions are the so-called hot bands (see Section 4.2).

For example, according to analysis by Sponer and co-workers,[3] the following two progressions appear in the benzene 260-mμ band:

$$M_0: \nu_{\max} \text{ (in cm}^{-1}) = \nu_{0\text{-}0} \text{ (absent)} + 520 + 923n;$$

$$M_1: \nu_{\max} \text{ (in cm}^{-1}) = \nu_{0\text{-}0} \text{ (absent)} - 606.4 + 923n;$$

where $n = 0, 1, 2, \ldots,$ and $\nu_{0\text{-}0} = 38{,}088.7$ cm^{-1}. As mentioned already (see Subsection 4.3.2), the benzene 260-mμ band is due to the symmetry-forbidden ${}^1A_{1g} \rightarrow {}^1B_{2u}$ transition, made allowed by coupling with an E_{2g} deformation vibration. 520 cm^{-1} is one quantum of the E_{2g} vibration in the excited electronic state ${}^1B_{2u}$, and 606.4 cm^{-1} is that in the ground electronic state ${}^1A_{1g}$. 923 cm^{-1} is one quantum of the totally symmetric pulsation vibration in the excited electronic state. Thus, the M_0 series, which appears strongly, is the progression due to transitions starting from the lowest vibronic state, and the M_1 series, which appears much more weakly, is the progression due to transitions starting from the first excited state of the E_{2g} vibration. As the temperature is lowered, the population ratio of the excited

[3] H. Sponer, G. Nordheim, A. L. Sklar, and E. Teller, *J. Chem. Phys.* **7**, 207 (1939).

vibrational state must decrease and hence the intensity of the M_1 progression must become smaller. Actually, as expected, the M_1 progression disappears in the spectrum of the solid at low temperature.

The spectrum of gaseous ethylene shows an intense absorption band in the 145–200 mμ region ($f \simeq 0.3$), which has a well-developed fine structure with the most intense peak at about 170 mμ (log $\epsilon_{max} = 4.2$).[4] This band is due to the excitation of an electron from the bonding π orbital ψ_{+1} to the antibonding π orbital ψ_{-1} without a change of spin. At room temperature this band has a pronounced shoulder on the long-wavelength side, which is much reduced in intensity at liquid nitrogen temperature (about 77°K).[5] This shoulder is almost certainly due to hot bands, that is, transitions from excited vibrational levels of the ground electronic state. As is well known, the equilibrium geometry of the ethylene molecule in the ground electronic state is planar. If the vibrational quantum number for the twisting vibration of the carbon-to-carbon double bond, which has a quantum of 1027 cm^{-1}, is denoted by $v(\theta)$, at room temperature not all the molecules are in the lowest vibronic state with $v(\theta) = 0$ and with the torsional amplitude of about 11°, but some portion of the molecules are in the first excited vibronic state with $v(\theta) = 1$ and with the torsional amplitude of about 19°. The shoulder observed on the long-wavelength side of the main band is probably largely due to transitions from the state with $v(\theta) = 1$. In the alkylated ethylenes the one quantum of the twisting vibration is smaller, and hence the population ratios of the first and higher twisting vibrational states are larger, so that the long-wavelength shoulder is more pronounced than in the case of ethylene.

5.2.2. Influence of Environment

Rotational, vibrational, and electronic energy levels of a molecule are sharp in an isolated molecule. Therefore, the most discrete spectrum with well-resolved individual spectral lines of a compound is usually obtained in the vapor state at low pressures. Collisions of molecules lead to the broadening of rotational energy levels. Hence in the vapor spectra at high pressures the broadening of the rotational structure occurs.

In condensed phases, molecules generally cannot rotate freely, and hence the rotational energy levels lose their meaning. As a consequence, in spectra of liquids, solutions, and solids, it is not possible to separate rotational lines. Furthermore, the vibrational levels are also broadened in liquids and

[4] V. J. Hammond and W. C. Price, *Trans. Faraday Soc.* **51,** 605 (1955); L. C. Jones, Jr., and L. W. Taylor, *Anal. Chem.* **27,** 228 (1955).

[5] W. J. Potts, Jr., *J. Chem. Phys.* **23,** 65 (1955).

solutions, because of varying interactions of molecules with their neighbors, which occur in a more or less random arrangement. This broadening is often sufficient to obscure most or all of the vibrational structure of bands. In the case of solutions, the extent of this broadening depends on the character of the solvent. The stronger the interaction between the solute and the solvent, the greater is the extent of the broadening. The cases are frequently encountered in which the vibrational structure observed in saturated hydrocarbon solvents is appreciably blurred in solvents such as benzene and carbon tetrachloride.

In the crystalline state, either pure or mixed, the environment of different molecules is much more uniform, and consequently spectra of solids and solid solutions generally show much better resolved vibrational structure.

5.3. Geometry of Molecules in Excited Electronic States

The Franck–Condon principle states that the geometry of a molecule remains almost completely unchanged at the moment of an electronic transition. This means that the geometry of a molecule corresponding to one of the turning points of vibration of the excited vibronic state at which the most probable vibrational component of an electronic transition terminates is almost identical with the equilibrium geometry of the molecule in the initial state, but, of course, it does not mean that the equilibrium geometry of the molecule in the excited state is almost identical with that in the initial state; that is, in an excited electronic state a molecule does not necessarily have the same equilibrium geometry as in the ground electronic state.

In some cases, the equilibrium geometry of a molecule in an excited electronic state can be inferred from theoretical considerations of the electronic structure of the state. However, it is generally difficult to determine experimentally the geometry of a molecule in an excited electronic state. The customary experimental techniques used for determination of the geometry of a molecule in its ground state cannot be used for determination of the geometry of the molecule in an excited state, since the lifetime of excited states is usually quite short, and hence ordinarily it is not possible to obtain the requisite quantities of molecules in the excited state for the time needed.

Most experimental information about the geometry of excited electronic states has been obtained for comparatively simple molecules from analysis of the vibrational and rotational structure of the corresponding absorption band in the vapor spectrum. Analysis of the vibrational structure permits identification of the vibrational modes and determination of the force constants of the vibrations in the excited electronic state, and analysis of the rotational structure permits determination of rotational constants of the excited molecule. Thus they can give information about the geometry of the molecule in the excited state.

Fortunately, the most prominent vibrational progression present in an absorption band frequently provides the most important information about the geometry of the molecule in the corresponding excited state. As mentioned in Subsection 5.1.3, the progression due to the vibration with respect to the normal coordinate whose equilibrium value is most greatly changed by the electronic transition appears prominently in the spectrum. Therefore, the prominent progression indicates the most important distortion which occurs upon the electronic excitation, and thus gives important clues to the geometry of the molecule in the excited state. A few examples will be given below.

As mentioned already, the spectrum of gaseous ethylene has an intense absorption band with structure in the 145–200 mμ region, which is considered to be due to the singlet–singlet π–π^* transition. In addition, ethylene has a very weak absorption band with structure in the 260–350 mμ region, whose maximum is at 270 mμ ($\epsilon_{\max} \simeq 10^{-4}$).[6] This band is intensified by the presence of oxygen molecules (see Subsection 4.3.1),[7] and is almost certainly due to the singlet–triplet π–π^* transition.[8]

In the excited ethylene molecule, the planar geometry is stabilized by one bonding π electron, but is destabilized by one antibonding π electron. A theoretical consideration[9] indicates that this results in a net destabilization of the planar geometry with respect to a geometry in which the two $2p\pi$ orbitals do not interact, and predicts that the equilibrium geometry of the ethylene molecule in the first excited singlet and triplet states is one in which the two methylene groups lie in mutually perpendicular planes. A more recent theoretical consideration[10] suggests that the methylene groups are bent and staggered.

An analysis[11] of the vibrational structure of the 170-mμ bands of ethylene and of tetradeuteroethylene in the vapor spectra supports a nonplanar geometry of the excited state. A progression with a spacing of 850 cm^{-1} appears,

[6] C. Reid, *J. Chem. Phys.* **18,** 1299 (1950).

[7] D. F. Evans, *J. Chem. Soc.* **1960,** 1735.

[8] R. S. Mulliken, *J. Chem. Phys.* **33,** 1596 (1960).

[9] R. S. Mulliken, *J. Chem. Phys.* **3,** 517 (1935); R. S. Mulliken and C. C. J. Roothaan, *Chem. Rev.* **41,** 219 (1947).

[10] A. D. Walsh, *J. Chem. Soc.* **1953,** 2325.

[11] P. G. Wilkinson and R. S. Mulliken, *J. Chem. Phys.* **23,** 1895 (1955).

which is due to the totally-symmetric C=C stretching vibration in the excited electronic state, and superimposed upon it are diffuse subbands due to the methylene twisting vibration, that is, the twisting vibration of the C=C bond. The appearance of the progression due to the twisting vibration suggests the nonplanarity of the equilibrium geometry of the excited state. The most intense peak of the vibrational structure corresponds to the vertical transition from the lowest vibronic state to the excited vibronic state with the vibrational quantum number for the stretching vibration of about 20. The equilibrium length of the C=C bond in the excited state is suggested to be about 1.69 Å, which is to be compared with the corresponding value of 1.333 Å in the ground state. In the S–T band at about 270 mμ a progression with a spacing of about 1000 cm^{-1} appears. This progression is attributed to the C=C stretching vibration in the excited state.

We have seen that excitation of a single electron in ethylene makes a large change in the bonding of the molecule, and hence a large change in the equilibrium geometry. However, in large conjugated molecules, excitation of one of many bonding π electrons to a low-energy antibonding π orbital will not so drastically change the bonding, and therefore a change in the equilibrium geometry upon the excitation will be much smaller.

In the acetylene molecule there are two sets of π orbitals, which may be designated as the π_y and the π_z orbitals, the molecular axis being taken as the x axis. There are two pairs of singly excited singlet electron configurations which arise from two pairs of electronic transitions, $\pi_y \rightarrow \pi_y{}^*$, $\pi_z \rightarrow \pi_z{}^*$, and $\pi_y \rightarrow \pi_z{}^*$, $\pi_z \rightarrow \pi_y{}^*$. The configuration interaction of the former pair gives rise to a high-energy "plus" state and a low-energy "minus" state.[12] The latter pair remains degenerate. The transition from the ground state to the plus state is allowed and polarized along the x axis, and probably this is responsible for an intense absorption band with the absorption maximum at about 152 mμ. Transitions to the other singly excited singlet states are forbidden. The doubly degenerate transitions, $\pi_y \rightarrow \pi_z{}^*$ and $\pi_z \rightarrow \pi_y{}^*$, are probably responsible for a moderately intense band near 182 mμ, and the transition to the minus state is probably responsible for a weak band appearing in the 210–250 mμ region ($f \simeq 10^{-4}$).

As for ethylene, the excited states of acetylene do not necessarily have the same equilibrium geometry as the ground state. The equilibrium geometry in the ground state is linear and has a C—C bond length of 1.205 Å. An analysis[13] of the vibrational structure of the weak band in the 210–250 mμ

[12] I. G. Ross, *Trans. Faraday Soc.* **48,** 973 (1952).

[13] C. K. Ingold and G. W. King, *J. Chem. Soc.* **1953,** 2702, 2704, 2708, 2725, 2745. See also K. K. Innes, *J. Chem. Phys.* **22,** 863 (1954).

region indicates that the equilibrium geometry of the molecule in the excited state is a planar *trans*-bent one in which the length of the C—C bond is 1.383 Å and the C—C—H angle is 120.2°, implying that the valence state of the carbon atoms is changed from the *sp* hybrid to the sp^2 hybrid upon the electronic excitation.

The cyano group is isoelectronic with the acetylene group, and is expected to have similar electronic transitions. The geometry of the hydrogen cyanide molecule in the ground state is linear. An analysis[14] of vibrational structure of two weak bands of this compound which appear in the 160–200 and 155–170 mμ regions shows that both the equilibrium geometries of the two excited states are bent, having the H—C—N bond angle of 125 and 114°, respectively.

As mentioned already (see Subsection 4.3.3), a weak absorption band of nonconjugated carbonyl compounds appearing with vibrational structure in the 270–300 mμ region ($\epsilon_{max} \simeq 10$) is due to the symmetry-forbidden singlet–singlet n–π^* transition. A very weak absorption band appearing near 400 mμ ($\epsilon_{max} \simeq 10^{-3}$) is attributed to the singlet–triplet n–π^* transition. An analysis[15] of the vibrational structure of the singlet–singlet n–π^* band of formaldehyde shows that the progression due to the out-of-plane bending vibration in the excited electronic state appears prominently, and indicates that the equilibrium geometry of the molecule in the excited state is a pyramidal one in which the angle between the plane of the methylene group and the carbonyl bond axis is about 20° and the carbonyl bond length is about 1.32 Å. It is shown that the equilibrium geometry of the molecule in the corresponding triplet state is also pyramidal.[16] Needless to say, in the ground state this molecule has a planar equilibrium geometry in which the carbonyl bond length is about 1.22 Å.

[14] G. Herzberg and K. K. Innes, *Can. J. Phys.* **35,** 842 (1957).
[15] J. C. D. Brand, *J. Chem. Soc.* **1956,** 858. See also G. W. Robinson, *Can. J. Phys.* **34,** 699 (1956).
[16] V. E. DiGiorgio and G. W. Robinson, *J. Chem. Phys.* **31,** 1678 (1959).

Chapter 6

Effect of Environment upon Electronic Absorption Spectra

6.1. Introduction

The electronic absorption spectra are affected by the environment of the molecules. As mentioned in Subsection 5.2.2, the shape of absorption bands markedly depends on the environment of the molecules. In addition, even the position and intensity of absorption maxima are affected by the environment. Therefore, when comparing the absorption spectra of related compounds, it is most desirable to compare the spectra measured in similar environments, or otherwise, it is necessary to make allowance for the effect of environment. In this chapter some main features of the effect of environment upon the electronic absorption spectra are discussed.

The energy levels of a liquid or solid are essentially a property of the whole system. However, in most cases, when treating molecules in a condensed state, it is more convenient to consider the perturbation of the energy levels of the isolated molecule by the interaction of the molecule with its neighbors. Needless to say, the weaker the interaction, the smaller is the perturbation, and hence the more closely will the spectrum of the condensed phase resemble the vapor spectrum. In fact, solutions in the least polar and least polarizable solvents, such as saturated hydrocarbons, with the lowest solvating power, give spectra most closely resembling vapor spectra.

The main interactions usually considered to operate between a molecule and its neighbors are electrostatic (dipole–dipole, dipole–induced dipole), dispersion, hydrogen-bonding, charge-transfer, and repulsive (orbital overlap) interactions. The effects of such intermolecular interactions upon the energies of electronic states of a molecule are generally different from one another. This affords a reason for the dependence of the positions of absorption bands on the environment.

If the intermolecular interaction stabilizes an excited electronic state more than it does the ground electronic state, the position of the absorption band due to the transition between the two states is shifted to longer wavelengths from the position in the vapor spectrum, and conversely, if the interaction stabilizes the excited state less than it does the ground state, the position of the band is shifted to shorter wavelengths.

It is to be noted that the arrangement of neighboring molecules of an excited molecule immediately after the electronic transition is not necessarily the optimum one, that is, the energetically most stable one, for that excited state. According to the Franck–Condon principle, motions of atomic nuclei are much slower than motions of electrons, so that the positions of atomic nuclei remain substantially unchanged during an electronic transition (see Section 5.1). For electronic transitions in molecules in condensed states, the principle means that not only the geometry of the molecule but also the arrangement of its neighbors before an electronic transition remains substantially unchanged immediately after the transition.

It seems to be convenient to explain here several words generally used in comparison of absorption spectra of a compound under different conditions or of absorption spectra of related compounds. When an absorption band is shifted toward longer wavelengths by a change of the measurement condition such as the environment and the temperature or by a change of a part of the molecular structure, for example, by introduction of a substituent, the shift is called a bathochromic shift or a red-shift. On the other hand, a shift of a particular band under discussion to shorter wavelengths is called a hypsochromic shift or a blue-shift. When a change of the measurement condition or a change of a part of the molecular structure results in an increase in the intensity of a particular band, the effect is called a hyperchromic effect. On the other hand, the effect resulting in a decrease in the intensity of a particular band is called a hypochromic effect.

6.2. Solvent Effect

6.2.1. Solvent Effect on Spectra of Nonpolar Molecules

For a nonpolar solute in a nonpolar solvent, the intermolecular interaction between the solute and the solvent arises only from dispersion forces, and is generally comparatively weak. In most cases, excited states of nonpolar molecules are also nonpolar. Therefore, the solvent effect upon absorption bands of nonpolar molecules is generally small. However, any electronic excitation in a molecule is accompanied with a displacement of charge in the molecule, nodes being introduced into the electronic wavefunction. Less work must be required to do this in a dielectric medium than in a vacuum. Therefore, it is predicted that an absorption band of a nonpolar molecule will shift to longer wavelengths or lower wave numbers with the increasing dielectric constant of the solvent, that is, with the increasing refractive index of the solvent. The larger the magnitude of the charge displacement, the greater will be the effect. As mentioned already, the magnitude of the transition moment can be taken as a measure of the magnitude of the charge displacement upon the transition, and its square is proportional to the total intensity of the corresponding absorption band. Thus, theory predicts that the shift of an absorption band by the solvent effect will increase approximately in proportion to the intensity of the band.

According to Bayliss,[1] the magnitude of the shift of an absorption maximum of a band of a nonpolar molecule to lower wave numbers in going from the vapor spectrum to the spectrum of the solution in a nonpolar solvent is approximately expressed as follows:

$$\Delta\nu = \frac{e^2M^2}{a^3hc} \cdot \frac{n^2 - 1}{2n^2 + 1} \doteqdot 10.71 \times 10^9 \frac{f}{\nu a^3} \cdot \frac{n^2 - 1}{2n^2 + 1}, \tag{6.1}$$

where $\Delta\nu$ is the magnitude of the bathochromic shift in wave numbers, ν the wave number of the absorption maximum, f the oscillator strength of the absorption band, M the magnitude of the transition moment of the corresponding electronic transition, n the refractive index of the solvent at the wave number ν, and a the radius (in angstrom units) of the cavity (assumed spherical) in the solvent in which the solute molecule is located. Too much reliance should not be placed on the quantitative meaning of this expression because of approximations used to derive this expression and especially because of inevitable arbitrarity in estimation of a. However, this expression has been found to represent qualitatively and even roughly quantitatively solvent effects for the spectra of *cata*-condensed aromatic hydrocarbons[2] and for other various systems.

Saturated hydrocarbon solvents, such as *n*-hexane, *n*-heptane, and cyclohexane, can be said to be good spectroscopic solvents, because they have comparatively low refractive indexes and hence give spectra nearly resembling vapor spectra. Perfluoroheptane has a much lower refractive index. At 225 mμ and room temperature, the refractive index of this compound is only 1.28 while that of *n*-heptane is 1.46. Solvent shifts of absorption bands

[1] N. S. Bayliss, *J. Chem. Phys.* **18,** 292 (1950).
[2] N. D. Coggeshall and A. Posefsky, *J. Chem. Phys.* **19,** 980 (1951).

(relative to the gas phase) have been shown by Evans[3] to be much less for perfluoroheptane than for *n*-heptane (see Table 6.1).

Benzene has a much higher refractive index than saturated hydrocarbons. Absorption bands measured with solutions in benzene are usually located at considerably longer wavelengths and are broader than those measured with solutions in saturated hydrocarbon solvents. For example, the band near

TABLE 6.1

SHIFTS ($\Delta\nu_{max}$, cm^{-1}) OF ABSORPTION BANDS IN GOING FROM VAPOR TO SOLUTION[a]

Compound	Naphthalene	Isoprene	Styrene	Benzene
Band	The band at $\lambda_{max} = 211$ mμ	The band at $\lambda_{max} = 216$ mμ	The band at $\lambda_{max} = 240$ mμ	Four strongest vibrational subbands of the 260-mμ band
ν_{max}, cm^{-1} (in vapor)	47510	46360	41660	—
ϵ_{max}	133×10^3	17×10^3	13×10^3	ca. 200
f	1.70	0.45	0.293	ca. 0.002
$\Delta\nu_{max}$(PFH), cm^{-1}	−1110	−470	−540	av. −40
$\Delta\nu_{max}$(H), cm^{-1}	−2220	−1560	−1290	av. −280

[a] Solvent: PFH, perfluoroheptane; H, heptane.

370 mμ (ϵ_{max} = ca. 9000, f = ca. 0.1) of anthracene and the band near 470 mμ (ϵ_{max} = ca. 12,500, f = ca. 0.08) of naphthacene are known to be shifted to lower wave numbers by about 300 cm^{-1} in going from solutions in saturated hydrocarbons to solutions in benzene.

In some cases, excited electronic states of nonpolar molecules are polar. For example, the lowest energy singlet–singlet transition in phenanthrene, which is responsible for the weak band in the 310–350 mμ region, is polarized along the symmetry axis of the molecule, and the corresponding excited state is considered to have a nonzero, though probably small, dipole moment along that direction. Such a transition is expected to shift to longer wavelengths in dielectric solvents, owing partly at least to the dipole–induced dipole interaction between the polar excited molecule and the solvent.

When the solvent is polar, the intermolecular interaction between the solute and the solvent arises mainly from dipole–induced dipole forces, which are generally stronger than dispersion forces. In most cases, the polarizability of an excited electronic state is larger than that of the ground state. Therefore, in this case again, bathochromic shifts of bands by the solvent are expected.

[3] D. F. Evans, *J. Chem. Phys.* **23**, 1429 (1955).

However, there can be some cases where the polarizability of the excited state is inferred to be smaller than that of the ground state. In such cases, the transition is expected to shift to shorter wavelengths in polar solvents. This appears to be the case for the first singlet–singlet band near 310 mμ of naphthalene and also for the third singlet–singlet band near 255 mμ of anthracene.

6.2.2. Solvent Effect on Spectra of Polar Molecules

When the solute is polar and the solvent is nonpolar, the predominant intermolecular interaction is of the dipole–induced dipole type. When the dipole moment of the solute molecule increases as a result of the charge displacement due to an electronic transition in the molecule, the excited electronic state will be more stabilized by solvation than the ground electronic state, and hence the corresponding absorption band will be shifted to longer wavelengths with increasing polarizability of the solvent. On the other hand, when the dipole moment of the solute molecule decreases upon an electronic transition, the corresponding absorption band will be shifted to shorter wavelengths with increasing polarizability of the solvent.

When both the solute and the solvent are polar, the predominant intermolecular interaction is of the dipole–dipole type. In this case the situation is considerably complicated, since the solvent molecules are strongly oriented around the solute molecule. However, in most cases, absorption bands due to electronic transitions which cause an increase in the dipole moment of the molecule can be expected to be shifted to longer wavelengths with increasing polarity of the solvent, and vice versa. A theoretical equation approximately representing the effect of the dipole–dipole interactions upon the position of absorption as well as emission bands has been given by Ooshika.[4] On the basis of his equation, it has been attempted to determine the dipole moments of polar molecules in excited electronic states from measurement of the solvent effect on the differences in wave number between absorption bands and the corresponding fluorescence bands.[5]

Table 6.2 shows data on effects of various solvents on the position of the first π–π^* and first n–π^* bands of mesityl oxide, $(CH_3)_2C{=}CHCOCH_3$.[6] In this table, $\Delta\nu$ represents the difference in wave number between the absorption

[4] Y. Ooshika, *J. Phys. Soc. Japan* **9**, 594 (1954).

[5] E. Lippert, *Z. Naturforsch.* **10a**, 541 (1955); *Z. Elektrochem.* **61**, 962 (1957). See also N. Mataga, Y. Kaifu, and M. Koizumi, *Bull. Chem. Soc. Japan* **29**, 465 (1956); W. W. Robertson, A. D. King, Jr., and O. E. Weigang, Jr., *J. Chem. Phys.* **35**, 464 (1961).

[6] A. E. Gillam and E. S. Stern, "An Introduction to Electronic Absorption Spectroscopy in Organic Chemistry," 2nd ed., p. 303. Arnold, London, 1957.

maximum in the spectrum of the solution in a solvent and that in the spectrum of the solution in hexane. As is seen, the π–π^* band shifts to lower frequencies, that is, to longer wavelengths, and the n–π^* band shifts to higher frequencies, that is, to shorter wavelengths, with increasing dielectric constant of the solvent. Needless to say, this compound, a conjugated enone, is polar, with the negative end at the oxygen atom of the carbonyl group. The one-electron transition from the highest occupied π orbital to the lowest vacant π orbital,

TABLE 6.2

SHIFTS ($\Delta\nu_{max}$, cm^{-1}) OF THE FIRST π–π^* AND FIRST n–π^* BANDS OF MESITYL OXIDE, $(CH_3)_2C{=}CHCOCH_3$, IN GOING FROM SOLUTION IN HEXANE TO SOLUTION IN OTHER SOLVENTS[a]

Solvent	Hexane			Ether	Ethanol	Methanol	Water
Dielectric constant		2.0		4.3	25.8	31	81
	λ_{max}, mμ	ν_{max}, cm^{-1}	log ϵ_{max}	$\Delta\nu_{max}$, cm^{-1}	$\Delta\nu_{max}$, cm^{-1}	$\Delta\nu_{max}$, cm^{-1}	$\Delta\nu_{max}$, cm^{-1}
π–π^* Band	229.5	43,570	4.1	−95	−1380	−1555	−2675
n–π^* Band	327.0	30,580	1.6	+95	+1165	+1470	(+2205)

[a] The data were taken from G. Scheibe, *Ber.* **58**, 586 (1925).

which is considered to be responsible for the π–π^* band, is a so-called intramolecular charge-transfer transition, accompanied with a charge displacement from the ethylenic bond to the carbonyl bond. Therefore, the excited state of this transition must be more polar than the ground state. On the other hand, in the n–π^* transition, one of two electrons in the n orbital centered at the oxygen atom transfers to the lowest vacant π orbital, which covers the whole conjugated system. Therefore, in this transition, a charge displacement occurs from the oxygen atom to the carbon atoms of the conjugated enone system, and hence the excited state of this transition is probably less polar than the ground state. This explains the shifts in opposed directions of the π–π^* and n–π^* bands. In general, the lowest energy π–π^* bands of conjugated enones and other polar conjugated systems shift to longer wavelengths, and the n–π^* bands of carbonyl, thiocarbonyl, azo compounds and others shift to shorter wavelengths, with increasing dielectric constant of the solvent.

Table 6.3 shows the positions of the absorption maximum of the n–π^* band of acetone in various solvents. It is noteworthy that the extent of the hypsochromic shift of the n–π^* band is particularly great in hydroxylic solvents such as alcohols and water. This effect of hydroxylic solvents is attributed at least partly to the hydrogen-bonding interaction between the solute molecule and the solvent molecule. The lone pair of electrons on the oxygen atom of the carbonyl group of the solute molecule forms a hydrogen bond with the hydrogen atom of the hydroxyl group of the solvent molecule. The formation of this hydrogen bond lowers the energy of the n orbital by an amount approximately equal to the energy of the hydrogen bond. In the excited state of the n–π^* transition, the hydrogen bond is almost completely broken or, at least, is largely weakened. Therefore, the energy of the n–π^* transition must

TABLE 6.3

THE EFFECTS OF SOLVENTS ON THE POSITION OF THE n–π^* BAND OF ACETONE

Solvent	λ_{max}, mμ
(Vapor)	280
Carbon tetrachloride	280
Benzene	280
Cyclohexane	280
Hexane	279
Heptane	279
Chloroform	277
(Acetone)	275
Acetonitrile	273
Ethanol	270
Methanol	269
Water	265

increase in the presence of hydroxylic solvents. This explains the great hypsochromic shift of the n–π^* band by hydroxylic solvents. The extent of the shift must increase with increasing hydrogen-bonding ability of the solvent. When the hydrogen bond is sufficiently strong, the n–π^* band will substantially disappear.

When the intermolecular interaction between the solute and the solvent is strong, the vibrational structure of absorption bands of the solute is generally blurred. If the solute molecule is nonpolar, the solvent molecules are substantially unoriented. If both the solute and the solvent are polar, however, the solvent molecules are oriented around the solute molecule, and after the electronic transition in the solute molecule the solvent molecules must rearrange to an equilibrium arrangement for the excited solute molecule. This relaxation of the solvent cage prevents establishment of vibrational quantization in the solute molecule, and leads to the blurring of the vibrational structure of absorption bands. Thus, the vibrational structure of a band is most distinctly observed in nonpolar solvents, and generally the vibrational structure of bands of polar molecules is almost completely blurred in polar solvents. For example, while the vibrational structure of n–π^* bands generally appears distinctly in nonpolar solvents, it is markedly blurred in polar solvents, especially in hydroxylic solvents.

We have so far considered only the case in which the chemical species of the absorbing substance is substantially not changed by the solvent. When the solvent changes the equilibrium of dissociation, association, or tautomeric change of the solute molecule or forms a molecular complex with the solute molecule, of course different types of solvent effect occur.

6.2.3. Solvent Effect on the Intensity of Bands

From consideration of classical oscillator theory Chako[7] concluded that the oscillator strength of an absorption band should increase by a factor $1/\gamma$ in going from the vapor state to dilute solution in a dielectric solvent. This factor, called the Lorentz–Lorenz correction factor, is given by

$$1/\gamma = f_{\mathrm{soln}}/f_{\mathrm{vap}} = (n^2 + 2)^2/9n, \tag{6.2}$$

where n is the refractive index of the solvent averaged over the wavelength range of the absorption band. This equation should be applied only to allowed transitions and to nonpolar solvents. For saturated hydrocarbon solvents the correction factor is about 1.2 to 1.3 in the ultraviolet region.

Chako's theory has been experimentally tested for several systems, and it has been found to be probably invalid for π–π^* transitions in conjugated hydrocarbons. For example, the correction factors $f_{\mathrm{soln}}/f_{\mathrm{vap}}$, that is, the ratios of the experimental oscillator strengths for solution and vapor, for the longest wavelength π–π^* S–S bands (near 240 mμ) of cyclopentadiene and cyclohexadiene in n-hexane were found to be 0.83 and 1.04, respectively, which are considerably smaller than the theoretical value, 1.30.[8] In Table 6.4 data of the longest wavelength π–π^* S–S bands of three conjugated dienes in vapor and in solution in n-heptane are shown. For all the three compounds the best values of the experimental correction factor $f_{\mathrm{soln}}/f_{\mathrm{vap}}$ is close to unity.[9] In view of their limited accuracy, these experimental data may be insufficient to draw a general conclusion. However, it seems to be true that the oscillator strengths of allowed π–π^* bands of conjugated hydrocarbons do not greatly change in going from the vapor spectrum to the solution spectrum, while the

[7] N. Q. Chako, *J. Chem. Phys.* **2,** 644 (1934).

[8] V. Henri and L. W. Pickett, *J. Chem. Phys.* **7,** 439 (1939); L. W. Pickett, E. Paddock, and E. Sackter, *J. Am. Chem. Soc.* **63,** 1073 (1941).

[9] L. E. Jacobs and J. R. Platt, *J. Chem. Phys.* **16,** 1137 (1948).

intensities of absorption maxima may vary depending on the phase and on the solvent. Jacobs and Platt[9] explained the deviation from Chako's theoretical value of the experimental value of the correction factor by saying that even in the vapor phase the electronic transition takes place in an effective dielectric due to the periphery of the molecule itself. A modification of Chako's theory was proposed by Schuyer,[10] using the Onsager model instead of the simple Lorentz–Lorenz one. However, Schuyer's theory gives also too high relative intensities for solution spectra.

TABLE 6.4

THE LONGEST WAVELENGTH π–π^* S–S BANDS (THE A BANDS) OF CONJUGATED DIENES IN VAPOR AND IN HEPTANE SOLUTION[a]

Compound		λ_{max}, mμ	ϵ_{max}, $\times 10^{-3}$	f	f_{soln}/f_{vap}
Isoprene (2-methyl-1,3-butadiene)	vapor	215	17	0.45 ± 0.02	1.07 ± 0.08
	solution	223	19	0.48 ± 0.02	
cis-Piperylene (*cis*-1,3-pentadiene)	vapor	216	20	0.57 ± 0.03	1.00 ± 0.07
	solution	226	22	0.57 ± 0.03	
trans-Piperylene (*trans*-1,3-pentadiene)	vapor	214	35	0.64 ± 0.03	0.95 ± 0.07
	solution	223	26	0.61 ± 0.03	

[a] The data were taken from Reference 9.

When the distribution of conformations of the molecules in the ground electronic state is changed in going from the vapor to solution, the f value of a band will naturally vary. According to Almasy and Laemmel,[11] the average of the f values of the π–π^* S–S band near 240 mμ of biphenyl in the vapor phase at 170, 260, 360, and 520°C is 0.311 ± 0.0035, while the f value of that band in the spectrum of the *n*-hexane solution is 0.411. Accordingly, the ratio of the f values for the solution and vapor is 1.32, a value which is in a fair agreement with the theoretical value of 1.28 calculated from Chako's equation. However, this agreement is probably only a result of coincidence, and the increase in the f value in going from the vapor to the solution probably should be attributed mainly to the difference in the conformation distribution between the different phases, as will be discussed in detail later on (see Chapter 12).

In contrast with allowed transitions, essentially forbidden transitions

[10] J. Schuyer, *Rec. trav. chim.* **72,** 933 (1953).

[11] F. Almasy and H. Laemmel, *Helv. Chim. Acta* **33,** 2092 (1950).

generally have intensities appreciably affected by change of the phase and of the solvent. As mentioned already (see Subsection 4.3.1), the intensities of singlet–triplet transitions are markedly influenced by the solvent. The intensities of n–π^* bands are also considerably influenced by the solvent. For example, the integrated intensity of the n–π^* band near 460 mμ (in n-heptane: λ_{max} = ca. 458 mμ, ϵ_{max} = ca. 20) of p-benzoquinone increases progressively when changing the solvent from n-heptane to benzene, dioxane, and to carbon disulfide.[12] Specially, the integrated intensity in carbon disulfide is much larger than in the other solvents. By the way, while the band shows well-developed vibrational fine structure in n-heptane, the structure is considerably blurred in the other solvents.

6.3. Absorption Spectra of Molecular Complexes

6.3.1. Charge-Transfer Complex

A complex is a loose reversible association of two or more distinct chemical components. It is well known that aromatic compounds and other unsaturated compounds form complexes with a wide variety of chemical species, such as picric acid, trinitrobenzene, quinones, tetracyanoethylene, halogens, and silver cation, and that amines and ethers form complexes with halogens.

The complex formation is usually accompanied with shift of absorption bands of the components and with appearance of new absorption bands. For example, the benzene–iodine complex shows a new intense absorption band at 296 mμ with ϵ_{max} = ca. 16,700 in n-heptane.[13] This complex is rather weak: the heat of formation $-\Delta H$ has been estimated at 1.1 kcal/mole.[14]

Mulliken[15] proposed a theory to explain the complex formation and the appearance of new absorption bands. According to his theory, in complexes, one component behaves as an electron donor (D) and the other as an electron acceptor (A). The complex is roughly described as a resonance hybrid between two states: a lower energy no-bond state $\Psi(D, A)$ and a higher energy ion-pair state $\Psi(D^+, A^-)$, which arises from one-electron transfer from D to A. For example, in complexes of aromatic hydrocarbons with picric acid, trinitrobenzene, quinones, tetracyanoethylene, etc., an electron

[12] A. Kuboyama, *Bull. Chem. Soc. Japan* **33,** 1027 (1960).

[13] H. A. Benesi and J. H. Hildebrand, *J. Am. Chem. Soc.* **71,** 2703 (1949); J. S. Ham, J. R. Platt, and H. McConnell, *J. Chem. Phys.* **19,** 1301 (1951); J. S. Ham, *J. Am. Chem. Soc.* **76,** 3875 (1954).

[14] J. A. A. Ketelaar, *J. phys. radium* **15,** 197 (1954).

[15] R. S. Mulliken, *J. Am. Chem. Soc.* **74,** 811 (1952).

is considered to transfer from a bonding π orbital of the former compounds to an antibonding π orbital of the latter compounds. These compounds are called π-donors and π-acceptors, respectively. In complexes of amines and ethers with iodine, the electron-transfer is considered to occur from a non-bonding lone-pair orbital of the amine or ether to an antibonding σ orbital of iodine. In this case, amines and ethers are called n-donors, and iodine is called a σ-acceptor.

The mixing of the two states, $\Psi(D, A)$ and $\Psi(D^+, A^-)$, produces a stabilized ground state Ψ_N and an excited state Ψ_E:

$$\Psi_N = a\Psi(D, A) + b\Psi(D^+, A^-); \quad (6.3)$$

$$\Psi_E = -b\Psi(D, A) + a\Psi(D^+, A^-), \quad (6.4)$$

where a and b are constants. In most cases $|b|$ is much smaller than $|a|$. The transition from Ψ_N to Ψ_E gives rise to a new absorption band characteristic of the complex at the wave number $(E_E - E_N)/hc$, where E_E and E_N are the energies of the excited and ground states, respectively. Since the ground state has primarily a character of the no-bond state and the excited state has primarily a character of the ion-pair or charge-transfer state, an electron displacement occurs from D to A in going from the ground state to the excited state. Hence this transition is called an intermolecular charge-transfer (abbreviated usually as CT) transition, and the absorption band due to this transition is called an intermolecular CT band. The transition moment of this transition is expected to be directed from D to A. Actually, it is known that in the crystal of the 1:1 complex of hexamethylbenzene and chloranil ($-\Delta H = 5.35$ kcal/mole)[16] the two components are stacked one upon the other in parallel planes and that the absorption band characteristic of this complex at 514 mμ is polarized in the direction perpendicular to the aromatic rings.[17]

The difference between the energy of the charge-transfer state, $E(D^+, A^-)$, and that of the no-bond state, $E(D, A)$ is given by

$$E(D^+, A^-) - E(D, A) = I_D - A_A - C, \quad (6.5)$$

where I_D is the ionization energy of the donor, A_A is the electron affinity of the acceptor, and C is the mutual electrostatic and dispersion energy of D^+ and A^- relative to that of D and A. If the stabilization energy of the ground state is represented by X, the energies of the ground and excited

[16] G. Briegleb and J. Czekalla, *Z. physik. Chem. (Frankfurt)* **24,** 37 (1960); *Z. Elektrochem.* **58,** 249 (1954).

[17] K. Nakamoto, *J. Am. Chem. Soc.* **74,** 1739 (1952).

states are expressed as follows:

$$E_{\mathrm{N}} = E(\mathrm{D}, \mathrm{A}) - X; \tag{6.6}$$

$$E_{\mathrm{E}} = E(\mathrm{D}^+, \mathrm{A}^-) + X = E(\mathrm{D}, \mathrm{A}) + I_{\mathrm{D}} - A_{\mathrm{A}} - C + X. \tag{6.7}$$

Accordingly, the energy of the CT transition is given by

$$E_{\mathrm{E}} - E_{\mathrm{N}} = I_{\mathrm{D}} - A_{\mathrm{A}} - C + 2X. \tag{6.8}$$

For weak complexes X is approximated by perturbation theory (see Section 2.4) as follows:

$$X = H^2/[E(\mathrm{D}^+, \mathrm{A}^-) - E(\mathrm{D}, \mathrm{A})], \tag{6.9}$$

where H represents the interaction matrix element between the two state functions, $\Psi(\mathrm{D}^+, \mathrm{A}^-)$ and $\Psi(\mathrm{D}, \mathrm{A})$. For a series of weak complexes formed from the same electron acceptor and various electron donors of a similar type, both $A_{\mathrm{A}} + C$ and H can be regarded as being approximately constant, and hence the energy of the CT transition can be approximately expressed as follows:

$$E_{\mathrm{E}} - E_{\mathrm{N}} = I_{\mathrm{D}} - C_1 + C_2/(I_{\mathrm{D}} - C_1), \tag{6.10}$$

where C_1 and C_2 are constants. It has been found that this approximate expression applies to most complexes in which the electron acceptor is iodine, chloranil, or trinitrobenzene.

The binding energies of complexes are usually rather small so that, in solution, many different geometries of the complex may exist in equilibrium with one another. Since the energy of the CT transition will differ for each geometry, the CT band is in most cases broad and structureless.

The charge-transfer excited state of a complex is generally much more polar than the ground state. Therefore, the intermolecular CT band is expected to shift to longer wavelengths with increasing dielectric constant of the solvent. This has been found to be the case for the CT bands of complexes formed from chloranil and various aromatic compounds.[18]

For weak complexes for which $a \gg b$, the transition moment length of the CT transition is approximately given by abR, where R is the distance of the electron displacement in going from the no-bond state to the CT state, that is, the distance between the center of the electron-donating orbital of D and that of the electron-accepting orbital of A. The ratio b/a is approximately given by $-H/[E(\mathrm{D}^+, \mathrm{A}^-) - E(\mathrm{D}, \mathrm{A})]$, and, needless to say, a and b are

[18] A. Kuboyama and S. Nagakura, *J. Am. Chem. Soc.* **77,** 2644 (1955); A. Kuboyama, *Nippon Kagaku Zasshi* **81,** 558 (1960).

related to each other by the normalization condition: $a^2 + b^2 = 1$. Accordingly, the greater the extent of the mixing of the no-bond and CT states, the more intense will be the CT transition; that is, the greater the stability of the complex, the higher will be the intensity of the CT band. The chloranil and trinitrobenzene complexes of aromatic compounds and the iodine–amine complexes give some support for this prediction, but the iodine complexes of aromatic compounds show a trend which is rather opposed to this prediction. This fact suggests the existence of another factor affecting the intensity of the CT band, which will be discussed later.

6.3.2. Charge-Resonance Complex

At the temperature of liquid nitrogen the *N*-ethylphenazyl radical forms the dimer, in which two radicals are situated in parallel planes.[19] This dimer shows an absorption band at 830 mμ, which is absent from the spectrum of the monomeric radical. This dimer is an example of an interesting type of complexes in which the two components are identical.

The electronic states of such a system have been interpreted as resonance hybrids of the no-bond structure RR and the two ion-pair structures R^+R^- and R^-R^+. Here R represents the monomer. Any state of the system must contain an equal contribution from R^+R^- and R^-R^+, and hence there cannot be any actual charge displacement from one component to the other upon an electronic transition. The ground state of the system is probably expressed by the wavefunction

$$\Psi_N = a\Psi(RR) + b[\Psi(R^+R^-) + \Psi(R^-R^+)]. \tag{6.11}$$

The absorption band of the *N*-ethylphenazyl dimer at 830 mμ is interpreted as being due to the transition from the ground state to the excited state expressed by the following wavefunction:

$$\Psi_E = (1/\sqrt{2})[\Psi(R^+R^-) - \Psi(R^-R^+)]. \tag{6.12}$$

The transition moment of this transition is in the direction perpendicular to the planes of the radicals.

A state of the type of $R^+R^- \pm R^-R^+$ is called a charge-resonance state, and a transition from the ground state to such a state is called a charge-resonance transition. A complex which is stabilized by a contribution of a charge-resonance state to the ground state is called a charge-resonance complex. The term "two-way charge-transfer" has also beeen used for the term "charge-resonance."

[19] K. H. Hausser and J. N. Murrell, *J. Chem. Phys.* **27**, 500 (1957).

6.3.3. Explanation of the Formation and Absorption Spectra of Complexes by the Simple MO Method

The formation and spectra of molecular complexes can also be explained by means of the simple MO method.[20] Here, the complex DA is regarded as formed by interaction between the molecular orbitals of D and those of A. Since the interaction is rather weak, it can be treated conveniently by perturbation method (see Section 2.4). Interactions between the occupied orbitals (d) of D and those (a) of A lead to no change in their total energy and to no

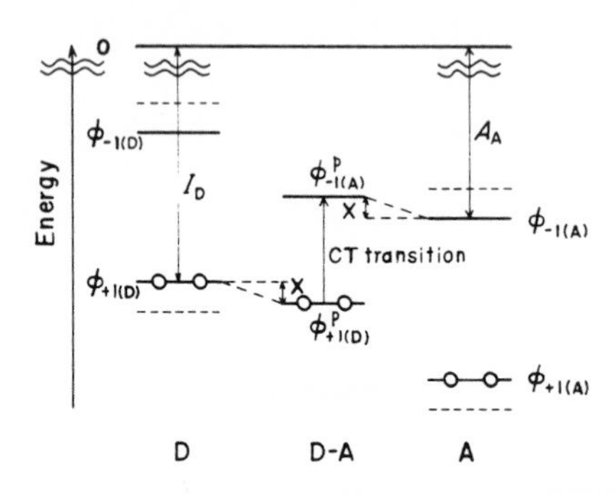

Fig. 6.1. An energy-level diagram explaining the formation of a charge-transfer complex and occurrence of an intermolecular charge-transfer transition in terms of interaction between molecular orbitals of the two components. D, an electron donor; A, an electron acceptor.

net transfer of charge between D and A. Interactions of the occupied orbitals (d) of D with the unoccupied orbitals (a^*) of A depress the former and raise the latter, leading to a net stabilization of the system with a simultaneous transfer of negative charge from D to A; interactions of the occupied orbitals (a) of A with the unoccupied orbitals (d^*) of D likewise lead to a net stabilization of the system with a simultaneous charge transfer in the opposite direction.

The extent of the interaction between two orbitals is inversely proportional to the difference in energy between the interacting orbitals, and is proportional to the square of the overlap integral between the orbitals. Figure 6.1 illustrates schematically the case in which the interaction between the highest occupied orbital of D, $\phi_{+1(D)}$, and the lowest unoccupied orbital of A, $\phi_{-1(A)}$, is most important. This case corresponds to the case in which the interaction

[20] M. J. S. Dewar and A. R. Lepley, *J. Am. Chem. Soc.* **83**, 4560 (1961); M. J. S. Dewar and H. Rogers, *ibid.* **84**, 395 (1962).

between the no-bond state (D, A) and the CT state (D^+, A^-) is predominant. By the interaction between the two orbitals, $\phi_{+1(D)}$ is lowered in energy by an amount X, giving rise to a perturbed orbital, $\phi^{p}_{+1(D)}$, and simultaneously $\phi_{-1(A)}$ is raised in energy by X, giving rise to a perturbed orbital, $\phi^{p}_{-1(A)}$. The perturbed orbitals are expressed as follows:

$$\phi^{p}_{+1(D)} = N(\phi_{+1(D)} + C\phi_{-1(A)}); \tag{6.13}$$

$$\phi^{p}_{-1(A)} = N(\phi_{-1(A)} - C\phi_{+1(D)}). \tag{6.14}$$

The coefficient C is usually much smaller than unity, and, according to perturbation theory, it is approximately given by

$$C \doteqdot -H/[\varepsilon_{-1(A)} - \varepsilon_{+1(D)}], \tag{6.15}$$

where H represents the matrix element of the perturbation operator (representing the interaction between D and A) between $\phi_{+1(D)}$ and $\phi_{-1(A)}$, and $\varepsilon_{-1(A)}$ and $\varepsilon_{+1(D)}$ represent the energies of $\phi_{-1(A)}$ and $\phi_{+1(D)}$, respectively. The energy X is approximately given by

$$X \doteqdot H^2/[\varepsilon_{-1(A)} - \varepsilon_{+1(D)}]. \tag{6.16}$$

The interaction matrix element H is usually a negative energy quantity, and is approximately proportional to the overlap integral between the two orbitals.

The occupied orbital $\phi^{p}_{+1(D)}$ has primarily a character of $\phi_{+1(D)}$, and the unoccupied orbital $\phi^{p}_{-1(A)}$ has primarily a character of $\phi_{-1(A)}$. Therefore, the one-electron transition from $\phi^{p}_{+1(D)}$ to $\phi^{p}_{-1(A)}$ is accompanied with a partial charge transfer from D to A. That is, this transition is a so-called intermolecular CT transition. In the simple MO approximation neglecting electronic repulsion, the ionization energy of D, I_D, is taken to be equal to $-\varepsilon_{+1(D)}$, and the electron affinity of A, A_A, is taken to be equal to $-\varepsilon_{-1(A)}$. Actually, it is known that there is a good linear relation between the ionization energy and the calculated energy of the highest occupied orbital, ϕ_{+1}, of aromatic hydrocarbons. Therefore, the energy of the CT transition from $\phi^{p}_{+1(D)}$ to $\phi^{p}_{-1(A)}$ can be expressed as follows:

$$\Delta E = \varepsilon_{-1(A)} - \varepsilon_{+1(D)} + 2X = I_D - A_A + 2X. \tag{6.17}$$

The transition moment of this transition is given by $\sqrt{2}N^2CR$, in which R represents the distance between the center of gravity of $\phi_{+1(D)}$ and that of $\phi_{-1(A)}$.

There may be more than one transition from orbitals arising from perturbation of occupied orbitals of one component to orbitals arising from perturbation of unoccupied orbitals of another component. In fact, it is known that there are cases where more than one intermolecular CT band is observed in the spectrum of a complex.

The absorption bands of the components are more or less markedly affected by the complex formation. For example, in the case illustrated in Figure 6.1, the absorption band of D due to the transition from $\phi_{+1(D)}$ to $\phi_{-1(D)}$ and the absorption band of A due to the transition from $\phi_{+1(A)}$ to $\phi_{-1(A)}$ will be shifted toward shorter wavelengths by the complex formation. It is well known that the absorption band near 500 mμ of iodine is shifted to shorter wavelengths and is intensified by the complex formation with aromatic compounds. The extraordinarily large solvent effect of carbon tetrachloride observed with the spectra of aromatic compounds and amines has been explained as being partly due to the complex formation of the solvent acting as a σ-acceptor with the solute acting as a π- or n-donor. In complexes involving heavy atoms, the spin selection rule is expected to be broken down by spin–orbit coupling. This must certainly occur in, for example, the iodine–benzene complex. The complex formation between the solute and the solvent may play an important role in the intensification of the singlet–triplet absorption bands of aromatic compounds dissolved in solvents containing heavy atoms like iodine.

In the case where $\phi_{-1(A)}$ interacts not only with $\phi_{+1(D)}$ but also with $\phi_{-1(D)}$, that is, in the case where $\phi^{p}_{-1(A)}$ has some contribution of $\phi_{-1(D)}$ in addition to the contribution of $\phi_{+1(D)}$, the intermolecular CT transition from $\phi^{p}_{+1(D)}$ to $\phi^{p}_{-1(A)}$ has a contribution of the local transition in D from $\phi_{+1(D)}$ to $\phi_{-1(D)}$. This case may be also described as the case where the charge-transfer state (D^+, A^-) mixes with the locally excited state (D*, A) as well as the no-bond state (D, A). If the local transition $\phi_{+1(D)} \rightarrow \phi_{-1(D)}$ is allowed, this transition contributes some intensity to the intermolecular CT transition. In addition, if $\phi^{p}_{+1(D)}$ has some contribution of $\phi_{+1(A)}$, and if the local transition in A from $\phi_{+1(A)}$ to $\phi_{-1(A)}$ is allowed, this transition also contributes some intensity to the CT transition. Thus, there are cases in which mixing of the CT state with locally excited states of the individual component molecules contributes additional intensity to the CT absorption.

In the halogen complexes of aromatic compounds the contribution of the local transitions in the aromatic compound to the intensity of the CT band is probably important. The most stable geometry of the halogen–benzene complex is the most symmetrical one, in which the halogen molecule lies on the sixfold symmetry axis of the benzene molecule.[21] In this geometry, the CT state cannot interact with any locally excited states of benzene, and hence

[21] O. Hassel, *Proc. Chem. Soc.* **1957,** 250; *Mol. Phys.* **1,** 241 (1958); O. Hassel and K. O. Stømme, *Acta Chem. Scand.* **12,** 1146 (1958).

the CT band cannot derive any intensity from the local transitions in benzene. These transitions can contribute to the intensity of the CT band only when the halogen molecule moves to give a less stable, unsymmetrical geometry. Since the halogen complexes of aromatic compounds are weak, there will be a large number of different geometries in equilibrium with one another. The weaker the complex, the greater its probability of assuming unsymmetrical geometries, and hence, the greater the contribution of the local transitions in the aromatic compound to the intensity of the CT band. This explains why in the iodine complexes of aromatic compounds the intensity of the CT band increases as the stability of the complex decreases. In general, in the case where the contribution of the local transitions to the intensity of the CT band is important, the intensity of the CT band is expected to increase as the stability of the complex decreases and as the temperature is raised. On the other hand, when the contribution of local transitions to the intensity of the CT band is not important, the intensity is expected to increase as the complex becomes more stable.

It is noteworthy that charge-transfer absorption is observed even in cases in which complex formation in any sense other than statistical collision pairing is dubious. For example, the solution of iodine in *n*-heptane shows an intense absorption band below 250 mμ, and this band is ascribed to a charge-transfer transition in which *n*-heptane and iodine behave as an electron donor and an electron acceptor, respectively, although these two compounds form no complex with each other.[22] In a statistical collision encounter between a donor and an acceptor there is no minimum in the potential energy curve, that is, no complex is formed, but the orbitals of the components may overlap adequately to give rise to an intermolecular CT transition. Such a transition is called a contact CT transition.[23] The occurrence of contact CT transitions may be interpreted in terms of the MO theory, as follows. For example, consider the case where $\phi_{-1(\mathrm{A})}$ does not interact with $\phi_{+1(\mathrm{D})}$ but interacts with $\phi_{-1(\mathrm{D})}$. This case corresponds to the case where the CT state (D^+, A^-) does not interact with the no-bond state (D, A) but interacts with the locally excited state in D (D*, A). In this case the ground state does not undergo any stabilization. However, since the perturbed unoccupied orbital $\phi^{\mathrm{p}}_{-1(\mathrm{A})}$ has a contribution of $\phi_{-1(\mathrm{D})}$, if the local transition in D from $\phi_{+1(\mathrm{D})}$ to $\phi_{-1(\mathrm{D})}$ is allowed, the intermolecular CT transition will be allowed by borrowing intensity from the local transition. Furthermore, in the case where $\phi_{-1(\mathrm{A})}$ and $\phi_{+1(\mathrm{D})}$ have different symmetries, as illustrated in Fig. 6.2, the overlap between these two orbitals cannot lead to any stabilization of the ground

[22] D. F. Evans, *J. Chem. Phys.* **23,** 1436 (1954).
[23] L. E. Orgel and R. S. Mulliken, *J. Am. Chem. Soc.* **79,** 4839 (1957).

state, but the transition from $\phi_{+1(D)}$ to $\phi_{-1(A)}$ will give rise to a CT band polarized in a plane perpendicular to the direction of the charge transfer.[24] It has been found that when oxygen is dissolved in various aromatic solvents, a new absorption occurs in the ultraviolet region.[25] When the solvent is benzene, the absorption appears near 230 mμ.[26] This absorption has been explained as being due to a contact CT transition.

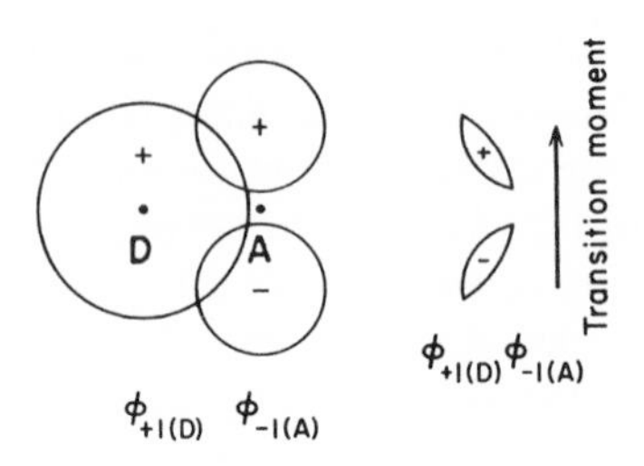

FIG. 6.2. Charge-transfer between orbitals of different symmetry.

6.3.4. Steric Factors in Complex Formation

As will be expected, the complex formation is highly sensitive to steric factors. Some examples indicating the importance of steric factors in complex formation will be given below.

Hexaethylbenzene invariably forms weaker complexes than does hexamethylbenzene. Thus, the heat of formation ($-\Delta H$) of the complex with iodine in carbon tetrachloride is 1.79 kcal/mole for hexaethylbenzene, while it is 3.73 kcal/mole for hexamethylbenzene.[27]

Biphenyl and its methylated derivatives form 1:1 complexes with 1,3,5-trinitrobenzene in carbon tetrachloride, which show CT bands in the region 300–380 mμ.[28] The complex stability increases as the biphenyl nucleus is increasingly methylated at *m*- and *p*-positions. On the other hand, as the biphenyl nucleus is increasingly methylated at *o*-positions, the complex stability decreases, in the order biphenyl ($-\Delta H = 2.33$–2.38 kcal/mole) $>$ 2-methylbiphenyl (1.2) $>$ 2,2′-dimethylbiphenyl (1.74–1.98) $>$ bimesityl

[24] J. N. Murrell, *Quart. Rev. (London)* **15,** 191 (1961).

[25] D. F. Evans, *J. Chem. Soc.* **1953,** 345; **1957,** 1351, 3885; **1959,** 2753; *Proc. Roy. Soc. (London)* **A255,** 55 (1960); A. U. Munck and R. N. Scott, *Nature* **177,** 587 (1956); H. Tsubomura and R. S. Mulliken, *J. Am. Chem. Soc.* **82,** 5966 (1960).

[26] E. C. Lim and V. L. Kowalski, *J. Chem. Phys.* **36,** 1729 (1962).

[27] L. J. Andrews and R. M. Keefer, *J. Am. Chem. Soc.* **74,** 4500 (1952); **77,** 2164 (1955).

[28] C. E. Castro, L. J. Andrews, and R. M. Keefer, *J. Am. Chem. Soc.* **80,** 2322 (1958).

(0.3–0.4) > 2,2′,6,6′-tetramethylbiphenyl (~0.2). A planar geometry of the biphenyl nucleus is required for maximum electronic interaction between the two components of the complex. The dependence of the complex stability on the number of methyl groups at *o*-positions is considered to indicate the dependence of the complex stability on the steric factor which prevents coplanarity of the two benzene rings in the biphenyl nucleus (see Chapter 12).

trans-Stilbene and related compounds form colored complexes with 1,3,5-trinitrobenzene. According to a qualitative study,[29] mixed solutions of the same concentrations of a series of phenylated stilbenes and trinitrobenzene in chloroform color as follows: *trans*-stilbene, intense yellow; triphenylethylene, yellow; tetraphenylethylene, very feeble yellow. Similarly, mixed solutions of a series of methylated stilbenes and trinitrobenzene in chloroform color as follows: *trans*-α-methylstilbene, yellow; *trans*-α,α′-dimethylstilbene, pale yellow. The coloration is certainly due to a CT transition in the complexes consisting of the stilbenes as electron donors and trinitrobenzene as an electron acceptor, and the degree of coloration is considered to indicate qualitatively the degree of the complex formation. It is evident that also in this case the degree of the complex formation is at its maximum for the planar molecular geometry and decreases with the increasing deviation from the planarity of the molecular geometry (see Chapter 14).

N,N-Dimethylaniline and its methylated derivatives form more or less stable complexes with iodine in *n*-heptane solutions, probably by the charge transfer from the *n* orbital of the nitrogen atom to the iodine molecule, and show intense CT bands in the near ultraviolet region. While the heats of formation of the iodine complexes of *N,N*-dimethylaniline and *N,N*-dimethyl-*p*-toluidine are very large (8.2 and 8.3 kcal/mole, respectively), those of the iodine complexes of *N,N*-dimethyl-*o*-toluidine and *N,N*-dimethyl-*o,o*′-xylidine (2.3 and 1.7 kcal/mole, respectively) are much smaller than the former two.[30] In the *ortho*-methylated *N,N*-dimethylanilines, the *ortho* methyl group approaches the lone-pair of electrons of the nitrogen atom, owing to the twist of the dimethylamino group (see Chapter 22). The decrease in the donor strength of the *ortho*-methylated *N,N*-dimethylanilines is therefore attributed to the steric effect of the *ortho* methyl group on the iodine molecule approaching the lone-pair of electrons of the nitrogen atom.

In connection with the importance of steric factors in complex formation, it is interesting to note that several cases are known in which pressure affects appreciably CT absorption bands of complexes. For example, the CT band of

[29] H. Ley and F. Rinke, *Ber.* **56,** 771 (1923).

[30] H. Tsubomura, *J. Am. Chem. Soc.* **82,** 40 (1960).

the benzene–iodine complex shifts to longer wavelengths under a pressure of 2000–50,000 atmospheres.[31,32] The CT band of quinhydrone and that of the chloranil–hexamethylbenzene complex also shift to lower wave numbers by about 2500 cm^{-1}, and increase in intensity by a factor of 1.2 and 1.7, respectively, under a pressure of 50,000 atm.[32]

6.4. Absorption Spectra of Molecules in the Solid State

6.4.1. Pure Crystals

The absorption spectrum of a molecule in the crystalline state is appreciably affected by the crystal field. Effects of the crystal field, or the intermolecular interaction in the crystal, upon the spectrum of the molecule may be summarized as follows.

(1) The symmetry selection rule is broken down by the intermolecular interaction (see Subsection 4.3.2). Thus the 0–0 component of a forbidden transition appears in the spectrum of the crystal. In the crystal field, a mixing of different excited states of different molecules can occur, and, as a result of this crystal-induced mixing of excited states, the intensities of electronic transitions in the isolated molecule are redistributed in the spectrum of the crystal.

(2) An excited state of a free molecule is split in the crystal into states of the number equal to the number (n) of molecules in a unit cell, and hence the transition from the ground state to the excited state is split into n components, polarized in different directions. This is the Davydov or factor group splitting, and can be explained in terms of the so-called exciton theory (see Chapter 20). The magnitude of the splitting is proportional to the dipole strength of the transition, and depends on the relative orientation of the molecules in a unit cell.

(3) The center of gravity of the Davydov components is usually at a longer wavelength than the corresponding band in the vapor spectrum. The magnitude of the shift is also proportional to the absorption intensity for similar geometry; in particular, for different electronic transitions in the same crystal.

In this way, the crystal spectrum has generally a considerably different feature than that of the corresponding solution and vapor spectra. A detailed description of the crystal spectra and their interpretations is beyond the scope

[31] J. S. Ham, *J. Am. Chem. Soc.* **76,** 3875, 3881 (1954).
[32] D. R. Stephens and H. G. Drickamer, *J. Chem. Phys.* **30,** 1518 (1959).

of this book. I will only give a few examples of the Davydov splitting and displacement of bands in the crystal spectra.

The $^1A_{1g} \rightarrow {}^1B_{2u}$ transition in benzene, which is responsible for the weak band near 260 mμ (ϵ_{max} = ca. 220, f = ca. 10^{-4} in solution), is forbidden by symmetry, but its 0–0 band appears in the crystal spectrum, probably by borrowing the intensity from the allowed $^1A_{1g} \rightarrow {}^1E_{1u}$ transition. The wave number of the 0–0 band has been determined by extrapolation to be 38,088.7 cm^{-1} in the vapor,[33] and 37,840 cm^{-1} in ethanol solution at 95°K.[34] Since the crystal of benzene contains four molecules in a unit cell, the $^1B_{2u}$ state of the molecule is split in the crystal into four states. The transition from the ground state to one of the four states is forbidden, and the transitions to the other three states are allowed, which give rise to three component bands polarized along the three crystal axes. According to Broude and his co-workers, the wave numbers of the 0–0 bands of these components polarized along the c, a, and b crystal axes in the spectrum of the crystal at 20°K are 37,803 cm^{-1} (strong), 37,839 cm^{-1} (strong), and 37,846 cm^{-1} (weak), respectively;[35] that is, the magnitude of splitting between the two strong (c and a) components is 36 cm^{-1}, and the average of the wave numbers of the two components is lower by about 270 cm^{-1} than the wave number of the 0–0 band in the vapor spectrum.

The longest wavelength π–π^* singlet–singlet band of naphthalene (λ_{max} = about 312 mμ, ϵ_{max} = 280, f = 0.002 in solution) is considered to be due to the $^1A_g \rightarrow {}^1B_{3u}^-$ transition, which is formally forbidden (see Subsection 10.3.3). The wave number of the 0–0 band of this transition is inferred to be 32,021 cm^{-1} in the vapor spectrum,[36] 31,750 cm^{-1} in the spectrum of ethanol solution at 95°K,[37] and 31,544 cm^{-1} in the spectrum of a mixed crystal with durene at 20°K.[36] In the pure crystal this transition becomes allowed probably by borrowing the intensity from the allowed $^1A_g \rightarrow {}^1B_{2u}$ transition. Since the pure crystal of naphthalene has two molecules in a unit cell, this transition is split into two components, one polarized in the ac plane and the other polarized along the b axis. The wave numbers of the 0–0 bands of these components are 31,476 and 31,642 cm^{-1}, respectively.[36] The b component is hundreds of times as strong as the ac component.

[33] H. Sponer, G. Nordheim, A. L. Sklar, and E. Teller, *J. Chem. Phys.* **7**, 207 (1939).

[34] H. C. Wolf, *in* "Solid State Physics" (F. Seits and D. Turnbull, eds.), Vol. 9, p. 1. Academic Press, New York, 1959.

[35] V. L. Broude, V. S. Medvedew, and A. F. Prikhotjko, *Optika i Spektroskopiya* **2**, 317 (1957).

[36] D. S. McClure, *J. Chem. Phys.* **23**, 1575 (1955).

[37] H. C. Wolf, *Z. Naturforsch.* **10**, 3 (1955).

The second π–π^* singlet–singlet band of naphthalene, an intense band with vibrational structure in the 240–290 mμ region, is due to the allowed $^1A_g \rightarrow {}^1B_{2u}$ transition, which is polarized along the short axis of the molecule, that is, the y axis (see Fig. 8.4). The wave number of the 0–0 band is 35,910 cm^{-1} in the vapor, 34,680 cm^{-1} in ethanol solution at 95°K, and 34,410 cm^{-1} in a mixed crystal with durene. This transition is also split in the pure crystal into two components. The 0–0 band of the a-polarized component is at 33,783 cm^{-1}, and that of the b-polarized component is at 33,610 or 33,460 cm^{-1}. The latter is much stronger than the former.[36]

Not only the spectrum of a single crystal, but the spectrum of microcrystals, as well, exhibit the splitting and shift of bands. According to Weigl, the intense band near 600 mμ (ϵ_{max} = about 10^5, f = about 0.7) of some cationic dyes such as Rhodamine B and Malachite Green is broader in the spectrum of a solid film consisting of microcrystals of the dye than in the spectrum of the solution in methanol, and in several cases the band in the solid spectrum is split into two maxima separated by about 1500 cm^{-1}.[38] This splitting has also been interpreted in terms of the Davydov theory.

6.4.2. Mixed Crystals

The spectrum of oriented molecules, not greatly affected by the intermolecular interaction, can be obtained by using a mixed crystal of the compound with another adequate inert compound in which the guest molecules are dispersed in a molecular manner, and are incorporated in an oriented way in the host crystal. The host crystal must be transparent in the spectral region of interest, and the guest molecule must have size, shape, and chemical properties similar to those of the host molecule. The host molecule cannot be entirely inert toward the guest molecule. However, if the electronic transition of the guest under consideration is separated in transition energy a sufficient amount from the transitions of the host, the perturbation of the spectrum of the guest by the intermolecular interaction between the host and the guest is small, so that the spectrum of an adequate dilute mixed crystal can be approximately regarded as the spectrum of the oriented but noninteracting molecules of the guest compound.

For example, in the spectrum of a mixed crystal of naphthalene (0.1%) with durene, which is transparent throughout the region of the first and in part of the second π–π^* singlet–singlet transitions of naphthalene, the first transition (the $^1A_g \rightarrow {}^1B_{3u}^-$ transition) of naphthalene is not greatly perturbed by the crystal field, that is, does not display the Davydov splitting, and its 0–0

[38] J. W. Weigl, *J. Chem. Phys.* **24**, 364 (1956).

band appears weakly as a band polarized along the long molecular axis (the x axis) almost exactly at the mean position of the Davydov components in the pure crystal (see Subsection 6.4.1).[39] As mentioned already, in the spectrum of the pure crystal, this transition shows an intensity ratio of the Davydov components which would be expected if the transition were polarized along the short molecular axis (the y axis), probably owing to the mixing with the ${}^1A_g \rightarrow {}^1B_{2u}$ transition, the second π–π^* singlet–singlet transition, which is allowed and polarized along the short molecular axis.

6.4.3. KCl Disks

In spectra of crystals, even in those measured with solid thin films prepared on quartz or glass plates by sublimation of crystals, by evaporation of thin layers of solutions, or by solidification of melted substances, the effect of the orientation of molecules in the crystal reveals itself markedly. This orientation effect is useful for the purpose of determining the direction of polarization of absorption bands,[40] but it makes the spectra unsuitable for direct comparison with the spectra of solutions and vapors, in which molecules are oriented at random. However, there are cases where it is desired to obtain the spectrum of a compound in the crystalline state suitable for direct comparison with the solution or vapor spectrum of the compound. This seems to be achieved most conveniently by the use of the pressed KCl disk technique.

The pressed KCl disk technique is quite analogous to the well-known pressed KBr disk technique developed originally for infrared spectroscopy. Thus the pure crystals of the organic compound under examination are mixed with pure potassium chloride in adequately dilute concentration, and the mixture is thoroughly ground so that the organic compound is evenly dispersed, and then is pressed under vacuum into a transparent disk. The absorption spectrum of the disk can easily be measured. While KBr starts to absorb appreciably below 240 mμ, KCl permits better transmission of light with wavelengths longer than 210 mμ. Hence KCl is more suitable than KBr for electronic absorption spectroscopy.

It may be somewhat doubtful whether or not the state of molecules in the disks is quite the same as that in the crystals. There may be a possibility that the compound is dispersed at least in part in a molecular manner, and that the state is rather similar to that of solid solutions. It seems, however, to be more probable that microcrystals are oriented at random in the disks. If

[39] D. S. McClure, *J. Chem. Phys.* **22,** 1668 (1954); **24,** 1 (1956).

[40] See, for example, A. C. Albrecht and W. T. Simpson, *J. Chem. Phys.* **23,** 1480 (1955).

it is so, it seems reasonable to consider that the spectra measured by this method are the spectra of the compounds in the crystalline state which are most suitable for direct comparison with the solution and vapor spectra.

When comparing the spectrum of a compound in a KCl disk with the corresponding solution spectrum, it is necessary to take account of the difference in the environment effect. If the disk spectrum is the spectrum of microcrystals oriented randomly, the effect of the crystal field must reveal itself in the spectrum. The environment effect in the KCl-disk spectra has not sufficiently been clarified for a wide variety of compounds. As far as the spectra measured hitherto (of compounds of rather restricted types) are concerned, the disk spectra resemble considerably the corresponding solution spectra on the whole. However, in most cases, vibrational structures of absorption bands become more diffuse and absorption maxima shift to longer wavelengths in going from the solution spectrum of the compound in an inert solvent to the KCl-disk spectrum.

In the case where the most probable geometry of the molecule in the solid state differs appreciably from that in solution, the difference between the KCl-disk spectrum and the solution spectrum, of course, must arise partly from the difference in the geometry of the molecule in the two states. In the case where the compound has substantially the same molecular geometry in the two states, the difference between the KCl-disk spectrum and the solution spectrum must, of course, be attributed wholly to the difference in the environment effect. The bathochromic shift observed with such a compound in going from the solution in an inert, saturated hydrocarbon solvent to the KCl disk was called by Dale[41] the normal red-shift. According to him, the magnitude of the normal red-shift, $\Delta\lambda$, of individual absorption maxima of naphthalene and anthracene increases regularly with the increasing wavelength of the absorption maximum, from about 2 mμ at 220 mμ to 8 mμ at 300 mμ, and to 19 mμ at 400 mμ (see Table 6.5). As far as the data shown in Table 6.5 are concerned, the regular relation between the magnitude of the normal red-shift and the wavelength of the band seems to hold regardless of the kind of compounds, the characters of bands, and the intensities of bands. However, when the comparison of the KCl-disk spectrum with the solution spectrum is extended over a more great variety of compounds, it seems safe to say that the magnitude of the normal red-shift depends on the type of the compound and on the character of the band. For example, the magnitude of the bathochromic shift, associated with the change from the heptane solution to the KCl disk, of absorption maxima of the

[41] J. Dale, *Acta Chem. Scand.* **11**, 650 (1957).

longest-wavelength singlet–singlet bands (A bands) of *trans*-stilbene and some related compounds increases fairly regularly with the wavelength of the absorption maximum, from about 2.5 mμ at about 265 mμ to about 5 mμ at about 300 mμ, and to about 9.5 mμ at about 350 mμ.[42] On the other hand,

TABLE 6.5

COMPARISON OF THE KCL-DISK SPECTRA AND HEXANE-SOLUTION SPECTRA OF NAPHTHALENE AND ANTHRACENE

Compound	KCl-disk λ_{max}, mμ	Hexane-solution λ_{max}, mμ	Shift $\Delta\lambda_{max}$, mμ	log ϵ_{max}[a]	f[b]	Transition and its polarization[c]
Naphthalene	292	285	7	3.64		$^1A_g \rightarrow {}^1B_{2u}$ (*p* band)
	281	275	6	3.82	0.11	Transverse
	272	267	5	3.89		(along the *y* axis)
						$^1A_g \rightarrow {}^1B_{3u}$ (β band)
	223	221	2	5.02	1.7	Longitudinal
						(along the *x* axis)
Anthracene	394	376	18	3.88		
	373	357	16	3.83		$^1A_g \rightarrow {}^1B_{2u}$ (*p* band)
	353	339	14	3.74	0.1	Transverse
	336	324	12	3.45		(along the *y* axis)
	318	309	9	3.17		
						$^1A_g \rightarrow {}^1B_{3u}$ (β band)
	256	252	4	5.25	2.3	Longitudinal
	249	246	3	—		(along the *x* axis)

[a] The values for the spectra of solutions in a mixture of methanol and ethanol, taken from E. Clar, *Spectrochim. Acta* **4,** 116 (1950).

[b] D. P. Craig and J. R. Walsh, *J. Chem. Soc.* **1958,** 1613; D. P. Craig and P. C. Hobbins, *ibid.* **1955,** 539.

[c] The coordinate axes are defined for naphthalene as shown in Fig. 8.4, and those are analogously defined for anthracene. Both naphthalene and anthracene molecules belong to point group D_{2h}.

the magnitudes of the bathochromic shifts of the second intense bands (B bands) of these compounds, appearing in the region about 220–250 mμ, are distributed over the range from about 1 to 12 mμ; for these bands there appears to be a very rough relation that the longer the conjugated system, the larger is the magnitude of the shift. The relations between the magnitude of the normal red-shift and the wavelength of the band for the compounds mentioned so far are shown graphically in Fig. 6.3. When comparing the KCl-disk spectrum with the solution spectrum it is necessary to allow for the normal red-shift.

[42] H. Suzuki, *Bull. Chem. Soc. Japan* **33,** 944 (1960).

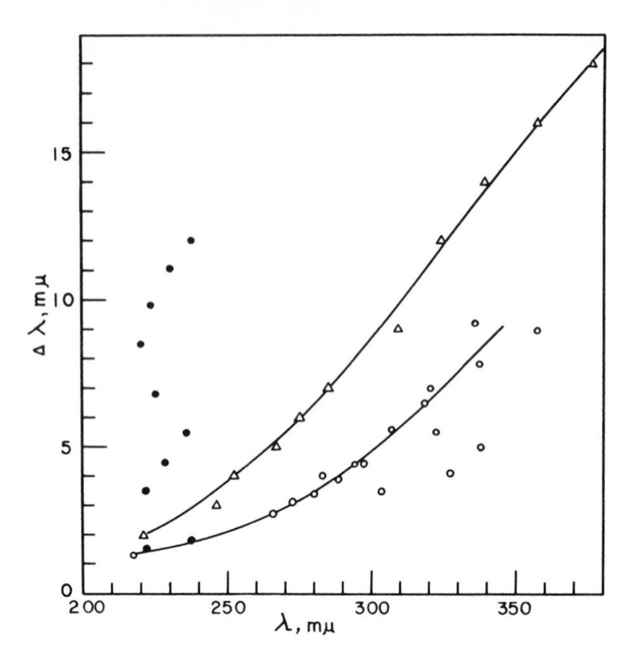

FIG. 6.3. The relation between the magnitude of bathochromic shift (Δλ) of the absorption maximum in going from the solution spectrum (in a saturated hydrocarbon solvent) to the KCl-disk spectrum and the wavelength (λ) of the absorption maximum in the solution spectrum. Open circle, for the A bands of *trans*-stilbene and sterically unhindered related compounds. Closed circle, for the B bands of *trans*-stilbene and sterically unhindered related compounds. Triangle, for the bands of naphthalene and anthracene (according to Dale's data, Reference 41).

It appears that the molar extinction coefficients in the KCl-disk spectrum are generally considerably smaller than the values in the corresponding solution spectrum.[42] Furthermore, with the A bands of *trans*-stilbene and related compounds it has been found that a marked redistribution of intensity occurs among the vibrational subbands in going from the solution spectrum to the KCl-disk spectrum (see Subsection 14.4.2).

Chapter 7

Simple LCAO MO Method

7.1. Elements of Simple LCAO MO Method

7.1.1. The LCAO MO Approximation

In molecular orbital (MO) theory an electronic state of a molecule is described in terms of molecular orbitals (MO's), that is, one-electron functions which in general extend over the whole molecule and to which the electrons are allocated so as to satisfy the Pauli exclusion principle (see Subsection 2.2.2). There are two principal approximate methods for obtaining the molecular orbitals: the free-electron (FE) method[1] and the LCAO MO method. The LCAO MO method will be almost exclusively used for interpretation of electronic absorption spectra throughout this book; therefore in the present chapter, elements of this method are briefly described.

The molecular orbitals are to be eigenfunctions of an effective one-electron Hamiltonian for the equilibrium nuclear configuration of the molecule, **h**:

$$\mathbf{h}\psi_m = \varepsilon_m \psi_m , \tag{7.1}$$

where the subscript *m* identifies the different molecular orbitals. When an electron gets close to a nucleus the influence of the potential field due to that nucleus becomes predominant. It is therefore reasonable to expect that in the regions close to the nuclei the molecular orbitals will resemble atomic orbitals (AO's). On the basis of this expectation, the molecular orbitals may be approximated as linear combinations of atomic orbitals. This approximation is known as the LCAO approximation to the molecular orbitals.

[1] J. R. Platt, *J. Chem. Phys.* **17**, 484 (1949); J. R. Platt *et al.*, "Free Electron Theory of Conjugated Molecules." Wiley, New York, 1964.

In the LCAO approximation the molecular orbitals are written as

$$\psi_m = \sum_i c_{mi}\chi_i , \tag{7.2}$$

where χ_i's are atomic orbitals. The orbital energies ε_m and the coefficients c_{mi} can be obtained by use of the variation method (see Section 2.3), if the matrix elements of the Hamiltonian and of unity between the atomic orbitals, that is, h_{ij}'s and S_{ij}'s, have been evaluated. The secular equations are given by

$$\sum_j c_j(h_{ij} - \varepsilon S_{ij}) = 0, \quad \text{for each } i. \tag{7.3}$$

If there are n atomic orbitals in the expression for the molecular orbitals, there are n secular equations of this type, and these will have nontrivial solutions only if the nth-order secular determinant is zero:

$$|h_{ij} - \varepsilon S_{ij}| = 0. \tag{7.4}$$

On expansion, this secular determinantal equation can be solved to give n solutions for the orbital energy ε, each of which can be substituted back into the secular equations to give the coefficients c of the atomic orbitals in the molecular orbital. This method of determining the molecular orbitals as linear combinations of atomic orbitals by use of the variation principle is called the simple LCAO MO method.

It is to be noted that in this method the two-electron terms representing repulsive interactions between electrons in the total electronic Hamiltonian have not been explicitly taken into account. Various methods of modifying the simple MO theory in order to explicitly allow for the interelectronic repulsions constitute the so-called advanced MO methods, which will be dealt with in a later chapter.

7.1.2. The π Electron Approximation

In this book we are interested mainly in π electronic states and π orbitals. In general, it can be assumed that interactions between π electrons and other electrons are comparatively small, and π electrons can be treated as being separated from other electrons. All the nuclei and all the electrons other than the π electrons, that is, all the inner electrons and all the electrons forming the σ bonds in a molecule, are said to form the core. Each of the π electrons can be treated as being in the potential field due to the core and other π electrons. Thus, if the potential acting on a single π electron i due to the core, namely, the core potential, is denoted by $\mathbf{V}_i^{\mathrm{C}}$, the electronic Hamiltonian for the π electrons can be written as follows:

$$\mathbf{H}^{\pi} = \sum_i (\mathbf{T}_i + \mathbf{V}_i^{\mathrm{C}}) + \sum_{i<j}\sum e^2/r_{ij} = \sum_i (\mathbf{T}_i + \mathbf{V}_i^{\mathrm{C}}) + \tfrac{1}{2}\sum_{i\neq j}\sum e^2/r_{ij}, \quad (7.5)$$

where i and j are running indices specifying π electrons, and the summation is carried out over all the π electrons. The last term in this expression represents the electrostatic repulsive interactions between π electrons. $\mathbf{T}_i + \mathbf{V}_i^{\mathrm{C}}$ is called the one-electron core Hamiltonian. When this is denoted by $\mathbf{h}_i^{\mathrm{C}}$, the effective one-electron Hamiltonian for π electrons can be expressed as follows:

$$\mathbf{h}_i = \mathbf{h}_i^{\mathrm{C}} + \mathbf{V}_i^{\pi}, \quad (7.6)$$

where $\mathbf{V}_i^{\pi}$ represents the effective potential acting on a π electron due to all the other π electrons. The π orbitals are considered to be eigenfunctions of this effective one-electron Hamiltonian. In the LCAO approximation, the π molecular orbitals are approximated as linear combinations of $p\pi$ atomic orbitals.

Ethylene provides the simplest example of π electronic systems. The ethylene molecule consists of two carbon atoms and four hydrogen atoms. Four C—H σ bonds and one C—C σ bond are said to form the σ-bond framework of the molecule; the six nuclei, the five pairs of electrons forming the σ-bond framework, and the two sets of two electrons in the inner shell, that is, the $1s$ atomic orbital of each carbon atom, are said to form the core. If the two remaining electrons, π electrons, are designated as 1 and 2, the electronic Hamiltonian for this π electronic system is given by

$$\mathbf{H}^{\pi} = \mathbf{T}_1 + \mathbf{T}_2 + \mathbf{V}_1^{\mathrm{C}} + \mathbf{V}_2^{\mathrm{C}} + e^2/r_{12} = \mathbf{h}_1^{\mathrm{C}} + \mathbf{h}_2^{\mathrm{C}} + e^2/r_{12}. \quad (7.7)$$

In the one-electron approximation the $\mathbf{H}^{\pi}$ is written as

$$\mathbf{H}^{\pi} = \mathbf{h}_1 + \mathbf{h}_2, \quad (7.8)$$

where $\mathbf{h}$'s are effective one-electron Hamiltonians, and are expressed as

$$\mathbf{h} = \mathbf{h}^{\mathrm{C}} + \mathbf{V}^{\pi}. \quad (7.9)$$

$\mathbf{h}^{\mathrm{C}}$ is the one-electron core Hamiltonian for the equilibrium nuclear configuration of the molecule, and $\mathbf{V}^{\pi}$ represents the effective potential acting on a π electron due to another π electron.

Each carbon atom in the ethylene molecule contributes one $2p\pi$ atomic orbital to the π molecular orbitals. That is, if the $2p\pi$ atomic orbitals of the two carbon atoms are denoted by χ_1 and χ_2, the two carbon atoms being designated as 1 and 2, as shown in Fig. 3.1, then the π molecular orbitals of the ethylene molecule are approximated as

$$\psi_m = c_{m1}\chi_1 + c_{m2}\chi_2 . \tag{7.10}$$

Thus our problem is now to find the orbital energies ε as eigenvalues of $\mathbf{h}$ and the best sets of values for the coefficients c by the variation method. The secular determinantal equation is evidently as follows:

$$\begin{vmatrix} h_{11} - \varepsilon S_{11} & h_{12} - \varepsilon S_{12} \\ h_{21} - \varepsilon S_{21} & h_{22} - \varepsilon S_{22} \end{vmatrix} = 0, \tag{7.11}$$

where, needless to say, h's are matrix elements of the effective one-electron Hamiltonian $\mathbf{h}$ between the $2p\pi$ carbon atomic orbitals and S's are the overlap integrals of these atomic orbitals. If χ_1 and χ_2 are real normalized orbitals, it follows that $h_{12} = h_{21}$, $S_{12} = S_{21}$, and $S_{11} = S_{22} = 1$. Furthermore, since χ_1 and χ_2 are identical atomic orbitals and undoubtedly are in a similar environment, it is expected that $h_{11} = h_{22}$.

7.1.3. The Hückel Approximation

The determination of appropriate numerical values of the matrix elements, h_{ij}'s and S_{ij}'s, remains as the only difficult problem. To evaluate these matrix elements, it is necessary to know the exact form of the Hamiltonian $\mathbf{h}$ and also the exact form of the atomic orbitals. However, by definition, the exact form of the effective one-electron Hamiltonian is not rigorously defined. Therefore, to make any progress, it is necessary to make a series of assumptions concerning the values of the matrix elements.

A few simplifying symbols are now introduced:

$$\alpha_i = h_{ii} ; \tag{7.12}$$

$$\beta_{ij} = h_{ij} = h_{ji} \quad (i \neq j). \tag{7.13}$$

The α_i is called the Coulomb integral of the atomic orbital χ_i, and has the physical significance of the energy of an electron which is localized in the atomic orbital χ_i and is acted by the effective potential in the effective one-electron Hamiltonian $\mathbf{h}$. Since the atomic orbital χ_i has a notable value only in the neighborhood of the ith atom, α_i is considered to be close to, and hence can be approximately taken to be equal to, the energy of the atomic orbital in the isolated atom in the appropriate valence state. The β_{ij} is called the resonance integral between the atomic orbitals χ_i and χ_j. This integral represents the energy of interaction of the two atomic orbitals. The most important contribution to this integral will come from the region of space where both χ_i and χ_j have large values. That is, the magnitude of this integral will depend on the extent of overlap of the two atomic orbitals χ_i and χ_j, and hence on the distance between the two atoms i and j. In fact, it has been shown by Mulliken[2] that the resonance integral β is approximately proportional to the overlap integral S. Both the Coulomb integrals and the resonance integrals are negative energy quantities. By the use of these symbols, and by assuming that all the atomic orbitals are normalized, the secular equations (7.3) can be rewritten as

$$\sum_{j \neq i} c_j(\beta_{ij} - \varepsilon S_{ij}) + c_i(\alpha_i - \varepsilon) = 0, \quad \text{for each } i. \tag{7.14}$$

When the π system consists entirely of carbon atoms, that is, when all the atomic orbitals in the expansion set of the π molecular orbitals are carbon $2p\pi$ atomic orbitals, the following set of approximations is frequently used:

(1) All the Coulomb intergals α_i are assumed to be equal and are replaced by a common symbol α, that is, it is assumed that $\alpha_1 = \alpha_2 = \cdots = \alpha$.

(2) The resonance integrals β_{ij} are assumed to be nonzero only when atoms i and j are neighbors to each other, that is, it is assumed that $\beta_{ij} \neq 0$ if $i \rightarrow j$ and $\beta_{ij} = 0$ if $i \not\rightarrow j$, where symbol $i \rightarrow j$ indicates that atoms i and j are neighbors and symbol $i \not\rightarrow j$ indicates that atoms i and j are nonneighbors. This approximation is called the neglect of nonneighbor interactions.

(3) All the resonance integrals for neighboring carbon atoms are assumed to have the same magnitude, and they are represented by a common symbol β, that is, it is assumed that $\beta_{ij} = \beta \neq 0$ if $i \rightarrow j$.

(4) All the overlap integrals S_{ij} for $i \neq j$ are neglected, that is, it is assumed that $S_{ij} = \delta_{ij}$, where δ is the Kronecker symbol. This approximation is called the neglect of overlap integrals or the zero-overlap approximation.

These approximations are called the Hückel approximations, or the set of these approximations is called simply the Hückel approximation.[3] The LCAO MO method involving the set of these approximations is called the Hückel MO (HMO) method, and molecular orbitals obtained by this method are called Hückel orbitals.

With the set of the Hückel approximations, the secular equations (7.14) reduces to

$$\sum_{j \rightarrow i} c_j\beta + c_i(\alpha - \varepsilon) = 0, \qquad \text{for each } i. \tag{7.15}$$

By dividing through Eq. (7.15) by β, and writing

$$(\alpha - \varepsilon)/\beta \equiv w, \tag{7.16}$$

[2] R. S. Mulliken, *J. chim. phys.* **46**, 675 (1949); *J. Phys. Chem.* **56**, 295 (1952).

[3] E. Hückel, *Z. Physik* **70**, 204 (1931); **76**, 628 (1932).

they are reduced to the following simple form:

$$\sum_{j \to i} c_j + wc_i = 0, \qquad \text{for each } i. \tag{7.17}$$

The secular determinant now has w everywhere on the diagonal, and 1 in the off-diagonal element of row i, column j if atoms i and j are neighbors, but zero otherwise. Thus we can easily set up the secular determinant for any π system consisting entirely of carbon π centers. For example, the secular determinantal equation for ethylene (7.11) can be written as

$$\begin{array}{c|cc|} & 1 & 2 \\ 1 & w & 1 \\ 2 & 1 & w \end{array} = 0. \tag{7.18}$$

Similarly, the secular determinantal equation in the Hückel approximation for benzene, with carbon atoms numbered as shown in Fig. 3.2, can be written as

$$\begin{array}{c|cccccc|} & 1 & 2 & 3 & 4 & 5 & 6 \\ 1 & w & 1 & 0 & 0 & 0 & 1 \\ 2 & 1 & w & 1 & 0 & 0 & 0 \\ 3 & 0 & 1 & w & 1 & 0 & 0 \\ 4 & 0 & 0 & 1 & w & 1 & 0 \\ 5 & 0 & 0 & 0 & 1 & w & 1 \\ 6 & 1 & 0 & 0 & 0 & 1 & w \end{array} = 0. \tag{7.19}$$

The nth-order secular determinantal equation can be solved to give n solutions for w. The energy of the mth molecular orbital is evidently given by

$$\varepsilon_m = \alpha + w_m(-\beta). \tag{7.20}$$

That is, w is the orbital energy as expressed in units of $-\beta$ with reference to α:

$$w_m = (\varepsilon_m - \alpha)/(-\beta). \tag{7.21}$$

Since $-\beta$ is a positive quantity, a negative value of w means a lower energy than α, and a positive value of w means a higher energy than α. A molecular orbital whose w value is negative is called a bonding orbital, and an orbital whose w value is positive is called an antibonding orbital. An orbital whose w value is zero is called a nonbonding orbital. In this book the following notation of molecular orbitals is principally used: the highest bonding orbital is designated by subscript $+1$, and lower bonding orbitals are numbered as $+2$, $+3$, ..., in the order of the decreasing energy; quite analogously, the lowest antibonding orbital is designated by subscript -1, and higher antibonding orbitals are designated by subscripts -2, -3, ..., in the order of the increasing energy; a nonbonding orbital is designated by subscript 0 or n. This notation will be referred to as the two-directional notation.

For each value of w the coefficients of atomic orbitals in the molecular orbital can be found by the usual procedure (see Subsection 2.3.2). In the zero-overlap approximation, the normalization condition is given by

$$\sum_i c_{mi}^2 = 1. \tag{7.22}$$

For example, by solving Eq. (7.18), the Hückel π orbitals of ethylene and their energies are obtained as follows:

$$\begin{aligned} w_{-1} &= +1, \qquad \varepsilon_{-1} = \alpha - \beta, \qquad \psi_{-1} = 2^{-1/2}(\chi_1 - \chi_2); \\ w_{+1} &= -1, \qquad \varepsilon_{+1} = \alpha + \beta, \qquad \psi_{+1} = 2^{-1/2}(\chi_1 + \chi_2). \end{aligned} \tag{7.23}$$

In the ground electron configuration the bonding π orbital ψ_{+1} is occupied by two electrons and the antibonding π orbital ψ_{-1} is vacant. Since the Coulomb integral α can be regarded approximately as the energy of the isolated carbon $2p\pi$ atomic orbital, the π bond energy in the ground electron configuration is approximately $2(-\beta)$. It is to be noted that the energy of the antibonding orbital lies as much above α as the energy of the bonding orbital lies below. It follows that the excited electron configuration resulting from promotion of one electron from the bonding to the antibonding orbital has no π bond energy. The value of $-\beta$ is maximal when the symmetry axes of the two carbon $2p\pi$ atomic orbitals are parallel, that is, when the molecule is planar, and must decrease as the symmetry axes deviate from the parallel arrangement by twist of the C—C bond. When the symmetry axes are perpendicular to each other, that is, when the two methylene groups lie in mutually perpendicular planes, the overlap integral between χ_1 and χ_2 is zero, and hence the value of $-\beta$ is also zero. This means that the ethylene molecule is most stable in the planar geometry, and that, as the C—C bond is twisted, the energy of the antibonding orbital is lowered and the energy of the bonding orbital is raised.

The Hückel approximations thus greatly simplify the calculation of the π orbitals of π-electron systems consisting entirely of carbon π centers. Only two integrals, α and β, remain unevaluated. These integrals are treated as empirical parameters, whose values are adjusted in order to produce agreement between experiment and calculation for one compound or one case, and then used in further calculations.

If a molecular orbital, ψ_m, has been normalized, the orbital energy can be expressed as follows:

$$\varepsilon_m = \int \psi_m \mathbf{h} \psi_m \, d\tau = \sum_i \sum_j c_{mi} c_{mj} h_{ij} \,. \tag{7.24}$$

By the use of the Hückel approximation (2), this equation is reduced to

$$\varepsilon_m = \sum_i c_{mi}^2 \alpha_i + \sum_{i \to j} \sum c_{mi} c_{mj} \beta_{ij} = \sum_i c_{mi}^2 \alpha_i + 2 \sum_{(i-j)} c_{mi} c_{mj} \beta_{ij} \,, \tag{7.25}$$

where $\sum_{(i-j)}$ means the summation over all the pairs of neighboring atomic orbitals. When the Hückel approximations (1) and (3) are used, this equation is further reduced to

$$\varepsilon_m = \alpha + 2 \sum_{(i-j)} c_{mi} c_{mj} \beta. \tag{7.26}$$

From comparison of this equation with (7.20), we obtain

$$w_m = -2 \sum_{(i-j)} c_{mi} c_{mj} \,. \tag{7.27}$$

7.1.4. Modifications of the Hückel Approximation

Some of the Hückel approximations may be removed by the use of some additional parameters within the framework of the simple LCAO MO method. For example, when applying the simple LCAO MO method to a π system containing a heteroatom, that is, an atom other than carbon, the difference of the Coulomb integral for the atomic orbital of the heteroatom from that of the carbon $2p\pi$ atomic orbital can be allowed for by expressing the Coulomb integral α_i as $\alpha + h_i\beta$, where α is a standard Coulomb integral and β is a standard resonance integral, and by assigning an appropriate value to the parameter h_i. Usually, the standard Coulomb integral is taken as the Coulomb integral for a carbon $2p\pi$ atomic orbital, and the standard resonance integral is taken as the resonance integral for two carbon $2p\pi$ atomic orbitals at a standard distance, for example, at the benzene bond distance. Since α and β are negative quantities, h_i is positive for an atom effectively more electronegative than the standard carbon.

The resonance integral β_{ij} must depend on the nature of the two atomic orbitals χ_i and χ_j, on the distance between the two atoms i and j, and on the relative arrangement of the two atomic orbitals. Such a variation of the resonance integral can be allowed for by expressing β_{ij} as $k_{ij}\beta$, where β is the standard resonance integral, and by assigning an appropriate value to the parameter k_{ij}.

Thus the Hückel approximations (1) and (3) can be modified by the use of the following new approximations:

$$(1') \quad \alpha_i = \alpha + h_i\beta; \tag{7.28}$$

$$(3') \quad \beta_{ij} = k_{ij}\beta. \tag{7.29}$$

If the expression (7.29) is used, the Hückel approximation (2) can be rewritten as:

$$(2') \quad k_{ij} = 0 \qquad \text{if} \quad i \nrightarrow j.$$

The set of approximations (1′), (2′), (3′), and (4) may be called the modified Hückel approximation, and the simple LCAO MO method using this set of approximations may be called the modified Hückel method. The parameters h and k will be referred to as the Coulomb parameter and resonance parameter, respectively.

When the modified Hückel approximation is used, the secular equations (7.3) can be rewritten as

$$\sum_{j \to i} c_j k_{ij}\beta + c_i(\alpha + h_i\beta - \varepsilon) = 0, \quad \text{for each } i. \tag{7.30}$$

Furthermore, when the expression (7.16) is used, these equations are reduced to

$$\sum_{j \to i} c_j k_{ij} + c_i(w + h_i) = 0, \quad \text{for each } i. \tag{7.31}$$

The secular determinant now has $w + h_i$ in the diagonal element of row i and column i, and k_{ij} in the off-diagonal element of row i and column j and in the off-diagonal element of row j and column i if atom i is directly bonded to j, but zero otherwise.

When this approximation is used, the expression (7.25) can be rewritten as

$$\begin{aligned} \varepsilon_m &= \sum_i c_{mi}^2(\alpha + h_i\beta) + 2\sum_{(i-j)} c_{mi}c_{mj}k_{ij}\beta \\ &= \alpha + \left(\sum_i c_{mi}^2 h_i + 2\sum_{(i-j)} c_{mi}c_{mj}k_{ij}\right)\beta. \end{aligned} \tag{7.32}$$

By comparing this expression with (7.20), we obtain

$$w_m = -\left(\sum_i c_{mi}^2 h_i + 2\sum_{(i-j)} c_{mi}c_{mj}k_{ij}\right). \tag{7.33}$$

The zero-overlap approximation might be thought to be rather drastic and questionable. In fact, when an explicit form of atomic orbitals is used, the overlap integrals of neighboring atomic orbitals in most π-electron systems have considerably large values. For example, the overlap integral of the two carbon $2p\pi$ atomic orbitals at an internuclear distance of about 1.39 Å is about 0.25. However, this approximation is not so drastic as it would seem, and is justified by the fact that, as will be seen below, inclusion of overlap integrals does not greatly change most of conclusions drawn on the basis of this approximation. The overlap integrals may be considered to have been implicitly allowed for in the empirical method by the vague definition and empirical nature of parameters α and β.

The zero-overlap approximation in the Hückel approximations may be replaced by the new assumption that all the overlap integrals S_{ij} for neighboring atomic orbitals have an equal nonzero value S. This new approximation is designated as (4′). The set of approximations (1), (2), (3), and (4′) is called the Wheland approximation, and molecular orbitals calculated on the basis of the Wheland approximation are called Wheland orbitals.[4] In this approximation, the secular equations (7.3) become

$$\sum_{j\to i} c_j{}^\gamma(\beta - \varepsilon^\gamma S) + c_i(\alpha - \varepsilon^\gamma) = 0, \quad \text{for each } i, \tag{7.34}$$

where the superscript γ indicates quantities to be obtained with inclusion of overlap integrals. By dividing through these equations by $(\beta - \varepsilon^\gamma S)$ and writting

$$(\alpha - \varepsilon^\gamma)/(\beta - \varepsilon^\gamma S) \equiv w, \tag{7.35}$$

these equations are reduced to the same form as the secular equations (7.17), which have been obtained for Hückel orbitals.

As mentioned previously, the resonance integral can be taken to be approximately proportional to the overlap integral. Accordingly, if the overlap integral for the standard π—π bond is denoted by S, the overlap integral S_{ij} can be expressed as follows:

$$(4'') \quad S_{ij} = k_{ij}S, \tag{7.36}$$

where k_{ij} is the resonance parameter for the i—j bond. The set of approximations (1′), (2′), (3′), and (4″) may be called the modified Wheland approximation. In this approximation the secular equations become

$$\sum_{j\to i} c_j{}^\gamma k_{ij}(\beta - \varepsilon^\gamma S) + c_i(\alpha + h_i\beta - \varepsilon^\gamma) = 0, \quad \text{for each } i. \tag{7.37}$$

By dividing by $(\beta - \varepsilon^\gamma S)$ and using expression (7.35) as before, these equations are evidently reduced to the same form as Eqs. (7.31), which have been derived in the modified Hückel approximation.

Thus, it is evident that whether the overlap integrals are included or not, the same form of the secular determinant results, and hence the same values of w and of the ratios between the coefficients of atomic orbitals in a molecular orbital must be obtained. However, while the orbital energy ε in the zero-overlap approximation, that is, in the Hückel or modified Hückel approximation, is given by Eq. (7.20), the corresponding orbital energy ε^γ in an approximation including overlap integrals, that is, in the Wheland or modified Wheland approximation, is given by

$$\varepsilon_m{}^\gamma = (\alpha - w_m\beta)/(1 - w_mS) = \alpha - (\beta - S\alpha)\, w_m/(1 - w_mS). \tag{7.38}$$

[4] G. W. Wheland, *J. Am. Chem. Soc.* **63,** 2025 (1941).

By setting

$$\beta - S\alpha \equiv \gamma \tag{7.39}$$

and

$$w_m/(1 - w_m S) \equiv w_m{}^\gamma, \tag{7.40}$$

Eq. (7.38) is rewritten as follows:

$$\varepsilon_m{}^\gamma = \alpha + w_m{}^\gamma(-\gamma). \tag{7.41}$$

Since β, S, and α are constants, γ is also a constant. This constant, like β, is a negative energy quantity, and is called the bond integral. Thus, while w represents the energy of the molecular orbital expressed in units of $-\beta$, w^γ represents the energy expressed in units of $-\gamma$. Since S is usually taken to be positive, it is evident that the inclusion of overlap integrals results in spreading antibonding energy levels and in compressing bonding energy levels. Furthermore, if a Hückel or modified Hückel molecular orbital, ψ_m, of a π system consisting entirely of carbon π centers, for which the Hückel approximation (1) can be used, has atomic orbital coefficients c_{mi}, it is easily verified that the corresponding molecular orbital calculated with inclusion of overlap integrals has the coefficients

$$c^\gamma_{mi} = (1 - w_m S)^{-1/2} c_{mi}\,. \tag{7.42}$$

Let us take ethylene as an example, again. The Hückel orbitals of ethylene and their energies were already given in expression (7.23). When the C—C bond length in the planar ethylene molecule is taken to be 1.34 Å, the value of the overlap integral between χ_1 and χ_2, S, is calculated to be about 0.27. With the use of this value of S, the Wheland orbitals and their energies can be calculated as follows:

$$\begin{aligned} &w^\gamma_{-1} \fallingdotseq +1.37, \quad &&\varepsilon^\gamma_{-1} \fallingdotseq \alpha - 1.37\,\gamma, \quad &&\psi^\gamma_{-1} \fallingdotseq 1.17 \times 2^{-1/2}(\chi_1 - \chi_2); \\ &w^\gamma_{+1} \fallingdotseq -0.79, \quad &&\varepsilon^\gamma_{+1} \fallingdotseq \alpha + 0.79\,\gamma, \quad &&\psi^\gamma_{+1} \fallingdotseq 0.89 \times 2^{-1/2}(\chi_1 + \chi_2). \end{aligned} \tag{7.43}$$

Thus, in the Wheland approximation, in contrast with the results of calculation in the Hückel approximation, the energy of the antibonding orbital lies above α by a greater amount than the energy of the bonding orbital lies below. It follows that the excited electron configuration $\psi^\gamma_{+1}\psi^\gamma_{-1}$ has a negative π bond energy. This means that in this excited electron configuration the planar structure is less stable than the nonplanar structure in which the two methylene groups lie in mutually perpendicular planes so that the two π electrons cannot interact with each other (see Section 5.3).

7.1.5. Energy and Moment of Electronic Transition in the Simple LCAO MO Approximation

In the simple MO theory, the transition energy of a one-electron transition from an occupied orbital, ψ_m, to an unoccupied orbital, ψ_n, is given by the difference between the energies of these orbitals. Thus, in the zero-overlap approximation, the transition energy is given by

$$\Delta E_{m\to n} = \varepsilon_n - \varepsilon_m = w_n - w_m \quad (\text{in } -\beta), \tag{7.44}$$

and, when the overlap integrals are included, it is given by

$$\Delta E_{m\to n} = \varepsilon_n{}^{\gamma} - \varepsilon_m{}^{\gamma} = w_n{}^{\gamma} - w_m{}^{\gamma} \quad (\text{in } -\gamma). \tag{7.45}$$

It should be noted that in such a simple theory, which neglects, or does not explicitly take into account, the electronic interaction, cannot account for the difference between the energy of a singlet–singlet transition and that of the corresponding singlet–triplet transition.

As stated previously, the transition moment of a singlet–singlet one-electron transition from a doubly occupied orbital, ψ_m, to an unoccupied orbital, ψ_n, is given by

$$\mathbf{M}_{m\to n} = \sqrt{2}\int \psi_m \mathbf{r} \psi_n \, d\tau, \tag{7.46}$$

where $\mathbf{r}$ is the vector defining the position of the promoted electron in the coordinate system fixed in the molecule. When the LCAO expressions of the molecular orbitals are substituted into this equation, the following equation is obtained:

$$\mathbf{M}_{m\to n} = \sqrt{2}\left[\sum_i c_{mi}c_{ni}\int \chi_i \mathbf{r} \chi_i \, d\tau + \sum_{i\neq j}\sum c_{mi}c_{nj}\int \chi_i \mathbf{r} \chi_j \, d\tau\right]. \tag{7.47}$$

Now the vector defining the position of the nucleus of atom i in the coordinate system fixed in the molecule is denoted by $\mathbf{r}_i$, and the vector defining the position of the electron in a new coordinate system in which the origin is at the position of the nucleus of atom i is denoted by $\mathbf{r}_{(i)}$. Then, we have

$$\mathbf{r} = \mathbf{r}_i + \mathbf{r}_{(i)}. \tag{7.48}$$

It follows that

$$\int \chi_i \mathbf{r} \chi_i \, d\tau = \int \chi_i (\mathbf{r}_i + \mathbf{r}_{(i)}) \chi_i \, d\tau = \mathbf{r}_i \int \chi_i^2 \, d\tau + \int \chi_i^2 \mathbf{r}_{(i)} \, d\tau. \tag{7.49}$$

Since χ_i^2 is symmetrical with respect to the nucleus of atom i unless χ_i is a hybrid orbital, the last term of this equation vanishes. Therefore, since χ_i is taken to be normalized, this equation reduces to

$$\int \chi_i \mathbf{r} \chi_i \, d\tau = \mathbf{r}_i. \tag{7.50}$$

When the Coulomb integrals of χ_i and χ_j are equal, that is, when $\alpha_i = \alpha_j$, the "hybrid" integral in expression (7.47) can be expressed as follows:

$$\int \chi_i \mathbf{r} \chi_j \, d\tau \quad (i \neq j) = S_{ij}(\mathbf{r}_i + \mathbf{r}_j)/2. \tag{7.51}$$

Therefore, when the zero-overlap approximation is adopted, all the hybrid integrals can be neglected. Even when the overlap integrals are not neglected, the sum of the terms in hybrid integrals in the expression (7.47) makes no contribution or only a negligibly small contribution to the transition moment. Therefore, the following simple expression can be given to the transition moment:

$$\mathbf{M}_{m\to n} = \sqrt{2}\sum_i c_{mi}c_{ni}\mathbf{r}_i \,. \tag{7.52}$$

The components of the transition moment along the coordinate axes fixed in the molecule are given by

$$M_{m\to n(q)} = \sqrt{2}\sum_i c_{mi}c_{ni}q_i \,, \tag{7.53}$$

where q is the general symbol for the coordinates x, y, z, and x_i, y_i, z_i are the coordinates of the position of the nucleus of atom i in the coordinate system fixed in the molecule. Thus, if the geometry of the molecule is known or is reasonably assumed and if the AO coefficients in the molecular orbitals are calculated, the transition moment of any one-electron transition can be easily calculated, and then, from the value of the transition moment, the value of the oscillator strength, that is, the f_{theor} value, of the corresponding absorption band can be calculated by means of Eq. (4.9).

Let us return to the study of ethylene. As mentioned previously, the intense absorption band appearing near 170 mμ can be considered as being, in the simple MO approximation, due to the singlet–singlet one-electron transition from the bonding π orbital ψ_{+1} to the antibonding π orbital ψ_{-1}. In the Hückel approximation the energy of this transition, ΔE_{11}, is evidently $2(-\beta)$. The transition moment of this transition lies along the C—C bond axis (that is, the x axis), and its magnitude is calculated as follows:

$$M_{11} = M_{11(x)} = \sqrt{2}(\tfrac{1}{2}x_1 - \tfrac{1}{2}x_2) = (1/\sqrt{2})R_{1-2} \,, \tag{7.54}$$

where x_1 and x_2 are the x coordinates of carbon atoms 1 and 2, respectively, and R_{1-2} is the C—C bond length. By taking a value of 1.34 Å for R_{1-2}, the magnitude of the dipole strength, that is, the square of the transition moment length, M_{11}^2, is calculated to be 0.90 (Å^2), and from this value, by taking a value of 58824 cm^{-1} (=170 mμ) for $\nu_{\max}$, the oscillator strength of this transition is calculated to be 0.57.

If the transition moment of the one-electron singlet–singlet transition from ψ_m to ψ_n in the approximation neglecting overlap integrals, that is, in the Hückel or modified Hückel approximation, is denoted by $\mathbf{M}_{m\to n}$, the transition moment in the approximation including overlap integrals, that is, in the Wheland or modified Wheland approximation, $\mathbf{M}^{\gamma}_{m\to n}$, evidently is given by

$$\mathbf{M}^{\gamma}_{m\to n} = (1 - w_m S)^{-1/2}(1 - w_n S)^{-1/2}\mathbf{M}_{m\to n}\,. \tag{7.55}$$

In the Wheland approximation where the S value of 0.27 is used, the energy of the one-electron transition from ψ^{γ}_{+1} to ψ^{γ}_{-1} in ethylene is calculated to be 2.16 $(-\gamma)$, and the dipole strength and oscillator strength of this transition are calculated to be 0.97 (Å^2) and 0.62, respectively.

7.1.6. The π-Electron Density and the π-Bond Order

If ψ_m is a normalized molecular π orbital expressed as $\sum_i c_{mi}\chi_i$ in the simple LCAO MO approximation with the neglect of overlap integrals, the probability that an electron in this molecular orbital will be found in the space region associated with the atomic orbital χ_i is considered to be given by c_{mi}^2. This quantity is called the partial π-electron density at atom i in the molecular orbital ψ_m, and is denoted by $p_{ii}^{(m)}$:

$$p_{ii}^{(m)} = c_{mi}^2\,. \tag{7.56}$$

The total π-electron density at atom i, P_{ii}, is defined as the sum of partial π-electron densities contributed by individual π electrons:

$$P_{ii} = \sum_m n_m p_{ii}^{(m)}, \tag{7.57}$$

where n_m is the number of electrons in the molecular orbital ψ_m, and the sum is taken over all the molecular π orbitals.

When atoms i and j are directly bonded to each other, the quantity $c_{mi}c_{mj}$ is called the partial π-bond order of the i—j bond in the molecular orbital ψ_m, and is denoted by $p_{ij}^{(m)}$:

$$p_{ij}^{(m)} = c_{mi}c_{mj}\,. \tag{7.58}$$

When the coefficients c_{mi} and c_{mj} are of like sign, the molecular orbital ψ_m is said to have no node between atoms i and j, and the partial π-bond order $p_{ij}^{(m)}$ is positive. When one of the coefficients, say c_{mi}, is zero, the molecular orbital ψ_m is said to have a node at the atom i, and the partial π-bond order $p_{ij}^{(m)}$ is zero. When the two coefficients are of opposite sign, the molecular orbital is said to have a node between the two atoms, and the partial π-bond order is negative. The partial π-bond order $p_{ij}^{(m)}$ may be construed as the partial π-electron density at the i—j bond (that is, the probability that an electron in the molecular orbital ψ_m will be found in the region of space between the two atoms i and j) and as representing the effect that a single electron in the molecular orbital ψ_m contributes to the binding power of the i—j bond: according as the partial π-bond order is positive, zero, or negative, an electron in the molecular orbital is considered to contribute a bonding, nonbonding, or antibonding effect to the i—j bond.

The total π-bond order of the i—j bond, P_{ij}, is defined as the sum of the partial π-bond orders contributed by individual π electrons:

$$P_{ij} = \sum_m n_m p_{ij}^{(m)}. \tag{7.59}$$

Let us take ethylene as an example again. The Hückel orbitals of ethylene were given in expression (7.23). The bonding π orbital ψ_{+1} has no node between the two carbon atoms, and the antibonding π orbital ψ_{-1} has a node between them. Evidently, the partial π-bond orders $p_{1,2}^{(+1)}$ and $p_{1,2}^{(-1)}$ are $+0.5$ and -0.5, respectively, and the partial π-electron densities $p_{1,1}^{(+1)}$, $p_{2,2}^{(+1)}$, $p_{1,1}^{(-1)}$, and $p_{2,2}^{(-1)}$ are all 0.5. In the ground electron configuration, in which two electrons with opposite spins occupy the bonding π orbital, the total π-electron density at each carbon atom is equal to unity, and the total π-bond order of the C—C bond is also equal to unity.

The connection between the orbital energy and the partial π-electron density as well as the partial π-bond order can be expressed mathematically. By substituting expressions (7.56) and (7.58) into Eqs. (7.25) and (7.33), we obtain

$$\varepsilon_m = \sum_i p_{ii}^{(m)} \alpha_i + 2 \sum_{(i-j)} p_{ij}^{(m)} \beta_{ij}; \tag{7.60}$$

$$w_m = -\left(\sum_i p_{ii}^{(m)} h_i + 2 \sum_{(i-j)} p_{ij}^{(m)} k_{ij} \right). \tag{7.61}$$

These equations indicate that the molecular orbital with the lesser number of nodes is the more stable. From these equations the following very useful relations are derived:

$$\frac{\partial \varepsilon_m}{\partial \alpha_i} = p_{ii}^{(m)}, \qquad \frac{\partial w_m}{\partial h_i} = -p_{ii}^{(m)}; \tag{7.62}$$

$$\frac{\partial \varepsilon_m}{\partial \beta_{ij}} = 2p_{ij}^{(m)}, \qquad \frac{\partial w_m}{\partial k_{ij}} = -2p_{ij}^{(m)}. \tag{7.63}$$

Hence, we can write to the first approximation, as follows:

$$\delta\varepsilon_m = p_{ii}^{(m)}\, \delta\alpha_i + \cdots, \qquad \delta w_m = -p_{ii}^{(m)}\, \delta h_i + \cdots; \tag{7.64}$$

$$\delta\varepsilon_m = 2p_{ij}^{(m)}\, \delta\beta_{ij} + \cdots, \qquad \delta w_m = -2p_{ij}^{(m)}\, \delta k_{ij} + \cdots. \tag{7.65}$$

When atom i is replaced by a more electronegative atom or when a substituent which exerts an electron-attracting inductive effect without conjugative effect is introduced to the atom i, the value of α_i becomes more negative.

Such a change in α_i is expressed by a positive value of δh_i, since both α and β are negative quantities. Equations (7.64) mean that the molecular orbital ψ_m will be stabilized by such a change in α_i, in proportion to the partial π-electron density $p_{ii}^{(m)}$. The value of k_{ij} for a fixed interatomic distance is at its maximum when the symmetry axes of the two $p\pi$ atomic orbitals χ_i and χ_j are parallel, and decreases as the i—j bond is twisted, that is, as the angle between the two axes, θ_{ij}, increases. A decrease in the value of k_{ij} is expressed by a negative value of δk_{ij}. Thus Eqs. (7.65) mean that the energy of the molecular orbital is raised or lowered by a twist of a bond according as the partial π-bond order of the twisted bond is positive or negative. For example, it is expected that a twist of the C—C bond of ethylene will result in destabilization of the bonding π orbital and stabilization of the antibonding π orbital.

In the simple MO theory neglecting the electronic repulsion, the total π-electron energy can be expressed as follows:

$$E = \sum_m n_m \varepsilon_m . \tag{7.66}$$

It can be easily verified that the total π-electron density and the total π-bond order are related to the total energy by the following equations:

$$\frac{\partial E}{\partial \alpha_i} = P_{ii} ; \tag{7.67}$$

$$\frac{\partial E}{\partial \beta_{ij}} = 2P_{ij} . \tag{7.68}$$

The transition energy of the one-electron transition from ψ_m to ψ_n is given by expression (7.44). The changes in the π-electron density at atom i and the π-bond order of the i—j bond upon this transition are simply given by

$$\Delta P_{ii} = p_{ii}^{(n)} - p_{ii}^{(m)} ; \tag{7.69}$$

$$\Delta P_{ij} = p_{ij}^{(n)} - p_{ij}^{(m)} . \tag{7.70}$$

Therefore, we can write to the first approximation, as follows:

$$\delta \, \Delta E_{m \to n} = -(p_{ii}^{(n)} - p_{ii}^{(m)}) \, \delta h_i = -\Delta P_{ii} \, \delta h_i ; \tag{7.71}$$

$$\delta \, \Delta E_{m \to n} = -2(p_{ij}^{(n)} - p_{ij}^{(m)}) \, \delta k_{ij} = -2 \, \Delta P_{ij} \, \delta k_{ij} . \tag{7.72}$$

Equation (7.71) means that when an atom is replaced by a more electronegative atom or when an electron-attracting substituent is introduced at an atom, the transition energy of an electronic transition will become smaller or greater, and hence the corresponding absorption band will be shifted toward longer or shorter wavelengths, according as the π-electron density at that atom increases or decreases upon that transition. Equation (7.72) means that when a bond is twisted, the transition energy of an electronic transition will increase or decrease, and hence a hypsochromic or bathochromic shift of the corresponding absorption band will result, depending on whether the π-bond order of the twisted bond increases or decreases upon that transition. These relations are very useful in discussion of the effect of a structural change upon the position of absorption bands.

7.2. The Pairing Property of π Orbitals of Alternant Hydrocarbons

A compound possessing a π-electron system which does not contain an odd-membered ring is said to be alternant.[5] Chain hydrocarbons such as ethylene, butadiene, and other conjugated polyenes, and ordinary aromatic hydrocarbons such as benzene and naphthalene, are alternant hydrocarbons

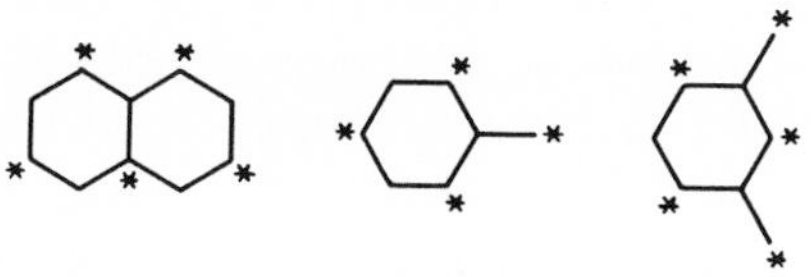

FIG. 7.1. The starring of carbon π centers in alternant hydrocarbons.

(frequently abridged as AH). On the other hand, a compound possessing a π-electron system which contains at least one odd-membered ring is said to be nonalternant. For example, azulene and fulvene are nonalternant hydrocarbons (frequently abridged as non-AH).

The carbon π centers in an alternant hydrocarbon can be divided into two sets, called starred and unstarred, such that no two members of the same set are directly linked. In most alternant hydrocarbons the numbers of starred and unstarred atoms are equal. In such cases the designation of the sets is arbitrary. If the numbers of atoms in the two sets are unequal, the more numerous set is starred. This starring is illustrated in Fig. 7.1 for naphthalene, benzyl radical, and a hypothetical molecule *m*-quinodimethane (*m*-dimethylenebenzene). It is clear that each starred atom has only unstarred atoms as

[5] C. A. Coulson and H. C. Longuet-Higgins, *Proc. Roy. Soc.* (*London*) **A192,** 16 (1947); H. C. Longuet-Higgins, *J. Chem. Phys.* **18,** 265 (1950).

neighbors and vice versa. In a nonalternant hydrocarbon such a division of π centers is not possible.

An alternant compound is further classed as "even" or "odd" according as the number of conjugated π centers in it is even or odd. A normal hydrocarbon is an even AH. An odd AH is necessarily a radical or ion (for example, allyl or benzyl radical, cation, and anion). For *m*-quinodimethane it is not possible to write structural formulas of Kekulé-type, in which each carbon atom is involved in two single bonds and one double bond. Such an AH is called a non-Kekulé AH.

By the way, two or more different conjugated compounds are said to be isoconjugate if they contain the same number of conjugated π centers in similar arrangements and the same number of π electrons. For example, benzene, pyridine, and pyrimidine are isoconjugate.

Many useful generalizations are applicable to the Hückel or modified Hückel π orbitals of alternant hydrocarbons, and they have been established as a series of formal theorems.[6] Some of these of interest to us are presented below.

Theorem 1. All the π orbital energies of an even alternant hydrocarbon occur in pairs, lying symmetrically above and below the nonbonding energy α. That is, if there is a bonding π orbital, ψ_{+m}, of energy $\alpha - w_{+m}\beta$, there is an antibonding π orbital ψ_{-m} of energy $\alpha - w_{-m}\beta$ where $w_{-m} = -w_{+m}$.

Theorem 2. The coefficients of the atomic orbitals in such pairing π orbitals are identical in absolute magnitude, and differ only in that the signs of the coefficients of one set of atomic orbitals of similar parity are inverted. That is, the coefficients $c_{+m,i}$ and $c_{-m,i}$ are equal if i is a starred atom, and they are equal in magnitude but have opposite signs if i is an unstarred atom:

$$c_{+m,i^*} = c_{-m,i^*}; \qquad c_{+m,j^\circ} = -c_{-m,j^\circ}, \tag{7.73}$$

where the superscript ° indicates unstarred. Therefore, if a bonding orbital, ψ_{+m}, is written as

$$\psi_{+m} = \sum_i^{*} c_{+m,i}\chi_i + \sum_j^{\circ} c_{+m,j}\chi_j, \tag{7.74}$$

where the summation $\overset{*}{\sum}$ is over starred atoms and the summation $\overset{\circ}{\sum}$ is over

[6] C. A. Coulson and G. S. Rushbrooke, *Proc. Cambridge Phil. Soc.* **36,** 193 (1940); C. A. Coulson and H. C. Longuet-Higgins, *Proc. Roy. Soc.* (*London*) **A191,** 39 (1947), **A192,** 16 (1947), **A193,** 447, 456 (1948), **A195,** 188 (1948); H. C. Longuet-Higgins, *J. Chem. Phys.* **18,** 265, 275, 283 (1950); M. J. S. Dewar, *J. Am. Chem. Soc.* **74,** 3341, 3345, 3350, 3353, 3355, 3357 (1952).

unstarred atoms, the paired antibonding orbital ψ_{-m} can be written as

$$\psi_{-m} = \sum_{i}^{*} c_{+m,i}\chi_i - \sum_{j}^{\circ} c_{+m,j}\chi_j . \qquad (7.75)$$

These two theorems, which are concerned with the pairing property of π orbitals of alternant hydrocarbons in the zero-overlap approximation, are sometimes called the MO correspondence rules in alternant hydrocarbons or the Coulson–Rushbrooke theorems. This pairing property can be easily proved. The secular equations for a hydrocarbon which are obtained on the basis of the zero-overlap approximation and of the approximation that all the Coulomb parameters h_i's are assumed to be zero [that is, the Hückel approximation (1)] are as follows [compare Eqs. (7.31)]:

$$wc_i + \sum_{j \to i} c_j k_{ij} = 0, \quad \text{for each } i. \qquad (7.76)$$

In an alternant hydrocarbon, if i is a starred atom, then j must be an unstarred atom. Evidently, if w_{+m}, $c_{+m,i}$, and $c_{+m,j}$ define one solution of these secular equations, the set of $-w_{+m}$, $c_{+m,i}$, and $-c_{+m,j}$ is also a solution.

In the LCAO MO approximation the number of π orbitals is equal to the number of the atomic orbitals in the expansion set. If there is an odd number of atoms in the conjugated system, an odd number of molecular π orbitals will occur. Thus, we have the following theorem:

Theorem 3. An odd alternant hydrocarbon has a nonbonding π orbital, whose energy is equal to α (that is, $w = 0$), in addition to pairs of bonding and antibonding π orbitals. A nonbonding molecular orbital (frequently abridged as NBMO) is denoted by ψ_0.

A non-Kekulé alternant hydrocarbon has two or more nonbonding molecular π orbitals. In general, if the number of double bonds in a principal resonance structure is T and the number of π centers in the conjugated system is N, there are at least $(N - 2T)$ nonbonding molecular π orbitals.

Concerning the nonbonding orbital in an odd alternant hydrocarbon we have the following theorems:

Theorem 4. In the nonbonding orbital all the coefficients of the unstarred atomic orbitals are zero. Accordingly, the nonbonding orbital can be expressed as

$$\psi_0 = \sum_{i}^{*} c_{0,i}\chi_i . \qquad (7.77)$$

Theorem 5. If j is an unstarred atom and i is a starred atom directly linked to j, the coefficients of atomic orbitals in the nonbonding orbital satisfy the following relation:

$$\sum_{i\to j} c_{0,i} k_{ij} = 0. \tag{7.78}$$

In the Hückel approximation, in which the k_{ij}'s for all the bonds are assumed to be unity, this relation reduces to

$$\sum_{i\to j} c_{0,i} = 0. \tag{7.79}$$

This theorem enables one to write down the coefficients of atomic orbitals in the LCAO expression of the nonbonding molecular orbital, without solving secular equations. Needless to say, both Theorems 4 and 5 can be derived from Eqs. (7.76) by putting $w = 0$.

From the pairing property of molecular π orbitals of an alternant hydrocarbon the following subsidiary theorems can be derived:

Theorem 6. In a neutral alternant hydrocarbon, which has the same number of π electrons as the number of the carbon π centers, that is, in an even AH molecule or in an odd AH radical, the π-electron density at each carbon atom in the ground electron configuration is unity. That is, for an even AH molecule,

$$P_{ii} = 2\sum_m c_{+m,i}^2 = 1; \tag{7.80}$$

for an odd AH radical,

$$P_{ii} = 2\sum_m c_{+m,i}^2 + c_{0,i}^2 = 1. \tag{7.81}$$

Since the molecular orbitals are to be orthonormal, the AO coefficients of the orbitals satisfy the relationship

$$\sum_m c_{m,i}^2 = 1, \tag{7.82}$$

where the summation is over all the molecular π orbitals. If we split up this sum into the contributions from occupied and vacant orbitals and use the pairing property that $c_{+m,i}^2 = c_{-m,i}^2$ [compare Eqs. (7.73)], we obtain the relation (7.80) or (7.81).

Theorem 7. The partial π-bond orders of each bond in pairing orbitals are identical in absolute magnitude, and differ in sign:

$$p_{ij}^{(+m)} = -p_{ij}^{(-m)} \tag{7.83}$$

This relation is derived immediately from Theorem 2, since two atoms directly bonded together belong to different sets. In an odd AH, the partial π-bond order of each bond in the nonbonding orbital is, of course, zero.

The energies of molecular π orbitals of a hydrocarbon in the modified Hückel approximation can be expressed as follows [compare Eqs. (7.33) and (7.58)]:

$$w_{+m} = -2 \sum_{(i-j)} p_{ij}^{(+m)} k_{ij} ; \tag{7.84}$$

$$w_{-m} = -2 \sum_{(i-j)} p_{ij}^{(-m)} k_{ij} . \tag{7.85}$$

Therefore, from relation (7.83), it follows that $w_{+m} = -w_{-m}$.

Theorem 8. In the ground electron configuration of an even alternant hydrocarbon, the "π-bond order" P_{ij} is zero if i and j are members of the same set of atoms (starred or unstarred):

$$P_{ij} = 2 \sum_{m} c_{+m,i} c_{+m,j} = 0 \tag{7.86}$$

if i and j are both starred or both unstarred. This theorem also follows from the orthonormality and pairing property of the molecular orbitals.

7.3. Applications of Perturbation Theory in the Simple LCAO MO Method

7.3.1. The Method of Polarizabilities

The principle of perturbation theory was mentioned already in Section 2.4. In the present section some important applications of this theory in the simple LCAO MO method will be described.

First, suppose that we wish to find the changes in the form and energy of π orbitals of a conjugated system when the Coulomb integral of an atomic orbital, say χ_r, is changed from α_r $(= \alpha + h_r\beta)$ to $\alpha_r + \delta\alpha_r$ $(= \alpha + (h_r + \delta h_r)\beta)$. It is assumed that the π orbitals of the original system and their energies are already given by

$$\psi_n = \sum_{s} c_{ns} \chi_s ; \qquad \varepsilon_n = \alpha + w_n(-\beta). \tag{7.87}$$

Now, the effective one-electron Hamiltonian for the original system is denoted by $\mathbf{h}^0$, and that for the new system is denoted by $\mathbf{h}^0 + \mathbf{h}'$. In the present problem, the exact form of the perturbation operator $\mathbf{h}'$ is of no interest, and use is made of the relation

$$\int \chi_r \mathbf{h}' \chi_r \, d\tau = \delta\alpha_r = \delta h_r \beta. \tag{7.88}$$

For the sake of simplicity, all the π orbitals of the unperturbed system are assumed to be nondegenerate. Then, by following Eq. (2.100), the change in the energy of a π orbital of the unperturbed system, say ψ_m, can be expressed as follows:

$$\delta\varepsilon_m = h'_{mm} + \sum_{n \neq m} (h'_{mn})^2/(\varepsilon_m - \varepsilon_n) + \cdots, \tag{7.89}$$

where

$$h'_{mn} = \int \psi_m \mathbf{h}' \psi_n \, d\tau = \sum_s \sum_t c_{ms} c_{nt} \int \chi_s \mathbf{h}' \chi_t \, d\tau = c_{mr} c_{nr} \, \delta\alpha_r . \tag{7.90}$$

By the use of the relation (7.90), this expression can be rewritten as

$$\delta\varepsilon_m = c_{mr}^2 \, \delta\alpha_r + (\delta\alpha_r)^2 \sum_{n \neq m} c_{mr}^2 c_{nr}^2 / (\varepsilon_m - \varepsilon_n) + \cdots. \tag{7.91}$$

The change in the total π-electron energy of a closed-shell electron configuration can therefore be expressed as follows:

$$\delta E = 2 \sum_m^{\text{occ}} c_{mr}^2 \, \delta\alpha_r + (\delta\alpha_r)^2 2 \sum_m^{\text{occ}} \sum_{n \neq m} c_{mr}^2 c_{nr}^2 / (\varepsilon_m - \varepsilon_n) + \cdots. \tag{7.92}$$

The coefficient of $\delta\alpha_r$ in the first term on the left-hand side of this equation is evidently the total π-electron density P_{rr}, and it is easily verified that the coefficient of $(\delta\alpha_r)^2$ in the second term is reduced to

$$2 \sum_m^{\text{occ}} \sum_n^{\text{unocc}} c_{mr}^2 c_{nr}^2 / (\varepsilon_m - \varepsilon_n).$$

Therefore, expression (7.92) can be rewritten as follows:

$$\delta E = P_{rr} \, \delta\alpha_r + \tfrac{1}{2} \pi_{rr} (\delta\alpha_r)^2 + \cdots, \tag{7.93}$$

where

$$\pi_{rr} = 4 \sum_m^{\text{occ}} \sum_n^{\text{unocc}} c_{mr}^2 c_{nr}^2 / (\varepsilon_m - \varepsilon_n). \tag{7.94}$$

The π_{rr} is evidently a negative quantity.

The π orbital of the perturbed system which reduces to ψ_m when the perturbation vanishes is denoted by $\psi_m{}^{\text{p}} = \sum_s c_{ms}^{\text{p}} \chi_s$. By the use of Eq. (2.101), this orbital can be approximately expressed as follows:

$$\sum_s c_{ms}^{\text{p}} \chi_s \doteqdot \sum_s c_{ms} \chi_s + \sum_s \sum_{n \neq m} [c_{mr} c_{nr} c_{ns} \, \delta\alpha_r / (\varepsilon_m - \varepsilon_n)] \chi_s + \cdots. \tag{7.95}$$

Comparison of the coefficients of χ_s on the two sides of this equation gives

$$c_{ms}^{\text{p}} \doteqdot c_{ms} + \sum_{n \neq m} c_{mr} c_{nr} c_{ns} \, \delta\alpha_r / (\varepsilon_m - \varepsilon_n) + \cdots. \tag{7.96}$$

Therefore, the total π-electron density at atom s in the perturbed system can be expressed as follows:

$$P_{ss}^{\mathrm{p}} \doteqdot 2\sum_{m}^{\mathrm{occ}}(c_{ms}^{\mathrm{p}})^2 = P_{ss} + \pi_{rs}\,\delta\alpha_r + \cdots, \tag{7.97}$$

where

$$\pi_{rs} = 4\sum_{m}^{\mathrm{occ}}\sum_{n}^{\mathrm{unocc}} c_{mr}c_{ms}c_{nr}c_{ns}/(\varepsilon_m - \varepsilon_n) = \partial P_{ss}/\partial\alpha_r\,. \tag{7.98}$$

If $s \neq r$, π_{rs} $(= \pi_{sr})$ is a quantity called the mutual polarizability of atoms r and s. If $s = r$, π_{rs} [that is, π_{rr}, defined by Eq. (7.94)] is called the self-polarizability (or autopolarizability) of atom r. The definition and formulation of these quantities, that is, the atom–atom polarizabilities, were made first by Coulson and Longuet-Higgins.[7] Following procedures similar to the above, they also defined and formulated the bond–atom, atom–bond, and bond–bond polarizabilities as follows:

$$\pi_{st,r} = \partial P_{st}/\partial\alpha_r = 2\sum_{m}^{\mathrm{occ}}\sum_{n}^{\mathrm{unocc}} c_{mr}c_{nr} \times (c_{ms}c_{nt} + c_{mt}c_{ns})/(\varepsilon_m - \varepsilon_n); \tag{7.99}$$

$$\pi_{r,st} = \partial P_{rr}/\partial\beta_{st} = 2\pi_{st,r}\,; \tag{7.100}$$

$$\pi_{rs,tu} = \partial P_{rs}/\partial\beta_{tu} = 2\sum_{m}^{\mathrm{occ}}\sum_{n}^{\mathrm{unocc}} (c_{mr}c_{ns} + c_{ms}c_{nr}) \times (c_{mt}c_{nu} + c_{mu}c_{nt})/(\varepsilon_m - \varepsilon_n); \tag{7.101}$$

$$\pi_{tu,rs} = \partial P_{tu}/\partial\beta_{rs} = \pi_{rs,tu}. \tag{7.102}$$

The quantities $\pi_{st,r}$ and $\pi_{r,st}$ are called the polarizability of bond s—t by atom r and the polarizability of atom r by bond s—t, respectively. The quantity $\pi_{tu,rs}$ $(= \pi_{rs,tu})$ is called the mutual polarizability of bonds r—s and t—u if r—$s \neq t$—u, and it is called the self-polarizability of bond r—s if r—$s = t$—u.

7.3.2. The Method of Building up Molecules from Fragments

As the next example of applications of perturbation theory, a method called the method of building up molecules from fragments or the method of composite molecule is outlined here. Now, suppose that a conjugated

[7] C. A. Coulson and H. C. Longuet-Higgins, *Proc. Roy. Soc. (London)* **A191,** 39 (1947); **A192,** 16 (1947).

system R—S is divided into two fragments R and S at bond r—s, where atoms r and s belong to R and S, respectively. For example, in the case of acetophenone, R and S may be taken as representing the benzene ring and the carbonyl bond, respectively. Then, the π orbitals of the composite molecule R—S can be approximated as linear combinations of the π orbitals of fragments R and S. According to this approximation, the π orbitals of the composite molecule R—S and their energies can be obtained by solving the proper secular equations based on the π orbitals of the separated fragments R and S. This approximation is called the LCMO (linear-combinations-of-molecular-orbitals) approximation.

Now, let the π orbitals of R, S, and R—S be $\phi_{m(\mathrm{R})}$'s, $\phi_{n(\mathrm{S})}$'s, and $\psi_{k(\mathrm{RS})}$'s, respectively. These orbitals are, of course, to be solutions of the equations

$$\mathbf{h}^{\mathrm{R}}\phi_{m(\mathrm{R})} = \eta_{m(\mathrm{R})}\phi_{m(\mathrm{R})}, \qquad \mathbf{h}^{\mathrm{S}}\phi_{n(\mathrm{S})} = \eta_{n(\mathrm{S})}\phi_{n(\mathrm{S})},$$

and

$$\mathbf{h}^{\mathrm{RS}}\psi_{k(\mathrm{RS})} = \varepsilon_{k(\mathrm{RS})}\psi_{k(\mathrm{RS})}, \tag{7.103}$$

where $\mathbf{h}^{\mathrm{R}}$, $\mathbf{h}^{\mathrm{S}}$, and $\mathbf{h}^{\mathrm{RS}}$ are effective one-electron Hamiltonian operators for R, S, and R—S, respectively, and can be expressed as follows:

$$\mathbf{h}^{\mathrm{R}} = \mathbf{T} + \mathbf{V}^{\mathrm{R}}; \qquad \mathbf{h}^{\mathrm{S}} = \mathbf{T} + \mathbf{V}^{\mathrm{S}}; \qquad \mathbf{h}^{\mathrm{RS}} = \mathbf{T} + \mathbf{V}^{\mathrm{R}} + \mathbf{V}^{\mathrm{S}}. \tag{7.104}$$

Here $\mathbf{V}^{\mathrm{R}}$ and $\mathbf{V}^{\mathrm{S}}$ are the effective potentials acting on an electron in R and S, respectively, and $\mathbf{T}$ is the kinetic energy operator for the electron. The effective one-electron Hamiltonian for R—S may also be expressed as

$$\mathbf{h}^{\mathrm{RS}} = \mathbf{h}^0 + \mathbf{h}', \tag{7.105}$$

where $\mathbf{h}^0$ is the Hamiltonian for the system of the two separated fragments R and S, and $\mathbf{h}'$ is the perturbation operator representing interaction between R and S. Since $\mathbf{V}^{\mathrm{R}}$ and $\mathbf{V}^{\mathrm{S}}$ are important only in the R and S regions, respectively, and $\phi_{m(\mathrm{R})}$'s and $\phi_{n(\mathrm{S})}$'s are localized in the R and S regions, respectively, the required matrix elements of the Hamiltonian $\mathbf{h}^{\mathrm{RS}}$ can be approximately given by

$$h^{\mathrm{RS}}_{m(\mathrm{R}),u(\mathrm{R})} = h^{\mathrm{R}}_{m(\mathrm{R}),u(\mathrm{R})} = \delta_{mu}\eta_{m(\mathrm{R})}, \tag{7.106}$$

$$h^{\mathrm{RS}}_{n(\mathrm{S}),v(\mathrm{S})} = h^{\mathrm{S}}_{n(\mathrm{S}),v(\mathrm{S})} = \delta_{nv}\eta_{n(\mathrm{S})}, \tag{7.107}$$

and

$$h^{\mathrm{RS}}_{m(\mathrm{R}),n(\mathrm{S})} = h'_{m(\mathrm{R}),n(\mathrm{S})} = b_{m(\mathrm{R}),r}b_{n(\mathrm{S}),s}\beta_{rs}, \tag{7.108}$$

where δ is the Kronecker symbol, $b_{m(\mathrm{R}),r}$ is the coefficient of atomic orbital χ_r in $\phi_{m(\mathrm{R})}$, $b_{n(\mathrm{S}),s}$ is the coefficient of atomic orbital χ_s in $\phi_{n(\mathrm{S})}$, and β_{rs}

is the resonance integral between χ_r and χ_s:

$$\beta_{rs} = \int \chi_r \mathbf{h}^{\mathrm{RS}} \chi_s \, d\tau = \int \chi_r \mathbf{h}' \chi_s \, d\tau. \tag{7.109}$$

The forms and energies of the molecular orbitals of R—S in the LCMO approximation can be approximately found by the use of perturbation theory, instead of solving the secular equations. Thus, if the relation

$$|h'_{m(\mathrm{R}),n(\mathrm{S})}| \ll |\eta_{m(\mathrm{R})} - \eta_{n(\mathrm{S})}| \tag{7.110}$$

holds good for any pair of $\phi_{m(\mathrm{R})}$ with one of $\phi_{n(\mathrm{S})}$'s, one of the molecular orbitals of R—S and its energy can be approximately expressed as follows [compare Eqs. (2.100) and (2.101)]:

$$\begin{aligned} \psi_{k(\mathrm{RS})} &\fallingdotseq N_k \left(\phi_{m(\mathrm{R})} + \sum_{n(\mathrm{S})} [h'_{m(\mathrm{R}),n(\mathrm{S})}/(\eta_{m(\mathrm{R})} - \eta_{n(\mathrm{S})})] \phi_{n(\mathrm{S})} \right) \\ &\equiv \phi^{\mathrm{p}}_{m(\mathrm{R})}\, ; \end{aligned} \tag{7.111}$$

$$\varepsilon_{k(\mathrm{RS})} \fallingdotseq \eta_{m(\mathrm{R})} + \sum_{n(\mathrm{S})} (h'_{m(\mathrm{R}),n(\mathrm{S})})^2/(\eta_{m(\mathrm{R})} - \eta_{n(\mathrm{S})}) \equiv \eta^{\mathrm{p}}_{m(\mathrm{R})}\, . \tag{7.112}$$

If we put

$$\eta_{m(\mathrm{R})} = \alpha + v_{m(\mathrm{R})}(-\beta), \qquad \eta_{n(\mathrm{S})} = \alpha + v_{n(\mathrm{S})}(-\beta),$$

$$\varepsilon_{k(\mathrm{RS})} = \alpha + w_{k(\mathrm{RS})}(-\beta) \equiv \eta^{\mathrm{p}}_{m(\mathrm{R})} = \alpha + v^{\mathrm{p}}_{m(\mathrm{R})}(-\beta),$$

and $\beta_{rs} = k_{rs}\beta$, and if we use relation (7.108), then we obtain

$$\phi^{\mathrm{p}}_{m(\mathrm{R})} = N^{\mathrm{p}}_{m(\mathrm{R})} \left(\phi_{m(\mathrm{R})} - \sum_{n(\mathrm{S})} [b_{m(\mathrm{R}),r} b_{n(\mathrm{S}),s} k_{rs}/(v_{m(\mathrm{R})} - v_{n(\mathrm{S})})] \phi_{n(\mathrm{S})} \right), \tag{7.113}$$

$$v^{\mathrm{p}}_{m(\mathrm{R})} = v_{m(\mathrm{R})} + \sum_{n(\mathrm{S})} b^2_{m(\mathrm{R}),r} b^2_{n(\mathrm{S}),s} k^2_{rs}/(v_{m(\mathrm{R})} - v_{n(\mathrm{S})}). \tag{7.114}$$

Thus, if S is regarded as a mesomeric substituent, by the use of perturbation theory we can express in an analytical form the changes in the energies and forms of the molecular orbitals of a molecule R produced by introduction of the substituent. As mentioned already in Section 6.3.4, a similar treatment is possible for molecular complexes: in this case, R and S are taken as representing the component molecules and R—S is taken as representing the molecular complex composed of R and S.

Chapter 8

Application of Simple LCAO MO Method to Some Simple π Systems

8.1. Simplification of Secular Determinants by the Use of Group Theory

The application of simple LCAO MO method to ethylene, the simplest example of π systems, was described in the preceding chapter. In the present chapter, general procedures of calculation of π orbitals by this method are illustrated for somewhat more complicated examples.

When the molecule or the skeleton of the π system belongs to a point group having one or more symmetry operations besides the identity operation, the secular determinant for determination of the π orbitals as linear combinations of $p\pi$ atomic orbitals can be factored into a product of lower-dimensional determinants, and hence solution of the secular equation can be simplified.

In general, any one of possible π MO's, ψ, as linear combinations of $p\pi$ AO's, χ, belongs to one of the irreducible representations of the point group to which the π system belongs, and the $p\pi$ AO's form a basis for a reducible representation, Γ_{red}, of that point group. The reducible representation Γ_{red} can be transformed to a set of irreducible representations by a similarity transformation (see Section 3.2), and correspondingly, it is possible to take the proper linear combinations of the AO's so that the combinations form bases for the irreducible representations. Such linear combinations, ϕ, are called intermediate orbitals or group orbitals (GO's). There are as many independent GO's as there are starting AO's. The π MO's can be determined not as linear combinations of AO's but as linear combinations of GO's. Since the one-electron Hamiltonian $\mathbf{h}$ is invariant under all the symmetry operations of the point group, that is, belongs to the totally symmetric representation, the element of the energy matrix $h_{ij} = \int \phi_i \mathbf{h} \phi_j \, d\tau$, as well as the overlap integral $S_{ij} = \int \phi_i \phi_j \, d\tau$, can be different from zero only if the functions ϕ_i and ϕ_j belong to the same irreducible representation. In other words, it can be said that there is no matrix element between two functions which belong to different symmetry species (see Subsection 3.4.3). Therefore, by finding the proper group orbitals and classifying them according to the representations to which they belong, it is possible to reduce the secular determinant to block form, that is, to the form of the product of lower-dimensional determinants.

The number of times the kth irreducible representation Γ_k occurs in the reducible representation Γ_{red}, a_k, can be given by

$$a_k = (1/h) \sum_R X_{\text{red}}(R) X_k(R), \tag{8.1}$$

where h is the order of the group, $X_{\text{red}}(R)$ and $X_k(R)$ are the characters for the operation R of representations Γ_{red} and Γ_k, respectively, and the sum is taken over all the operations in the group (see Subsection 3.4.2). If the dimension of Γ_k is denoted by l_k, the number of independent group orbitals belonging to Γ_k is given by $a_k \times l_k$.

When a symmetry operation R of the point group is applied to an AO χ_i, the resultant function $\mathbf{R}\chi_i$ is $+\chi_j$ or $-\chi_j$, in which χ_j is one of the AO's. The AO χ_j is called a partner of χ_i or an AO equivalent to χ_i. In general, the AO's found in this manner by applying each symmetry operation in the point group to a given AO are called the partners of, or the AO's equivalent to, the starting AO. The group orbitals belonging to an irreducible representation Γ_k, that is, the proper linear combinations of AO's which belong to Γ_k and out of which the MO's belonging to Γ_k are to be built, can be constructed as linear combinations of each AO and its partners by means of the following equation:

$$\phi_{i(k)} = N_{i(k)} \sum_R X_k(R) \mathbf{R} \chi_i, \tag{8.2}$$

where the summation is taken over all the operations of the group, and $N_{i(k)}$ is the normalization constant.

The procedure of simplification of the secular equation by the use of group theory is illustrated by taking 1,3-butadiene (referred to hereafter simply as butadiene) as an example. In the usual LCAO approximation neglecting nonneighbor interactions, the π system of butadiene can be treated by considering a linear model for the carbon skeleton. If the carbon atoms are numbered as in Fig. 8.1, the four π MO's are to be expressed as follows:

$$\psi_m = c_{m,2'}\chi_{2'} + c_{m,1'}\chi_{1'} + c_{m,1}\chi_1 + c_{m,2}\chi_2, \tag{8.3}$$

where, of course, the χ's are carbon $2p\pi$ AO's. According to the symmetry of the system, evidently $\alpha_{2'} = \alpha_2$, $\alpha_{1'} = \alpha_1$, and $\beta_{1'2'} = \beta_{12}$. Therefore,

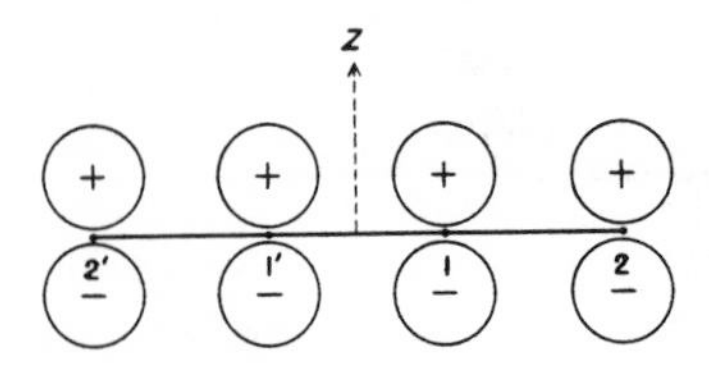

FIG. 8.1. An abstract linear model of the skeleton of 1,3-butadiene and the carbon $2p\pi$ atomic orbitals.

when it is assumed that all the α's are equal and overlap integrals are neglected, the secular equation for determination of the π MO's can be written as follows, each column and each row of the determinant being labeled by the number of the corresponding AO:

$$\begin{array}{c} \\ 2' \\ 1' \\ 1 \\ 2 \end{array} \begin{array}{c} \begin{array}{cccc} 2' & 1' & 1 & 2 \end{array} \\ \begin{vmatrix} w & k_{12} & 0 & 0 \\ k_{12} & w & k_{11'} & 0 \\ 0 & k_{11'} & w & k_{12} \\ 0 & 0 & k_{12} & w \end{vmatrix} \end{array} = 0, \tag{8.4}$$

where w is the orbital energy defined by expression (7.16), and the k's are the resonance parameters, defined by expression (7.29). Thus it is necessary to solve this fourth-order determinantal secular equation. However, when group theory is used, this fourth-order equation can be reduced to two second-order equations, as will be shown below, and hence the determination of the MO's can be greatly simplified.

It is convenient to take the midpoint of the 1—1′ bond as the origin of the coordinate system for the linear "abstract" model of the carbon skeleton, and to take the direction of the symmetry axes of the four parallel carbon $2p\pi$ AO's as the direction of the z axis. This linear model has evidently many symmetry elements. However, the sufficient simplification of the secular determinant can be achieved by assuming that this model belongs to point group C_2, which has only two symmetry operations, the identity operation (E) and the rotation by 180° about the z axis ($C_2(z)$), operations other than these two producing no new results. That is, it can be said that the assignment of point group C_2 to this system gives optimum use of the symmetry of the system. The character table for point group C_2 is given in Table 8.1.

When the identity operation E is applied to the AO's, of course each AO

TABLE 8.1

CHARACTER TABLE OF POINT GROUP C_2

C_2	E	$C_2(z)$	
A	1	1	z, R_z
B	1	-1	x, y, R_x, R_y

remains unchanged. When the operation $C_2(z)$ is applied, the AO's $\chi_{2'}$, $\chi_{1'}$, χ_1, and χ_2 are transformed to χ_2, χ_1, $\chi_{1'}$, and $\chi_{2'}$, respectively. Therefore, when the transformation of the set of the AO's by application of an operation R is expressed as

$$\begin{bmatrix} \mathbf{R}\chi_{2'} \\ \mathbf{R}\chi_{1'} \\ \mathbf{R}\chi_1 \\ \mathbf{R}\chi_2 \end{bmatrix} = \mathbf{A}(R) \begin{bmatrix} \chi_{2'} \\ \chi_{1'} \\ \chi_1 \\ \chi_2 \end{bmatrix}, \tag{8.5}$$

the matrices of transformation for operations E and C_2 are expressed as follows:

$$\mathbf{A}(E) = \begin{bmatrix} 1 & 0 & 0 & 0 \\ 0 & 1 & 0 & 0 \\ 0 & 0 & 1 & 0 \\ 0 & 0 & 0 & 1 \end{bmatrix}; \qquad \mathbf{A}(C_2) = \begin{bmatrix} 0 & 0 & 0 & 1 \\ 0 & 0 & 1 & 0 \\ 0 & 1 & 0 & 0 \\ 1 & 0 & 0 & 0 \end{bmatrix}. \tag{8.6}$$

This set of four-dimensional matrices makes up a reducible representation, Γ_{red}, of point group C_2, and the set of the four AO's is said to form a basis for this reducible representation. Evidently, the character of this representation, $\mathrm{X}_{\text{red}}(R)$, is 4 for operation E, and is 0 for operation C_2. Therefore, since h is 2, by the use of Eq. (8.1) we obtain

$$\begin{aligned} a_A &= \tfrac{1}{2}(4 \times 1 + 0 \times 1) = 2; \\ a_B &= \tfrac{1}{2}(4 \times 1 - 0 \times 1) = 2. \end{aligned} \tag{8.7}$$

Thus it is found that two sets of two group orbitals can be formed from the four AO's, one set belonging to symmetry species A, and another to B.

By the use of Eq. (8.2), the proper linear combinations of AO's belonging to A can be found to be the following, aside from the normalization factors:

$$\begin{aligned} &\sum_R \mathrm{X}_A(R)\mathbf{R}\chi_2 = \chi_2 + \chi_{2'}; \qquad \sum_R \mathrm{X}_A(R)\mathbf{R}\chi_1 = \chi_1 + \chi_{1'}; \\ &\sum_R \mathrm{X}_A(R)\mathbf{R}\chi_{1'} = \chi_{1'} + \chi_1; \qquad \sum_R \mathrm{X}_A(R)\mathbf{R}\chi_{2'} = \chi_{2'} + \chi_2. \end{aligned} \tag{8.8}$$

Of these combinations only two are independent: after normalization these give two group orbitals belonging to A:

$$\begin{aligned} \phi_{1(a)} &= \sqrt{\tfrac{1}{2}}(\chi_1 + \chi_{1'}); \\ \phi_{2(a)} &= \sqrt{\tfrac{1}{2}}(\chi_2 + \chi_{2'}). \end{aligned} \tag{8.9}$$

Quite analogously, the two independent group orbitals belonging to B are found to be as follows:

$$\begin{aligned} \phi_{1(b)} &= \sqrt{\tfrac{1}{2}}(\chi_1 - \chi_{1'}); \\ \phi_{2(b)} &= \sqrt{\tfrac{1}{2}}(\chi_2 - \chi_{2'}). \end{aligned} \tag{8.10}$$

Since matrix elements between functions belonging to different irreducible representations of the group vanish, our problem reduces to solving separately the following two second-order determinantal secular equations and determining the MO's as linear combinations of the GO's belonging to the same representation:

$$\begin{array}{c|cc|l} & 1(a) & 2(a) & \\ 1(a) & w + k_{11'} & k_{12} & \\ 2(a) & k_{12} & w & \end{array} = 0; \tag{8.11}$$

$$\begin{array}{c|cc|l} & 1(b) & 2(b) & \\ 1(b) & w - k_{11'} & k_{12} & \\ 2(b) & k_{12} & w & \end{array} = 0. \tag{8.12}$$

This means that the original fourth-order secular determinant (8.4) has been transformed to block form, or in other words, has been partially diagonalized. The process of this transformation can be shown as follows:

$$\begin{array}{c|cccc|} & 2' & 1' & 1 & 2 \\ 2' & w & k_{12} & 0 & 0 \\ 1' & k_{12} & w & k_{11'} & 0 \\ 1 & 0 & k_{11'} & w & k_{12} \\ 2 & 0 & 0 & k_{12} & w \end{array} \rightarrow$$

$$\begin{array}{c|cccc|}
 & 1(a) & 2(a) & 1(b) & 2(b) \\
2' & \sqrt{\tfrac{1}{2}}k_{12} & \sqrt{\tfrac{1}{2}}w & -\sqrt{\tfrac{1}{2}}k_{12} & -\sqrt{\tfrac{1}{2}}w \\
1' & \sqrt{\tfrac{1}{2}}(w + k_{11'}) & \sqrt{\tfrac{1}{2}}k_{12} & \sqrt{\tfrac{1}{2}}(k_{11'} - w) & -\sqrt{\tfrac{1}{2}}k_{12} \\
1 & \sqrt{\tfrac{1}{2}}(w + k_{11'}) & \sqrt{\tfrac{1}{2}}k_{12} & \sqrt{\tfrac{1}{2}}(w - k_{11'}) & \sqrt{\tfrac{1}{2}}k_{12} \\
2 & \sqrt{\tfrac{1}{2}}k_{12} & \sqrt{\tfrac{1}{2}}w & \sqrt{\tfrac{1}{2}}k_{12} & \sqrt{\tfrac{1}{2}}w
\end{array} \rightarrow$$

$$\begin{array}{c|cccc|}
 & 1(a) & 2(a) & 1(b) & 2(b) \\
1(a) & w + k_{11'} & k_{12} & 0 & 0 \\
2(a) & k_{12} & w & 0 & 0 \\
1(b) & 0 & 0 & w - k_{11'} & k_{12} \\
2(b) & 0 & 0 & k_{12} & w
\end{array} . \tag{8.13}$$

8.2. Application of Simple LCAO MO Method to Butadiene

8.2.1. The π Orbitals and Their Energies

The four π orbitals of butadiene are denoted by ψ_{+2}, ψ_{+1}, ψ_{-1}, and ψ_{-2} in the order of increasing energy. Then, the energies of these orbitals are found by solving Eqs. (8.11) and (8.12) to be as follows:

$$\begin{aligned} w_{\mp 2} &= \pm\tfrac{1}{2}[(k_{11'}^2 + 4k_{12}^2)^{1/2} + k_{11'}]; \\ w_{\mp 1} &= \pm\tfrac{1}{2}[(k_{11'}^2 + 4k_{12}^2)^{1/2} - k_{11'}]. \end{aligned} \tag{8.14}$$

ψ_{+2} and ψ_{-1} belong to symmetry species A, and ψ_{+1} and ψ_{-2} belong to symmetry species B. Of course, ψ_{+2} and ψ_{+1} are bonding orbitals, and ψ_{-1} and ψ_{-2} are antibonding orbitals. The ratio of the coefficients of group orbitals in a molecular orbital ψ_m, $C_{m,2}/C_{m,1}$, is given by $-k_{12}/w_m$. Therefore, since $w_{+1} = -w_{-1}$ and $w_{+2} = -w_{-2}$, the molecular orbitals can be expressed as follows:

$$\begin{aligned} \psi_{\mp 2} &= [2(k_{12}^2 + w_{-2}^2)]^{-1/2}(k_{12}\chi_2 \mp w_{-2}\chi_1 + w_{-2}\chi_{1'} \mp k_{12}\chi_{2'}); \\ \psi_{\mp 1} &= [2(k_{12}^2 + w_{-1}^2)]^{-1/2}(k_{12}\chi_2 \mp w_{-1}\chi_1 - w_{-1}\chi_{1'} \pm k_{12}\chi_{2'}). \end{aligned} \tag{8.15}$$

Since k_{12}, w_{-2}, and w_{-1} are positive, it is evident that ψ_{+2}, ψ_{+1}, ψ_{-1}, and ψ_{-2} have, respectively, 0, 1, 2, and 3 nodal planes perpendicular to the C—C bonds, in addition to the common nodal plane coinciding with the molecular plane.

In the Hückel approximation, in which it is assumed that $k_{12} = k_{11'} = 1$, the molecular orbitals and their energies are calculated to be as follows:

$$\begin{aligned} w_{\mp 2} &= \pm 1.618, \quad & \psi_{\mp 2} &= 0.372\chi_2 \mp 0.601\chi_1 \\ & & &\quad + 0.601\chi_{1'} \mp 0.372\chi_{2'}\,; \\ w_{\mp 1} &= \pm 0.618, \quad & \psi_{\mp 1} &= 0.601\chi_2 \mp 0.372\chi_1 \\ & & &\quad - 0.372\chi_{1'} \pm 0.601\chi_{2'}\,. \end{aligned} \tag{8.16}$$

The partial π-bond orders and partial π-electron densities calculated for these orbitals are summarized in Table 8.2.

TABLE 8.2

ENERGIES, PARTIAL π-BOND ORDERS, AND PARTIAL π-ELECTRON DENSITIES OF THE π MO'S OF 1,3-BUTADIENE IN THE HÜCKEL APPROXIMATION

MO	Energy	Partial π-bond order		Partial π-electron density	
ψ_m	w_m	$p_{12}^{(m)} = p_{1'2'}^{(m)}$	$p_{11'}^{(m)}$	$p_{22}^{(m)} = p_{2'2'}^{(m)}$	$p_{11}^{(m)} = p_{1'1'}^{(m)}$
$\psi_{\mp 2}$	± 1.618	∓ 0.224	∓ 0.361	0.138	0.361
$\psi_{\mp 1}$	± 0.618	∓ 0.224	± 0.138	0.361	0.138

The ground electron configuration can be expressed as $\psi_{+2}^2\psi_{+1}^2$, and its energy is given by

$$E_0 = 2\varepsilon_{+2} + 2\varepsilon_{+1} = 4\alpha + 2(w_{+2} + w_{+1})(-\beta). \tag{8.17}$$

In the Hückel approximation this energy is calculated to be $4\alpha + 4.472\beta$. If the π-π interaction between atoms 1 and 1′ were absent, that is, if $k_{11'}$ were zero, the ground electron configuration would have the energy of $4\alpha + 4\beta$. Therefore, the extra-resonance energy (delocalization energy) of butadiene is calculated to be $0.472(-\beta)$.

In the Hückel approximation, the total π-bond orders of the 1—2 (as well as 1′—2′) bond and of the 1—1′ bond in the ground electron configuration are calculated to be 0.896 and 0.446, respectively. Of course, the total π-electron density at each carbon atom in the ground electron configuration is equal to unity. Bonds such as the 1—2 and 1′—2′ bonds of butadiene are called "essential" double bonds, and bonds such as the 1—1′ bond of butadiene are called "essential" single bonds. The essential double bond and the essential single bond are defined as follows[1]: the essential double (single) bond is a bond which is double (single) in all principal resonance structures of the molecule. Principal resonance structures are structures which contain the greatest possible number of double bonds and do not involve formal charges or "long" bonds. In the ground electron configurations, most essential double bonds have π-bond orders higher than 0.8, and most essential single bonds have π-bond orders lower than 0.5. The essential double bond and the essential single bond are frequently written as the "double" bond and the "single" bond, respectively.

8.2.2. The Transition Energies and the Transition Moments

The singlet–singlet one-electron transition from the highest occupied π orbital ψ_{+1} to the lowest vacant π orbital ψ_{-1} is considered, to a first approximation, to be responsible for the first intense absorption band of butadiene ($\lambda_{\text{max}} \simeq 217$ mμ, $\epsilon_{\text{max}} \simeq 2 \times 10^4$, in *n*-hexane). In the Hückel approximation, the energy of this transition is calculated to be 1.236 $(-\beta)$. Upon this transition the π-bond order of the essential double bonds (1—2 and 1′—2′ bonds) decreases from 0.896 to 0.448 and that of the essential single bond (1—1′ bond) increases from 0.446 to 0.722. Therefore, it is expected that the first intense band of butadiene will be shifted to longer wavelengths by a twist of the essential double bonds and to shorter wavelengths by a twist of the essential single bond (see Subsection 7.1.6).

As shown up to this point, for calculating the π orbitals of butadiene and their energies in the approximation neglecting nonneighbor interactions the abstract linear model of the carbon skeleton sufficed and further specification of the molecular geometry was unnecessary. However, consideration of the selection rule and calculation of the transition moments must be based on the actual geometry of the molecule.

There are two planar conformations for the carbon skeleton of butadiene, *s-trans* and *s-cis*, which are distinguished by the geometry around the essentially single 1—1′ bond. With regard to the numbering of the carbon atoms and the coordinate system, the conventions illustrated in Figs. 8.2 and 8.3 are adopted. That is, the midpoint of the 1—1′ bond is taken as the origin of the coordinate system, the direction of the 1—1′ bond as the y axis, the molecular plane as the xy plane, and the direction of the parallel symmetry axes of the carbon $2p\pi$ atomic orbitals as the z axis. The planar *s-trans* form belongs to point group C_{2h}, and the planar *s-cis* form to point group C_{2v}.

[1] H. C. Longuet-Higgins, *J. Chem. Phys.* **18**, 265 (1950).

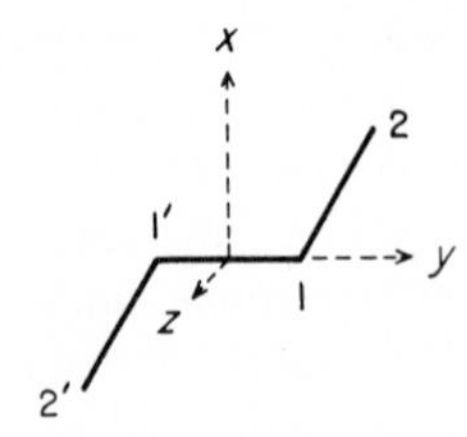

FIG. 8.2. The skeleton of the planar *s-trans* form of 1,3-butadiene.

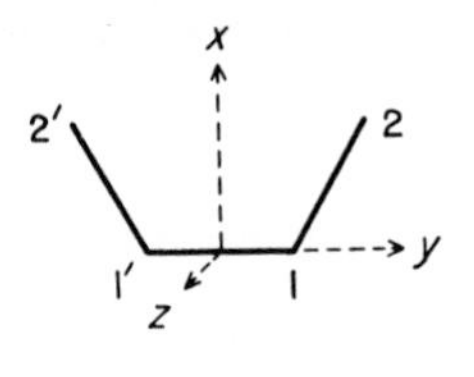

FIG. 8.3. The skeleton of the planar *s-cis* form of 1,3-butadiene.

The character tables for these point groups are shown as Table 8.3 and Table 8.4, respectively. It is to be noted that in the present case the twofold symmetry axis of point group C_{2v} is taken as the x axis, although in the usual notation it is taken as the z axis.

The $2p\pi$ atomic orbital at each carbon atom is antisymmetric with respect to the reflection in the plane of the three σ bonds by which that atom is linked to three neighboring atoms. Therefore, π orbitals of planar molecules are invariably antisymmetric with respect to the reflection in the molecular

TABLE 8.3

CHARACTER TABLE OF POINT GROUP C_{2h}

C_{2h}	E	$C_2(z)$	$\sigma_h(xy)$	i	
A_g	1	1	1	1	R_z
A_u	1	1	−1	−1	z
B_g	1	−1	−1	1	R_x, R_y
B_u	1	−1	1	−1	x, y

TABLE 8.4

CHARACTER TABLE OF POINT GROUP C_{2v}

$C_{2v}(z)$ $C_{2v}(x)$	E E	$C_2(z)$ $C_2(x)$	$\sigma_v(zx)$ $\sigma_v(zx)$	$\sigma_v(yz)$ $\sigma_v(xy)$	$C_{2v}(z)$	$C_{2v}(x)$
A_1	1	1	1	1	z	x
A_2	1	1	−1	−1	R_z	R_x
B_1	1	−1	1	−1	x, R_y	z, R_y
B_2	1	−1	−1	1	y, R_x	y, R_z

plane. Thus the π orbitals of butadiene in the planar *s-trans* form must belong to either of symmetry species A_u and B_g of point group C_{2h}, and those of butadiene in the planar *s-cis* form must belong to either of symmetry species A_2 and B_1 of point group C_{2v}. Actually, it can be easily verified that ψ_{+2} and ψ_{-1} belong to symmetry species A_u of point group C_{2h} in the planar *s-trans* form and to symmetry species B_1 of point group C_{2v} in the planar *s-cis* form, and that ψ_{+1} and ψ_{-2} belong to symmetry species B_g of point group C_{2h} in the planar *s-trans* form and to symmetry species A_2 of point group C_{2v} in the planar *s-cis* form.

TABLE 8.5

THE SYMMETRY OF SINGLY-EXCITED ELECTRON CONFIGURATIONS AND THE SELECTION PROPERTY OF ONE-ELECTRON S–S TRANSITIONS IN 1,3-BUTADIENE

Transition	Configuration	Planar *s-trans* form, C_{2h}	Planar *s-cis* form, $C_{2v}(x)$
$\psi_{+1} \rightarrow \psi_{-1}$	V_1^{11}	B_u, allowed (x, y)	B_2, allowed (y)
$\psi_{+1} \rightarrow \psi_{-2}$	V_1^{12}	A_g, forbidden	A_1, allowed (x)
$\psi_{+2} \rightarrow \psi_{-1}$	V_1^{21}	A_g, forbidden	A_1, allowed (x)
$\psi_{+2} \rightarrow \psi_{-2}$	V_1^{22}	B_u, allowed (x, y)	B_2, allowed (y)

The ground electron configuration is, of course, totally symmetric, that is, it belongs to A_g of point group C_{2h} in the *s-trans* form and to A_1 of point group C_{2v} in the *s-cis* form. The representation to which the excited electron configuration Ψ_1^{mn} arising from the one-electron transition from an occupied orbital ψ_{+m} to a vacant orbital ψ_{-n} belongs is given by the direct product of the representations to which orbitals ψ_{+m} and ψ_{-n} belong. For example, the lowest energy singlet excited configuration, V_1^{11}, which arises from the singlet–singlet one-electron transition from ψ_{+1} to ψ_{-1}, belongs to B_u of point group C_{2h} in the *s-trans* form and to B_2 of point group C_{2v} in the *s-cis* form. Therefore, it is expected that this transition is symmetry-allowed, having nonzero components of the transition moment along the x and y axes in the *s-trans* form and along the y axis in the *s-cis* form. The symmetry species of the singly excited electron configurations and the selection properties of the transitions from the ground electron configuration to these excited electron configurations are summarized in Table 8.5. In this table the letters in parentheses following "allowed" indicate the coordinate axes along which the components of the transition moment can be nonzero.

It is noteworthy that two transitions which are forbidden in the *s-trans* form, $\psi_{+1} \rightarrow \psi_{-2}$ and $\psi_{+2} \rightarrow \psi_{-1}$, become allowed in the *s-cis* form. This fact can be explained most conveniently by expressing the transition moment as the vector sum of two "partial" transition moments. The transition moment of the singlet–singlet one-electron transition from ψ_{+m} to ψ_{-n} can be expressed as

$$\mathbf{M}_{mn} = \sqrt{2} \sum_i c_{+m,i} c_{-n,i} \mathbf{r}_i = \mathbf{m}_{mn} + \mathbf{m}'_{mn}, \tag{8.18}$$

where $c_{+m,i}$ and $c_{-n,i}$ are the coefficients of atomic orbital χ_i in ψ_{+m} and ψ_{-n}, respectively, $\mathbf{r}_i$ is the position vector of atom i, and $\mathbf{m}_{mn}$ and $\mathbf{m}'_{mn}$ are vector quantities called the partial transition moments corresponding to two halves

TABLE 8.6

ENERGIES (ΔE) AND ELECTRIC MOMENTS (M) OF ONE-ELECTRON S–S TRANSITIONS IN 1,3-BUTADIENE CALCULATED BY THE USE OF THE HÜCKEL APPROXIMATION[a]

		Planar *s-trans* form			Planar *s-cis* form		
Transition	ΔE in $-\beta$	M_x in R	M_y in R	M^2 in R^2	M_x in R	M_y in R	M^2 in R^2
$\psi_{+1} \rightarrow \psi_{-1}$	1.236	+0.885	+0.826	1.465	0	+0.826	0.682
$\psi_{+1} \rightarrow \psi_{-2}$	2.236	0	0	0	+0.548	0	0.300
$\psi_{+2} \rightarrow \psi_{-1}$	2.236	0	0	0	+0.548	0	0.300
$\psi_{+2} \rightarrow \psi_{-2}$	3.236	+0.339	−0.118	0.129	0	−0.118	0.014

[a] R: the length of each C—C bond.

of the molecule. In the case of butadiene, $\mathbf{m}_{mn}$ represents the sum of the terms $\sqrt{2} c_{+m,i} c_{-n,i} \mathbf{r}_i$ for $i = 1$ and $i = 2$, and $\mathbf{m}'_{mn}$ represents the sum of the terms for $i = 1'$ and $i = 2'$. Evidently, $\mathbf{m}_{mn}$ and $\mathbf{m}'_{mn}$ have the same magnitude, but they do not always have the same direction. For example, in the *s-trans* form of butadiene the partial transition moments of the transition from ψ_{+1} to ψ_{-2} have opposed directions and cancel each other. On the other hand, in the *s-cis* form, while the y components of these partial transition moments have opposed directions and cancel each other, the x components have the same direction and give a nonzero resultant in the direction of the x axis.

The transition moments of singlet–singlet one-electron transitions in butadiene, calculated from the Hückel orbitals (8.16) by assuming that all the valence angles are 120° and that all the C—C bonds have an equal bond length R, are summarized in Table 8.6. In this table M_x and M_y represent x and y components of the transition moment $\mathbf{M}$. The z component M_z is zero for all the transitions.

8.3. Application of Simple LCAO MO Method to Some Aromatic Hydrocarbons

8.3.1. BENZENE

The secular determinant for determination of the six π orbitals of benzene as linear combinations of the six carbon $2p\pi$ atomic orbitals in the Hückel approximation was already given in Eq. (7.19). This sixth-order determinant can be factorized on the basis of the symmetry of the molecule. As mentioned

TABLE 8.7

TRANSFORMATION OF THE $2p\pi$ AO'S OF BENZENE UNDER THE SYMMETRY OPERATIONS OF POINT GROUP D_2

Operation (R)	E	$C_2(z)$	$C_2(y)$	$C_2(x)$
	χ_1	χ_4	$-\chi_1$	$-\chi_4$
	χ_2	χ_5	$-\chi_6$	$-\chi_3$
	χ_3	χ_6	$-\chi_5$	$-\chi_2$
	χ_4	χ_1	$-\chi_4$	$-\chi_1$
	χ_5	χ_2	$-\chi_3$	$-\chi_6$
	χ_6	χ_3	$-\chi_2$	$-\chi_5$
Character, $X_{red}(R)$	6	0	−2	0

already (see Subsection 3.3.2), the equilibrium geometry of the benzene molecule in the ground state belongs to point group D_{6h}. However, the use of symmetry elements such as threefold and higher symmetry axes in this point group makes it rather difficult to find the group orbitals forming bases for the irreducible representations of the group. The sufficient simplification of the secular determinant can be achieved by using only twofold symmetry axes. Thus, at the present stage, we treat the benzene molecule as belonging to point group D_2. In general, no error is made by assuming that a molecule has lower symmetry than it actually has.

The character table for point group D_2 was shown already as Table 3.2. With regard to the numbering of the carbon atoms and the Cartesian coordinate system, the conventions illustrated in Fig. 3.2 are used. When the six carbon $2p\pi$ atomic orbitals are subjected to the symmetry operations of point group D_2, they are transformed as shown in Table 8.7, thus forming a basis for a reducible sixfold degenerate representation of this point group, Γ_{red}. Evidently, χ_1 and χ_4 are equivalent, and χ_2, χ_3, χ_5, and χ_6 are also equivalent. The numbers at the bottom of the table below the line are the characters

$X_{red}(R)$ of the reducible representation Γ_{red} for the symmetry operations at the head of the respective columns.

By the use of Eq. (8.1) it is found that one A, two B_1, one B_2, and two B_3 group orbitals can be formed as linear combinations of the six atomic orbitals. The use of Eq. (8.2) gives the following set of group orbitals:

$$\begin{aligned}
&A && \phi_{2(a)} = \tfrac{1}{2}(\chi_2 - \chi_3 + \chi_5 - \chi_6) \\
&B_1 && \begin{cases} \phi_{1(b1)} = \sqrt{\tfrac{1}{2}}(\chi_1 + \chi_4) \\ \phi_{2(b1)} = \tfrac{1}{2}(\chi_2 + \chi_3 + \chi_5 + \chi_6) \end{cases} \\
&B_2 && \phi_{2(b2)} = \tfrac{1}{2}(\chi_2 + \chi_3 - \chi_5 - \chi_6) \\
&B_3 && \begin{cases} \phi_{1(b3)} = \sqrt{\tfrac{1}{2}}(\chi_1 - \chi_4) \\ \phi_{2(b3)} = \tfrac{1}{2}(\chi_2 - \chi_3 - \chi_5 + \chi_6). \end{cases}
\end{aligned} \tag{8.19}$$

Accordingly, the secular determinant in Eq. (7.19) is factorized into two first-order and two second-order determinants:

$$\begin{array}{c} \\ 2(a) \\ 1(b1) \\ 2(b1) \\ 2(b2) \\ 1(b3) \\ 2(b3) \end{array}
\begin{array}{c} \begin{array}{cccccc} 2(a) & 1(b1) & 2(b1) & 2(b2) & 1(b3) & 2(b3) \end{array} \\
\left| \begin{array}{cccccc}
w-1 & 0 & 0 & 0 & 0 & 0 \\
0 & w & \sqrt{2} & 0 & 0 & 0 \\
0 & \sqrt{2} & w+1 & 0 & 0 & 0 \\
0 & 0 & 0 & w+1 & 0 & 0 \\
0 & 0 & 0 & 0 & w & \sqrt{2} \\
0 & 0 & 0 & 0 & \sqrt{2} & w-1
\end{array} \right| \end{array} = 0. \tag{8.20}$$

By solving the two first-order and two second-order equations separately, the energies and forms of the six π orbitals are found to be as follows:

$$\begin{aligned}
&A: && w_{-2} = +1, && \psi_{-2} = \phi_{2(a)}; \\
&B_1: && \begin{cases} w_{-1} = +1, \\ w_{+3} = -2, \end{cases} && \begin{aligned} &\psi_{-1} = (1/\sqrt{3})[\phi_{2(b1)} - \sqrt{2}\,\phi_{1(b1)}], \\ &\psi_{+3} = (1/\sqrt{3})[\sqrt{2}\,\phi_{2(b1)} + \phi_{1(b1)}]; \end{aligned} \\
&B_2: && w_{+2} = -1, && \psi_{+2} = \phi_{2(b2)}; \\
&B_3: && \begin{cases} w_{+1} = -1, \\ w_{-3} = +2, \end{cases} && \begin{aligned} &\psi_{+1} = (1/\sqrt{3})[\phi_{2(b3)} + \sqrt{2}\,\phi_{1(b3)}], \\ &\psi_{-3} = (1/\sqrt{3})[\sqrt{2}\,\phi_{2(b3)} - \phi_{1(b3)}]. \end{aligned}
\end{aligned} \tag{8.21}$$

These molecular π orbitals written as linear combinations of group orbitals can be easily rewritten as linear combinations of atomic orbitals, as shown in Table 3.5. It is to be noted that also in this case the number of nodes in a π orbital increases in the order of increasing orbital energy. Thus, the lowest bonding orbital (ψ_{+3}), the degenerate highest bonding orbitals (ψ_{+2} and ψ_{+1}), the degenerate lowest antibonding orbitals (ψ_{-1} and ψ_{-2}), and the highest antibonding orbital (ψ_{-3}) have, respectively, 0, 1, 2, and 3 nodal planes perpendicular to the molecular plane, in addition to the common nodal plane which coincides with the molecular plane.

In general, the energies and AO coefficients of the Hückel MO's of a conjugated ring system consisting of n carbon π centers can be written in the following analytic forms:

$$w_k = -2\cos(2k\pi/n); \tag{8.22}$$

$$c_{kj} = (1/\sqrt{n})\exp[2\pi ik(j-1)/n], \tag{8.23}$$

where k is the index specifying the molecular orbital, j is the index specifying the atomic orbital, and i represents $\sqrt{-1}$. The k takes the values $0, \pm 1, \pm 2, \ldots, \pm(n-1)/2$ when n is odd, and takes the values $0, \pm 1, \pm 2, \ldots, \pm n/2$ when n is even. When n is even the MO for $k = +n/2$ and that for $k = -n/2$ are identical. The j takes the values $1, 2, \ldots, n$. For $j = 1, j-1$ is taken to represent n. It is to be noted that the coefficients are given in complex form.

For benzene, n is 6, and k can take the values $0, \pm 1, \pm 2$, and 3. The π orbitals of benzene and their energies obtained by the use of Eq. (8.22) and (8.23) can be expressed as follows, being distinguished from the real functions given in Table 3.5 by attached primes:

$$w_0' = -2, \quad w'_{\pm 1} = -1, \quad w'_{\pm 2} = +1, \quad w_3' = +2; \tag{8.24}$$

$$\psi_k' = (1/\sqrt{6})\sum_j \exp[2\pi ik(j-1)/6]\chi_j; \quad j = 1, 2, \ldots, 6. \tag{8.25}$$

These complex functions (8.25) are related to the already obtained real functions (given in Table 3.5) as follows:

$$\begin{aligned} &\psi_{-3} = -\psi_3'; \\ &\psi_{-2} = (1/i\sqrt{2})(\psi'_{+2} - \psi'_{-2}), \qquad \psi_{-1} = -(1/\sqrt{2})(\psi'_{+2} + \psi'_{-2}); \\ &\psi_{+1} = (1/\sqrt{2})(\psi'_{+1} + \psi'_{-1}), \qquad \psi_{+2} = (1/i\sqrt{2})(\psi'_{+1} - \psi'_{-1}); \\ &\psi_{+3} = \psi_0'. \end{aligned} \tag{8.26}$$

Needless to say, linear combinations of degenerate eigenfunctions of a Hamiltonian are also eigenfunctions of the Hamiltonian. The absolute value of k for a complex function coincides with the number of nodal planes in the corresponding real function.

In the ground electron configuration all the three bonding π orbitals are occupied each by two electrons. In this electron configuration, the π-bond order of each C—C bond is $\frac{2}{3}$ ($\doteqdot$ 0.667), the π-electron density at each carbon atom is unity, and the resonance energy relative to the energy of three isolated ethylenic bonds is $2(-\beta)$.

The ground electron configuration is, of course, singlet and belongs to the totally symmetric species, that is, A of point group D_2, or A_{1g} of point group D_{6h}. One-electron transitions from the highest occupied degenerate π orbitals to the lowest vacant degenerate π orbitals give rise to four degenerate lowest energy excited singlet configurations V_1^{11}, V_1^{12}, V_1^{21}, and V_1^{22} and the corresponding triplet configurations, T_1^{11}, T_1^{12}, T_1^{21}, and T_1^{22}. V_1^{11}, V_1^{22}, and their triplet counterparts belong to symmetry species B_2 of point group D_2, and V_1^{12}, V_1^{21}, and their triplet counterparts belong to symmetry species B_3 of point group D_2. Therefore, the transitions from the ground configuration, V_0, to V_1^{11} and V_1^{22} are expected to be allowed and polarized along the y axis, and those to V_1^{12} and V_1^{21} are expected to be allowed and polarized along the x axis. The transition moments of these transitions are calculated to be as follows:

$$
\begin{aligned}
M_{11} &= M_{11(y)} = -(1/\sqrt{2})R, \qquad & M_{22} &= M_{22(y)} = +(1/\sqrt{2})R, \\
M_{12} &= M_{12(x)} = +(1/\sqrt{2})R, \qquad & M_{21} &= M_{21(x)} = +(1/\sqrt{2})R,
\end{aligned} \tag{8.27}
$$

where M_{11} and $M_{11(y)}$, for example, represent the transition moment of the transition from V_0 to V_1^{11} and its component along the y axis, respectively, and R represents the length of each C—C bond.

Actually, the benzene molecule belongs to point group D_{6h}, and, as mentioned already (see Subsection 3.4.2), the degenerate lowest energy excited configurations give rise to the electronic states belonging to symmetry species B_{2u}, B_{1u}, and E_{1u} of that point group (see Table 3.7). The transitions from the ground state to the α and p states are expected to be forbidden, and the transitions to the degenerate β and β' states are expected to be allowed. In fact, the transition moments of these transitions are calculated as follows:

$$
\begin{aligned}
M_\alpha &= (1/\sqrt{2})(M_{12} - M_{21}) = 0, \\
M_p &= (1/\sqrt{2})(M_{11} + M_{22}) = 0, \\
M_\beta &= (1/\sqrt{2})(M_{12} + M_{21}) = +R(x), \\
M_{\beta'} &= (1/\sqrt{2})(M_{11} - M_{22}) = -R(y).
\end{aligned} \tag{8.28}
$$

8.3.2. Naphthalene

With regard to the numbering of the carbon atoms in the naphthalene molecule and the coordinate system assigned to the molecule, the conventions illustrated in Fig. 8.4 are used. The equilibrium nuclear configuration of the naphthalene molecule in the ground electronic state belongs to point group D_{2h}. However, the sufficient simplification of the secular determinant for determination of the π orbitals as linear combinations of 10 carbon $2p\pi$ atomic orbitals can be achieved by the use of the character table for point group D_2. The character tables for point groups D_2 and D_{2h} were shown already as Tables 3.2 and 3.3, respectively. By the usual procedure the tenth-order secular determinant can be factorized into two second-order and two third-order determinants.

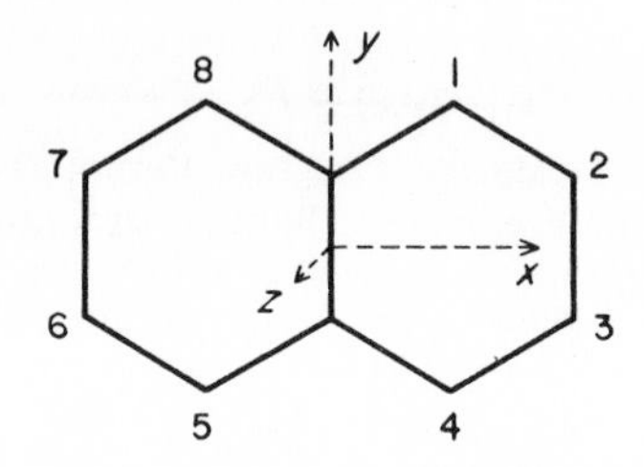

FIG. 8.4. The skeleton of naphthalene.

The energies and symmetry species of the 10 Hückel orbitals of naphthalene are as follows:

$$\begin{aligned}
b_{3g} &: w_{-5} = +2.303; & b_{1u} &: w_{+5} = -2.303; \\
a_{u} &: w_{-4} = +1.618; & b_{2g} &: w_{+4} = -1.618; \\
b_{1u} &: w_{-3} = +1.303; & b_{3g} &: w_{+3} = -1.303; \\
b_{3g} &: w_{-2} = +1.000; & b_{1u} &: w_{+2} = -1.000; \\
b_{2g} &: w_{-1} = +0.636; & a_{u} &: w_{+1} = -0.636.
\end{aligned} \tag{8.29}$$

The two lowest antibonding and two highest bonding π orbitals can be expressed as follows:

$$\begin{aligned}
b_{3g} &: \psi_{-2} = 0.408(\chi_2 - \chi_3 - \chi_6 + \chi_7 - \chi_9 + \chi_{10}); \\
b_{2g} &: \psi_{-1} = 0.268(\chi_2 + \chi_3 - \chi_6 - \chi_7) - 0.422(\chi_1 + \chi_4 - \chi_5 - \chi_8); \\
a_{u} &: \psi_{+1} = 0.268(\chi_2 - \chi_3 + \chi_6 - \chi_7) + 0.422(\chi_1 - \chi_4 + \chi_5 - \chi_8); \\
b_{1u} &: \psi_{+2} = 0.408(\chi_2 + \chi_3 + \chi_6 + \chi_7 - \chi_9 - \chi_{10}).
\end{aligned} \tag{8.30}$$

We may say that these π orbitals have forms resembling those of the corresponding π orbitals of benzene. It is noteworthy that the π orbitals of naphthalene are nondegenerate, while those of benzene are degenerate.

The ground electron configuration of naphthalene, in which all the five bonding π orbitals are occupied each by two electrons, belongs to symmetry species A_g of point group D_{2h}. In this electron configuration, the resonance energy relative to the energy of five isolated ethylenic bonds is $3.720(-\beta)$, the π-electron density at each carbon atom is unity, and the π-bond orders of the 1—2, 2—3, 1—9, and 9—10 bonds are 0.725, 0.603, 0.555, and 0.518, respectively. C—C bonds in aromatic ring systems are neither essential double bonds nor essential single bonds. Such bonds are classified as "aromatic" bonds. Most aromatic bonds have π-bond orders of 0.5–0.8.

TABLE 8.8

ENERGIES (ΔE) AND ELECTRIC MOMENTS (M) OF LOW-ENERGY ONE-ELECTRON TRANSITIONS IN NAPHTHALENE IN THE HÜCKEL APPROXIMATION[a]

Transition	Configuration	ΔE, in $-\beta$	M, in R
$\psi_{+1} \rightarrow \psi_{-1}$	V_1^{11}, $^1B_{2u}$	1.272	$-0.803(y)$
$\psi_{+1} \rightarrow \psi_{-2}$	V_1^{12}, $^1B_{3u}$	1.636	$+1.073(x)$
$\psi_{+2} \rightarrow \psi_{-1}$	V_1^{21}, $^1B_{3u}$	1.636	$+1.073(x)$
$\psi_{+2} \rightarrow \psi_{-2}$	V_1^{22}, $^1B_{2u}$	2.000	$+0.707(y)$

[a] R: the length of each C—C bond.

Excited singlet electron configurations V_1^{11} and V_1^{22} belong to symmetry species B_{2u}. Therefore, the transitions from the ground electron configuration to these configurations are expected to be allowed and polarized in the direction of the y axis. On the other hand, V_1^{12} and V_1^{21}, which are degenerate, belong to symmetry species B_{3u}, and hence the transitions from the ground configuration to these configurations are expected to be allowed and polarized along the x axis. Table 8.8 shows the transition moment lengths of these transitions calculated by assuming that all the valence angles are 120° and that all the C—C bonds have an equal bond length R.

8.3.3. AZULENE

As mentioned in Section 7.2, and as have been seen for ethylene, butadiene, benzene, and naphthalene, the Hückel π orbitals of alternant hydrocarbons have the pairing property. On the other hand, the Hückel π orbitals of nonalternant hydrocarbons have not the pairing property. This fact can be illustrated by taking azulene as an example.

The carbon skeleton of the azulene molecule is illustrated in Fig. 8.5, together with the numbering of the carbon atoms and the coordinate system. This skeleton belongs to point group C_{2v}, in which the rotational symmetry axis is taken as the x axis. The π orbitals must be antisymmetric with respect to the reflection in the molecular plane (xy plane), that is, they must belong to either of symmetry species A_2 and B_1 of point group C_{2v}. The tenth-order

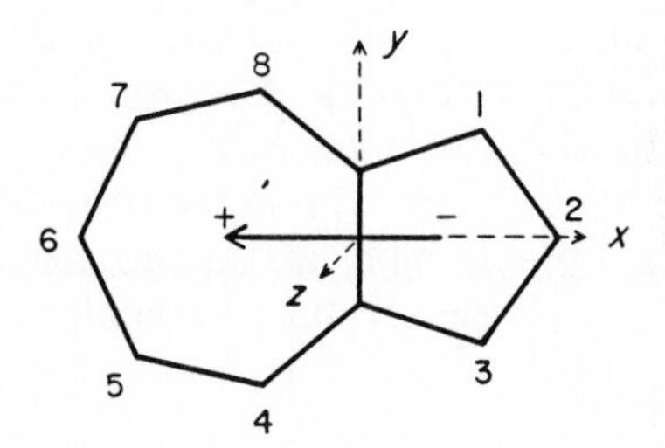

FIG. 8.5. The skeleton of azulene. The solid arrow indicates the direction of the dipole moment of the molecule in the ground state.

secular determinant for determination of the π orbitals as linear combinations of the carbon $2p\pi$ atomic orbitals can be factorized by the usual procedure into one fourth-order determinant for a_2 orbitals and one sixth-order determinant for b_1 orbitals. The values of w of the ten Hückel orbitals are calculated to

TABLE 8.9

THE TWO LOWEST ANTIBONDING AND TWO HIGHEST BONDING π MO'S OF AZULENE IN THE HÜCKEL APPROXIMATION[a]

MO	Sym.	w	(χ_2)	$(\chi_1 \pm \chi_3)$	$(\chi_9 \pm \chi_{10})$	$(\chi_4 \pm \chi_8)$	$(\chi_5 \pm \chi_7)$	(χ_6)
ψ_{-2}	a_2	+0.738	0.000	−0.299	0.221	−0.357	0.484	0.000
ψ_{-1}	b_1	+0.400	0.316	−0.063	−0.290	0.470	0.102	−0.511
ψ_{+1}	a_2	−0.477	0.000	0.543	0.259	0.160	0.336	0.000
ψ_{+2}	b_1	−0.887	0.583	0.259	−0.354	−0.219	0.160	0.360

[a] The coefficients $c_{m,i}$'s are listed under the corresponding AO (χ_i). The signs + and − in $(\chi_i \pm \chi_j)$ refer to MO's of symmetry b_1 and a_2, respectively.

be as follows: $-2.310(b_1)$, $-1.652(b_1)$, $-1.356(a_2)$, $-0.887(b_1)$, $-0.477(a_2)$, $+0.400(b_1)$, $+0.738(a_2)$, $+1.579(b_1)$, $+1.869(b_1)$, and $+2.095(a_2)$.[2] The energies and forms of the two lowest antibonding and two highest bonding π orbitals in the Hückel approximation are shown in Table 8.9. It is evident that the π orbitals are not in pairs.

[2] R. Pariser, *J. Chem. Phys.* **25,** 1112 (1956).

In the Hückel approximation the π-electron densities at carbon atoms in the ground electron configuration are as follows: 1.047 at atom 2, 1.173 at each of atoms 1 and 3, 1.027 at each of atoms 9 and 10, 0.855 at each of atoms 4 and 8, 0.986 at each of atoms 5 and 7, and 0.870 at atom 6. This means that the π electrons are more concentrated in the five-membered ring than in the seven-membered ring. Therefore, it is expected that the azulene molecule has a dipole moment in the direction of the x axis, as illustrated by an arrow in Fig. 8.5. Actually, it is known that the azulene molecule has a dipole moment of 1.0 ± 0.05 D.[3] By the way, fulvene, a nonalternant hydrocarbon, has a dipole moment of about 1.1 D.[4] It is noteworthy that nonalternant hydrocarbons have nonzero dipole moments in contrast with alternant hydrocarbons, which are completely, or almost completely, nonpolar.

[3] G. W. Wheland and D. E. Mann, *J. Chem. Phys.* **17**, 264 (1949).
[4] J. Thiec and J. Wiemann, *Bull. soc. chim. France* **1956,** 177.

Chapter 9

Evaluation of Parameters Used in Simple LCAO MO Method

9.1. Introduction

As mentioned in Subsection 7.1.4, in the modified or refined Hückel method the variation of the Coulomb and resonance integrals is taken into account by expressing the integrals as

$$\alpha_i = \alpha + h_i\beta, \tag{9.1}$$

$$\beta_{ij} = k_{ij}\beta, \tag{9.2}$$

and by assigning appropriate values to parameters h_i and k_{ij}. Usually, the Coulomb integral for a carbon $2p\pi$ atomic orbital is taken as the standard Coulomb integral α, and the resonance integral for a pair of carbon $2p\pi$ atomic orbitals of a C—C bond in benzene is taken as the standard resonance integral β.

The Coulomb and resonance integrals are defined in terms of an effective one-electron Hamiltonian, which is assumed to include the electronic interactions in an averaged manner (see Subsection 7.1.2). By definition, the effective one-electron Hamiltonian cannot be written explicitly. Therefore, the Coulomb and resonance integrals cannot be calculated theoretically, and for the same reason, the Coulomb and resonance parameters, h_i and k_{ij}, cannot be calculated *ab initio*. There are, however, some principles establishing scales of relative values of the parameters. In practice, it is important to choose parameter values which are in a reasonable range of magnitude and in a reasonable relative order. In this chapter the principles used in this book for choosing the parameter values are described and some others are referred to.

9.2. Evaluation of Coulomb Parameters

9.2.1. Relation of the Coulomb Integral to the Electronegativity

As mentioned above, the variation of the Coulomb integral is allowed for by assigning appropriate values to the Coulomb parameter h in expression (9.1). In a rough approximation we can equate the Coulomb integral for an atomic orbital with the energy of the orbital, or at least, we can say that the former must be closely related to the latter. Since both α and β are negative energy quantities, $p\pi$ atomic valence-shell orbitals of atoms that are more electron-attracting than the carbon atom are expected to have more negative Coulomb integrals than α and hence to have positive h values. It seems to be reasonable to relate the Coulomb integral (α_X) or the Coulomb parameter (h_X) to the electronegativity (x_X) of the atom or to the effective nuclear charge*[,1] (Z_X') for the orbital. It has been suggested that α_X is proportional to the electronegativity, x_X;[2] and also that h_X is proportional to the difference in electronegativity between the atom X and the carbon atom, $x_X - x_C \equiv \Delta x_X$.[3–5] The constant of proportionality between h_X and Δx_X has frequently been taken as unity:[6]

$$h_X = x_X - x_C, \tag{9.3}$$

where the x's are electronegativities in Pauling's scale.[7] By the use of this equation the h's for $p\pi$ atomic valence-shell orbitals can be evaluated from Pauling's electronegativity values. In Table 9.1 some of the electronegativity values are shown, together with some of Slater's effective nuclear charge values for a $2p$ valence electron. The assumption of the proportionality between

* The effective nuclear charge for a Slater atomic orbital is given by

$$Z' = Z - s = (n - d)\mu,$$

where Z is the nuclear charge, n the principal quantum number of the orbital, s the screening constant dependent on n, μ the so-called Slater μ-parameter, and $(n - d)$ the effective principal quantum number, which is equal to 2 for orbitals with the principal quantum number of 2.

[1] J. C. Slater, *Phys. Rev.* **36**, 57 (1930).

[2] C. Sandorfy, *Bull. soc. chim. France* **1949**, 615.

[3] R. S. Mulliken, *J. chim. phys.* **46**, 497, 675 (1949).

[4] E. Gyoerffy, *Compt. rend.* **232**, 515 (1951).

[5] C. A. Coulson, "Valence." Oxford Univ. Press, London and New York, 1952, p. 242; C. A. Coulson and J. de Heer, *J. Chem. Soc.* **1952**, 483.

[6] A. Laforgue, *J. chim. phys.* **46**, 568 (1949).

[7] L. Pauling, *J. Am. Chem. Soc.* **54**, 3570 (1932); "The Nature of the Chemical Bond," 3rd edition, Cornell University Press, Ithaca, New York (1960).

α_X and x_X and that between h_X and Δx_X with the proportionality constant of unity imply that the ratio of the standard Coulomb integral to the standard resonance integral, α/β, is equal to the electronegativity of the carbon atom, 2.5.

Account should be taken of one important principle. For example, in pyridine the nitrogen atom contributes one electron to the π system, while in pyrrole it contributes two. In the former case, the heteroatom contributes an effective charge of +1 to the core potential, while in the latter the contributed effective charge is +2. Thus, from the point of view of participation in the π-electron system, heteroatoms may be classified into two different groups according to the number of electrons that they contribute to the system. Atoms contributing one π electron are essentially atoms bound by a double bond to an adjacent atom in the structural formulas; among the representatives of this group are: the pyridine-type nitrogen, the nitrogen of the C=N and N=N bonds, the oxygen of the C=O bond, the sulfur of the C=S bond, the phosphorus of the P=O bond, etc. Atoms contributing two π electrons are essentially atoms with lone pair electrons, bound only by single bonds to adjacent atoms in the structural formulas; among the representatives of this group are: the pyrrole or aniline-type nitrogen, the furan or phenol-type oxygen, the thiophene or thiophenol-type sulfur, the halogen atoms directly linked to an aromatic or ethylenic carbon atom, etc. The Coulomb integral for a heteroatom that contributes two electrons must be more negative than that for the same heteroatom that contributes one electron: if a heteroatom X that contributes one electron and the same atom that contributes two electrons are represented as $\dot{X}$ and $\ddot{X}$, respectively, $h_{\ddot{X}}$ must be more positive than $h_{\dot{X}}$.

TABLE 9.1

PAULING'S ELECTRONEGATIVITY (x) AND SLATER'S EFFECTIVE NUCLEAR CHARGE (Z') FOR A $2p$ VALENCE ELECTRON

Atom(X)	Z_X'	x_X	$x_X - x_C$
B	2.60	2.0	−0.5
C	3.25	2.5	0
N	3.90	3.0	+0.5
N^+	4.25	(3.27)	(+0.77)
O	4.55	3.5	+1.0
O^+	4.90	(3.77)	(+1.27)
F	5.20	4.0	+1.5

Table 9.1 indicates that the electronegativity x is proportional to the effective nuclear charge Z'. On the basis of this proportionality, electronegativity values for singly positively charged atoms, x_{X^+}, can be estimated from the corresponding Z' values. Such values of x_{X^+} are shown in parentheses in Table 9.1. If, as mentioned above, $\alpha_{\dot{X}}$ is assumed to be proportional to x_X, it may be assumed that $\alpha_{\ddot{X}}$ be proportional to the mean of x_X and x_{X^+}. This means that if Eq. (9.3) is rewritten as

$$h_{\dot{X}} = x_X - x_C , \tag{9.4}$$

the following equation may be written:

$$h_{\ddot{X}} = (x_X + x_{X^+})/2 - x_C . \tag{9.5}$$

According to these assumptions, $h_{\dot{N}}$ and $h_{\ddot{N}}$, for example, are evaluated as +0.5 and +0.63, respectively. Furthermore, the h for a singly positively charged heteroatom, $h_{\dot{X}^+}$, may be assumed to be equal to $x_{X^+} - x_C$. According to this assumption, $h_{\dot{N}^+}$, the h-parameter for the singly positively charged nitrogen atom, for example, in pyridinium ion, is evaluated as +0.77.

9.2.2. Relation of the Coulomb Integral to the Atomic Ionization Energy

It has so far been assumed that all the atomic orbitals which are incorporated into π systems are purely p valence orbitals. However, there is a possibility that hybrid orbitals having some contribution of s orbitals are incorporated into π systems. The value of the h-parameter for a hybrid valence orbital must be different from that for the p valence orbital of the same atom. Since in general an s orbital of an atom has a lower energy than a p orbital with the same principal quantum number of the same atom, it is expected that the Coulomb integral for a valence orbital will become more negative, and hence the h-parameter will become more positive, as the contribution of the s orbital to that valence orbital becomes greater.

In order to take account of the dependence of the Coulomb integral and h-parameter on the hybridization of the atomic orbital, it is assumed that the Coulomb integral be proportional to the ionization energy of an electron occupying the atomic orbital in the pertinent valence state of the atom. This assumption is considered to be almost equivalent to the foregoing assumption that the Coulomb integral be proportional to the electronegativity, in view of the following facts. Mulliken[8] defined the electronegativity (absolute electroaffinity) of an atom as the mean value of its valence-state

[8] R. S. Mulliken, *J. Chem. Phys.* **2**, 782 (1934).

ionization energy I and its valence-state electron affinity A, both measured in electron volts. The magnitudes of the electron affinities are usually much smaller than the magnitudes of the ionization energies, and the variations of the former are much smaller than the variations of the latter. Therefore, Mulliken's electronegativity, x^{M}, is approximately proportional to the ionization energy. In addition, x^{M} has been shown to be roughly proportional to Pauling's electronegativity, x^{P}:[9]

$$x^{\mathrm{P}} \fallingdotseq x^{\mathrm{M}}/3.15. \tag{9.6}$$

Pritchard and Skinner[9] presented extensive tables of energies of various valence states of various atoms, and of ionization energies of atoms in various valence states. The data for C, N, and O selected from their tables are shown in Table 9.2. In this table energies are given in units of electron volts. The figures in parentheses following the symbols for electron configurations represent the number of electrons whose spins are independent of one another, or in other words, the number of singly occupied orbitals. Thus, for example, the valence state s^2xy (2) is to be regarded as a mixture in the ratio 3:1 of the triplet and singlet states of electron configuration $(2s)^2(2p_x)^1(2p_y)^1$. The symbol p (or s) following figures representing ionization energies indicates that the ionization energy is the energy required to remove one electron from a singly occupied p (or s) orbital in the corresponding valence state. The data which are not presented in Pritchard and Skinner's tables and have been derived in the following manner from the pertinent valence state energies and ionization energies are shown in parentheses. The valence state s^2xy of C has the energy of 0.32 ev above the ground state, in which the two $2p$ electrons have parallel spins, and the energy required to remove one $2p$ electron from this valence state, giving the valence state s^2x of C^+, which is the ground state of C^+, is 10.94 ev. Accordingly, the energy of the s^2x state of C^+ $(+e^-)$ relative to the ground state of C is calculated to be 11.26 ev.

Energies of hybrid valence states can be calculated as means of energies of pertinent valence states, and from those the ionization energies of atoms in hybrid valence states can be calculated. For example, if the trigonal (sp^2) hybrid orbital is expressed as tr, the valence state $(tr)^2(tr)^1(tr)^1(p)^1$ of N is calculated as the mean of energies of the valence states s^2xyz, sx^2yz, and sxy^2z: $(1.19 + 2 \times 14.23)/3 = 9.88\dot{3}$ (ev). Analogously, the energy of the valence state $(tr)^2(tr)^1(tr)^1$ of N^+ relative to the ground state of N is calculated as the mean of energies of the valence states s^2xy, sx^2y, and sxy^2:

$$(15.02 + 2 \times 28.72)/3 = 24.15\dot{3} \text{ (ev)}.$$

[9] H. O. Pritchard and H. A. Skinner, *Chem. Rev.* **55**, 745 (1955).

TABLE 9.2

ENERGIES OF VALENCE STATES AND IONIZATION ENERGIES OF C, N, AND O (UNITS: EV)

	Neutral atom			Positively charged atom X^+		
Atom (X)	Valence state	Energy[a]	Ionization energy (I)	Valence state	Energy[b]	Energy[c]
C	s^2xy (2)	0.32	10.94 p	s^2x (1)	0	(11.26)
	s^2x^2 (0)	1.74	—	—	—	—
	$sxyz$ (4)	8.26	11.42 p	sxy (3)	8.42	(19.68)
			21.43 s	xyz (3)	(18.43)	(29.69)
	sx^2y (2)	9.85	(11.59 p)	sx^2 (1)	10.18	(21.44)
N	—	—	—	Ground state	0	(14.54)
	s^2xyz (3)	1.19	13.83 p	s^2xy (2)	0.48	(15.02)
	s^2x^2y (1)	2.98	(14.18 p)	s^2x^2 (0)	2.62	(17.16)
	sx^2yz (3)	14.23	14.49 p	sx^2y (2)	14.18	(28.72)
			27.5 s	x^2yz (2)	(27.19)	(41.73)
				$sxyz$ (4)	11.95	(26.49)
O	—	—	—	Ground state	0	(13.61)
	s^2x^2yz (2)	0.50	17.28 p	s^2x^2y (1)	4.17	(17.78)
	sx^2y^2z (2)	17.65	17.76 p	sx^2y^2 (1)	21.80	(35.41)
			35.30 s	x^2y^2z (1)	(39.34)	(52.95)
	$s^2x^2y^2$ (0)	2.71	—	—	—	—
				s^2xyz (3)	1.66	(15.27)
				sx^2yz (3)	19.17	(32.78)

[a] Energy above the ground state of the atom.
[b] Energy above the ground state of the positively charged atom X^+.
[c] Energy of X^+ (in the valence state) $+ e^-$ above the ground state of the atom X.

Therefore, the ionization energy of an electron in the $2p\pi$ orbital of N in the trigonal valence state, in which the n orbital is an sp^2 hybrid orbital, is calculated to be 14.27 ev ($= 24.15\dot{3} - 9.88\dot{3}$). The energy of the valence state $(tr)^1(tr)^1(tr)^1(p)^1$ of N^+ is equal to the energy of the valence state $sxyz$ of N^+, 26.49 ev. Therefore, the energy required to remove one of the two electrons occupying the sp^2 n orbital of N in the trigonal valence state is calculated to be 16.61 ev ($= 26.49 - 9.88$). Furthermore, since the energy of the valence state $(tr)^1(tr)^1(tr)^1(p)^2$ of N is equal to the energy of the valence state sx^2yz of N, 14.23 ev, the energy required to remove one of the two electrons occupying the $2p\pi$ orbital of N in the trigonal valence state is calculated to be 12.26 ev (= 26.49 − 14.23). The state energies and ionization energies of various hybrid valence states of interest to us, calculated in this manner, are summarized in Table 9.3, where *di* represents the *sp* hybrid orbital.

The energies of ionization of singly positively charged atoms to doubly positively charged atoms, that is, the energies required for the processes

TABLE 9.3

ENERGIES OF HYBRID VALENCE STATES (E), IONIZATION ENERGIES (I), AND VALUES OF THE COULOMB PARAMETER (h)

Atom	Valence state	E,[a] ev	Ionization process	I, ev	h	
C	$(tr)^1(tr)^1(tr)^1(p)^1$	8.26	$(p)^1 \rightarrow (p)^0$	11.42	0	$(h_{\dot{C}(p)})$
			$(tr)^1 \rightarrow (tr)^0$	14.76	0.70	
N	$(tr)^2(tr)^1(tr)^1(p)^1$	9.88	$(p)^1 \rightarrow (p)^0$	14.27	0.60	$(h_{\dot{N}(p)})$
			$(tr)^1 \rightarrow (tr)^0$	18.61	1.51	
			$(tr)^2 \rightarrow (tr)^1$	16.61	1.09	
	$(tr)^1(tr)^1(tr)^1(p)^2$	14.23	$(p)^2 \rightarrow (p)^1$	12.26	0.18	
N^+	$(tr)^1(tr)^1(tr)^1(p)^1$	26.49	$(tr)^1 \rightarrow (tr)^0$	21.05	2.02	
			$(p)^1 \rightarrow (p)^0$	15.95	0.95	$(h_{\dot{N}^+(p)})$
O	$(tr)^2(tr)^2(tr)^1(p)^1$	6.22	$(p)^1 \rightarrow (p)^0$	17.44	1.27	
			$(tr)^1 \rightarrow (tr)^0$	23.29	2.49	
			$(tr)^2 \rightarrow (tr)^1$	20.73	1.96	
	$(tr)^2(tr)^1(tr)^1(p)^2$	11.93	$(p)^2 \rightarrow (p)^1$	15.01	0.75	
	$(di)^2(di)^1(p)^2(p)^1$	9.08	$(p)^1 \rightarrow (p)^0$	17.52	1.28	$(h_{\dot{O}(p)})$
			$(p)^2 \rightarrow (p)^1$	14.95	0.74	
O^+	$(tr)^2(tr)^1(tr)^1(p)^1$	26.94	$(tr)^1 \rightarrow (tr)^0$	26.10	3.08	
			$(p)^1 \rightarrow (p)^0$	19.30	1.66	
N	$(tr)^2(tr)^1(tr)^1(p)^1$	9.88	$(tr)^2 \rightarrow (tr)^0$	37.66	1.56	$(h_{\ddot{N}(tr)})$
	$(tr)^1(tr)^1(tr)^1(p)^2$	14.23	$(p)^2 \rightarrow (p)^0$	28.21	0.56	$(h_{\ddot{N}(p)})$
O	$(tr)^2(tr)^2(tr)^1(p)^1$	6.22	$(tr)^2 \rightarrow (tr)^0$	46.83	2.52	
	$(tr)^2(tr)^1(tr)^1(p)^2$	11.93	$(p)^2 \rightarrow (p)^0$	34.31	1.21	$(h_{\ddot{O}(p)})$
	$(di)^2(di)^1(p)^2(p)^1$	9.08	$(p)^2 \rightarrow (p)^0$	34.25[b]	1.20	$(h_{\ddot{O}(p)})$

[a] Energy relative to the ground state of the neutral atom.

[b] The energy required for the process $(p)^1 \rightarrow (p)^0$ from O^+ in the valence state $(di)^2(di)^1(p)^1(p)^1$ was assumed to be equal to the energy required for the process $(p)^1 \rightarrow (p)^0$ from O^+ in the valence state $(tr)^2(tr)^1(tr)^1(p)^1$.

$X^+ \to X^{2+}$, cannot be obtained from Pritchard and Skinner's tables. However, with the neutral C, N, and O atoms in the trigonal valence state, it is found that the estimated ionization energies for the process $(p)^1 \to (p)^0$ and those for the process $(tr)^1 \to (tr)^0$ have separately an approximately linear relationship with the squares of the effective nuclear charges, Z', for the neutral atoms. Therefore, by the use of this approximate linear relationship, the ionization energies of X^+ may be estimated from the values of Z' for the charged atoms X^+. The values estimated in this manner are also shown in Table 9.3.

On the basis of the assumption of proportionality of the Coulomb integral to the valence state ionization energy, the value of h for a given singly occupied orbital of atom X in a given valence state, $h_{\dot{X}}$, can be calculated by the use of the following equation:

$$h_{\dot{X}} = [(I_{\dot{X}} - I_{\dot{C}(p)})/I_{\dot{C}(p)}](\alpha/\beta), \tag{9.7}$$

where $I_{\dot{X}}$ is the ionization energy required to remove one electron from the singly occupied orbital of X in the valence state, and $I_{\dot{C}(p)}$ is the ionization energy required to remove one electron from the $2p\pi$ orbital of C in the trigonal valence state, 11.42 ev. The value of h for a doubly occupied orbital, $h_{\ddot{X}}$, is estimated as the mean of the h values corresponding to the pertinent ionization processes $\ddot{X} \to \dot{X}^+$ and $\dot{X}^+ \to X^{2+}$. For example, the value of h for the doubly occupied $2p\pi$ orbital of N in the trigonal valence state, $h_{\ddot{N}(p)}$, is estimated as the mean of the h value corresponding to the ionization process $(p)^2 \to (p)^1$ in the valence state $(tr)^1(tr)^1(tr)^1(p)^2$ of N and the h value corresponding to the ionization process $(p)^1 \to (p)^0$ in the valence state $(tr)^1(tr)^1(tr)^1(p)^1$ of N^+. Matsen[10] has given values of -7.2 and -3.0 ev to the standard α and β, respectively, implying that the ratio α/β is 2.4. The values of h estimated by the use of this value for α/β are shown in the last column of Table 9.3.

9.3. Evaluation of Resonance Parameters

9.3.1. The Proportionality of the Resonance Integral to the Overlap Integral

There are two principal methods for evaluating the resonance parameters, k_{ij} in expression (9.2). In one of them the resonance integral is related to the bond energy. Lennard-Jones[11] has proposed that the π-π resonance integral

[10] F. A. Matsen, *J. Am. Chem. Soc.* **72,** 5243 (1950).
[11] J. E. Lennard-Jones, *Proc. Roy. Soc.* (*London*) **A158,** 280 (1937).

β_{ij} be considered as representing half the difference between the energy of the i—j double bond of a given length, $E_{i=j}$, and that of the i—j single bond of the same length, E_{i-j} :

$$\beta_{ij} = (E_{i=j} - E_{i-j})/2. \tag{9.8}$$

According to this relation, k_{ij} can be expressed as follows:

$$k_{ij} = (E_{i=j} - E_{i-j})/(E_{C=C} - E_{C-C}), \tag{9.9}$$

where $E_{C=C}$ is the energy of the C—C double bond of a standard length and E_{C-C} is that of the C—C single bond of the same length.

In the other important method the resonance integral is related to the overlap integral. Mulliken[12,13] has proposed that the resonance integral be proportional to the overlap integral as long as the same type of bonds is concerned:

$$\beta_{ij} = KS_{ij}, \tag{9.10}$$

where K is a constant. In order to simplify the calculations involved in molecules containing heteroatoms when overlap integrals are included in the simple LCAO MO method, Wheland[14] has assumed that the value of K be the same for all types of bonds (see Subsection 7.1.4). This assumption implies that the ratio of two resonance integrals is given by the ratio of the corresponding overlap integrals, and in particular that we may write

$$k_{ij} = S_{ij}/S, \tag{9.11}$$

where S is the standard π–π overlap integral, that is, the π–π overlap integral for a standard C—C bond, which is usually taken as a C—C bond in benzene. By the use of this relation, if the overlap integrals have been calculated, the resonance parameters can easily be evaluated.

In this book we are interested mainly in the variation of β_{ij} associated with changes in the bond length and in the angle of twist of the two $p\pi$ atomic orbitals, χ_i and χ_j, from coplanarity. With respect to this problem, the use of Eq. (9.11) is much more convenient than the use of Eq. (9.9). Therefore, in this book Eq. (9.11) is exclusively used for evaluating resonance parameters. As will be mentioned below, it is somewhat doubtful that the proportionality constant K in expression (9.10) is the same for all types of bonds. Nevertheless, it is assumed, for simplicity, that the K is the same,

[12] R. S. Mulliken, *J. chim. phys.* **46,** 497 (1949).
[13] R. S. Mulliken, *J. Phys. Chem.* **56,** 295 (1952).
[14] G. W. Wheland, *J. Am. Chem. Soc.* **64,** 900 (1942).

and hence Eq. (9.11) can be used for different types of bonds as long as both the principal quantum numbers of atomic orbitals χ_i and χ_j are 2. It has been suggested that the assumption of the proportionality of the resonance integral to the overlap integral is not appropriate with a change in principal quantum number.[15]

The validity of relation (9.10) has been verified for C—C bonds by comparison of the variation of the resonance integral estimated by the Lennard-Jones method as a function of bond length with the corresponding variation of the overlap integral for two Slater $2p\pi$ orbitals of carbon. This relation is probably also valid for any other type of bonds. It is, however, not certain that the proportionality constant K is the same for different types of bonds. When a heteroatom X is more electronegative than C it is expected that the core potential due to C—X be stronger than that due to C—C. Accordingly, even when the π–π overlap integral for the C—X bond is smaller than the standard S, it may be possible that the absolute value of the π–π resonance integral for the C—X bond is higher than that of the standard β, and hence that the resonance parameter k for the C—X bond is higher than 1. For example, as shown in Fig. 9.2, the curve of the π–π overlap integral for the C—O bond as a function of the interatomic distance R lies below the corresponding curve for the C—C bond. Therefore, the assumption of the same value for the proportionality constant K in Eq. (9.10) would lead to a k value smaller than 1 for the carbonyl bond. In fact, values higher than 1 have been advocated by many workers for the k-parameter for the carbonyl bond.

In order to allow for the possible dependence of the proportionality constant K on the type of bond, relation (9.10) may be refined in the following manner. Mulliken[12] has proposed an approximation that an electron distribution $\chi_i\chi_j$ is replaced by the distribution $S_{ij}(\chi_i^2 + \chi_j^2)/2$. Following this approximation, which is called the Mulliken approximation, it may be assumed that β_{ij} is equal to, or at least, is proportional to, $S_{ij}(\alpha_i + \alpha_j)/2$:

$$\beta_{ij} = K'S_{ij}(\alpha_i + \alpha_j)/2, \tag{9.12}$$

where K' is a constant. By the use of expression (9.1) this relation can be rewritten as

$$\beta_{ij} = K'S_{ij}\alpha\,\{1 + [(h_i + h_j)/2](\beta/\alpha)\}. \tag{9.13}$$

According to this assumption, k_{ij} can be written as

$$k_{ij} = (S_{ij}/S)\,\{1 + [(h_i + h_j)/2](\beta/\alpha)\}. \tag{9.14}$$

Instead of Eq. (9.11) this equation may be used for evaluation of k_{ij}.

It is to be noted that the value of the overlap integral S_{ij} must be dependent on whether the involved atomic orbitals are one-electron contributors or two-electron contributors. When the orbital χ_i is a one-electron contributor, that is, when the ith atom X contributes one electron in the orbital χ_i to the π system, the orbital χ_i is usually taken as the orbital in the neutral atom X. On the other hand, when the orbital χ_i is a two-electron contributor, that is, when the ith atom X contributes a pair of electrons in the orbital χ_i to the π system, it seems to be reasonable to take the orbital χ_i as the orbital in the partially positively charged atom $X^{\delta+}$, where δ represents a fractional number smaller than unity. The value of δ may be taken to be $\frac{1}{2}$. Since the effective nuclear charge must be greater in the

[15] A. Streitwieser, Jr., *J. Am. Chem. Soc.* **82,** 4123 (1960).

positively charged atom $X^{\delta+}$ than in the neutral atom X, an atomic orbital in $X^{\delta+}$ must be contracted compared with the orbital of the same type in X. Therefore, when the orbital χ_j is, for example, a $2p\pi$ orbital of a carbon atom and the orbital χ_i is, for example, a $2p\pi$ orbital of a heteroatom X, the value of the overlap integral S_{ij} at a fixed internuclear distance must be smaller in the case where the orbital χ_i is a two-electron contributor than in the case where the orbital χ_i is a one-electron contributor. If the other factors are assumed to be the same, this means that the value of the resonance parameter k_{ij} must be smaller in the former case than in the latter case.

9.3.2. THE DEPENDENCE OF THE OVERLAP INTEGRAL ON THE BOND LENGTH

Mulliken and his co-workers[16] presented extensive tables of overlap integrals for various pairs of Slater atomic orbitals at various interatomic distances. On the basis of Mulliken's tables, in Fig. 9.1 the values of overlap

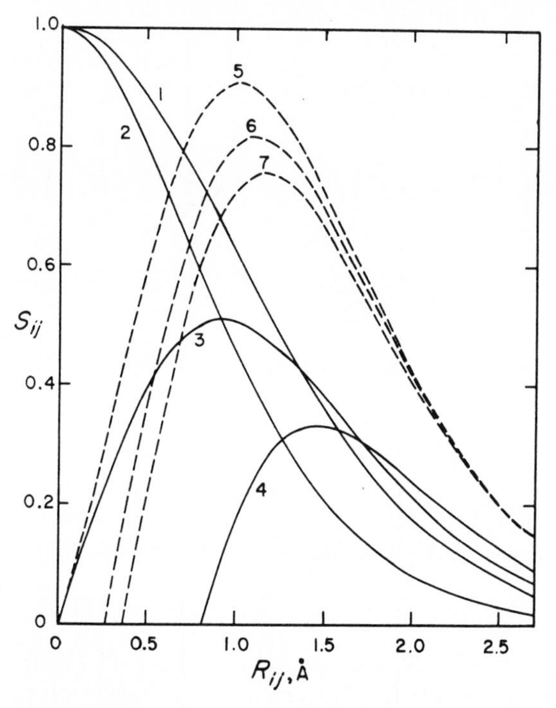

FIG. 9.1. Overlap integrals (S) for various pairs of carbon Slater atomic orbitals as functions of the interatomic distance R. Curve 1, $2s$–$2s$; curve 2, $2p\pi$–$2p\pi$; curve 3, $2s$–$2p\sigma$; curve 4, $2p\sigma$–$2p\sigma$; curve 5, $2sp\sigma$–$2sp\sigma$; curve 6, $2sp^2\sigma$–$2sp^2\sigma$; curve 7, $2sp^3\sigma$–$2sp^3\sigma$.

[16] R. S. Mulliken, C. A. Rieke, D. Orloff, and H. Orloff, *J. Chem. Phys.* **17,** 1248 (1949).

integrals for various pairs of carbon atomic orbitals are shown as functions of the interatomic distance R. Overlap integrals between a σ orbital on one atom and a π orbital on another atom, for example, the $2s$–$2p\pi$ and $2p\sigma$–$2p\pi$ overlap integrals, are zero. Similarly, when the bond axis is taken as the x axis, the $2p_y$–$2p_z$ overlap integral is zero.

In Fig. 9.2 the values of the $2p\pi$–$2p\pi$ overlap integrals for C—C, C—N, C—O, and N—N bonds as functions of the interatomic distance are shown.

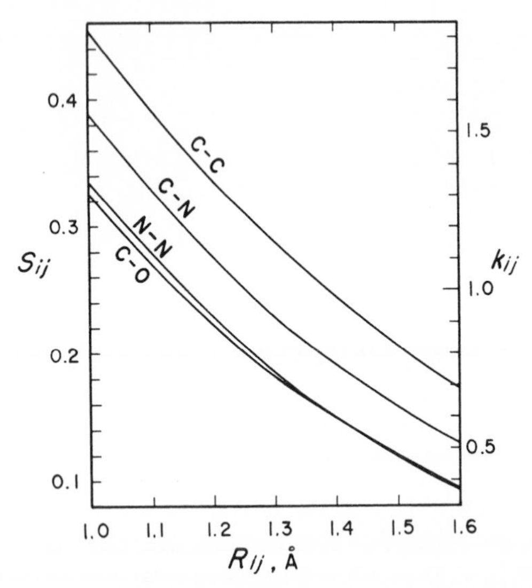

FIG. 9.2. The $2p\pi$–$2p\pi$ overlap integrals (S) for C—C, C—N, C—O, and N—N bonds as functions of the interatomic distance R. The k_{ij} is the resonance parameter calculated from Eq. (9.11) by the use of the value of 0.249 for the standard overlap integral.

It is seen that the value of the $2p\pi$–$2p\pi$ overlap integral for the C—C bond varies approximately linearly with the interatomic distance in the region between 1.44 and 1.54 Å. The relation can be approximately expressed as

$$S(R) = 0.192 + (1.540 - R) \times 0.360, \tag{9.15}$$

where R is the interatomic distance in units of Å. The values of the C—C $2p\pi$–$2p\pi$ overlap integral for the interatomic distances of 1.39, 1.397, and 1.40 Å are 0.249, 0.246, and 0.244, respectively. When the value of 0.249 is taken as the value of the standard overlap integral S, and when relations (9.11) and (9.15) are used, the resonance parameter for the C—C π—π bond at the interatomic distance R, $k(R)$, can be approximately expressed as follows, as long as R is in the region between 1.44 and 1.54 Å:

$$k(R) = 0.771 + (1.540 - R) \times 1.446. \tag{9.16}$$

In Tables 9.4 and 9.5 the values of S for C—C, C—N, C—O, and N—N $2p\pi$—$2p\pi$ bonds of lengths of interest to us and the values of k for these bonds calculated by the use of Eq. (9.11) are shown. The values of k calculated by using Eq. (9.14) with a value of $1/2.4$ for β/α and with values listed in Table 9.3

TABLE 9.4

VALUES OF THE π–π OVERLAP INTEGRAL (S) AND RESONANCE PARAMETER (k) FOR C—C BONDS[a]

R, Å	S	k[b]
1.20	0.337	1.353
1.33	0.275	1.102
1.34	0.270	1.084
1.39	0.249	1.000
1.44	0.228	0.916
1.48	0.214	0.858
1.50	0.206	0.827
1.54	0.192	0.770

[a] R: the bond length.
[b] Calculated by the use of Eq. (9.11).

for the appropriate h-parameters are also shown in Table 9.5. Since C—C "double" and "single" bonds in most conjugated systems have lengths of about 1.34 and 1.44–1.54 Å, respectively, the appropriate values of k for these bonds are about 1.1 and 0.8–0.9, respectively.

9.3.3. OVERLAP INTEGRALS OF NONPARALLEL p ORBITALS AND HYBRID ORBITALS

In the preceding subsection the $2p\pi$–$2p\pi$ overlap integral refers to the overlap integral of two $2p\pi$ atomic orbitals whose symmetry axes are parallel. When the bond is twisted and the symmetry axes of the two atomic orbitals become nonparallel, the overlap integral must vary depending on the angle of twist. The overlap integral of two p orbitals whose symmetry axes are not parallel to each other can be estimated in the following manner.

Consider first the case of two adjacent $2p$ orbitals, $\chi_i(p_z)$ and $\chi_j(p_{z'})$, whose symmetry axes lie in parallel planes at the distance R and are canted

TABLE 9.5

VALUES OF THE π–π OVERLAP INTEGRAL (S) AND RESONANCE PARAMETER (k) FOR C—N, C—O, AND N—N BONDS[a]

Bond	R, Å	S	k^b	k^c
$\dot{C}$—$\dot{N}$	1.27 (C═NOH)	0.243	0.975	1.096
	1.35 (pyridine)	0.209	0.841	0.946
	1.41 (═C—N═, *trans*-azobenzene)	0.187	0.749	0.843
	1.47 (═C—N═, *cis*-azobenzene)	0.166	0.668	0.751
$\dot{C}$—$\ddot{N}$	1.42 (pyrrole)	0.183	0.736	0.822
	1.43 (Ar—$\ddot{N}$, C_6H_5—$NHCOCH_3$)	0.180	0.722	0.806
	1.47 (Ar—$\ddot{N}$, *p*-HOC_6H_4—NH_2)	0.166	0.668	0.746
$\dot{C}$—$\dot{O}$	1.15 (C═O, *p*-quinones)	0.245	0.984	1.246
	1.23 (C═O)	0.210	0.844	1.069
$\dot{C}$—$\ddot{O}$	1.36 (phenols)	0.161	0.645	0.808
	1.37 (furan)	0.158	0.633	0.792
	1.47 (phenols, *p*-$H_2NC_6H_4$—OH)	0.128	0.513	0.642
$\dot{N}$—$\dot{N}$	1.23 (N═N)	0.215	0.862	1.077

[a] R: the bond length.

[b] Calculated by the use of Eq. (9.11).

[c] Calculated by the use of Eq. (9.14) with a value of 1/2.4 for β/α. In the calculation the following h values were used: $h_{\dot{N}(p)} = 0.60$; $h_{\ddot{N}(p)} = 0.56$; $h_{\dot{O}(p)} = 1.28$; $h_{\ddot{O}(p)} = 1.21$.

with respect to each other by the angle θ_{ij}, as illustrated in Fig. 9.3. The Slater $2p$ atomic orbitals are expressed as follows:

$$\chi(2p_x) = N_{2p}xe^{-\mu r}; \qquad \chi(2p_y) = N_{2p}ye^{-\mu r}; \qquad \chi(2p_z) = N_{2p}ze^{-\mu r}, \tag{9.17}$$

where N_{2p} is the normalization constant equal to $(\mu^5/\pi)^{1/2}$, μ is the effective nuclear charge divided by the effective principal quantum number, and r is the distance from the atomic nucleus. Evidently, $2p$ orbitals, more generally, p orbitals, can be treated as vectors. Since the symmetry axis (z') of the $2p_{z'}$ orbital on atom j is in the yz plane and makes the angle θ_{ij} with the z axis, the orbital can be resolved into y and z components by the following equation:

$$\chi_j(p_{z'}) = \chi_j(p_z) \cos \theta_{ij} + \chi_j(p_y) \sin \theta_{ij}. \tag{9.18}$$

Since the overlap integral between $\chi_i(p_z)$ and $\chi_j(p_y)$ is zero, the overlap integral between $\chi_i(p_z)$ and $\chi_j(p_{z'})$ reduces to the overlap integral between the two parallel p orbitals $\chi_i(p_z)$ and $\chi_j(p_z)$, multiplied by $\cos \theta_{ij}$:

$$S_{ij}(p_z\text{–}p_{z'}) = S_{ij}(p\pi\text{–}p\pi) \cos \theta_{ij}. \tag{9.19}$$

That is, when the overlap integral between two p orbitals with the interatomic distance R and the angle of twist θ is expressed as $S(R, \theta)$, we can write as follows:

$$S(R, \theta) = S(R, 0°) \cos \theta. \quad (9.20)$$

Therefore, by the use of relation (9.11), the resonance parameter k for the twisted p orbitals can be expressed as follows:

$$k(R, \theta) = S(R, \theta)/S = [S(R, 0°)/S] \cos \theta = k(R, 0°) \cos \theta. \quad (9.21)$$

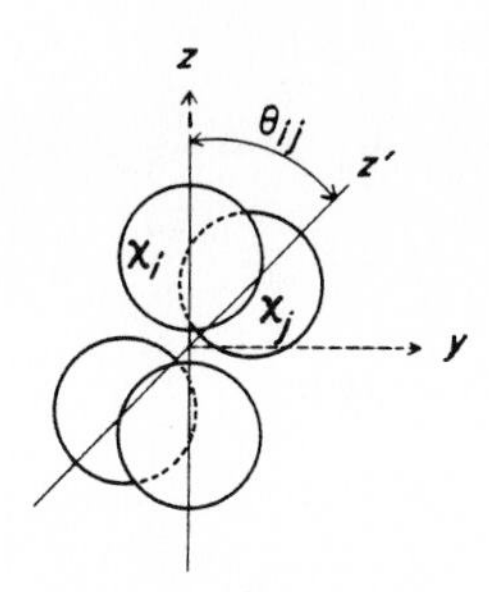

FIG. 9.3. The overlap between two $p\pi$ atomic orbitals on adjacent atoms i and j when the i—j bond is twisted by an angle θ_{ij} (end view).

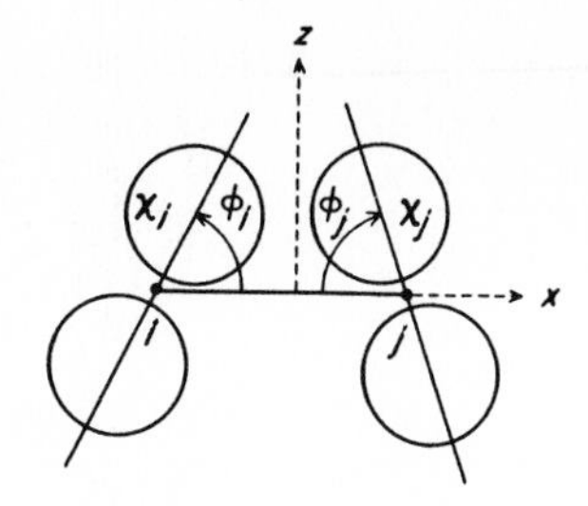

FIG. 9.4. The overlap between two p atomic orbitals whose symmetry axes lie in a plane and are slanted towards each other.

In a similar manner, any two oriented p orbitals may be resolved into appropriate p_x, p_y, and p_z orbitals for calculation of the overlap integral. For example, in the case of two p orbitals χ_i and χ_j, whose symmetry axes lie in a plane (xz plane) (that is; $\theta_{ij} = 0°$) but are slanted toward each other at the angles ϕ_i and ϕ_j with respect to the internuclear line (the x axis), as illustrated in Fig. 9.4, the orbitals can be resolved as

$$\chi_i = \chi_i(p_x) \cos \phi_i + \chi_i(p_z) \sin \phi_i \quad (9.22)$$

and

$$\chi_j = \chi_j(p_x) \cos \phi_j + \chi_j(p_z) \sin \phi_j, \quad (9.23)$$

and hence the overlap integral between the two orbitals can be resolved into p_x–p_x (that is, $p\sigma$–$p\sigma$) and p_z–p_z (that is, $p\pi$–$p\pi$) contributions:

$$S_{ij} = S_{ij}(p\sigma–p\sigma) \cos \phi_i \cos \phi_j + S_{ij}(p\pi–p\pi) \sin \phi_i \sin \phi_j. \quad (9.24)$$

If the orbitals in addition are canted in such a way that $\theta_{ij} \neq 0°$, evidently the correction factor $\cos \theta_{ij}$ should be introduced into the term with the $p\pi$–$p\pi$ overlap integral:

$$S_{ij} = S_{ij}(p\sigma–p\sigma) \cos \phi_i \cos \phi_j + S_{ij}(p\pi–p\pi) \sin \phi_i \sin \phi_j \cos \theta_{ij}. \quad (9.25)$$

Such cases are frequently encountered when treating nonneighbor interactions in nonplanar π systems.

In a similar manner the overlap integral between a p orbital and a hybrid orbital or between two hybrid orbitals can be resolved into $p\sigma$–$p\sigma$, $p\pi$–$p\pi$, and s–$p\sigma$ contributions. For example, let us consider the case illustrated in Fig. 9.5, in which χ_i is a trigonal (sp^2) hybrid orbital whose symmetry axis is in the xy plane and makes an angle of 120° with the internuclear axis (x axis), and χ_j is a p orbital whose symmetry axis is in the yz plane and makes an angle θ with the z axis. In this case, the orbitals may be resolved as follows:

$$\chi_i(sp^2) = 3^{-1/2}\,\chi_i(s) - 6^{-1/2}\,\chi_i(p_x) + 2^{-1/2}\,\chi_i(p_y); \qquad (9.26)$$

$$\chi_j(p) = \chi_j(p_z)\cos\theta + \chi_j(p_y)\sin\theta. \qquad (9.27)$$

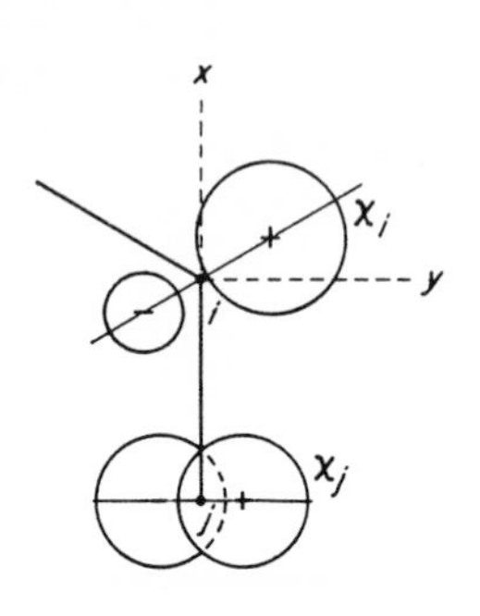

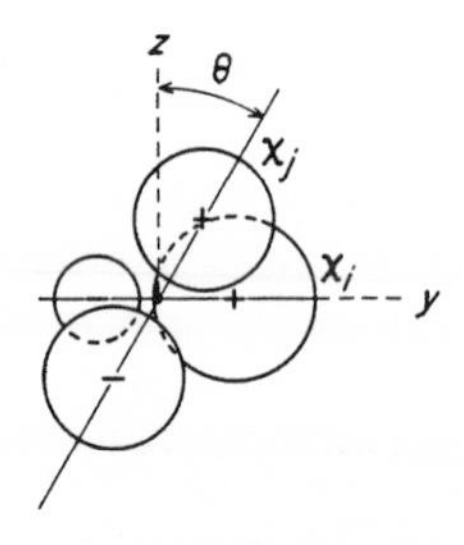

FIG. 9.5. The overlap between an sp^2 hybrid orbital on atom i (χ_i) whose symmetry axis is in the xy plane and makes an angle of 120° with the i—j bond axis (x axis) and a p orbital on atom j (χ_j) whose symmetry axis is in the yz plane and makes an angle θ with the z axis. A projective figure on the xy plane and a projective figure on the yz plane (end view).

Since s–$p\pi$, $p\sigma$–$p\pi$, and p_y–p_z overlap integrals are zero, the overlap integral of χ_i and χ_j reduces to the following simple form:

$$S_{ij} = 2^{-1/2}\,S_{ij}(p\pi\text{–}p\pi)\sin\theta. \qquad (9.28)$$

Such a case is encountered in treatment of the interaction between n orbitals and π orbitals, for example, in nonplanar *cis*-azobenzene (compare Chapter 23).

Lastly, we consider the case where both two orbitals χ_i and χ_j are sp^2 hybrid orbitals whose symmetry axes lie in a plane (xy plane) and make angles of 120° with the internuclear axis (x axis) either in the *trans* or in the *cis* arrangement, just as the two n orbitals on the two nitrogen atoms in the *trans* and *cis* azo groups (compare Chapter 23, Fig. 23.1). If atoms i and j are of the same element, the overlap integral of χ_i and χ_j can evidently be expressed as follows:

$$S_{ij} = (1/3)S(s\text{–}s) - (\sqrt{2}/3)S(s\text{–}p\sigma) + (1/6)S(p\sigma\text{–}p\sigma) \pm (1/2)S(p\pi\text{–}p\pi). \quad (9.29)$$

The sign in front of the last term is + for the *cis* arrangement and is − for the *trans* arrangement.

9.3.4. Variation of Bond Length

When the symmetry axes of two adjacent $p\pi$ atomic orbitals are twisted around the bond axis from each other by the angle θ, the π–π interaction across the bond is reduced in the manner expressed by Eq. (9.20). This reduction of the π–π interaction must be accompanied with a change, probably an increase, in the bond length. Therefore, in order to relate the value of k to the twist angle θ by Eq. (9.21), it is necessary to take account of the variation of the bond length R associated with the twist.

The C—C bond length is known to vary from about 1.20 Å for a triple bond to about 1.54 Å for a single bond. Usually, the variation of the bond length is related to the variation of the π-bond order. The C—C bonds in ethane, ethylene, and acetylene have π-bond orders of 0, 1, and 2, respectively, and there is a monotonic relationship between these π-bond orders and the corresponding bond lengths, 1.54, 1.34, and 1.20 Å. Bonds of intermediate lengths are associated with fractional π-bond orders. For example, the lengths of C—C bonds in graphite and in benzene are 1.4210 ± 0.0001 Å and 1.397 ± 0.003 Å, respectively, and the π-bond orders of these bonds calculated by the Hückel method are 0.535 and 0.667, respectively. It is possible to draw a smooth π-bond order–length correlation curve through these five fundamental points for ethane, graphite, benzene, ethylene, and acetylene. Coulson[17] expressed this correlation by the following equation:

$$R = R_s - (R_s - R_d)/[1 + K(1 - P)/P], \quad (9.30)$$

where P is the π-bond order, R_s and R_d are the "natural" single bond length (that is, the bond length for $P = 0$) and the "natural" double bond length (that is, the bond length for $P = 1$), respectively, and K is a constant. A value of 0.765 was assigned to K. The lengths of most C—C bonds between trigonal carbon atoms in aromatic rings are in the range of 1.36–1.46 Å. When the observed bond length is plotted against the π-bond order calculated

[17] C. A. Coulson, *Proc. Roy. Soc. (London)* **A169,** 413 (1939).

by the Hückel method, the points for a wide variety of aromatic molecules all lie within about 0.02 Å of the correlation line.

There are some cases where the plot of the observed bond length against the calculated π-bond order is considerably deviated from the correlation line. For example, the π-bond orders of the essential double bonds and essential single bond in butadiene are 0.896 and 0.446, respectively, in the Hückel approximation (see Subsection 8.2.1). The observed length of the essential single bond of 1.483 ± 0.01 Å[18] is considerably longer than expected from the π-bond order of 0.446, and the observed length of the essential double bonds of 1.337 ± 0.005 Å[18] is shorter than expected from the π-bond order of 0.896, and is not lengthened compared with the length of the double bond in ethylene (1.333,[19] 1.337,[20] or 1.339 Å[21]). Part of this discrepancy comes from the assumption of equality of all β's in the Hückel method. In addition, Mulliken[22] has shown that more advanced calculations including electronic interactions yield a lower π-bond order (about 0.2) for the central "single" bond in butadiene and that only a small lengthening of the essential double bonds as compared with the ethylenic double bond, less than the present experimental uncertainties, is to be expected. In general, inclusion of electronic interactions appears to result in a greater localization of double bonds than in the simple MO method.[23] In the case of aromatic hydrocarbons, it has been pointed out that refinement of the simple MO method to advanced MO methods by inclusion of electronic interactions appears to be equivalent to giving more weight to the principal resonance structures in which the number of benzenoid rings is maximum.[24] Interestingly, however, it also has been pointed out that advanced MO methods are in fact not more accurate in prediction of bond length than the simple method.[25]

The bond length has so far been related solely to the π-bond order. However, it is certain that there must be some other factors affecting the bond length. Coulson[26] has pointed out that, since a bond in a conjugated system may be regarded as the superposition of a σ bond and a fractional π bond, its length will depend not only on the π-bond order but also on the nature of the σ bond. The dependence of the bond length on the nature of the σ bond is shown very clearly by the lengths of C—H bonds of various types: CH_3—H, 1.093 Å; CH_2=CH—H, 1.071 Å; CH≡C—H, 1.058 Å.

[18] (By electron diffraction method) A. Almenningen, O. Bastiansen, and M. Traetteberg, *Acta Chem. Scand.* **12,** 1221 (1958).

[19] (By electron diffraction method) L. S. Bartell and R. A. Bonham, *J. Chem. Phys.* **27,** 1414 (1957); **31,** 400 (1959).

[20] (By infrared spectroscopy) H. C. Allen, Jr. and E. K. Plyler, *J. Am. Chem. Soc.* **80,** 2673 (1958).

[21] (By Raman spectroscopy) J. M. Dowling and B. P. Stoicheff, *Can. J. Phys.* **37,** 703 (1959).

[22] R. S. Mulliken, *Tetrahedron* **6,** 68 (1959).

[23] M. J. S. Dewar and C. E. Wulfman, *J. Chem. Phys.* **29,** 158 (1958).

[24] M. Randić, *J. Chem. Phys.* **34,** 693 (1961).

[25] H. O. Pritchard and F. H. Sumner, *Proc. Roy. Soc.* (*London*) **A226,** 128 (1958).

[26] C. A. Coulson, *in* "Theoretical Organic Chemistry, Proceedings and Discussions of the Kekulé Symposium," p. 49. Butterworths, London and Washington, D.C., 1959.

In this case there is no complicating π-electron contribution. Evidently, the σ-bond radius of the carbon atom depends on the type of hybridization: there are contradictions in the radius which amount to 0.022 and 0.013 Å in the change in hybridization from sp^3 to sp^2 and from sp^2 to sp, respectively. Accordingly, it is expected that, for example, the natural sp^2–sp^2 C—C single bond length, that is, the length of the sp^2–sp^2 C—C bond with the π-bond order of zero, is not equal to the length of the sp^3–sp^3 C—C single bonds in saturated hydrocarbons, and hence probably the foregoing single order–length correlation curve for C—C bonds should be replaced by a family of curves, each member relating to a particular hybridization in the σ bond.

In usual conjugated systems bonds between trigonal carbon átoms, that is, sp^2–sp^2 C—C bonds, are the commonest, and hence we are most interested in this type of bond. The natural sp^2–sp^2 C—C double bond length, that is, the length of the sp^2–sp^2 C—C bond with the π-bond order of 1, R_d, is of course the length of the isolated or nonconjugated ethylenic bond, and is generally taken from ethylene as 1.34 or 1.335 Å (average). On the other hand, the natural sp^2–sp^2 C—C single bond length, R_s, is not yet known with absolute reliability.

Coulson[27] proposed a value of 1.50 Å for R_s, but later suggested that a shorter length might be better. Dewar and Schmeising[28] have taken as R_s a value of 1.48 Å, which is equal to the length of the "single" C—C bonds in some conjugated systems such as butadiene and biphenyl, and have advanced a view that variations of bond length are due almost entirely to changes in hybridization in σ bonds and that effects of π conjugation are negligible in many conjugated systems. In view of the fact that the length of the sp^2–sp^2 C—C bond varies from about 1.34 Å for the isolated ethylenic bond to about 1.40 Å for aromatic bonds, to about 1.44–1.50 Å for "single" bonds in planar conjugated systems, and to about 1.51–1.55 Å for "single" bonds in some sterically hindered conjugated systems (see below), the view of Dewar and Schmeising is evidently invalid, overemphasizing the effect of σ hybridization in determining the bond length. Mulliken[29] and Bak and Hansen-Nygaard[30] have carefully examined the effect of σ hybridization in determining the bond length and have affirmed the importance of π-bond order. Mulliken has proposed a value of 1.51 ± 0.01 Å, and Bak and Hansen-Nygaard a value of 1.517 Å, for R_s.

[27] C. A. Coulson, *Proc. Roy. Soc.* (*London*) **A207,** 91 (1951); *J. Phys. Chem.* **56,** 311 (1952); Special Publication No. 12, p. 85. The Chemical Society, London, 1958.

[28] M. J. S. Dewar and H. N. Schmeising, *Tetrahedron* **5,** 166 (1959); **11,** 96 (1960).

[29] R. S. Mulliken, *Tetrahedron* **6,** 68 (1959).

[30] B. Bak and L. Hansen-Nygaard, *J. Chem. Phys.* **33,** 418 (1960).

When Coulson's equation (9.30) is solved for the unknowns by the use of the values of R and P for ethylene ($R = 1.335$ Å; $P = 1$), benzene ($R = 1.397$ Å; $P = 0.667$), and graphite ($R = 1.421$ Å; $P = 0.535$), the values of R_s and K are found to be 1.515 Å and 1.05, respectively. When the value of K is assumed to be 1, the equation reduces to the following simple form:

$$R = R_s - (R_s - R_d)P. \tag{9.31}$$

This means that the length of the sp^2–sp^2 C—C bond varies approximately linearly with the π-bond order. Schmeising[31] has proposed values of 1.511 and 0.173 Å for R_s and $(R_s - R_d)$, respectively:

$$R \text{ (in Å)} = 1.511 - 0.173P. \tag{9.32}$$

Another approach to the bond length problem has been made by Costain and Stoicheff.[32] They have compiled accurate values of C—C and C—H bond lengths in many simple compounds, most of which have probably an accuracy of at least ± 0.005 Å, and deduced a generalization that the bond lengths vary linearly with the number of adjacent bonds, irrespective of the bond multiplicity and type of the adjacent bonds. However, this generalization seems to be consistent with the view so far discussed that the C—C bond length depends primarily on the σ-bond hybridization and the π-bond order.

Thus the most values proposed hitherto for the natural sp^2–sp^2 C—C single bond length R_s are about 1.51 Å. However, there are many sp^2–sp^2 C—C bonds whose lengths have been reported to be longer than 1.51 Å. For example, the lengths of the essential single bonds joining together benzene rings in hexaphenylbenzene, tetraphenylene, and triphenylene have been reported to be 1.52,[33] 1.52,[34] and about 1.53 Å,[35] respectively, and the average length of the six C—C *peri*-bonds joining naphthalenic residues in quaterrylene has been reported to be 1.53 ± 0.01 Å.[36] Furthermore, the length of the coannular "single" bond in *m*-tolidine (4,4′-diamino-2,2′-dimethylbiphenyl) has been reported to be as long as 1.55 Å.[37] These unusually long bond lengths of sp^2–sp^2 C—C bonds in sterically overcrowded molecules suggest that the intramolecular repulsive interaction between nonbonded atoms affects the bond length. Bartell[38] has emphasized the importance of this steric effect.

[31] H. N. Schmeising, presented at International Symposium on Molecular Structure and Spectroscopy, Tokyo, September, 1962.

[32] C. C. Costain and B. P. Stoicheff, *J. Chem. Phys.* **30,** 777 (1959).

[33] A. Almenningen, O. Bastiansen, and P. N. Skancke, *Acta Chem. Scand.* **12,** 1215 (1958).

[34] I. L. Karle and L. O. Brockway, *J. Am. Chem. Soc.* **66,** 1974 (1944).

[35] T. H. Goodwin, *J. Chem. Soc.* **1960,** 485.

[36] H. N. Shrivastava and J. C. Speakman, *Proc. Roy. Soc.* (*London*) **A257,** 477 (1960).

[37] F. Fowweather, *Acta Cryst.* **5,** 820 (1952).

[38] L. S. Bartell, *J. Chem. Phys.* **32,** 827 (1960); L. S. Bartell and R. A. Bonham, *ibid.* **32,** 824 (1960).

9.3.5. CORRELATION OF THE sp^2–sp^2 C—C BOND LENGTH TO THE ANGLE OF TWIST AND EVALUATION OF k FOR TWISTED sp^2–sp^2 C—C BONDS

As discussed in the preceding subsection, there are many factors affecting bond lengths. Besides the foregoing factors, that is, the hybridization of σ bond, the π-bond order, and the intramolecular repulsive interaction between nonbonded atoms, there may be additional factors, such as the ionicity of the bond, the formal charge on the bonded atoms, and the intermolecular interaction. Furthermore, the bond length will depend, though probably to a small extent, on the temperature owing to the anharmonicity of the stretching vibration of the bond. However, as far as sp^2–sp^2 C—C bonds are concerned, it may be said that the bond length is determined primarily by the π-bond order and the steric strain around the bond. In the absence of the steric strain, the sp^2–sp^2 C—C bond length varies approximately linearly with the π-bond order. As will be mentioned later, the π-bond order of the essential single bond is approximately proportional to the resonance parameter k, which in turn is proportional to the π–π overlap integral S.

When an sp^2–sp^2 C—C bond is twisted, the k value and π-bond order will decrease approximately in proportion to the π-π overlap integral, and consequently the bond will be lengthened. The following relation has been tentatively assumed between the twist angle θ and the bond length R:

$$[S(R, 0°) - S(R_{90}, 0°)]/[S(R_0, 0°) - S(R_{90}, 0°)] = S(R, \theta)/S(R, 0°) = \cos\theta, \tag{9.33}$$

where $S(R, \theta)$ represents, as mentioned already, the π–π overlap integral for the bond with the bond length R and the twist angle θ, and R_0 and R_{90} are the lengths of the bond when θ is 0° and when θ is 90°, respectively.[39] In view of the approximate linear relationship between P, k, and S, this assumed relation seems to be reasonable. According to this relation, the resonance parameter $k(R, \theta)$, which is equal to $k(R, 0°)\cos\theta$ [compare Eq. (9.20)], can be expressed as follows:

$$\begin{aligned} k(R, \theta) &= k(R_{90}, 0°)\cos\theta + [k(R_0, 0°) - k(R_{90}, 0°)]\cos^2\theta \\ &= k(R_0, 0°)\cos\theta - [k(R_0, 0°) - k(R_{90}, 0°)]\cos\theta\cdot(1 - \cos\theta). \end{aligned} \tag{9.34}$$

The value of $[k(R_0, 0°) - k(R_{90}, 0°)]$ is considered to be 0.15 at the most. Therefore, in a rough approximation, the terms containing this quantity may be neglected.

[39] H. Suzuki, *Bull. Chem. Soc. Japan* **25,** 145 (1952).

TABLE 9.6

THE LENGTH (R) AND TWIST ANGLE (θ) OF THE COANNULAR ESSENTIAL SINGLE BOND OF BIPHENYL AND RELATED COMPOUNDS

Compound	R, Å	θ, deg.
Biphenyl	1.497	ca. 0
Biphenyl	1.48	0
p-Terphenyl	1.48	0
p-Quaterphenyl	1.48	0
4,4′-Dinitrobiphenyl	1.485	0
4-Hydroxybiphenyl	1.49	0
1,3,5-Triphenylbenzene	1.49 ± 0.03	28.3 ± 2
4,4′-Difluorobiphenyl	1.52	44 ± 5
o-Terphenyl	1.52 ± 0.04	45
3,3′,5,5′-Tetrabromobiphenyl	1.50	49
2-Fluorobiphenyl	1.52	49 ± 5
3,3′-Dichlorobenzidine	1.50	52 ± 10
3,3′-Dibromobiphenyl	1.49	54 ± 5
2,2′-Difluorobiphenyl	1.52	60 ± 5
m-Tolidine hydrochloride	1.525 ± 0.02	70.6
2,2′-Dichlorobenzidine	1.53 ± 0.02	ca. 72
m-Tolidine	1.55 ± 0.05	86
Hexaphenylbenzene	1.52 ± 0.01	90 ± 10

As mentioned already, the overlap integral $S(R, 0°)$ decreases approximately linearly with the increase in the bond length R over the range from about 1.44 to about 1.54 Å. Therefore, from Eq. (9.33) the following relation is deduced:

$$R = R_{90} - (R_{90} - R_0) \cos \theta. \tag{9.35}$$

The value of R_0 depends on the conjugated system. For example, as will be discussed later, the value of R_0 may be taken as 1.483 Å for the central "single" bond in the conjugated diene system, as 1.48 or 1.50 Å for the coannular "single" bond in the biphenyl system, and as 1.445 Å for the "single" bonds in the stilbene system. The value of R_{90} should not be taken to be equal to the so-called natural sp^2–sp^2 C—C single bond length R_s. As mentioned in the preceding subsection, in sterically hindered molecules having large values of θ the sp^2–sp^2 C—C bond appears to be lengthened beyond R_s.

In Table 9.6 values of R and θ for the coannular "single" bonds in biphenyl and related compounds, determined by X-ray or electron diffraction, are collected.[40] The plot of these values in Fig. 9.6 shows a trend that R increases from about 1.48 Å to about 1.52 or 1.55 Å as θ increases from 0 to nearly 90°. This fact suggests that the value of R_{90} may safely be assumed as 1.54 Å.

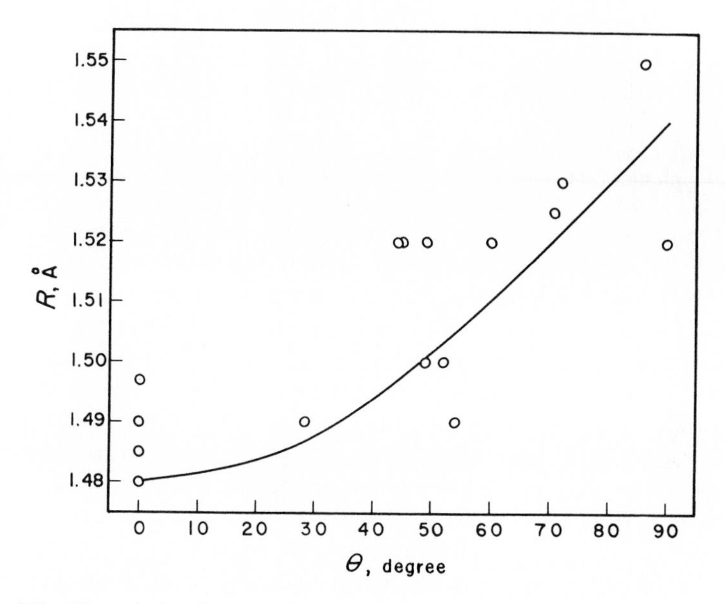

FIG. 9.6. The relation between the twist angle (θ) and the bond length (R) for the coannular essential single bonds in biphenyl and related compounds.

When the value of R_{90} is assumed to be 1.54 Å, R is related to θ by the equation

$$R = 1.54 \text{ Å} - (1.54 \text{ Å} - R_0) \cos \theta. \tag{9.36}$$

In this case, since $k(R_{90}, 0°) = 0.771$ and $k(R_0, 0°) - k(R_{90}, 0°) = 1.446 \times (1.54 - R_0)$ [compare Eq. (9.16)], Eq. (9.34) reduces to the following:

$$k(R, \theta) = 0.771 \cos \theta + 1.446(1.54 - R_0) \cos^2 \theta. \tag{9.37}$$

When the values of R_{90} and R_0 are assumed to be 1.54 and 1.48 Å, respectively, R can be related to θ by the equation

[40] Most of the data were taken from the following: Tables of Interatomic Distances and Configuration in Molecules and Ions, Special Publication No. 11. The Chemical Society, London, 1958.

$$R \text{ (in Å)} = 1.54 - (1.54 - 1.48) \cos \theta = 1.54 - 0.06 \cos \theta. \quad (9.38)$$

In this case, since the value of $k(1.48 \text{ Å}, 0°)$ is 0.858 (see Table 9.4), the parameter k is related to θ by the following equation:

$$\begin{aligned} k(R, \theta) &= 0.771 \cos \theta + 0.087 \cos^2 \theta \\ &= 0.858 \cos \theta - 0.087 \cos \theta \cdot (1 - \cos \theta). \end{aligned} \quad (9.39)$$

The R–θ correlation curve according to Eq. (9.38) is also shown in Fig. 9.6.

9.4. Supplementary Remarks

By the use of the methods developed in the preceding sections, we can evaluate the parameters h and k for various atomic orbitals and pairs of atomic orbitals. Such parameter values are used, in principle, throughout this book. It is, however, to be noted that these methods are not the only possible ones and that the individual numerical values of the parameters estimated by these methods are not the unconditionally precise ones. As emphasized previously, the effective one-electron Hamiltonian is an operator whose exact form cannot be written, and it must vary from molecule to molecule. Since the parameters h and k are defined in terms of the effective one-electron Hamiltonian, it is essentially impossible to assign precise numerical values to the individual parameters. The methods developed above should be considered to fix only the relative values of the parameters for different atomic orbitals and pairs of atomic orbitals and, possibly, the order of magnitude of the parameter values.

Usually, the parameter values are chosen so as to produce good correlation between some experimental property and the corresponding calculated quantity. In recent years, a theoretical procedure has become partially available, in which the parameter values are chosen so as to reproduce in the simple MO approximation the results obtained by the SCF MO method. Depending on the method used for evaluation, on the experimental property and type of compounds used for adjustment, and on the degree of approximation of the MO calculation, many different sets of parameter values have been proposed and used. Some of the parameter values recommended by Streitwieser,[41] those used by Pullman and Pullman,[42] and those selected by

[41] A. Streitwieser, Jr., "Molecular Orbital Theory for Organic Chemists," Chap. 5. Wiley, New York, 1961.

[42] B. Pullman and A. Pullman, "Quantum Biochemistry," Chap. III, Sect. V. Wiley (Interscience), New York, 1963.

Yonezawa and others[43] are shown in Table 9.7. Compare these values with the corresponding values in Tables 9.3–9.5.

One should use a homogeneous set of parameter values. The set of parameter values which obey the rules fixing their relative values and, at the same time, give the best agreement between theoretical predictions and experimental observations is considered as the most satisfactory. When

TABLE 9.7

PARAMETER VALUES CHOSEN BY SOME AUTHORS

	Coulomb parameter h_X			Resonance parameter $k_{\dot{C}-X}$		
Atom (X)	S[a]	P[b]	Y[c]	S	P	Y
$\dot{N}$	0.5	0.4	0.6	1	1	1
$\ddot{N}$	1.5	1	1	0.8	0.9	1
$\dot{N}^+$	2	2	—	—	1	—
$\dot{O}$	1	1.2	2	1	2	$2^{1/2}$
$\ddot{O}$	2	2	2	0.8	0.9	0.6
$\dot{S}$	—	0	0.9	—	1.2	1.2
$\ddot{S}$	—	0	0.9	—	0.6	0.5
$\ddot{F}$	3	—	2.1	0.7	—	1.25
$\ddot{Cl}$	2	—	1.8	0.4	—	0.8
$\ddot{Br}$	1.5	—	1.4	0.3	—	0.7
$\ddot{I}$	—	—	1.2	—	—	0.6

[a] S: A. Streitwieser, Jr., Reference 41; $k_{\dot{C}-\dot{C}} = 0.9$ for $R = 1.47$ Å (essential single bond); $k_{\dot{C}-\dot{C}} = 1$ for $R = 1.40$ Å (aromatic bond); $k_{\dot{C}-\dot{C}} = 1.1$ for $R = 1.34$ Å (essential double bond).

[b] P: B. Pullman and A. Pullman, Reference 42.

[c] Y: T. Yonezawa et al., Reference 43, $h_{\ddot{N}H_2} = 0.4$; $k_{\dot{C}-\ddot{N}H_2} = 0.6$; $h_{\ddot{O}H} = 0.6$; $k_{\dot{C}-\ddot{O}H} = 0.7$; $h_{\ddot{O}CH_3} = 0.5$; $k_{\dot{C}-\ddot{O}CH_3} = 0.6$; $h_{\ddot{S}H} = 0.55$; $k_{\dot{C}-\ddot{S}H} = 0.6$.

drawing a conclusion from the MO calculation, one should make sure that the conclusion is not altered qualitatively when the parameter values are varied within reasonable limits. Fortunately, as long as the parameter values used are in an acceptable range of magnitude and their relative order is a reasonable one, and as long as the results of the MO calculation are used for comparative studies of related properties in series of related molecules,

[43] T. Yonezawa, C. Nagata, H. Kato, A. Imamura, and K. Morokuma, "Ryoshikagaku Nyumon (An Introduction to Quantum Chemistry)," Chap. 2. Kagakudojin, Kyoto, 1963.

the conclusions obtained in most cases are largely independent of the individual numerical values of the parameters.

Sometimes, an inductive effect of a heteroatom X on an attached carbon atom is allowed for by assigning to the carbon atom a h value, $h_{C'}$. The $h_{C'}$ is usually taken to be some given fraction, d, of the value of h for the heteroatom:

$$h_{C'} = d \cdot h_X . \tag{9.40}$$

Values ranging from about 0.1 to about 0.3 have been proposed for this "auxiliary inductive parameter" d. Streitwieser[41] has suggested that if $d = 0.1$ is suitable for $\dot{X}$, a value closer to $d = 0.05$ may be better for the corresponding $\ddot{X}$. The inductive effect is considered to be rapidly damped, to be quite small two bonds away from the heteroatom, and unmeasurable thereafter.

Chapter 10

Usefulness and Limitations of Simple LCAO MO Method in Interpretation of Electronic Absorption Spectra

10.1. General

The calculation of π orbitals of π-electron systems can be easily carried out by simple LCAO MO method in the manner illustrated in Chapter 8. Regretfully, however, it is not possible to interpret satisfactorily detailed features of the electronic absorption spectra by this method. For example, no distinction between singlet and triplet states can be made in this method. Furthermore, for example, as mentioned already, the four degenerate lowest energy excitations in benzene in the simple MO approximation should actually split into three singlet–singlet transitions of different energies, two nondegenerate and one doubly degenerate, and their singlet–triplet counterparts.

The failure of the simple method to give an overall correct picture of electronic spectra is due to the neglect of the electronic repulsive interactions. Inclusion of the electronic interactions constitutes the so-called advanced MO methods, which will be dealt with in the succeeding chapter. In the first part of the present chapter limitations of the simple method in interpreting electronic spectra and effects of inclusion of the electronic interactions are briefly discussed.

The situation that the simple method cannot give a satisfactory interpretation of electronic spectra would appear to be very discouraging. The simple method is, however, very useful for correlating the position and intensity of related absorption bands in a series of related compounds with the structure of molecules. In the latter part of this chapter the usefulness of the simple method in interpretation of electronic spectra is illustrated.

10.2. Singlet–Triplet Splitting

A singly-excited electron configuration, in which two orbitals are singly occupied, can be either singlet or triplet. A triplet state is invariably lower in energy than its singlet counterpart. The energy separation of the corresponding singlet and triplet states is called the singlet–triplet splitting. This splitting may be qualitatively interpreted as follows. In a triplet state the parallel spins of the two electrons in the singly occupied orbitals prevent them from approaching closely, so that the average distance between the electrons with the same spin is larger than that between the corresponding electrons with opposed spins in the singlet counterpart. Accordingly, a triplet state has lower energy than its singlet counterpart, owing to the smaller repulsive Coulombic interactions between the electrons.

As mentioned already, in the simple MO method no distinction between singlet and triplet states is made. The singlet–triplet splitting can be elucidated only by explicitly taking account of the electronic interaction term in the Hamiltonian operator.

For simplicity, only two electrons are considered, being designated as 1 and 2, and other electrons, if any, are assumed to contribute only to the core potential. It is assumed that in the ground state the two electrons are in an orbital ψ_k. Then, the antisymmetrized function of the ground state can be expressed as

$$V_0 = \psi_k(1)\psi_k(2)S, \tag{10.1}$$

and the antisymmetrized functions of the singly-excited singlet and triplet states which arise from the ground state by promoting one electron from ψ_k to a higher orbital ψ_r can be expressed as

$$V_1^{kr} = 2^{-1/2}[\psi_k(1)\psi_r(2) + \psi_k(2)\psi_r(1)]S; \tag{10.2}$$

$$T_1^{kr} = 2^{-1/2}[\psi_k(1)\psi_r(2) - \psi_k(2)\psi_r(1)]T. \tag{10.3}$$

In these expressions, S represents the singlet spin state function for two electrons, given in expression (2.34), and T represents the triplet spin state function for two electrons, given in expression (2.33) (see Subsection 2.2.4). The energy of a state Ψ is given by

$$E(\Psi) = \int \Psi^* \mathbf{H} \Psi \, d\tau, \tag{10.4}$$

where $\mathbf{H}$ is the electronic Hamiltonian. In the present case the $\mathbf{H}$ can be expressed as

$$\mathbf{H} = \mathbf{h}_1{}^{\mathrm{C}} + \mathbf{h}_2{}^{\mathrm{C}} + e^2/r_{12}, \tag{10.5}$$

where $\mathbf{h}_1{}^{\mathrm{C}}$ and $\mathbf{h}_2{}^{\mathrm{C}}$ are the one-electron core Hamiltonians for electrons 1 and 2, respectively, and e^2/r_{12} is the term representing the repulsive interaction between the two electrons, r_{12} being the distance between them (compare Subsection 7.1.2). The operator **H** has no effect on spin functions. If ψ_k and ψ_r are assumed to be orthonormal, since the spin state functions are also orthonormal, the energies of V_0, V_1^{kr}, and T_1^{kr} can evidently be expressed as

$$E(V_0) = 2\varepsilon_k{}^{\mathrm{C}} + J_{kk}; \tag{10.6}$$

$$E(V_1^{kr}) = \varepsilon_k{}^{\mathrm{C}} + \varepsilon_r{}^{\mathrm{C}} + J_{kr} + K_{kr}; \tag{10.7}$$

$$E(T_1^{kr}) = \varepsilon_k{}^{\mathrm{C}} + \varepsilon_r{}^{\mathrm{C}} + J_{kr} - K_{kr}. \tag{10.8}$$

In these expressions the following notation is used:

$$\varepsilon_k{}^{\mathrm{C}} \equiv \int \psi_k{}^*(1)\mathbf{h}_1{}^{\mathrm{C}}\psi_k(1)\, d\tau; \tag{10.9}$$

$$\begin{aligned} J_{kr} &\equiv \int \psi_k{}^*(1)\psi_r{}^*(2)(e^2/r_{12})\psi_k(1)\psi_r(2)\, d\tau \\ &= \int \psi_k{}^*(1)\psi_k(1)(e^2/r_{12})\psi_r(2)\psi_r{}^*(2)\, d\tau \equiv [kk \mid rr]; \end{aligned} \tag{10.10}$$

$$\begin{aligned} K_{kr} &\equiv \int \psi_k{}^*(1)\psi_r{}^*(2)(e^2/r_{12})\psi_r(1)\psi_k(2)\, d\tau \\ &= \int \psi_k{}^*(1)\psi_r(1)(e^2/r_{12})\psi_k(2)\psi_r{}^*(2)\, d\tau \equiv [kr \mid kr]. \end{aligned} \tag{10.11}$$

The $\varepsilon_k{}^{\mathrm{C}}$ may be called the core energy of the orbital ψ_k. J_{kr}, or $[kk \mid rr]$, represents the energy of the electrostatic repulsive interaction between an electron in ψ_k and another electron in ψ_r, and is called an MO Coulomb repulsion integral. K_{kr}, or $[kr \mid kr]$, can be interpreted as representing the energy of the electrostatic repulsive interaction between the two charge distributions $e\psi_k{}^*(1)\psi_r(1)$ and $e\psi_r{}^*(2)\psi_k(2)$, and is called an MO exchange repulsion integral.

In this way, we have

$$E(V_1^{kr}) - E(T_1^{kr}) = 2K_{kr}. \tag{10.12}$$

Even when more than two electrons are taken into consideration, if V_1^{kr} and T_1^{kr} are taken to represent the singlet and triplet states that arise from a closed-shell electron configuration by promoting one electron from an orbital ψ_k to an orbital ψ_r, relation (10.12) is always valid. In general, K_{kr} is a positive energy quantity. Therefore, a triplet state is lower in energy than the corresponding singlet state, and hence an S–T band appears at longer wavelength than the corresponding S–S band. For example, as mentioned previously, ethylene exhibits the π–π^* S–S absorption band at about 160 mμ and the corresponding S–T band at about 270 mμ. The singlet–triplet splitting of n–π^* transitions is, in general, comparatively small. For example, while the singlet–triplet splitting of the π–π^* transition in monoolefins is about 24,000 cm^{-1}, that of the n–π^* transition in simple carbonyl compounds is only about 3000 cm^{-1}. This fact can be explained as follows: since the overlap of the n orbital (ψ_k) and the π^* orbital (ψ_r) is small, the charge distribution $e\psi_k\psi_r$ is small everywhere, and hence the MO exchange repulsion integral K_{kr} is small.

The calculated transition energy by the simple MO method can be associated with the mean of the energies of the S–S and S–T transitions. In a series of related compounds the singlet–triplet splitting of a given type transition diminishes progressively as the size of the conjugated system increases, and in most cases there is a high degree of proportionality between the singlet–triplet splitting and the overall transition energy. Therefore, in most cases, in a series of related compounds the transition energy calculated by the simple MO method can be correlated with the position of the observed S–S band.

10.3. Effects of Configuration Interaction

10.3.1. General

As has been emphasized, in the simple MO method the total electronic Hamiltonian is approximated as the sum of one-electron Hamiltonians. The antisymmetrized functions of electron configurations made up of the orthonormal set of orbitals that are eigenfunctions of the one-electron Hamiltonian are eigenfunctions of the approximate total electronic Hamiltonian. If the electronic repulsion terms are included in the Hamiltonian, the antisymmetrized configuration functions are no longer eigenfunctions of the Hamiltonian: there can be nonzero matrix elements of the Hamiltonian between different configuration functions. Better approximations to the true state functions are built as appropriate linear combinations of the configuration functions by solving the appropriate secular determinant. This method is called the method of configuration interaction (frequently abbreviated as CI).

The total electronic Hamiltonian [see expression (2.19) or (7.5)] does not involve the electronic spin coordinate, and it is invariant under all the symmetry operations of the point group to which the molecule belongs. Therefore, the matrix elements can be nonzero only between configurations of the same spin multiplicity and of the same symmetry species. Furthermore, since the Hamiltonian contains only one- and two-electron terms, if two configurations differ by three or more spin-orbitals, the matrix element between them is zero because of the orthogonality of the spin-orbitals.

In general, the larger the absolute value of the matrix element, and the smaller the energy separation, the stronger is the interaction between the two configurations. Since the energy separations between the ground and the excited configurations are generally comparatively large, the interactions between them are generally of small importance. Important interactions occur between singly-excited configurations. The interaction between two configurations of the same energy is usually very effective, and as a result of it, the degenerate pair splits into two (plus and minus) states of different energies. Such an interaction is called a first-order configuration interaction. The interaction between configurations of different energies is usually less effective. As a result of the interaction, the higher energy level is raised and the lower is lowered. The interaction between configurations of different energies is called a second-order configuration interaction.

Sometimes, the sequence of the energy levels of excited electron configurations in the simple MO approximation is reversed in part by introducing the configuration interactions. Usually, in a series of related compounds, the effect of the configuration interaction of a similar type diminishes progressively as the size of the conjugated system becomes larger.

10.3.2. Configuration Interaction in Butadiene and Polyenes

Butadiene is first taken as an example for illustrating the effect of the configuration interaction. The usual notation for electron configurations is used: the ground configuration is denoted by V_0, and singly-excited singlet and triplet configurations that are obtained from the ground configuration by promoting one electron from a bonding π orbital, say ψ_{+k}, to an antibonding π orbital, say ψ_{-r}, are denoted by V_1^{kr} and T_1^{kr}, respectively.

As mentioned already, configuration interaction can occur only between configurations of the same spin multiplicity and of the same symmetry. It is evident from Table 8.5 that in the case of butadiene configuration interaction will occur within each of the following sets: V_0, V_1^{12}, V_1^{21}; V_1^{11}, V_1^{22}; T_1^{12}, T_1^{21}; T_1^{11}, T_1^{22}. Since V_1^{12} and V_1^{21} are degenerate, they undergo a first-order configuration interaction and give rise to two states of different energies: a plus state, $2^{-1/2}(V_1^{12} + V_1^{21}) \equiv {}^+V_1^{12}$, and a minus state, $2^{-1/2}(V_1^{12} - V_1^{21}) \equiv {}^-V_1^{12}$. Similarly, T_1^{12} and T_1^{21} undergo a first-order configuration interaction.

In general, in alternant hydrocarbons, because of the pairing properties of the π orbitals, V_1^{kr} and V_1^{rk} have the same symmetry and the same energy even with the inclusion of electronic interactions, and hence they undergo a first-order configuration interaction and give rise to a plus and a minus state:

$$2^{-1/2}(V_1^{kr} + V_1^{rk}) \equiv {}^+V_1^{kr}; \tag{10.13}$$

$$2^{-1/2}(V_1^{kr} - V_1^{rk}) \equiv {}^-V_1^{kr}. \tag{10.14}$$

Similarly, T_1^{kr} and T_1^{rk} undergo a first-order configuration interaction:

$$2^{-1/2}(T_1^{kr} + T_1^{rk}) \equiv {}^+T_1^{kr}; \tag{10.15}$$

$$2^{-1/2}(T_1^{kr} - T_1^{rk}) \equiv {}^-T_1^{kr}. \tag{10.16}$$

The matrix element of the total π-electronic Hamiltonian between V_0 and V_1^{kr} and that between V_0 and V_1^{rk} have the same absolute value but are opposed in sign. Therefore, while minus states can interact with the ground configuration, plus states cannot interact with it. Furthermore, in alternant hydrocarbons, the electron densities $\psi_{+k}\psi_{-r}$ and $\psi_{+r}\psi_{-k}$ are identical. This means that the electric moments of the transitions from the ground configuration to the degenerate configurations V_1^{kr} and V_1^{rk} are equal and act in the same direction. It follows that in the transition to the minus state ${}^-V_1^{kr}$ these transition moments exactly cancel one another. In general, transitions from the ground state to minus states are forbidden.

As a result of the configuration interaction, the electronic states of butadiene can be expressed as follows:

$$\begin{array}{ll}
\text{Singlet states} & \text{Triplet states} \\
{}^+[V]_1^{22} = -{}^1bV_1^{11} + {}^1BV_1^{22}; & {}^+[T]_1^{22} = -{}^3bT_1^{11} + {}^3BT_1^{22}; \\
{}^-[V]_1^{12} = {}^1aV_0 + {}^1A\,{}^-V_1^{12}; & {}^-[T]_1^{12} = {}^-T_1^{12}; \\
{}^+[V]_1^{12} = {}^+V_1^{12}; & {}^+[T]_1^{12} = {}^+T_1^{12}; \\
{}^+[V]_1^{11} = {}^1BV_1^{11} + {}^1bV_1^{22}; & {}^+[T]_1^{11} = {}^3BT_1^{11} + {}^3bT_1^{22}. \\
{}^-[V]_0 = {}^1AV_0 - {}^1a\,{}^-V_1^{12}. &
\end{array} \tag{10.17}$$

The coefficients of configuration functions in these expressions are all positive, and in planar molecular forms the following relations exist between them:

$$^1B \gg {}^1b; \qquad {}^1A \gg {}^1a; \qquad {}^3B \gg {}^3b. \tag{10.18}$$

According to the actual calculation, which will be shown in a later chapter, the coefficients written in capital letters, ^1A, ^1B, and ^3B, are greater than 0.9.

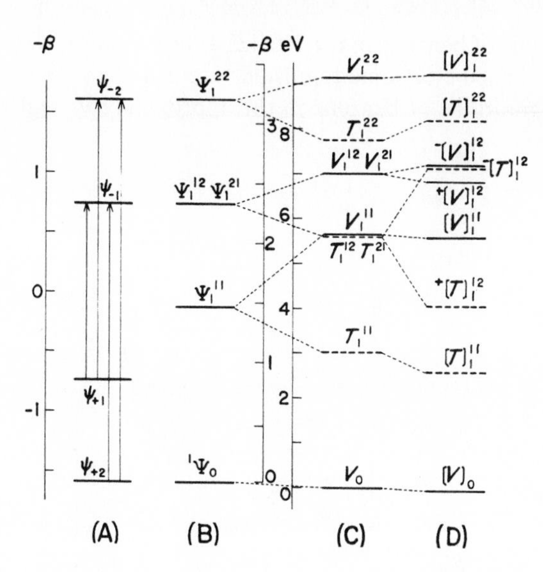

FIG. 10.1. The energy level diagram of the planar *s-trans* form of 1,3-butadiene. (A) the orbital energy levels in the simple LCAO MO approximation; (B) the term energy levels in the simple LCAO MO approximation; (C) the energy levels of the electron configurations calculated with inclusion of electronic interactions; (D) the energy levels of the electronic states calculated with inclusion of configuration interaction.

This means that, for example, the state $^-[V]_0$ has almost exclusively a character of V_0 and the state $^+[V]_1^{11}$ has almost exclusively a character of V_1^{11}. The energy level diagrams for the planar *s-trans* form of butadiene before and after inclusion of the configuration interaction are shown in Fig. 10.1.

It is noteworthy that the sequence of the energy levels remains unchanged after inclusion of the configuration interaction, except for the splitting of the degenerate level, and that the lowest excited singlet state is $^+[V]_1^{11}$, which has a predominant contribution of V_1^{11}. Thus it can be said that the lowest energy S–S transition has mainly a character of the one-electron transition from ψ_{+1}

to ψ_{-1} in the simple MO approximation. This is generally the case for conjugated linear polyenes. The longest wavelength intense S–S band of a polyene can be attributed to the one-electron transition from the highest bonding π orbital to the lowest antibonding π orbital in the simple MO approximation.

10.3.3. Configuration Interaction in Benzene, Naphthalene, and Other Aromatic Hydrocarbons

The π orbitals of benzene in the Hückel approximation were already shown in Table 3.5. When electronic repulsions are included, the degenerate lowest excited configurations of benzene in the approximation neglecting electronic

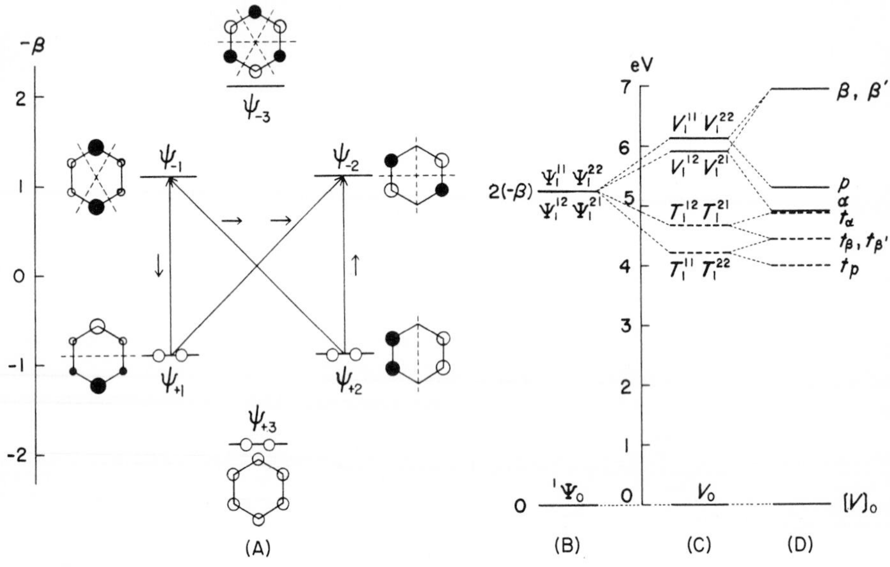

FIG. 10.2. The energy level diagram of benzene. (A) the orbital energy levels in the Hückel approximation. The short arrows indicate the direction of the transition moment. (B) the term energy levels in the Hückel approximation; (C) the energy levels of the electron configurations; (D) the energy levels of the electronic states after inclusion of the configuration interaction.

repulsions split into the following four pairs of degenerate configurations: V_1^{11}, V_1^{22}; V_1^{12}, V_1^{21}; T_1^{11}, T_1^{22}; T_1^{12}, T_1^{21}. Configuration interaction occurs within each pair and gives rise to the state functions shown in Table 3.7. In Fig. 10.2 the energy level diagrams before and after inclusion of the configuration interaction are shown. The β' state ($^+[V]_1^{22}$) and the β state ($^+[V]_1^{12}$) form a degenerate pair, which belongs to symmetry species E_{1u} of point group D_{6h}. The transition from the ground state to this degenerate state is allowed. The p state $({}^{+}[V]_1^{11})$ belongs to B_{1u}, and the α state $({}^{-}[V]_1^{12})$ belongs to B_{2u}. The transitions from the ground state to these states are forbidden.

Clar[1] has classified the absorption bands of benzenoid aromatic hydrocarbons into three types, mainly by their intensity and by their vibrational structure. Bands of type 1 are weak bands ($\epsilon = 10^2$–10^3, $f = 10^{-3}$–10^{-2}) often possessing a complicated vibrational structure. The first band of the spectrum

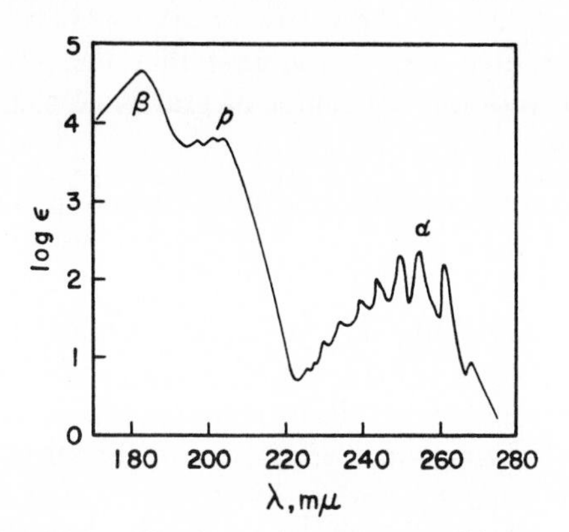

FIG. 10.3. The absorption spectrum of benzene.

is often of this type. If such a band appears to be absent, one assumes that it underlies more intense bands which are present. Bands of type 2 are moderately intense bands ($\epsilon \approx 10^4$, $f \approx 10^{-1}$) usually showing a very regular vibrational structure. A band of this type is either the first or the second band in the spectrum, depending on whether the band of type 1 is seen or not. Bands of type 3 are intense bands ($\epsilon \approx 10^5$, $f \approx 1$) which have rather less vibrational structure than either of the previous bands. Clar has designated the three types of band 1, 2, and 3 by the symbols α, p, and β, respectively, distinguishing different bands of the same type by primed superscripts.

The absorption spectrum of benzene is shown in Fig. 10.3. The weak band near 250 mμ ($\epsilon_{\max} = 200$, $f = 0.001$), the moderately intense band near 200 mμ ($\epsilon_{\max} = 7400$, $f = 0.04$), and the intense band near 180 mμ ($\epsilon_{\max} = 46{,}000$, $f = 0.88$) are α, p, and β bands, respectively, and are attributed to the transitions from the ground state (${}^1A_{1g}$) to the foregoing excited singlet states, ${}^1B_{2u}$, ${}^1B_{1u}$, and ${}^1E_{1u}$, respectively. Benzene exhibits also a very weak band near 325 mμ. This band is attributed to the first S–T transition, that is, the

[1] E. Clar, "Aromatische Kohlenwasserstoffe." Springer Verlag, Berlin, 1952.

transition from the ground state to the lowest excited triplet state, $^{+}[T]_1^{11}$ (t_p or $^3B_{1u}$ state).

When only S–S transitions are taken into consideration, the transition energy of the lowest energy degenerate transitions in the simple MO approximation, $2(-\beta)$ or $2.133\,(-\gamma)$, may be considered to correspond to the mean of the energies of the foregoing three S–S transitions. This mean of the transition energies is called the center of gravity of singlets.[2] The wave numbers of the 0–0 components of the α, p, and β bands of benzene in solution in saturated hydrocarbons are 38,200, 48,000, and 53,000 cm^{-1}, respectively.[3] Accordingly, by considering that the β band is due to a doubly degenerate transition, the center of gravity of singlets of benzene is calculated as follows:

$$\text{c.g. of singlets of benzene} = (38{,}200 + 48{,}000 + 2 \times 53{,}000)/4 \doteqdot 48{,}000 \text{ cm}^{-1}. \tag{10.19}$$

The two lowest antibonding and two highest bonding π orbitals of naphthalene in the Hückel approximation were shown in expression (8.30), and characteristics of the four excited electron configurations resulting from one-electron transitions between these orbitals were shown in Table 8.8. As is generally the case, V_1^{12} and V_1^{21} undergo first-order configuration interaction, and give rise to the plus state (β state) and the minus state (α state):

$$^{+}V_1^{12} \doteqdot {}^{+}[V]_1^{12} \equiv \Phi_\beta; \tag{10.20}$$

$$^{-}V_1^{12} \doteqdot {}^{-}[V]_1^{12} \equiv \Phi_\alpha. \tag{10.21}$$

The plus state is higher in energy than the minus state. The transition from the ground state to the plus state is allowed, having the nonzero transition moment in the direction of the long molecular axis (x axis). On the other hand, the transition to the minus state is forbidden. V_1^{11} and V_1^{22} have different energies, in contrast with the corresponding electron configurations of benzene. Therefore, they undergo second-order configuration interaction, and give rise to two states, β' and p states:

$$-bV_1^{11} + BV_1^{22} \doteqdot {}^{+}[V]_1^{22} \equiv \Phi_{\beta'}; \tag{10.22}$$

$$+BV_1^{11} + bV_1^{22} \doteqdot {}^{+}[V]_1^{11} \equiv \Phi_p. \tag{10.23}$$

Both the coefficients B and b are positive, and B is larger than b. The β' state, $^{+}[V]_1^{22}$, which has a predominant contribution of V_1^{22}, is higher in energy than

[2] J. R. Platt, *J. Chem. Phys.* **18,** 1168 (1950).
[3] H. B. Klevens and J. R. Platt, *J. Chem. Phys.* **17,** 470 (1949).

the p state, ${}^{+}[V]_1^{11}$, which has a predominant contribution of V_1^{11}. In contrast with the case of benzene, the moments of the transitions from V_0 to V_1^{11} and V_1^{22} do not cancel each other in the transition from the ground state to the p state. Both the transitions from the ground state to the β' and p states are allowed, having nonzero transition moments in the direction of the short molecular axis (y axis).

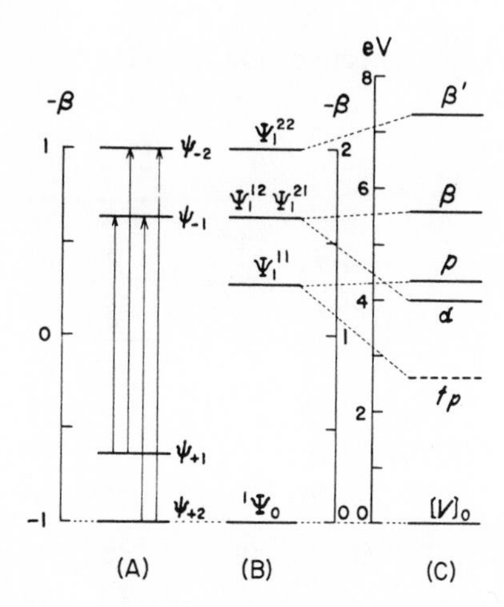

FIG. 10.4. The energy level diagram of naphthalene. (A) the orbital energy levels in the Hückel approximation; (B) the term energy levels in the Hückel approximation; (C) the energy levels of the electronic states.

Energy levels of naphthalene are shown schematically in Fig. 10.4. It is noteworthy that in contrast with the case of butadiene, in the case of naphthalene, the state ${}^{-}[V]_1^{12}$ (α state) is lower in energy than the state ${}^{+}[V]_1^{11}$. It is also noteworthy that the β and β' states of naphthalene have different energies, while the corresponding states of benzene are degenerate. Naphthalene has three main band systems in the near ultraviolet region. The longest wavelength band, which appears near 310 mμ, has the lowest intensity ($\epsilon_{max} = 280$, $f = 0.002$), and is the α band, which is attributed to the forbidden transition from the ground state to ${}^{-}[V]_1^{12}$. The shorter wavelength bands have successively higher intensity, as in the case of benzene. The moderately intense band near 290 mμ ($\epsilon_{max} = 9300$, $f =$ ca. 0.2) and the very intense band at about 220 mμ ($\epsilon_{max} = 133{,}000$, $f = 1.70$) are the p and the β band, respectively, which are attributed to the allowed transitions to ${}^{+}[V]_1^{11}$ and ${}^{+}[V]_1^{12}$, respectively. The band due to the allowed transition to ${}^{+}[V]_1^{22}$, the β' band, is considered to occur in the far ultraviolet region.

The spectra of most benzenoid hydrocarbons can be satisfactorily interpreted in a similar manner.[4] The reversal of the sequence of the energy levels of the states ${}^{-}[V]_1^{12}$ and ${}^{+}[V]_1^{11}$, which occurs in naphthalene, also occurs in most polyphenes and *peri*-condensed benzenoid hydrocarbons: in these compounds, just as in benzene and naphthalene, the α band is at longer wavelength than the p band. This reversal is considered to be due to the fact that in the "round" π systems the first-order configuration interaction between V_1^{12} and V_1^{21} is highly effective. On the other hand, in polyacenes higher

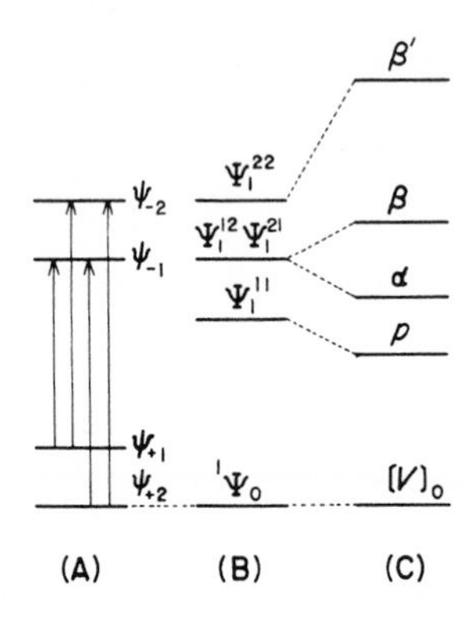

FIG. 10.5. The energy level diagram of a polyacene. (A) the orbital energy levels; (B) the term energy levels; (C) the energy levels of the electronic states.

than anthracene, as in butadiene and conjugated linear polyenes, the p state, ${}^{+}[V]_1^{11}$, is lower than the α state, ${}^{-}[V]_1^{12}$, that is, the p band is the first S–S band (see Fig. 10.5). In these "long" π systems, the first-order configuration interaction between V_1^{12} and V_1^{21} is not so effective, and V_1^{11} is much lower than V_1^{12} and V_1^{21}.

10.3.4. Configuration Interaction in Azulene

The two lowest antibonding and two highest bonding π orbitals of azulene in the Hückel approximation were shown in Table 8.9. Since azulene is a nonalternant hydrocarbon, the π orbitals are not paired, and hence electron configurations V_1^{kr} and V_1^{rk} are not degenerate. However, since V_1^{12} and V_1^{21} are nearly degenerate and belong to the same symmetry species (A_1), they undergo a quasi-first-order configuration interaction, and give rise to two states of considerably different energies. On the analogy of the case of

[4] M. J. S. Dewar and H. C. Longuet-Higgins, *Proc. Phys. Soc.* (*London*) **A67,** 795 (1954).

alternant hydrocarbons, the higher of these states is called the β state, and the lower the α state. V_1^{11} and V_1^{22} belong to the same symmetry species (B_2). Therefore, they undergo a second-order configuration interaction, and give rise to two states. The lower state (p state) has a predominant contribution of V_1^{11}, and the higher state (β' state) has a predominant contribution of V_1^{22}: even when interaction of all singly-excited configurations is taken into account, the coefficient of V_1^{11} in the p state function, as well as the coefficient of V_1^{22} in the β' state function, is larger than 0.9.[5]

The calculation by Pariser[5] shows that the sequence of the energy levels is as follows: $\beta > \beta' > \alpha > p$. Transitions from the ground state, $^1A_1$, to other $^1A_1$ states, for example, the α and the β state, have the transition moment in the direction of the symmetry axis of the molecule, the x axis, and transitions to $^1B_2$ states, for example, the p and the β' state, have the transition moment in the direction of the y axis (see Fig. 8.5). The magnitude of the transition moment is very small for the transitions to the p and the α state, is small for the transition to the β' state, and is large for the transition to the β state.

The weak band near 690 mμ ($\epsilon_{\max} = 300$, $f = 0.009$), the moderately intense band near 340 mμ ($\log \epsilon_{\max} = 4.0$, $f = 0.08$), and the intense band near 270 mμ ($\log \epsilon_{\max} = 5.1$, $f = 1.10$) in the azulene spectrum are assigned as p, α, and β bands, respectively. The first band, the p band, is responsible for the purple color of azulene. The bands which are assigned here as the p and the α band are the bands which were called the α and the p band, respectively, from the empirical standpoint. The β' band is considered to be at about 310 mμ and to be submerged under the neighboring band systems. The calculation starting with the Hückel orbitals and including configuration interaction leads to numerical results in good agreement with the experimental spectral data both with respect to energy and intensity of transition.

10.4. Correlation between Results of Simple MO Calculation and Spectral Data in Conjugated Linear Polyenes

10.4.1. Simple MO Calculation on Conjugated Linear Polyenes

We have seen that the simple MO method, based on a one-electron Hamiltonian, fails to explain the general features of the spectra. However, as mentioned previously, the simple MO method is very useful for correlating the position and intensity of certain types of band in a restricted class of

[5] R. Pariser, *J. Chem. Phys.* **25**, 1112 (1956).

compound. Hereafter some examples of the correlation between results of simple MO calculation and experimental spectral data are given.

The conjugated linear polyene system is taken as the first example. The secular determinant for determination of the Hückel MO's of a conjugated linear polyene can be easily written: all the diagonal elements, a_{ii}'s, are w, all the off-diagonal elements at the both sides of the diagonal elements, $a_{i,i-1}$'s and $a_{i,i+1}$'s, are 1, and all the other off-diagonal elements are zero. The secular determinant for a conjugated linear polyene consisting of n carbon π centers is denoted by D_n. When expanding D_n with respect to the first row, we obtain

$$D_n = wD_{n-1} - D_{n-2}. \tag{10.24}$$

In this manner, the expanded form of the secular determinant, that is, the secular polynomial, can be easily obtained.

When n is even, only terms involving w with even exponents and a simple numerical term of $+1$ or -1 occur in the secular polynomial. Therefore, in this case, all the roots for w are symmetrically disposed about the zero value, that is, all the roots occur in $n/2$ pairs (compare Theorem 1, Section 7.2). On the other hand, when n is odd, only terms involving w with odd exponents occur in the secular polynomial. In this case, there is necessarily a root of zero, and the other roots occur in $(n - 1)/2$ pairs (see Theorem 3, Section 7.2).

For the continuous chain system, the energies of the Hückel MO's, w, and the coefficients of the carbon $2p\pi$ AO's in the Hückel MO's, c, can be written in the following analytic forms[6]:

$$w_r = -2 \cos [r\pi/(n + 1)]: \qquad r = 1, 2, 3, \ldots, n; \tag{10.25}$$

$$c_{rj} = [2/(n + 1)]^{-1/2} \sin [jr\pi/(n + 1)]: \qquad j = 1, 2, 3, \ldots, n, \tag{10.26}$$

where r is an index specifying MO's. The carbon π centers are numbered as $1, 2, 3, \ldots, n$ from one end of the carbon chain, and c_{rj} is the coefficient of the $2p\pi$ AO at the jth carbon atom, χ_j, in the rth MO, ψ_r. The MO index r takes the increasing integer from 1 for the lowest MO to n for the highest MO in the order of increasing energy. This notation of MO's will be referred to as the one-directional notation, thus being distinguished from the two-directional notation, which is mainly used in this book. Evidently, in the case where n is even, the values of r for the lowest antibonding MO (ψ_{-1}) and the highest bonding MO (ψ_{+1}) are $n/2 + 1$ and $n/2$, respectively, and, in the case where n is odd, the values of r for the lowest antibonding MO (ψ_{-1}), the

[6] C. A. Coulson, *Proc. Roy. Soc.* (*London*) **A169**, 413 (1939).

nonbonding MO (ψ_0), and the highest bonding MO (ψ_{+1}) are $(n + 1)/2 + 1$, $(n + 1)/2$, and $(n + 1)/2 - 1$, respectively.

A π MO specified by the index r has $(r - 1)$ nodal points along the conjugated chain, in addition to the nodal plane common to all the π MO's, which coincides with the molecular plane. In continuous chain systems consisting of even π centers the nodel points occur only between atoms. For example, the highest bonding MO has $(n/2 - 1)$ nodal points at alternate bonds, that is, at the essential single bonds. This means that the partial π-bond orders of the "single" bonds for this MO are negative and those of the "double" bonds are positive. On the other hand, the lowest antibonding MO has $n/2$ nodal points at alternate bonds, that is, at the essential double bonds. For this orbital, the partial π-bond orders of the "single" bonds are positive, and those of the "double" bonds are negative (see Theorem 7, Section 7.2). In continuous chain systems consisting of odd π centers, the nodal points in some of the MO's coincide with nuclear positions. In the nonbonding MO, for which r is $(n + 1)/2$, all the nodal points are at all the even-numbered atoms, and hence the partial π-bond orders of all the bonds for this orbital are zero.

As in the case of butadiene (see Section 8.1), in the approximation neglecting interactions between nonneighboring AO's, a conjugated linear polyene can be treated in general as belonging to point group C_2 (or C_{2h}), by assuming that all the carbon π centers are aligned on a straight line. Of course, the midpoint of the straight chain is the center of symmetry, and the axis that is through this point and parallel to the symmetry axes of the carbon $2p\pi$ AO's is taken as the principal rotational axis, the z axis. Then, if the π MO's are arranged in the order of energy, alternate MO's belong to different symmetry species. Thus, the lowest MO ($r = 1$) and the other MO's with odd values of r belong to symmetry species A of C_2 (or A_u of C_{2h}), and the MO's with even values of r belong to symmetry species B of C_2 (or B_g of C_{2h}).

10.4.2. Correlation between the Observed Frequency of the First S–S Band and the Calculated Transition Energy

In a conjugated linear polyene, V_1^{11}, V_1^{22}, etc., belong to the same symmetry species (for the abstract straight model, B of C_2, or B_u of C_{2h}), and V_1^{12}, V_1^{21}, etc., belong to the other same symmetry species (for the abstract straight model, A of C_2, or A_g of C_{2h}). Configuration interaction occurs within a set of configurations belonging to the same symmetry species. However, as in butadiene (see Subsection 10.3.2), generally, in conjugated polyenes, the sequence of energy levels of configurations is not reversed by the configuration interaction, and the lowest singlet excited state has a predominant contribution of V_1^{11}. Therefore, the first or longest wavelength S–S band of a conjugated polyene can be considered as being due to the one-electron transition from ψ_{+1} to ψ_{-1}, which is allowed by symmetry and has the transition moment directed along the long axis of the conjugated chain.

TABLE 10.1

CORRELATION BETWEEN THE OBSERVED WAVE NUMBER OF THE FIRST S–S BAND (THE A BAND) AND THE CORRESPONDING HMO ENERGY DIFFERENCE Δw_{11} IN H—(CH=CH)$_{n/2}$—H[a-d]

$n/2$	λ_{max}, mμ	ϵ_{max}	ν_{max}, cm^{-1}	Δw_{11}	$\nu_{max}/\Delta w_{11}$
1	162.5	—	61,500	2.000	30,750
2	217	21,000	46,080	1.236	37,280
3	268	34,600	37,310	0.890	41,920
4	304	—	32,900	0.695	47,330
5	334	121,000	29,940	0.569	52,620
6	364	138,000	27,470	0.482	57,000
7	390	—	25,640	0.418	61,340
8	410	—	24,390	0.370	65,920
9	—	—	—	0.330	—
10	447	—	22,370	0.299	74,820

[a] The spectral data for $n/2 = 1$: J. R. Platt, H. B. Klevens, and W. C. Price, *J. Chem. Phys.* **17**, 466 (1949).

[b] The spectral data for $n/2 = 2$: in hexane; W. C. Price and A. D. Walsh, *Proc. Roy. Soc. (London)* **A174**, 220 (1940).

[c] The spectral data for $n/2 = 4$: in cyclohexane; G. F. Woods and L. H. Schwartzman, *J. Am. Chem. Soc.* **71**, 1396 (1949).

[d] The other spectral data: in 2,2,4-trimethylpentane; F. Sondheimer, D. A. Ben-Efraim, and R. Wolovsky, *J. Am. Chem. Soc.* **83**, 1675 (1961).

Since the values of the index r for ψ_{-1} and ψ_{+1} of a conjugated chain with an even value of n are $n/2 + 1$ and $n/2$, respectively, when the analytic expression for the orbital energy, Eq. (10.25), is used, the energy difference between ψ_{-1} and ψ_{+1}, Δw_{11}, in units of $-\beta$ can be expressed as follows:

$$\Delta w_{11} = 4 \sin [\pi/2(n + 1)]. \tag{10.27}$$

It is evident that the value of Δw_{11} becomes smaller with the increasing n. As will be verified later, the magnitude of the transition moment of the one-electron transition from ψ_{+1} to ψ_{-1} becomes larger as the length of the polyene chain becomes longer.

In Table 10.1 the data on the first S–S bands of a series of conjugated linear polyenes, H—(CH=CH)$_{n/2}$—H, are listed, together with the corresponding Δw_{11} values. Although the compounds listed in this table may be not stereochemically pure, the *all-trans* configuration probably predominates. Evidently, both λ_{max} and ϵ_{max} increase with the increasing length of the conjugated polyene chain. It can be said that a nearly linear relationship exists

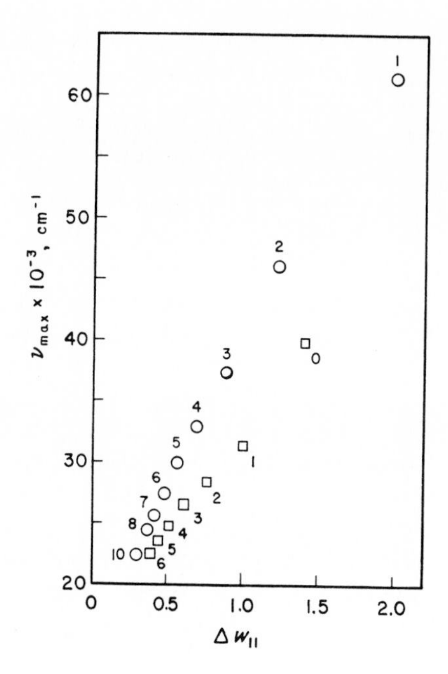

FIG. 10.6. Correlation of experimental values of ν_{max} of the first S–S bands (the A bands) of conjugated polyenes with values of Δw_{11} calculated by the Hückel method. Open circle, points for H—(CH=CH)$_{n/2}$—H, based on the data in Table 10.1. Square, points for C_6H_5—(CH=CH)$_{n/2}$—C_6H_5, based on the data in Table 10.2. The numerals denote $n/2$.

between λ^2_{max} and n as well as between ϵ_{max} and n. It is known that similar linear relationships exist in related series, such as the α-phenylpolyene, the α,ω-diphenylpolyene, and the α,α,ω-triphenylpolyene series.[7] The plot of the values of ν_{max} of the first S–S bands of the series of conjugated polyenes listed in Table 10.1 against the corresponding values of Δw_{11} shows the existence of a nearly linear relationship (compare Fig. 10.6). In Table 10.2 the

[7] H. M. Walborsky and J. F. Pendleton, *J. Am. Chem. Soc.* **82**, 1405 (1960).

data on the first S–S bands of a series of conjugated polyenes terminated by phenyl groups, $C_6H_5—(CH═CH)_{n/2}—C_6H_5$, and the corresponding values of Δw_{11} calculated by the Hückel method, are listed. Also in this series, there exists a nearly linear relationship between ν_{max} and Δw_{11} (see Fig. 10.6). The existence of such an excellent linear correlation between ν_{max} and Δw_{11}, that is, between the observed and the calculated transition energy, is rather

TABLE 10.2

CORRELATION BETWEEN THE OBSERVED WAVE NUMBER OF THE FIRST S–S BAND (THE A BAND) AND THE CORRESPONDING HMO ENERGY DIFFERENCE Δw_{11} in $C_6H_5—(CH═CH)_{n/2}—C_6H_5$[a]

$n/2$	λ_{max}, mμ	$\epsilon_{max} \times 10^{-3}$	ν_{max}, cm^{-1}	Δw_{11}	$\nu_{max}/\Delta w_{11}$
0	251.5	18	39,760	1.408	28,240
1	319	24	31,350	1.011	31,010
2	352	40	28,410	0.772	36,800
3	377	75	26,530	0.620	42,790
4	404	86	24,750	0.520	47,600
5	424	94	23,580	0.444	53,110
6	445	113	22,470	0.392	57,320

[a] The spectral data are for benzene solutions; K. W. Hausser, R. Kuhn, and A. Smakula, *Z. physik. Chem.* **B29**, 384 (1935).

surprising, in consideration of many approximations and assumptions involved in the simple MO method. This excellent correlation must mean that, as mentioned already, there is a high degree of proportionality within a series of related conjugated systems between the effect of configuration interaction and singlet–triplet splitting and the overall transition energy.

As is seen in Fig. 10.6, both the nearly linear correlation lines for the polyene and the diphenylpolyene series do not encounter the origin. This means that there is no proportionality between the calculated and the observed transition energy. In other words, this means that the value of $-\beta$ varies regularly from compound to compound in either series. In the last column of Table 10.1, as well as that of Table 10.2, the values of $-\beta$ (that is, the values of $\nu_{max}/\Delta w_{11}$) for the individual compounds are listed. These values show that the value of $-\beta$ increases regularly with the elongation of the conjugated system. In general, the effect of configuration interaction diminishes as the length of the conjugated system increases. That is, the extent of the lowering of the lowest singlet excited state by the configuration interaction becomes relatively smaller as the length of the conjugated system increases. Thus the regular increase in the value of $-\beta$ associated with the increase in the length of the conjugated system can be considered to reflect the regular decrease in the importance of the effect of electronic repulsions associated with the increase in the length of the conjugated system.

It can be said that if a constant value of $-\beta$ is to be used, the transition energy calculated by the Hückel method decreases too rapidly as the length of the conjugated polyene chain increases. Since for small angles $\sin\theta$ can be approximated as θ, the Hückel theory predicts that for a long polyene the λ_{max} of the first band will be proportional to n [see Eq. (10.27)], and will increase to infinity as the length of the conjugated chain increases to infinity. However, in fact, as mentioned already, the λ_{max} increases in a rough proportion rather to the square root of n, and appears to approach an asymptotic value.

This fault of the Hückel method can be mended by allowing for an alternation of bond lengths of essential single and essential double bonds in the framework of the simple LCAO MO method. It has been shown on theoretical grounds that an alternation of bond lengths, long and short, of essential single and essential double bonds will always occur in a conjugated polyene chain even if the chain is long enough.[8-10] In the Hückel approximation the π orbital energies of the infinite polyene are continuous. A bond length alternation opens up a gap between the antibonding and the bonding π orbitals, so that the energy of the first transition approaches a nonzero limit as the length of the conjugated system becomes infinite. The bond alternation can be allowed for by assigning different resonance integrals to the essential single and double bonds, β_s and β_d, respectively ($\beta_s/\beta_d < 1$).[11] It has been found that the best fit of the calculated transition energy with the observed one is obtained for the first S–S band of a series of α,ω-dimethylpolyenes, CH_3—$(CH{=}CH)_{n/2}$—CH_3, by using the values $-\beta_d = 4.00$ eV and $\beta_s/\beta_d = 0.72$.[12] By using these values the limiting wavelength for the infinite polyene is calculated to be 553 mμ.

In the cumulenes, $R_2(C{=})_nR_2$, there is no bond length alternation, since all the C—C bonds in the conjugated chain are essential double bonds. It is of interest that the λ_{max} of the first S–S band of the cumulenes is proportional to n, as is predicted by the Hückel theory.[13]

10.4.3. Intensity of Absorption Bands of Conjugated Linear Polyenes

Conjugated linear polyenes can assume various geometries. In Table 10.3 are given the values of the dipole strength, M^2, and its x and y components, $M^2_{(x)}$ and $M^2_{(y)}$, calculated by the Hückel method for the one-electron transitions $\psi_{+1} \rightarrow \psi_{-1}$ and $\psi_{+1} \rightarrow \psi_{-2}$ in planar geometries of some lower members of the conjugated linear polyene series. The calculation

[8] H. Labhart, *J. Chem. Phys.* **27,** 957 (1957).
[9] Y. Ooshika, *J. Phys. Soc. Japan* **12,** 1238, 1246 (1957).
[10] H. C. Longuet-Higgins and L. Salem, *Proc. Roy. Soc.* (*London*) **A251,** 172 (1959).
[11] M. J. S. Dewar, *J. Chem. Soc.* **1952,** 3544.
[12] J. N. Murrell, "The Theory of the Electronic Spectra of Organic Molecules," p. 72. Methuen, London, 1963.
[13] H. Krauch, *J. Chem. Phys.* **28,** 898 (1958).

has been carried out by assuming that all the valence angles are 120° and that all the C—C bonds have an equal bond length R. The values of M^2, $M^2_{(x)}$, and $M^2_{(y)}$ are expressed in units of R^2. Except for the 3-*cis* configuration of the polyene with $n = 8$, the midpoint of the central bond of each polyene has been taken as the origin of the coordinate system, the direction of that bond

TABLE 10.3

THE ENERGIES (Δw) AND DIPOLE STRENGTHS (M^2, IN UNITS OF R^2) OF ONE-ELECTRON TRANSITIONS IN CONJUGATED LINEAR POLYENES H—(CH═CH)$_{n/2}$—H CALCULATED BY THE HÜCKEL METHOD[a]

		$\psi_{+1} \rightarrow \psi_{-1}$ transition				$\psi_{+1} \rightarrow \psi_{-2}$ (as well as $\psi_{+2} \rightarrow \psi_{-1}$) transition			
n	Form	Δw	$M^2_{(x)}$	$M^2_{(y)}$	M^2	Δw	$M^2_{(x)}$	$M^2_{(y)}$	M^2
2		2.000	0	0.500	0.500	—	—	—	—
4	*s-trans*	1.236	0.783	0.682	1.465	2.236	0	0	0
	s-cis	1.236	0	0.682	0.682	2.236	0.300	0	0.300
6	*all-trans*	0.890	0.285	2.661	2.946	1.692	0	0	0
	3-*cis*	0.890	0	2.661	2.661	1.692	0.056	0	0.056
8	*all-trans*	0.695	1.961	2.953	4.914	1.347	0	0	0
	3-*cis*	0.695	0.105	4.293	4.398	1.347	0.092	0	0.092
10	*all-trans*	0.569	1.082	6.295	7.377	1.115	0	0	0
	3-*cis*	0.569	0.409	6.295	6.704	1.115	0.094	0	0.094
	5-*cis*	0.569	0	6.295	6.295	1.115	0.238	0	0.238
	3,5,7-*cis*	0.569	0	6.295	6.295	1.115	0.015	0	0.015

[a] R: the length of each C—C bond.

as the direction of the y axis, and the molecular plane as the xy plane. All π–π^* transitions in planar molecules have no z component of the transition moment. The configuration *all-trans* refers to the form with *trans* configuration around all the C—C bonds. The configuration 3-*cis*, for example, refers to the form with *cis* configuration around the 3—4 bond only, and *trans* configuration around all the other bonds. For the 3-*cis* configuration of the polyene with 8 carbon π centers, the direction of the 3—4 bond has been taken as the direction of the y axis. The dipole strength of the transition $\psi_{+2} \rightarrow \psi_{-1}$ is invariably equal to that of the transition $\psi_{+1} \rightarrow \psi_{-2}$.

The one-electron transition from ψ_{+1} to ψ_{-1} has the transition moment approximately directed along the long axis of the conjugated system. Therefore, the transition moment of this transition is the largest in the most elongated form, that is, the *all-trans* configuration of a polyene, and becomes larger as the length of the conjugated polyene is increased by adding another C—C double bond. This trend has been actually observed with the first S–S bands of carotenoids.[14]

In Table 10.4 the data on the first S–S bands of *all-trans-α,ω*-dimethylpolyenes, $CH_3—(CH{=}CH)_{n/2}—CH_3$, are shown, together with some relevant data. The f_{theor} represents the value of the oscillator strength calculated from

TABLE 10.4

COMPARISON OF THE EXPERIMENTAL AND THEORETICAL VALUES OF THE OSCILLATOR STRENGTH OF THE FIRST S–S BAND (THE A BAND) IN THE SERIES OF *all-trans-α,ω*-DIMETHYLPOLYENES $CH_3—(CH{=}CH)_{n/2}—CH_3$

$n/2$	Solvent	λ_{max}, mμ	ν_{max}, cm^{-1}	f_{exp}	f_{theor}	f_{theor}/f_{exp}
2	Methanol[a]	226	44,250	0.32[c]	1.378	4.31
3	Methanol[a]	274	36,500	0.98	2.287	2.33
	Chloroform[b]	279	35,840	0.74	2.246	3.04
4	Methanol[a]	314	31,850	1.11	3.328	3.00
	Chloroform[b]	316.5	31,600	1.11	3.303	2.98
5	Chloroform[b]	349	28,650	1.66	4.494	2.70

[a] F. Bohlmann and H. J. Mannhardt, *Chem. Ber.* **89,** 1307 (1956).
[b] P. Nayler and M. C. Whiting, *J. Chem. Soc.* **1955,** 3037.
[c] This value seems to be too low. The f value of the A band (λ_{max} = 223 mμ) of *trans*-piperylene $CH_2{=}CH—CH{=}CH—CH_3$ in heptane has been reported to be 0.61 ± 0.03 (L. E. Jacobs and J. R. Platt, *J. Chem. Phys.* **16,** 1137 (1948)).

the value of M^2 of the transition $\psi_{+1} \rightarrow \psi_{-1}$ in the *all-trans* form listed in Table 10.3 and the experimental value of ν_{max} by using a value of 1.40 Å for R. It may be said that a fairly good correlation exists between the observed oscillator strength, f_{exp}, and the calculated one, f_{theor}. It is to be noted that the value of f_{theor} is larger than the corresponding value of f_{exp} by a factor of about 2–3. This is a general tendency in the simple MO calculation of the oscillator strength of the first S–S transition, and is inevitable when a single electron configuration is used without inclusion of configuration interaction.[15,16] In some cases the ratio of the calculated to the observed oscillator strength, f_{theor}/f_{exp}, amounts to as much as 5.

Planar *all-trans* configurations of conjugated linear polyenes belong to point group C_{2h}, having the center of symmetry. When the number of

[14] L. Zechmeister, *Chem. Rev.* **34,** 267 (1944).
[15] R. S. Mulliken and C. A. Rieke, *Rept. Progr. in Phys.* **8,** 231 (1941).
[16] C. A. Coulson and J. Jacobs, *Proc. Roy. Soc.* (*London*) **A206,** 287 (1951).

essential double bonds, $n/2$, is odd, ψ_{+1}, ψ_{-2}, etc., belong to representation A_u, and ψ_{+2}, ψ_{-1}, etc., belong to B_g. On the other hand, when $n/2$ is even, ψ_{+1}, ψ_{-2}, etc., belong to B_g, and ψ_{+2}, ψ_{-1}, etc., belong to A_u. In either case, the transitions $\psi_{+1} \rightarrow \psi_{-2}$ and $\psi_{+2} \rightarrow \psi_{-1}$ are symmetry-forbidden. When one or more bonds become *cis*, the center of symmetry is destroyed, and these transitions become allowed (see Table 10.3). This means that, when one or more bonds become *cis*, the transition from the ground state to the plus state, ${}^{+}V_1^{12}$ or ${}^{+}[V]_1^{12}$, becomes allowed.

Zechmeister and Polgár[17] have observed in their study of the spectra of carotenoids that the first S–S band has the highest intensity in the *all-trans* isomer and decreases in intensity when one or more bonds become *cis*, and that a shorter wavelength band which is absent from the spectrum of the *all-trans* isomer appears in the spectra of the *cis* isomers. This second band, which is called the *cis* band, is considered to be due to the transition to the plus state ${}^{+}[V]_1^{12}$. The intensity of the *cis* band depends on the location of the *cis* bond in the conjugated polyene chain: the more centrally located the *cis* bond, the more intense the *cis* band.

10.4.4. π-Bond Orders of Conjugated Linear Polyenes

It seems to be convenient to refer in this place to the π-bond orders of conjugated linear polyenes, for the sake of reference in a later chapter. In Tables 10.5 and 10.6 are shown the values of *RE*, P_{ij}, ΔP_{ij}, and Δw calculated by the Hückel method for some lower members of a series of the conjugated linear systems consisting of n carbon π centers. *RE* represents the resonance energy (in units of $-\beta$) of the ground electron configuration, and P_{ij} represents the π-bond order of the i—j bond in the ground electron configuration. ΔP_{ij} represents the change in the π-bond order of the i—j bond associated with the lowest energy excitation, that is, the one-electron transition from ψ_{+1} to ψ_{-1} in the case of systems with even n values, and the one-electron transition from ψ_0 to ψ_{-1} as well as that from ψ_{+1} to ψ_0 in the case of systems with odd n values. Δw represents the energy of that transition, that is, Δw_{11} in the former case, and Δw_{01}, as well as Δw_{10}, in the latter case. As mentioned already (compare Section 7.2), the following relations exist: $p_{ij}^{(+1)} = -p_{ij}^{(-1)}$, $p_{ij}^{(0)} = 0$, $w_{+1} = -w_{-1}$, and $w_0 = 0$. Therefore, for systems where n is even, $\Delta P_{ij} = p_{ij}^{(-1)} - p_{ij}^{(+1)} = 2p_{ij}^{(-1)}$, and $\Delta w = \Delta w_{11} = w_{-1} - w_{+1} = 2w_{-1}$; and for systems where n is odd, $\Delta P_{ij} = p_{ij}^{(-1)} - p_{ij}^{(0)} = p_{ij}^{(0)} - p_{ij}^{(+1)} = p_{ij}^{(-1)}$, and $\Delta w = w_{-1} - w_0 = w_0 - w_{+1} = w_{-1}$.

[17] L. Zechmeister and A. Polgár, *J. Am. Chem. Soc.* **65**, 1522 (1943).

In systems where n is even, the essential double bonds have large P values, and the essential single bonds have small P values. The extent of alternation of π-bond order decreases gradually as the value of n increases. The values of ΔP for the essential double bonds are negative, and those for the essential single bonds are positive. This suggests that deviation of the geometry of the

TABLE 10.5

RESULTS OF THE CALCULATION BY THE HÜCKEL METHOD ON CONJUGATED LINEAR SYSTEMS CONSISTING OF n CARBON π CENTERS, IN WHICH n IS EVEN[a]

n		Bond	1–2	2–3	3–4	4–5	5–6
2	$RE = 0$	P	1.000	—	—	—	—
	$\Delta w = 2$	ΔP	−1.000	—	—	—	—
4	$RE = 0.472$	P	0.896	0.446	—	—	—
	$\Delta w = 1.236$	ΔP	−0.448	+0.276	—	—	—
6	$RE = 0.988$	P	0.871	0.483	0.785	—	—
	$\Delta w = 0.890$	ΔP	−0.242	+0.194	−0.349	—	—
8	$RE = 1.517$	P	0.862	0.495	0.758	0.529	—
	$\Delta w = 0.695$	ΔP	−0.150	+0.132	−0.247	+0.184	—
10	$RE = 2.054$	P	0.858	0.500	0.747	0.544	0.730
	$\Delta w = 0.569$	ΔP	−0.101	+0.093	−0.179	+0.149	−0.208

[a] RE: The resonance energy (in units of $-\beta$), that is, the difference in π-electronic energy between the system consisting of $n/2$ isolated ethylenic bonds and the conjugated system consisting of n carbon π centers. Δw: the energy (in units of $-\beta$) of the lowest energy one-electron transition, that is, the energy difference between ψ_{-1} and ψ_{+1}. P: the π-bond order. ΔP: the change in the π-bond order on the lowest energy one-electron transition, the one-electron transition from ψ_{+1} to ψ_{-1}.

conjugated system from planarity by twist of an essential double bond will cause bathochromic shift of the first S–S band and that deviation by twist of an essential single bond will cause hypsochromic shift of the band (see Subsection 7.1.6).

Systems with odd values of n are positive ions, radicals, or negative ions according as the number of electrons in the nonbonding orbital ψ_0 is 0, 1, or 2. Since $w_0 = 0$ and $p_{ij}^{(0)} = 0$, evidently the ions and radical with the same value of n have the same values of RE, P_{ij}, ΔP_{ij}, and Δw. In these systems RE is the resonance energy relative to the energy of the system having $(n - 1)/2$ isolated double bonds and one isolated carbon $2p\pi$ atomic orbital.

In systems with odd values of n the alternation of π-bond order is not so prominent as in systems with even values of n. It is noteworthy that the value of ΔP, that is, the value of $p^{(-1)}$ or $-p^{(+1)}$, of the central bonds is zero for the systems where $(n + 1)/2$ is odd, and is negative for the systems where $(n + 1)/2$ is even. This suggests that twist of the central bonds of systems with odd values of $(n + 1)/2$ will exert no effect on the position of the first S–S band and that twist of the central bonds of systems with even values of $(n + 1)/2$ will exert a bathochromic effect on the first band.

TABLE 10.6

RESULTS OF THE CALCULATION BY THE HÜCKEL METHOD ON CONJUGATED LINEAR SYSTEMS CONSISTING OF n CARBON π CENTERS, IN WHICH n IS ODD[a]

n		Bond	1–2	2–3	3–4	4–5	5–6
3	$RE = 0.828$	P	0.707	—	—	—	—
	$\Delta w = 1.414$	ΔP	−0.354	—	—	—	—
5	$RE = 1.464$	P	0.789	0.577	—	—	—
	$\Delta w = 1.000$	ΔP	−0.250	0	—	—	—
7	$RE = 2.055$	P	0.816	0.545	0.653	—	—
	$\Delta w = 0.765$	ΔP	−0.163	+0.068	−0.096	—	—
9	$RE = 2.628$	P	0.828	0.531	0.682	0.616	—
	$\Delta w = 0.618$	ΔP	−0.112	+0.069	−0.112	0	—
11	$RE = 3.191$	P	0.834	0.524	0.697	0.599	0.644
	$\Delta w = 0.518$	ΔP	−0.081	+0.059	−0.102	+0.037	−0.043

[a] RE: The resonance energy (in units of $-\beta$), that is, the difference in π-electronic energy between the system, consisting of $(n - 1)/2$ isolated ethylenic bonds and one isolated carbon π center, and the conjugated system consisting of n carbon π centers. Δw: the energy (in units of $-\beta$) of the lowest energy one-electron transition, that is, the energy of the one-electron transition from ψ_0 to ψ_{-1} or from ψ_{+1} to ψ_0. P: the π-bond order. ΔP: the change in the π-bond order on the lowest energy one-electron transition.

The fact that the value of ΔP of the central bonds of the odd-membered conjugated chain is zero or negative according as the value of $(n + 1)/2$ is odd or even is understandable from the following consideration. In the abstract linear model of an odd-membered system the center of symmetry is at the central or $[(n + 1)/2]$th atom. The values of index r for ψ_{-1}, ψ_0, and ψ_{+1} are $(n + 1)/2 + 1$, $(n + 1)/2$, and $(n + 1)/2 - 1$, respectively. When $(n + 1)/2$ is odd the central atom is starred (compare Section 7.2). In this case, ψ_0, having an odd value of r, is symmetric, and ψ_{+1} and ψ_{-1}, having even values of r, are antisymmetric with respect to the rotation around the z axis, $C_2(z)$, or the reflection in the plane that contains the z axis and is perpendicular to the conjugated chain, $\sigma_v(xz)$. This implies that the coefficients of the central $2p\pi$ AO in ψ_{+1} and ψ_{-1} are zero and, therefore, that the values of $p^{(+1)}$ and $p^{(-1)}$ for the central bonds are zero. On the other hand, when $(n + 1)/2$ is even, the central atom is unstarred. In this case, ψ_0, having an even r value, is antisymmetric, and ψ_{+1} and ψ_{-1}, having odd r values, are symmetric with respect to $C_2(z)$ or $\sigma_v(xz)$. This implies that the coefficients of the central AO in ψ_{+1} and ψ_{-1} can be nonzero and hence the values of $p^{(+1)}$ and $p^{(-1)}$ for the central bonds çan be nonzero. ψ_{-1} must have $(n + 1)/2$ nodes along the conjugated system. These nodes are at the alternate bonds from both the terminal bonds to the central bonds. The values of $p^{(-1)}$ for the central bonds are thus negative.

10.5. Correlation between Results of Simple MO Calculation and Spectral Data in Aromatic Hydrocarbons

10.5.1. Empirical Correlation between Spectral Data and Molecular Structure

As mentioned already, main absorption bands of aromatic hydrocarbons are classified into four types: α, p, β, and β'. In a series of *cata*-condensed linear polycyclic benzenoid hydrocarbons, namely, acenes, as well as in a series of *cata*-condensed angular (or bent) polycyclic benzenoid hydrocarbons, namely, phenes, all the bands are progressively displaced toward longer wavelengths as the number of benzene rings increases.

In the acene series, the p band is more sensitive to the increase in the number of benzene rings than is the α band. While the α band is at longer wavelength than the p band, that is, $\lambda_\alpha > \lambda_p$, in benzene as well as in naphthalene, the p band shifts so rapidly toward longer wavelengths with progressive annellation that the α band is overtaken by the p band in going along the series, naphthalene, anthracene, tetracene (naphthacene), pentacene, and so on. Thus, in anthracene the α band is already swamped by the p band, and in pentacene both bands are clearly distinguishable with $\lambda_\alpha < \lambda_p$. In the phene series, the p band does not shift so rapidly as it does in the acene series; in phenes, as in benzene and naphthalene, $\lambda_\alpha > \lambda_p > \lambda_\beta$.

The α and the β band appear to be related: in both series these bands undergo parallel moderate bathochromic shifts with progressive annellation. The ratio of the wavelength of the α band to that of the β band, $\lambda_\alpha/\lambda_\beta$, has a nearly constant value of 1.35 in many hydrocarbons.[1] In the acene series the β band increases in intensity almost linearly with length of the acene.

In the spectra of larger hydrocarbons a very intense band (β' band) is found on the shorter wavelength side of the β band. The β' band has almost constant intensity in acenes.

In the spectra of most aromatic compounds, a weak S–T band (t band) has been observed on the longer wavelength side of the S–S band system. The p and the t band appear to be related: on progressive annellation these bands undergo parallel shifts which are large or small bathochromic according as the annellation is linear or angular. The ratio of the wavelength of the p band to that of the t band, λ_p/λ_t, is roughly constant (about 0.6) in the phene series, but falls progressively in the acene series.

10.5.2. Interpretation of Spectra of Aromatic Hydrocarbons by the MO Method Including Configuration Interaction

The spectra of aromatic hydrocarbons can be satisfactorily interpreted by the MO method including configuration interaction. As mentioned in Subsection 10.3.3, the α and the β band are interpreted to be due to the transitions from the ground state ${}^-[V]_0$ to excited states ${}^-[V]_1^{12}$ and ${}^+[V]_1^{12}$, respectively, which arise primarily from the configuration interaction between V_1^{12} and V_1^{21}, and the p and the β' band are interpreted to be due to the transitions from the ground state to excited states ${}^+[V]_1^{11}$ and ${}^+[V]_1^{22}$, respectively, which arise primarily from the configuration interaction between V_1^{11} and V_1^{22}. The t band is interpreted to be the triplet counterpart of the p band, that is, to be due to the transition from the ground state to triplet state ${}^+[T]_1^{11}$.

In benzene and, in general, in compounds with D_{3h} or D_{6h} symmetry, such as triphenylene and coronene, both the highest bonding level and the lowest antibonding level are doubly degenerate, that is, $w_{+2} = w_{+1}$ and $w_{-2} = w_{-1}$, and hence V_1^{11} and V_1^{22} have the equal energies and make equal contributions to the p state (${}^+[V]_1^{11}$) as well as to the β' state (${}^+[V]_1^{22}$). However, in other compounds, V_1^{11} is lower in energy than V_1^{22}, and hence these two electron configurations contribute differently to the p and the β' state. Thus, V_1^{11} contributes predominantly to the lower state, the p state, and V_1^{22} contributes predominantly to the higher state, the β' state. Accordingly, in the simple MO approximation, the transition responsible for the p band can be approximated as the one-electron transition from ψ_{+1} to ψ_{-1}, and the transition giving rise to the β' band can be approximated as the one-electron transition from ψ_{+2} to ψ_{-2}. Acenes belong to point group D_{2h}, and these transitions in acenes are ${}^1A_{1g} \rightarrow {}^1B_{2u}$ transitions and have nonzero transition moments directed along the y axis, that is, the short symmetry axis of the molecule.

Actually, the p and the β' band in acenes are found experimentally to be polarized in that direction.[18,18a-b]

In alternant hydrocarbons, V_1^{12} and V_1^{21} have the same energy, and hence the interaction between these electron configurations is of the so-called first order. In acenes, the α state $({}^-[V]_1^{12})$ and the β state $({}^+[V]_1^{12})$ belong to symmetry species B_{3u} of point group D_{2h}. Since the α state is a minus state, the transition to the α state is forbidden, to a first approximation. The transition to the β state is predicted to have the nonzero transition moment in the direction of the long symmetry axis (x axis) of the molecule, the direction found experimentally.

The interpretation that the α and the β band are due to transitions to the states arising from the same set of electron configurations accounts for the similar trends in the shifts of these bands on progressive annellation. In general, as the size of the conjugated system becomes larger, the value of Δw_{12} and that of Δw_{21} become smaller, and, at the same time, the splitting of energy levels due to configuration interaction becomes smaller. Therefore, the approximate constancy of the ratio $\lambda_\alpha/\lambda_\beta$ is understandable.

Furthermore, the MO treatment accounts for the different trends in the shift of the p band on linear and angular annellation. It has been shown that the energies of ψ_{-1} and ψ_{+1}, that is, the values of w_{-1} and w_{+1}, converge to a common zero with the infinite annellation in the acene series, and that they converge to a finite energy separation in the phene series.[19] In both series, however, ψ_{-2} and ψ_{+2} converge to a finite energy separation. These behaviors of the orbitals on annellation account for the moderate increase in the wavelength of the α and the β band associated with either linear or angular annellation, and for the rapid bathochromic shift of the p band associated with linear annellation in the acene series.

10.5.3. Correlation between the Observed Frequencies of Bands and Transition Energies Calculated by the Hückel Method

Data on α, p, and β bands of a great number of *cata*-condensed and *peri*-condensed aromatic hydrocarbons have been compiled by Clar,[1] and data of the corresponding energy differences of Hückel orbitals have been compiled

[18] D. S. McClure, *J. Chem. Phys.* **22,** 1256, 1668 (1954); *ibid.* **24,** 1 (1956); also *in* "Solid State Physics" (F. Seitz and D. Turnbull, eds.), Vol. 8, p. 1. Academic Press, New York, 1959.

[18a] J. W. Sidman, *J. Chem. Phys.* **25,** 115, 122 (1956).

[18b] H. C. Wolf, *in* "Solid State Physics" (F. Seitz and D. Turnbull, eds.), Vol. 9, p. 1. Academic Press, New York, 1959.

[19] C. A. Coulson, *Proc. Phys. Soc.* (*London*) **A60,** 257 (1948).

TABLE 10.7

ABSORPTION SPECTRA OF BENZENOID HYDROCARBONS AND ENERGIES OF ONE-ELECTRON TRANSITIONS CALCULATED BY THE HÜCKEL METHOD[a]

Compound	λ_t	λ_p	λ_α	λ_β	λ_p/λ_t	$\lambda_\alpha/\lambda_\beta$	Δw_{11}	Δw_{12}
Benzene	340	206.8	264	183	0.61	1.44	2.000	2.000
Naphthalene	469.5	288.5	314.5	221	0.61	1.42	1.236	1.618
[Acene]								
Anthracene	670	378.5	—	254.5	0.57	—	0.828	1.414
Naphthacene	975.5	471	—	274	0.48	—	0.590	1.072
Pentacene	1300	575.5	428	310	0.44	1.38	0.439	0.838
[Phene]								
Phenanthrene	463	294.5	344.5	254.7	0.64	1.35	1.210	1.374
Chrysene	505	319	360	267	0.63	1.35	1.040	1.312
Benz[*a*]anthracene	606	359	385	290	0.59	1.33	0.905	1.167
Benzo[*c*]phenanthrene	—	315	372	281	—	1.32	1.135	1.230
Picene	—	328.5	376	286.5	—	1.31	1.004	1.182
Dibenz[*a,h*]anthracene	546	351	395	300	0.64	1.32	0.946	1.158
Dibenz[*a,j*]anthracene	—	351	395	304	—	1.30	0.983	1.110
Dibenzo[*b,h*]phenanthrene	—	359	423	317	—	1.33	0.874	0.958
Dibenzo[*c,g*]phenanthrene (pentahelicene)	—	329	395	310	—	1.27	1.071	1.192

[a] λ_t, λ_p, and λ_α are the wavelengths (in mμ) of the origins of *t*, *p*, and α bands, respectively, and λ_β is the wavelength (in mμ) of the maximum of the β band. The solvent is usually an alcohol or benzene. When available, data for alcohol-glass solutions at -170°C are given.

by Streitwieser.[20] In addition, data on *t* bands of some aromatic compounds have been compiled by Mason.[21] Some of these data are listed in Table 10.7.

It has been found that there is an excellent linear correlation between the observed wave number of the *p* band, ν_p, and the energy difference between ψ_{-1} and ψ_{+1}, Δw_{11}:[20]

$$\nu_p\,(\text{cm}^{-1}) = (19{,}020 \pm 330)\,\Delta w_{11} + (10{,}520 \pm 340). \tag{10.28}$$

This correlation holds for a wide variety of aromatic compounds. Considered from the fact that V_1^{22} and other electron configurations of the same symmetry make some contributions to the *p* state, although the contribution of V_1^{11}

[20] A. Streitwieser, Jr., "Molecular Orbital Theory for Organic Chemists." Chap. 8. Wiley, New York, 1961.

[21] S. F. Mason, *Quart. Rev.* (*London*) **15**, 287 (1961).

is predominant, this excellent linear correlation is rather surprising. Especially, it is noteworthy that this relation holds for benzene, in which the configuration interaction between V_1^{11} and V_1^{22} is of the first order.

It is of interest that there is an excellent linear relationship between the frequency of the p band and the polarographic half-wave reduction potential.[22,22a] In the simple MO approximation, the reduction potential is considered as the energy required to place one electron into the lowest unoccupied orbital of the molecule, and hence should be linearly related to w_{-1}. On the other hand, as mentioned above, the frequency of the p band is linearly related to Δw_{11}, which is equal to $2w_{-1}$ in alternant hydrocarbons.

The t band is the S–T counterpart of the p band. The ratio ν_t/ν_p $(= \lambda_p/\lambda_t)$ has roughly a constant value of about 0.6 in the phene series. Since the decrease in ν_p associated with annellation is approximately proportional to the decrease in Δw_{11}, this means that in phenes the decreases in $\nu_p - \nu_t$ and in ν_t are also approximately proportional to the decrease in Δw_{11}. The following linear relationship between ν_t and Δw_{11} holds approximately good for the phene series and also for some *peri*-condensed aromatic hydrocarbons:

$$\nu_t\,(\text{cm}^{-1}) \doteqdot 11{,}500\,\Delta w_{11} + 7000. \tag{10.29}$$

The value of Δw_{11} decreases so rapidly with linear annellation, that the ratio ν_t/ν_p decreases progressively in the acene series, from 0.61 for naphthalene to 0.44 for pentacene (see Table 10.7). The following linear relationship between ν_t and Δw_{11} holds approximately good for the acene series:

$$\nu_t\,(\text{cm}^{-1}) \doteqdot 16{,}600\,\Delta w_{11} + 700. \tag{10.30}$$

In alternant hydrocarbons, as mentioned already, V_1^{12} and V_1^{21} are degenerate, and the α and the β state are considered to arise primarily from the first-order interaction of these two electron configurations. If the splitting of the two states due to the configuration interaction is linearly related to the mean of the energies of the two states relative to the ground state, we may expect correlations between Δw_{12} $(= \Delta w_{21})$ and the wave numbers of the α and the β band. Actually, the following linear correlations hold approximately good for a wide variety of aromatic compounds:

$$\nu_\alpha\,(\text{cm}^{-1}) \doteqdot 14{,}400\,\Delta w_{12} + 8800; \tag{10.31}$$

$$\nu_\beta\,(\text{cm}^{-1}) \doteqdot 23{,}750\,\Delta w_{21} + 6600. \tag{10.32}$$

[22] A. Maccoll, *Nature* **163,** 178 (1949).
[22a] A. T. Watson and F. A. Matsen, *J. Chem. Phys.* **18,** 1305 (1950).

In nonalternant hydrocarbons, V_1^{12} and V_1^{21} are no longer degenerate. Nevertheless, fluoranthene derivatives fit these correlations satisfactorily if the wave number of the α band is related to Δw_{12} and the wave number of the β band is related to Δw_{21}.[20]

Thus it can be said that the simple MO method is sufficiently useful even for correlation and prediction of the rather complicated spectra of aromatic hydrocarbons.

10.6. Effects of Replacement of Carbon π Centers by Heteroatom π Centers and of Introduction of Nonmesomeric Substituents on Spectra of Conjugated Hydrocarbons

10.6.1. THE EFFECTS IN ALTERNANT HYDROCARBONS

The spectral effect of replacement of a carbon π center, say the ith π center, in a carbon π system by a heteroatom π center, and the effect of introduction of a nonmesomeric substituent onto the ith carbon atom, can be treated within the framework of the simple LCAO MO method, to a first approximation, by assigning an appropriate altered value of the Coulomb integral, $\alpha_i = \alpha + h_i\beta$, to the ith π AO of the parent π system (see Subsections 7.1.4 and 7.1.6, and also Chapter 9). The value of parameter h_i is positive for heteroatoms which are more electronegative than carbon and for electron-attracting substituents. The change in the energy separation between MO's ψ_n and ψ_m, $\Delta E_{m\to n}$ $(= w_n - w_m)$, associated with the change in the Coulomb integral for the π AO at the ith atom from α to $\alpha + h_i\beta$ can be expressed approximately as follows [compare Eq. (7.71)]:

$$\delta\Delta E_{m\to n} = -(c_{n,i}^2 - c_{m,i}^2)h_i = -(p_{ii}^{(n)} - p_{ii}^{(m)})h_i = -\Delta P_{ii}h_i, \quad (10.33)$$

where $c_{n,i}$ is the coefficient of the ith π AO in MO ψ_n of the parent π system, $p_{ii}^{(n)}$ is the partial π-electron density of ψ_n at the ith atom, and ΔP_{ii} is the change in the π-electron density at the ith atom associated with one-electron transition from ψ_m to ψ_n. Although this expression is of a rough approximation, this is useful to some extent for obtaining a qualitative or semiquantitative explanation of the effects of the foregoing structural changes on the spectra of π systems.

In even alternant hydrocarbons, needless to say, $c_{-k,i}^2$ and $c_{+k,i}^2$ are identical. Therefore, in such systems it is expected that replacement of a carbon π center by a heteroatom π center and introduction of a nonmesomeric substituent will cause no change in the position of the absorption bands due to transitions $\psi_{+1} \rightarrow \psi_{-1}$, $\psi_{+2} \rightarrow \psi_{-2}$, etc. This expectation is substantiated by the following examples.

Data on the first main (π–π^* S–S) bands of butadiene and related compounds and those of styrene and related compounds are shown in Table 10.8 and in Table 10.9, respectively. It will be seen that the replacements of

TABLE 10.8

THE FIRST MAIN BANDS (THE A BANDS) OF 1,3-BUTADIENE AND RELATED COMPOUNDS

Compound	Solvent	λ_{max}, mμ	ϵ_{max}
$CH_2{=}CH{-}CH{=}CH_2$	Hexane	217	20,900[a]
$CH_2{=}CH{-}CH{=}CH{-}CH_3$	Heptane	223	26,000[b]
	Hexane	223.5	25,500[c]
	Ethanol	223.5	23,000[c]
$CH_2{=}C(CH_3){-}CH{=}CH_2$	Hexane	220	23,900[a]
$CH_2{=}C(CH_3){-}C(CH_3){=}CH_2$	Hexane	225	20,400[a]
	Hexane	226	21,400[d]
$CH_3{-}CH{=}CH{-}CH{=}CH{-}CH_3$	Hexane	227	25,500[c]
	Ethanol	227	22,500[c]
$CH_3{-}CH{=}CH{-}CH{=}N{-}C_4H_9$	Ethanol	220	23,000[e]
$CH_3{-}CH{=}CH{-}CH{=}O$	Ethanol	217	15,650[f]

[a] A. Smakula, *Angew. Chem.* **47,** 657 (1934).
[b] L. E. Jacobs and J. R. Platt, *J. Chem. Phys.* **16,** 1137 (1948).
[c] H. Booker, L. K. Evans, and A. E. Gillam, *J. Chem. Soc.* **1940,** 1453.
[d] G. Scheibe, *Ber.* **59,** 1321 (1926).
[e] H. C. Barany, E. A. Braude, and M. Pianka, *J. Chem. Soc.* **1949,** 1898.
[f] K. W. Hausser, R. Kuhn, A. Smakula, and M. Hoffer, *Z. physik. Chem.* **B29,** 371 (1935).

=CH— by =N— and by =O cause substantially no change in the position of the first main band of each parent compound.

Replacement of one methine group, =CH—, in benzene by =N— affords pyridine. The replacement of the =CH— group at position 1 and that at position 2 in naphthalene afford quinoline and isoquinoline, respectively, and the replacement of the =CH— group at position 9 in anthracene yields acridine. As is expected, the spectra of these mono-aza compounds are closely similar to those of the respective parent hydrocarbons (compare Table 10.10).

The most prominent difference of the spectra of the aza-aromatic compounds from the spectra of the corresponding aromatic hydrocarbons lies in the greater intensity and partial loss of vibrational structure of the α band.

TABLE 10.9

THE FIRST MAIN BANDS (THE A BANDS) OF STYRENE AND RELATED COMPOUNDS

Compound	Solvent	λ_{max}, mμ	ϵ_{max}
C_6H_5—CH=CH_2	Heptane	248	14,760[a]
	Ethanol	248	14,620[b]
C_6H_5—CH=CH—CH_3	Ethanol	250	17,300[c]
C_6H_5—CH=N—CH_3	Ethanol	247	17,000[d]
C_6H_5—CH=O	Heptane	237	14,100[e]
	Ethanol	246	13,200[f]

[a] D. F. Evans, *J. Chem. Phys.* **23,** 1429 (1955).
[b] C. G. Overberger and D. Tanner, *J. Am. Chem. Soc.* **77,** 369 (1955).
[c] R. Y. Mixer, R. F. Hech, S. Winstein, and W. G. Young, *J. Am. Chem. Soc.* **75,** 4094 (1953).
[d] G. E. McCasland and E. C. Horswill, *J. Am. Chem. Soc.* **73,** 3923 (1951).
[e] H. Suzuki, *Bull. Chem. Soc. Japan* **33,** 613 (1960).
[f] R. P. Mariella and R. R. Raube, *J. Am. Chem. Soc.* **74,** 521 (1952).

TABLE 10.10

ABSORPTION SPECTRA OF SOME AROMATIC HYDROCARBONS AND THEIR MONO-AZA ANALOGS

Compound	Solvent	α band		*p* band		β band	
		λ_{max}, mμ	ϵ_{max}	λ_{max}, mμ	ϵ_{max}	λ_{max}, mμ	ϵ_{max}
Benzene	Hexane	255[a]	250[a]	198[a]	8000[a]	—	—
Pyridine	Hexane	250[a]	2000[a]	195[a]	7500[a]	—	—
Naphthalene	Ethanol	311[a]	320[a]	275[a]	5600[a]	221[b]	95,500[b]
Quinoline	Ethanol	314[c]	3000[c]	278[c]	3500[c]	225[c]	35,000[c]
	Hexane	311[a]	6300[a]	275[a]	4500[a]	—	—
Isoquinoline	Ethanol	320[c]	2700[c]	267[c]	3700[c]	218[b]	79,500[b]
	Hexane	317[a]	3500[a]	262[a]	3700[a]	—	—
Anthracene	Ethanol	—	—	380[a]	6500[a]	252[a]	160,000[a]
Acridine	Ethanol	—	—	347[a]	8000[a]	252[a]	170,000[a]

[a] A. E. Gillam and E. S. Stern, "An Introduction to Electronic Absorption Spectroscopy in Organic Chemistry," 2nd ed., p. 153. Arnold, London, 1957.
[b] R. A. Friedel and M. Orchin, "Ultraviolet Spectra of Aromatic Compounds," Wiley, New York and London, 1951.
[c] J. M. Hearn, R. A. Morton, and J. C. E. Simpson, *J. Chem. Soc.* **1951,** 3318.

Both these facts are attributed to the reduced symmetry of the aza-aromatic compounds compared with the parent hydrocarbons. As mentioned previously, the α band of alternant aromatic hydrocarbons is forbidden because of the pairing property of the molecular orbitals, and appears as a result of coupling with vibration. On the aza substitution, this pairing property is lost, and hence the α band becomes allowed. Probably, the energies of electron configurations V_1^{12} and V_1^{21} of an aromatic hydrocarbon are raised and lowered, respectively, by the aza substitution, and hence in the aza compound these two configurations contribute with different weights to the α state.

The nitrogen atom has a lone pair of electrons, and hence in the aza compounds n–π^* transitions must occur. In the solution spectra of pyridine the n–π^* bands cannot be distinctly observed, but on the long-wavelength edge of the α band there are a series of shoulders which probably arise from an n–π^* transition. This n–π^* band becomes more distinctly observable in the vapor spectra of pyridine[23] and picolines (methylpyridines),[24] although still partly submerged under the α band. Now we take the benzene π orbitals in real form as given in Table 3.5, and suppose that the methine group at position 1 is replaced by an aza group. Then, since nitrogen is more electronegative than carbon, ψ_{-1} will be lowered by the aza substitution because it has a nonzero partial electron density at position 1, and ψ_{-2} will be unaffected to a first approximation because its partial electron density at position 1 is zero. The lowest energy n–π^* transition of pyridine is considered to have as the acceptor the perturbed orbital arising from ψ_{-1} of benzene. Since the n orbital of pyridine is a hybrid orbital, the n–π^* transition of pyridine is allowed and polarized perpendicularly to the molecular plane (see Subsection 4.3.3). The magnitude of the transition moment depends on the amount of s character in the n orbital. If it is a pure sp^2 hybrid there is $\frac{1}{3}$ s character in it. Experimentally, the oscillator strength of the n–π^* band of pyridine is estimated to be about 0.003,[25] which is about ten times as great as that of the normal carbonyl n–π^* band, which is forbidden.

The n–π^* band of an aza-aromatic compound shifts to longer wavelengths as more nitrogen atoms are introduced into the molecule. This is in marked contrast to the behavior of the π–π^* bands, which, as have been seen above, are rather insensitive to aza substitution. According to first-order perturbation theory, the lowering of the lowest unoccupied π orbital of a parent aromatic hydrocarbon on aza substitution will depend on the partial electron density of the orbital at the substitution position. If aza substitution is made

[23] M. Kasha, *Discussions Faraday Soc.* **9**, 14 (1950).

[24] J. H. Rush and H. Sponer, *J. Chem. Phys.* **20**, 1847 (1952).

[25] H. P. Stephenson, *J. Chem. Phys.* **22**, 1077 (1954).

at positions i, and if the AO coefficients of the lowest unoccupied orbital at positions i are c_i, the lowering of this orbital will be proportional to $\Sigma\, c_i^2$. This means that the decrease in the energy of the lowest energy n–π^* transition on aza substitution will be proportional to $\Sigma\, c_i^2$. Actually, for aza derivatives of benzene and naphthalene it has been found that, unless two nitrogen atoms are adjacent to each other, the wave number of the absorption onset (at $\epsilon = 20$) of the longest wavelength n–π^* band can be expressed approximately as follows:[26,26a–b]

$$\nu(\mathrm{cm}^{-1}) = K_H - 15{,}000 \sum c_i^2, \tag{10.34}$$

where K_H is a constant for a given parent hydrocarbon, which has values of about 39,110 and 32,150 for benzene and naphthalene, respectively. If two

TABLE 10.11

COMPARISON OF ABSORPTION SPECTRA OF BENZENE, ANILINIUM ION, AND ANILINE

Compound	λ_{max}, mμ	ϵ_{max}	λ_{max}, mμ	ϵ_{max}
C_6H_6	255[a]	200	200	5000
C_6H_5—NH_3^+	254[a]	160	203	7400
C_6H_5—NH_2	287.5	1860	234	87,100

[a] Approximate center of fine-structure bands.

nitrogen atoms are adjacent to each other, the n–π^* band occurs approximately 6000 cm^{-1} below the wave number expected from the value of $\Sigma\, c_i^2$ by this equation. This is because the two neighboring nitrogen n orbitals overlap each other. This overlap gives one orbital of higher energy (by about 6000 cm^{-1}) than an isolated n orbital (the antisymmetric combination of the two n orbitals) and one of lower energy (the symmetric combination), and it is the higher energy orbital which gives rise to the observed n–π^* band.

The spectra of the salts of aromatic amines, Ar—$NH_3^+X^-$, are remarkably similar to the spectra of the respective parent hydrocarbons, while the spectra of aromatic amines are considerably different from the spectra of the parent hydrocarbons.[27] For example, the spectrum of anilinium ion is similar to that of benzene, while the spectrum of aniline is considerably different from that of benzene (see Table 10.11). Since the —NH_3^+ group

[26] S. F. Mason, *J. Chem. Soc.* **1959,** 1240.

[26a] J. N. Murrell, "The Theory of the Electronic Spectra of Organic Molecules," Sect. 8.5. Methuen, London, 1963.

[26b] L. Goodman and R. W. Harrell, *J. Chem. Phys.* **30,** 1131 (1959).

[27] D. Peters, *J. Chem. Soc.* **1957,** 4182.

is considered to be nonmesomeric or purely inductive while the amino group is mesomeric, these facts provide a further confirmation of the simple theory.

10.6.2. THE EFFECTS OF SUBSTITUENTS IN NONALTERNANT HYDROCARBONS

In nonalternant hydrocarbons, the π MO's have no pairing property, and hence any π–π^* transition is accompanied with change in the π-electron density at each π center. Therefore, on the basis of Eq. (10.33), it is expected that introduction of a nonmesomeric substituent into a nonalternant hydrocarbon will exert a bathochromic or a hypsochromic effect on a band, depending on the electronegativity of the substituent, through the value of h_i, and on the position of substitution, through the value of ΔP_{ii}. This expectation has been borne out experimentally by the observation of effects of methyl substitution at various positions on the longest wavelength S–S band (p band) of azulene.[28,28a]

Methyl and other alkyl groups attached to π centers are usually considered to exert hyperconjugative and electron-donating inductive effects. The hyperconjugative effect usually reduces the energies of electronic transitions in both alternant and nonalternant hydrocarbons by extending the conjugated system. In alternant hydrocarbons, the inductive effect is of subsidiary importance owing to the pairing property of π orbitals, and the hyperconjugative small bathochromic effect is dominant. On the other hand, in nonalternant hydrocarbons, the inductive effect is expected to reveal itself if the π-electron density change at the position of substitution associated with the electronic transition under consideration is large. The electron-donating inductive effect of an alkyl substituent can be treated by taking the value of h_i as negative. If we consider the alkyl substituent as exerting the inductive effect only, it is expected that the substitution will give rise to a hypsochromic or a bathochromic shift of an absorption band according as ΔP_{ii} for the corresponding transition is positive or negative, and that the magnitude of the shift will be proportional to the absolute value of ΔP_{ii}.

The p state of azulene has a predominant contribution of V_1^{11} (compare Subsection 10.3.4). Therefore, the p band can be considered approximately as being due to the one-electron transition from ψ_{+1} to ψ_{-1}. The values of ΔP_{ii}'s for this transition in the Hückel approximation and the corresponding values for the transition from the ground state to the p state calculated with inclusion of configuration interaction are shown in Table 10.12. This table

[28] P. A. Plattner and E. Heilbronner, *Helv. Chim. Acta* **30,** 910 (1947).

[28a] H. C. Longuet-Higgins and R. G. Sowden, *J. Chem. Soc.* **1952,** 1404.

TABLE 10.12

THE CHANGES (ΔP_{ii}) IN π-ELECTRON DENSITIES AT VARIOUS POSITIONS IN AZULENE ON THE ELECTRONIC TRANSITIONS RESPONSIBLE FOR THE p AND α BANDS AND THE SHIFTS ($\Delta\lambda$, mμ) OF THE p BAND (λ_{max} = 697 mμ) OF AZULENE BY INTRODUCTION OF A METHYL SUBSTITUENT AT VARIOUS POSITIONS

Band	Position (i)	2	1 (3)	9 (10)	4 (8)	5 (7)	6
p	ΔP_{ii}, Hückel MO[a]	+0.0997	−0.2907	+0.0172	+0.1952	−0.1021	+0.2610
	ΔP_{ii}, Hückel MO + CI[a]	+0.1388	−0.2412	+0.0114	+0.2016	−0.1214	+0.1601
	$\Delta\lambda_{obs}$, mμ[b]	−21	+41	—	−17	+18	−16
	$\Delta\lambda_{calc}$, mμ	−22	+41	—	−31	+20	−25
α	ΔP_{ii}, Hückel MO + CI[a]	−0.0996	−0.0975	−0.0230	+0.1341	+0.0281	+0.0171

[a] Reference 5.
[b] Reference 28.

also shows the observed magnitudes of shift of the *p* band by methyl substitution at various positions, $\Delta\lambda_{obs}$, and the corresponding expected magnitudes, $\Delta\lambda_{calc}$, calculated from the values of ΔP_{ii} for the transition from the ground state to the *p* state and calibrated for the substitution at position 1 (for the numbering of the carbon atoms in azulene, see Fig. 8.5). Evidently, the calculated shifts are in a fairly good agreement with the observed shifts both with respect to the direction and magnitude.

For di- and poly-substitution, the shift of the *p* band is approximately additive, that is, approximately equals the sum of the shifts induced by each methyl group. The approximate additivity appears to hold generally for small inductive and hyperconjugative effects of nonmesomeric substituents on electronic absorption spectra, while the effects of mesomeric or conjugative substituents are usually large and are not additive.

The azulene α state is a state that has predominant contributions of V_1^{12} and V_1^{21}. The values of ΔP_{ii} for the transition from the ground state to the α state are shown in the last row of Table 10.12. It is expected from these values that the azulene α band (the second S–S band) would undergo a hypsochromic shift on methyl substitution at position 4, 5, or 6 and a bathochromic shift on methyl substitution at position 1 or 2. In fact, however, experiment appears to indicate small bathochromic shifts for all positions.[29] The bathochromic shifts for positions 2, 4, and 6 appear to be larger than those for positions 1 and 5. In this case the hyperconjugative effect probably outweighs the inductive effect.

10.6.3. The Small Bathochromic Effect of Alkyl Substituents in Alternant Hydrocarbons

The bathochromic shifts of π–π^* bands of alternant hydrocarbons by an alkyl substituent are usually small and rarely exceed 10 mμ. For example, as is seen in Table 10.8, the first S–S band of butadiene undergoes a bathochromic shift of about 6 mμ on introduction of one methyl group, and further, a bathochromic shift of about 4 mμ on introduction of the second methyl group. Furthermore, as is seen in Table 10.9, the first main band of styrene undergoes a bathochromic shift of about 2 mμ on methyl substitution at the β position. In addition to the bathochromic effect, introduction of a methyl group exerts usually a small hyperchromic effect on π–π^* bands, that is, intensifies slightly π–π^* bands.

In Table 10.13 data showing the bathochromic effect of alkyl substituents on the spectrum of benzene are collected. Introduction of one alkyl group

[29] P. A. Plattner and E. Heilbronner, *Helv. Chim. Acta* **31**, 804 (1948).

TABLE 10.13

ABSORPTION SPECTRA OF BENZENE AND ITS ALKYLATED DERIVATIVES

A. The α band

Benzene	The most intense max.[a] (in heptane) λ_{max}, mμ	ϵ_{max}	The first max.[b] (vapor) λ_{max}, mμ	The first max.[c] (vapor) λ_{max}, mμ
Unsubstituted	255	230	262.54	262.46
Methyl-	261	300	266.78	266.75
Ethyl-	261	220	—	266.43
n-Propyl-	261	245	—	266.52
Isopropyl-	—	—	—	265.87
n-Butyl-	260	235	—	266.63
1,3-Dimethyl-	266	400	—	—
1,3,5-Trimethyl-	266	305	274.06	—

B. The p and β bands (in heptane)[d]

Benzene	The p band λ_{max}, mμ	ϵ_{max}	f	The β band λ_{max}, mμ	ϵ_{max}	f
Unsubstituted	195.0	6900	0.10	183.5	46,000	0.69
Methyl-	200.0	8100	0.12	188.5	55,000	0.97
Ethyl-	200	7400	0.11	189	57,000	1.00
1,3-Dimethyl-	203	10,500	0.18	193	76,000	1.28

[a] A. E. Gillam and E. S. Stern, "An Introduction to Electronic Absorption Spectroscopy in Organic Chemistry," 2nd ed., p. 134. Arnold, London, 1957.

[b] K. F. Herzfeld, *Chem. Rev.* **41,** 233 (1947).

[c] F. A. Matsen, *in* "Technique of Organic Chemistry, Vol. IX, Chemical Applications of Spectroscopy" (W. West, ed.), p. 677. Wiley (Interscience), New York, 1956.

[d] J. R. Platt and H. B. Klevens, *Chem. Rev.* **41,** 301 (1947).

causes a bathochromic shift of about 5 mμ of each benzene band. Introduction of the second methyl group to the *ortho* or the *para* position of toluene causes a bathochromic shift of about 4 mμ of the β band, and that to the *meta* position causes a bathochromic shift of about 6 mμ of the same band. The greater bathochromic effect of the methyl substituent at the *meta* position as compared with the bathochromic effect of the same substituent at the *ortho* or the *para* position can be also recognized in comparison of the

TABLE 10.13 (*Contd.*)

C. The β band (in solution)[e]

Benzene	λ_{max}, mμ	ϵ_{max}
Unsubstituted	184	50,000
Methyl-	188	56,000
t-Butyl-	188.7	80,000
1,2-Dimethyl-	192	53,000
1,3-Dimethyl-	194	50,000
1,4-Dimethyl-	192	54,000
1-Methyl-2-ethyl-	192.5	55,000
1-Methyl-3-ethyl-	194.5	57,000
1-Methyl-4-ethyl-	193.5	55,000
1,2,3-Trimethyl-	195	55,000
1,2,4-Trimethyl-	196	51,000
1,3,5-Trimethyl-	199	55,000
1,2,3,5-Tetramethyl-	199.5	55,000
1,2,4,5-Tetramethyl-	197.5	50,000

[e] D. W. Turner, *in* "Determination of Organic Structures by Physical Method" (F. C. Nachod and W. D. Phillips, eds.), Vol. 2, p. 384. Academic Press, New York, 1962.

spectral data of positional isomers of trimethylbenzene as well as of tetramethylbenzene. The bathochromic effect of methyl substitution on absorption bands of benzene is said to be approximately additive.[30] However, more strictly speaking, the effect of one methyl substituent appears to become gradually smaller as methyl substituents are accumulated. Treatments of the hyperconjugative bathochromic effect of the methyl substituent by the simple MO method will be described later.

[30] K. F. Herzfeld, *Chem. Rev.* **41,** 233 (1947).

Chapter 11

Advanced MO Methods

11.1. Introduction

Molecular orbital methods in which explicit consideration is given to the electronic repulsion terms in the total electronic Hamiltonian are called advanced MO methods. The earliest one of advanced MO methods is a method called the method of antisymmetrized products of molecular orbitals or the antisymmetrized molecular orbital method (ASMO method), developed largely by Goeppert-Mayer and Sklar.[1] In this method, the individual molecular orbitals are taken to be the same as those obtained by the simple LCAO MO method, the product wavefunctions for electron configurations are made antisymmetric with respect to each interchange of electron labels in fulfillment of the requirement of the Pauli principle (see Subsection 2.2.4), and the energies associated with the antisymmetrized electron configuration functions are calculated by the use of the total electronic Hamiltonian. Craig[2] introduced configuration interaction (CI) into this method. The ASMO method including configuration interaction is called the ASMO CI method. In this method, more nearly correct wavefunctions of electronic states are constructed as linear combinations of appropriate antisymmetrized electron configuration functions. Another important one of advanced MO methods is the so-called self-consistent-field molecular orbital method (SCF MO method), developed largely by Roothaan[3] (see Subsection 2.2.2).

In these methods, the required integrals are calculated or estimated by the use of assumed forms of the atomic orbitals (usually taken as Slater atomic orbital functions), and no empirical parameter is used. These methods are

[1] M. Goeppert-Mayer and A. L. Sklar, *J. Chem. Phys.* **6,** 645 (1938).
[2] D. P. Craig, *Proc. Roy. Soc.* (*London*) **A200,** 474 (1950).
[3] C. C. J. Roothaan, *Rev. Modern Phys.* **23,** 69 (1951).

therefore called nonempirical methods. In their complete form the non-empirical methods are difficult to apply: benzene is the most complex molecule on which the methods have been fully tested.

Pariser and Parr[4,5] have developed a method which incorporates empirical elements into the ASMO CI method and is applicable to more complex molecules. This semiempirical method is called the Pariser–Parr method. This method is different from the nonempirical ASMO CI method chiefly in the following two features: the use of the so-called zero-differential-overlap approximation[6] and the determination of some required integral values by empirical or semiempirical procedures. This method is said to combine the advantages of the conventional simple LCAO MO method with the advantages of the purely theoretical ASMO CI method, and has been widely adopted. On the other hand, Pople[7] has introduced into the SCF MO method a set of simplifying approximations closely related to the Pariser–Parr approximations in the ASMO CI method. Sometimes, the Pariser–Parr method and the Pople method are combined together, and are called the Pariser–Parr–Pople method or the P-method.

In the present chapter we chiefly describe the Pariser–Parr method, with some reference to the Pople method. In order to maintain a uniform notation, the notation used in the original papers of Pariser and Parr and of Pople will be partly changed.

11.2. Matrix Elements of the Total π-Electronic Hamiltonian between Antisymmetrized Electron Configuration Functions

The total π-electronic Hamiltonian for a π-electron system can be expressed as follows [see Eq. (7.5)]:

$$\mathbf{H} = \sum_i \mathbf{h}_i^{\mathrm{C}} + \sum_{ij} \mathbf{G}_{ij}, \tag{11.1}$$

where $\mathbf{h}_i^{\mathrm{C}}$ is the one-electron core Hamiltonian for π electron i and $\mathbf{G}_{ij}$ is the two-electron repulsion operator e^2/r_{ij}, representing the repulsion between π electrons i and j:

$$\sum_{ij} \mathbf{G}_{ij} = \tfrac{1}{2} \sum_i \sum_{j \neq i} e^2/r_{ij}. \tag{11.2}$$

If the π-electron system has n $(= 2m)$ π electrons, the antisymmetrized function of the ground electron configuration, in which the lowest m π orbitals

[4] R. Pariser and R. G. Parr, *J. Chem. Phys.* **21**, 466 (1953).
[5] R. Pariser and R. G. Parr, *J. Chem. Phys.* **21**, 767 (1953).
[6] R. G. Parr, *J. Chem. Phys.* **20**, 1499 (1952).
[7] J. A. Pople, *Trans. Faraday Soc.* **49**, 1375 (1953).

from ψ_a through ψ_m are fully occupied, is given by expression (2.46). Promotion of one electron from an occupied π orbital ψ_k to an unoccupied π orbital ψ_r will give rise to singlet and triplet electron configurations, V_1^{kr} and T_1^{kr}, whose antisymmetrized functions are given by expressions (2.47) and (2.48), respectively. If the orbitals ψ are orthonormal eigenfunctions of an effective one-electron Hamiltonian, these configuration functions are eigenfunctions of the approximate Hamiltonian as the sum of the effective one-electron Hamiltonians, but are not eigenfunctions of the correct total π-electronic Hamiltonian, **H** given by expression (11.1). That is, the off-diagonal matrix elements of the Hamiltonian **H** between electron configurations can be nonzero. More nearly correct functions of electronic states can be constructed as linear combinations of antisymmetrized electron configuration functions by diagonalizing the matrix of the Hamiltonian **H**.

The Hamiltonian matrix elements between antisymmetrized electron configuration functions built up from orthonormal orbitals can all be reduced to a sum of one- and two-electron integrals.

Consider first the diagonal matrix element for the ground electron configuration, that is, the energy associated with the ground electron configuration:

$$E(V_0) \equiv \langle V_0 | \mathbf{H} | V_0 \rangle = \langle |\psi_a{}^\alpha \psi_a{}^\beta \cdots \psi_m{}^\alpha \psi_m{}^\beta \| \mathbf{H} \| \psi_a{}^\alpha \psi_a{}^\beta \cdots \psi_m{}^\alpha \psi_m{}^\beta | \rangle. \tag{11.3}$$

The Slater determinant has a normalization factor $(1/n!)^{1/2}$. When the unnormalized determinant is used [see Eq. (2.43)], expression (11.3) can be rewritten as

$$E(V_0) = (1/n!) \langle [\psi_a{}^\alpha \psi_a{}^\beta \cdots \psi_m{}^\alpha \psi_m{}^\beta] | \mathbf{H} | [\psi_a{}^\alpha \psi_a{}^\beta \cdots \psi_m{}^\alpha \psi_m{}^\beta] \rangle. \tag{11.4}$$

If the first determinant in the matrix element is expanded, one of the terms will be

$$(1/n!) \langle \psi_a{}^\alpha(1) \psi_a{}^\beta(2) \cdots \psi_m{}^\alpha(n-1) \psi_m{}^\beta(n) | \mathbf{H} | [\psi_a{}^\alpha \psi_a{}^\beta \cdots \psi_m{}^\alpha \psi_m{}^\beta] \rangle. \tag{11.5}$$

There are $n!$ terms of this type. Since there is nothing special about this particular allocation of electrons to orbitals, all these terms must have the same value. Therefore, the factor $n!$ cancels out the normalization factors of the Slater determinants, and we have

$$E(V_0) = \langle \psi_a{}^\alpha(1) \psi_a{}^\beta(2) \cdots \psi_m{}^\alpha(n-1) \psi_m{}^\beta(n) | \mathbf{H} | [\psi_a{}^\alpha \psi_a{}^\beta \cdots \psi_m{}^\alpha \psi_m{}^\beta] \rangle. \tag{11.6}$$

In general, when one of the determinants in the matrix element between two Slater determinants is expanded, the normalization factors of the determinants disappear. Thus, after substitution of the explicit expression of **H**, we can write

$$E(V_0) = \left\langle \psi_a{}^\alpha(1)\psi_a{}^\beta(2) \cdots \psi_m{}^\alpha(n-1)\psi_m{}^\beta(n) \left| \left(\sum_i \mathbf{h}_i{}^{\mathrm{C}} + \sum_{ij} \mathbf{G}_{ij} \right) \right| \right.$$

$$\left. \times \sum_{\mathrm{P}} (-1)^{\mathrm{P}} \mathbf{P}[\psi_a{}^\alpha(1)\psi_a{}^\beta(2) \cdots \psi_m{}^\alpha(n-1)\psi_m{}^\beta(n)] \right\rangle, \qquad (11.7)$$

where $\mathbf{P}$ is the permutation operator, which interchanges electron labels [compare expression (2.44)].

We concentrate first on the terms with one-electron operators, $\mathbf{h}_i{}^{\mathrm{C}}$'s. Because of orthogonality of the orbitals, a term containing $\mathbf{h}_i{}^{\mathrm{C}}$ as an only operator can be nonzero only if electron i has the same spin function and each electron other than i is in the same spin-orbital on both sides of the operator. When this requirement is fulfilled, the integration over all the coordinates of electrons other than i gives only a factor of unity since all the spin-orbitals are normalized, and, if electron i is in an orbital ψ_c on one side of the operator and in an orbital ψ_d on the other side, the term reduces to the following one-electron integral:

$$\langle \psi_c(i) | \, \mathbf{h}_i{}^{\mathrm{C}} \, | \psi_d(i) \rangle \equiv h^{\mathrm{C}}_{[cd]} \, . \qquad (11.8)$$

This integral is called the core integral over MO's ψ_c and ψ_d. In $E(V_0)$ such one-electron integrals can arise only from the diagonal term of the determinant in expression (11.6), or, in other words, only when there is no permutation of electron labels in (11.7); if there is a permutation, at least one electron other than i is in different spin-orbitals on the two sides of the operator $\mathbf{h}_i{}^{\mathrm{C}}$. The nonzero terms arising from $\mathbf{h}_i{}^{\mathrm{C}}$'s in $E(V_0)$ are thus $2 \sum_c{}^m \varepsilon_c{}^{\mathrm{C}}$ in all, in which $\varepsilon_c{}^{\mathrm{C}}$ is the so-called core energy of MO ψ_c, defined as

$$\varepsilon_c{}^{\mathrm{C}} \equiv h^{\mathrm{C}}_{[cc]} = \langle \psi_c(i) | \, \mathbf{h}_i{}^{\mathrm{C}} \, | \psi_c(i) \rangle, \qquad (11.9)$$

and $\sum^m$ represents the summation over all the occupied orbitals.

Now we consider the terms with two-electron operators, $\mathbf{G}_{ij}$'s. For a term containing $\mathbf{G}_{ij}$ as an only operator to be nonvanishing, each electron other than i and j must be in the same spin-orbital and each of two electrons i and j must have the same spin function on the both sides of the operator. When this requirement is fulfilled, and if electrons i and j are in orbitals ψ_c and ψ_d, respectively, on one side of the operator $\mathbf{G}_{ij}$, and are in orbitals ψ_e and ψ_f, respectively, on the other side, the term reduces to the following two-electron integral:

$$\langle \psi_c(i)\psi_d(j) | \, \mathbf{G}_{ij} \, | \psi_e(i)\psi_f(j) \rangle = \int \psi_c{}^*(i)\psi_e(i)(e^2/r_{ij})\psi_f(j)\psi_d{}^*(j) \, d\tau$$

$$\equiv [ce \mid fd]. \qquad (11.10)$$

This integral represents the energy of electrostatic repulsive interaction between two charge distributions $e\psi_c{}^*\psi_e$ and $e\psi_f\psi_d{}^*$. Such an integral is called an electronic repulsion integral over MO's. In $E(V_0)$, only integrals of the following two types can occur:

$$[cc \mid dd] \equiv J_{cd}\,; \tag{11.11}$$

$$[cd \mid cd] \equiv K_{cd}\,. \tag{11.12}$$

J_{cd} is the so-called MO Coulomb repulsion integral, and K_{cd} is the so-called MO exchange repulsion integral (see Section 10.2). In $E(V_0)$, integrals of type J_{cd} occur only when there is no permutation in expression (11.7), and hence there is one integral of this type for each pair of spin-orbitals. On the other hand, integrals of type K_{cd} occur only when two electron labels assigned to spin-orbitals having the same spin function are interchanged, and hence there is one integral of this type for each pair of orbitals having the same spin function. The nonzero terms arising from $\mathbf{G}_{ij}$'s in $E(V_0)$ are thus

$$2\sum_{c}^{m}\sum_{d\neq c}^{m} J_{cd} + \sum_{c}^{m} J_{cc} - \sum_{c}^{m}\sum_{d\neq c}^{m} K_{cd}\,. \tag{11.13}$$

Since $J_{cc} = K_{cc}$, this expression can be rewritten as

$$\sum_{c}^{m}\sum_{d}^{m}(2J_{cd} - K_{cd}). \tag{11.14}$$

We now sum up all the contributions to $E(V_0)$ and obtain

$$E(V_0) = 2\sum_{c}^{m} h_{[cc]}^{\mathrm{C}} + \sum_{c}^{m}\sum_{d}^{m}(2J_{cd} - K_{cd}). \tag{11.15}$$

A quantity $F_{[cf]}$ is defined as follows:

$$F_{[cf]} \equiv h_{[cf]}^{\mathrm{C}} + \sum_{d}^{m}\{2[cf \mid dd] - [cd \mid fd]\}. \tag{11.16}$$

This quantity is called the **F** matrix element between MO's ψ_c and ψ_f. When $f = c$, we have

$$F_{[cc]} = h_{[cc]}^{\mathrm{C}} + \sum_{d}^{m}(2J_{cd} - K_{cd}) \equiv \varepsilon_c{}^{\mathrm{S}}. \tag{11.17}$$

The $\varepsilon_c{}^{\mathrm{S}}$ is called the zeroth-order SCF orbital energy of ψ_c. Expression (11.15) can be rewritten as

$$E(V_0) = \sum_{c}^{m}(h_{[cc]}^{\mathrm{C}} + F_{[cc]}) = \sum_{c}^{m}(\varepsilon_c{}^{\mathrm{C}} + \varepsilon_c{}^{\mathrm{S}}). \tag{11.18}$$

As is easily verified in a similar manner, the diagonal matrix elements for singly-excited electron configurations V_1^{kr} and T_1^{kr} are given by

$$\left.\begin{array}{l}\langle V_1^{kr}|\,\mathbf{H}\,|V_1^{kr}\rangle \equiv E(V_1^{kr})\\ \langle T_1^{kr}|\,\mathbf{H}\,|T_1^{kr}\rangle \equiv E(T_1^{kr})\end{array}\right\} = 2\sum_{c\neq k}^{m} h_{[cc]}^{\mathrm{C}} + h_{[kk]}^{\mathrm{C}} + h_{[rr]}^{\mathrm{C}}$$

$$+ \sum_{c\neq k}^{m}\sum_{d\neq k}^{m}(2J_{cd} - K_{cd})$$

$$+ \sum_{c\neq k}^{m}(2J_{kc} - K_{kc} + 2J_{rc} - K_{rc}) + J_{kr} \pm K_{kr}, \tag{11.19}$$

where the upper and the lower sign in front of the last term are valid for $E(V_1^{kr})$ and $E(T_1^{kr})$, respectively. Therefore, we obtain

$$E(V_1^{kr}) - E(T_1^{kr}) = 2K_{kr}. \tag{11.20}$$

The energies of the singly-excited configurations relative to the energy of the ground configuration can be expressed as follows:

$$E(V_1^{kr}) - E(V_0) = F_{[rr]} - F_{[kk]} - J_{kr} + 2K_{kr}\,; \tag{11.21}$$

$$E(T_1^{kr}) - E(V_0) = F_{[rr]} - F_{[kk]} - J_{kr}. \tag{11.22}$$

The off-diagonal matrix elements can also be reduced to sums of one- and two-electron integrals in a similar manner. Evidently, as mentioned already (see Subsection 10.3.1), for a matrix element between different configurations to be nonzero, the configurations must have the same spin multiplicity, must belong to the same symmetry species, and must not differ by more than three spin-orbitals. The matrix elements involving singly-excited configurations are given by the following expressions:[3,8]

$$\langle V_1^{kr}|\,\mathbf{H}\,|V_0\rangle = \sqrt{2}F_{[kr]}\,; \tag{11.23}$$

$$\langle V_1^{kr}|\,\mathbf{H}\,|V_1^{ls}\rangle - \delta_{kl}\,\delta_{rs}E(V_0) = \delta_{kl}F_{[rs]} - \delta_{rs}F_{[kl]} - [kl\,|\,rs] + 2[kr\,|\,ls]; \tag{11.24}$$

$$\langle T_1^{kr}\,|\mathbf{H}|\,T_1^{ls}\rangle - \delta_{kl}\,\delta_{rs}E(V_0) = \delta_{kl}F_{[rs]} - \delta_{rs}F_{[kl]} - [kl\,|\,rs]. \tag{11.25}$$

In these expressions, δ is the Kronecker symbol, and F's are $\mathbf{F}$ matrix elements between MO's, defined by expression (11.16). Needless to say, when $l = k$ and $s = r$, expressions (11.24) and (11.25) reduce to (11.21) and (11.22), respectively.

[8] J. A. Pople, *Proc. Phys. Soc.* (*London*) **A68,** 81 (1955).

Similar formulas for the **H** matrix elements involving doubly-excited configurations have been given.[9] For example:

$$E(V_2^{kr}) - E(V_0) = (\varepsilon_r{}^{\mathrm{C}} - \varepsilon_k{}^{\mathrm{C}}) + (\varepsilon_r{}^{\mathrm{S}} - \varepsilon_k{}^{\mathrm{S}}) - 2J_{kr} + K_{kr} + J_{rr}\,; \quad (11.26)$$

$$\langle V_2^{kr} |\, \mathbf{H} \,| V_0 \rangle = K_{kr}\,, \quad (11.27)$$

where V_2^{kr} is the doubly-excited singlet configuration which arises from the ground configuration by promoting two electrons from ψ_k to ψ_r.

In this way, all the matrix elements of **H** are expressible in terms of one-electron integrals of type $h_{[kl]}^{\mathrm{C}}$, that is, the core integrals over MO's, and two-electron integrals of type $[kl \mid rs]$, that is, the electronic repulsion integrals over MO's. When the MO's are LCAO MO's, these integrals over MO's can be expanded into sums of integrals over AO's for numerical evaluation, as will be mentioned in the following section.

11.3. The Semiempirical LCAO ASMO CI Method (The Pariser–Parr Method)

11.3.1. Expansion of Electronic Repulsion Integrals over MO's into Sums of Electronic Repulsion Integrals over AO's and the Zero-Differential-Overlap Approximation

The MO's ψ_k will further be taken to be orthonormal linear combinations of π AO's χ_p on the several π centers:

$$\psi_k = \sum_p c_{k,p}\chi_p\,, \quad (11.28)$$

where the coefficients $c_{k,p}$ are taken to be real.

An electronic repulsion integral over MO's, say $[kr \mid ls]$, can be expanded through integrals over AO's as follows:

$$[kr \mid ls] = \sum_p \sum_t \sum_q \sum_u c_{k,p}c_{r,t}c_{l,q}c_{s,u}(pt \mid qu). \quad (11.29)$$

In this expression, $(pt \mid qu)$ is an electronic repulsion integral over AO's, defined by

$$(pt \mid qu) = \int \chi_p{}^*(i)\chi_t(i)(e^2/r_{ij})\chi_q(j)\chi_u{}^*(j)\, d\tau. \quad (11.30)$$

This quantity can be interpreted as representing the energy of Coulomb interaction between the charge distributions $e\chi_p{}^*\chi_t$ and $e\chi_q\chi_u{}^*$.

[9] J. N. Murrell and K. L. McEwen, *J. Chem. Phys.* **25**, 1143 (1956).

The so-called zero-differential-overlap approximation is the neglect of differential overlap:

$$\chi_p{}^*\chi_t = 0 \quad \text{for } p \neq t. \tag{11.31}$$

In this approximation we take

$$(pt \mid qu) = \delta_{pt}\,\delta_{qu}(pp \mid qq); \tag{11.32}$$

that is, all of the electronic repulsion integrals over AO's, except those of type $(pp \mid pp)$ and $(pp \mid qq)$, are taken to be zero. Expression (11.32) does in fact follow from an earlier approximation suggested by Mulliken[10]

$$(pt \mid qu) = (S_{pt}S_{qu}/4)[(pp \mid qq) + (pp \mid uu) + (tt \mid qq) + (tt \mid uu)], \tag{11.33}$$

if the overlap integrals S are taken to be zero. The integral $(pp \mid pp)$ represents the energy of Coulomb repulsion between two electrons in the same atomic orbital χ_p, and the integral $(pp \mid qq)$ in which $p \neq q$ represents the energy of Coulomb repulsion between two electrons in the different atomic orbitals χ_p and χ_q. Integrals of type $(pp \mid pp)$ are called one-center Coulomb repulsion integrals, and integrals of type $(pp \mid qq)$ are called two-center Coulomb repulsion integrals.

According to the zero-differential-overlap approximation, we can write

$$[kr \mid ls] = \sum_p \sum_q c_{k,p}c_{r,p}c_{l,q}c_{s,q}(pp \mid qq). \tag{11.34}$$

In particular,

$$J_{kr} = [kk \mid rr] = \sum_p \sum_q c_{k,p}^2 c_{r,q}^2 (pp \mid qq); \tag{11.35}$$

$$K_{kr} = [kr \mid kr] = \sum_p \sum_q c_{k,p}c_{r,p}c_{k,q}c_{r,q}(pp \mid qq). \tag{11.36}$$

The evaluation of electronic repulsion integrals over MO's is thus reduced to evaluation of a relatively small number of Coulomb repulsion integrals over AO's.

11.3.2. Evaluation of Coulomb Repulsion Integrals over AO's

Pariser and Parr[5] have assumed that the value of the one-center Coulomb integral $(pp \mid pp)$ is given empirically as the difference between the appropriate valence state atomic ionization energy I_p (that is, the energy required to remove one electron from the singly occupied orbital χ_p of the atom p in the appropriate valence state) and the appropriate valence state atomic electron affinity A_p (that is, the energy released by adding one electron into the singly

[10] R. S. Mulliken, *J. Chim. Phys.* **46**, 497 (1949).

occupied orbital χ_p of the atom p in the appropriate valence state):

$$(pp \mid pp) = I_p - A_p . \tag{11.37}$$

According to Mulliken,[11] $I = 11.22$ ev and $A = 0.69$ ev for the $2p\pi$ AO on the sp^2 hybridized carbon atom, so that the derived value of $(pp \mid pp)$ for the $2p\pi$ AO on the sp^2 hybridized carbon atom is 10.53 ev. (According to Pritchard and Skinner,[12] $I = 11.42$ ev and $A = 0.58$ ev for the carbon $2p\pi$ AO. From these values, a value of 10.84 ev is derived for $(pp \mid pp)$ for the carbon $2p\pi$ AO.) Similarly, a value of 12.27 (= 14.63 − 2.36) ev has been assigned to the $(pp \mid pp)$ for the $2p\pi$ AO on the pyridine-type nitrogen atom. If the atom p contributes two electrons in the AO χ_p to the π-electron system, I_p and A_p must be replaced by I_{p^+} and A_{p^+}, respectively. I_{p^+} and A_{p^+} represent, respectively, the energy required to remove one electron from, and the energy released by adding one electron into, the singly occupied orbital χ_p of the singly positively charged atom p in the appropriate valence state.

The two-center Coulomb integral $(pp \mid qq)$ is a function of the interatomic distance R_{pq}. The values of $(pp \mid qq)$ are determined from the uniformly charged sphere approximation, in which the charge densities $e\chi_p^2$ are replaced by tangent uniformly charged nonconducting spheres of diameter

$$r_p = (4.597/Z_p') \text{ Å}, \tag{11.38}$$

where Z_p' is the Slater effective nuclear charge for χ_p, and the values of $(pp \mid qq)$ are computed as the energy of repulsion of these charged spheres by means of classical electrostatic theory. The values of $(pp \mid qq)$ for $R_{pq} \geq 2.80$ Å are thus determined by the equation

$$\begin{aligned}(pp \mid qq) = (7.1975/R_{pq})\{&[1 + (1/2R_{pq})^2(r_p - r_q)^2]^{-1/2} \\ &+ [1 + (1/2R_{pq})^2(r_p + r_q)^2]^{-1/2}\} \text{ ev},\end{aligned} \tag{11.39}$$

where R_{pq}, r_p, and r_q are given in angstrom units. With $Z' = 3.25$, the values of $(pp \mid qq)$ for two carbon $2p\pi$ AO's with $R_{pq} \geq 2.80$ Å are given by

$$(pp \mid qq)_{\text{CC}} = (7.1975/R_{pq})[1 + (1 + 2.0007/R_{pq}^2)^{-1/2}] \text{ ev}, \tag{11.40}$$

where the letter C indicates the carbon π center. For $R_{pq} \leq 2.80$ Å, the values of $(pp \mid qq)$ for two carbon $2p\pi$ AO's are given by

$$(pp \mid qq)_{\text{CC}} = (pp \mid pp)_{\text{CC}} - 2.625R_{pq} + 0.2157R_{pq}^2 \text{ ev}, \tag{11.41}$$

in which $(pp \mid pp)_{\text{CC}} = 10.53$ (ev) and the constants have been chosen to fit

[11] R. S. Mulliken, *J. Chem. Phys.* **2**, 782 (1934).

[12] H. O. Pritchard and H. A. Skinner, *Chem. Rev.* **55**, 745 (1955).

TABLE 11.1

VALUES OF THE COULOMB REPULSION INTEGRAL $(pp \mid qq)_{CC}$ FOR VARIOUS VALUES OF THE INTERATOMIC DISTANCE R_{pq}

R_{pq}, Å	$(pp \mid qq)_{CC}$, ev		R_{pq}, Å	$(pp \mid qq)_{CC}$, ev
0.00	10.53	$(pp \mid pp)_{CC}$	3.00	4.57
1.34	7.40	Ethylenic bond	4.00	3.50
1.39	7.30	Benzene, 1–2 (*ortho*)	5.00	2.82
2.00	6.14		6.00	2.37
2.41	5.46	Benzene, 1–3 (*meta*)	8.00	1.79
2.78	4.90	Benzene, 1–4 (*para*)	10.00	1.43

the values of $(pp \mid qq)_{CC}$ calculated from equation (11.40) for $R_{pq} = 2.80$ Å and $R_{pq} = 3.70$ Å.

In all the applications of the Pariser–Parr method in this book, the value of $(pp \mid pp)_{CC}$ of 10.53 ev and values of $(pp \mid qq)_{CC}$ calculated from Eqs. (11.40) and (11.41) will be used. Such values of $(pp \mid qq)_{CC}$ for some values of R_{pq} are shown in Table 11.1 (compare Fig. 11.1).

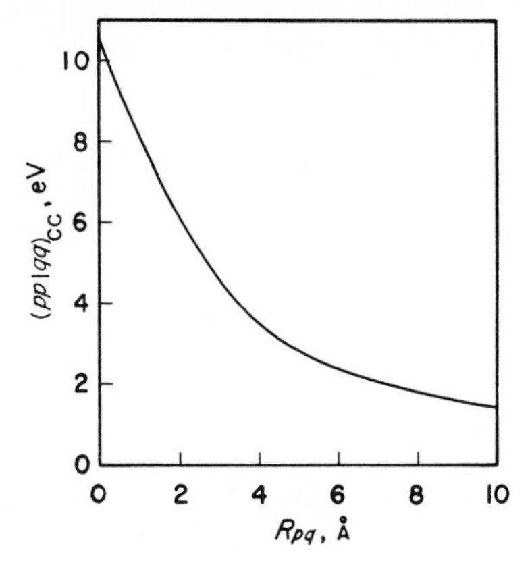

FIG. 11.1. The Coulomb repulsion integral $(pp \mid qq)_{CC}$ as a function of the interatomic distance R_{pq}.

A later development was to treat the larger of the Coulomb repulsion integrals as parameters to be determined empirically. For example, in order to get the best fit of the calculated energies of the lowest triplet and the lowest three singlet excited states of benzene to the experimental with the use of a value of −2.371 eV for β^{C} (see the succeeding subsection), Pariser[13] has assigned a value of 10.959 eV to $(pp \mid pp)_{CC}$ and values of 6.895,

[13] R. Pariser, *J. Chem. Phys.* **24**, 250 (1956).

5.682, and 4.978 ev to $(pp \mid qq)_{CC}$ at $R_{pq} = 1.390$, 2.407, and 2.780 Å, respectively, and has shown that these empirical values give very satisfactory calculations of the energy levels of larger aromatic hydrocarbons of the acene series. By the way, Pariser has used the values of $(pp \mid qq)_{CC}$ for $R_{pq} \geq 2.78$ Å calculated from Slater carbon $2p$ orbitals with $Z' = 3.18$, and has shown that these values are simply obtained exact to at least three decimal places from the largest three terms of the theoretical equation:

$$(pp \mid qq) = a/b - 3a/b^3 + 81a/4b^5, \tag{11.42}$$

where $a = Z'e^2/2a_0$, $b = Z'R_{pq}/2a_0$, and a_0 is the Bohr radius. However, these values are substantially the same as the values calculated from Eq. (11.40).

A more drastic approximation has been made which makes the calculations much simpler but still gives good results. This approximation is to neglect all the Coulomb repulsion integrals over AO's except the one-center integrals $(pp \mid pp)$ and the two-center integrals $(pp \mid qq)$ for neighboring AO's.[14,15]

11.3.3. Expansion of Core Integrals over MO's in Terms of Core Integrals over AO's and Evaluation of the Latter

In the **H** matrix elements core integrals over MO's of type $h^{C}_{[kr]}$ appear, in addition to electronic repulsion integrals over MO's. The core integrals over MO's can also be expanded into integrals over AO's:

$$h^{C}_{[kr]} = \sum_{p}\sum_{q} c_{k,p} c_{r,p} h^{C}_{(pq)}, \tag{11.43}$$

where $h^{C}_{(pq)}$ is a core integral over AO's, defined by

$$h^{C}_{(pq)} \equiv \langle \chi_p(i) | \, \mathbf{h}_i^{C} \, | \chi_q(i) \rangle. \tag{11.44}$$

The core Coulomb integrals $\alpha_p{}^{C}$ and the core resonance integrals β^{C}_{pq} are defined by

$$\alpha_p{}^{C} \equiv h^{C}_{(pp)} = \langle \chi_p(i) | \, \mathbf{h}_i^{C} | \chi_p(i) \rangle; \tag{11.45}$$

$$\beta^{C}_{pq} \equiv h^{C}_{(pq)} = \langle \chi_p(i) | \, \mathbf{h}_i^{C} \, | \chi_q(i) \rangle \quad \text{for } p \neq q. \tag{11.46}$$

Then Eq. (11.43) can be rewritten as

$$h^{C}_{[kr]} = \sum_{p} c_{k,p} c_{r,p} \alpha_p{}^{C} + \sum_{p}\sum_{q \neq p} c_{k,p} c_{r,q} \beta^{C}_{pq}. \tag{11.47}$$

It is to be recalled that in Hückel method two similar parameters, α and β, were used. However, these parameters are integrals involving an effective one-electron Hamiltonian, an operator which cannot be precisely defined. On the other hand, core integrals α^{C} and β^{C} involve the core one-electron Hamiltonian, an operator which is precisely defined. Therefore, in principle, their values can be calculated.

[14] H. C. Longuet-Higgins and L. Salem, *Proc. Roy. Soc.* (*London*) **A257**, 445 (1960).
[15] J. N. Murrell and L. Salem, *J. Chem. Phys.* **34**, 1914 (1961).

The core one-electron Hamiltonian for electron i, $\mathbf{h}_i^{\mathrm{C}}$, consists of the kinetic energy operator $\mathbf{T}(i)$ and the core potential field $\mathbf{V}^{\mathrm{C}}(i)$, that is, the potential field due to the molecule which has been stripped of all its π electrons [see Subsection 7.1.2, especially Eq. (7.5)]. This potential field can be written as a sum of contributions from each atom, that is, the core Hamiltonian can be written as

$$\mathbf{h}_i^{\mathrm{C}} = \mathbf{T}(i) + \sum_p \mathbf{V}_p(i) + \sum_v \mathbf{V}_v^*(i), \tag{11.48}$$

where the atoms p are charged in the core, that is, are π centers (for example, the carbon atoms in benzene) and the atoms v are uncharged in the core (for example, the hydrogen atoms in benzene), and where the asterisk connotes a potential due to a neutral atom.

The potential due to a distant atom of unit charge, say $\mathbf{V}_q$, is equal to the potential due to the neutral atom, $\mathbf{V}_q^*$, minus the potential due to the electron density $e\chi_q^*\chi_q$. Accordingly, we can write

$$\mathbf{V}_q(i) = \mathbf{V}_q^*(i) - z_q \int \chi_q^*(j)\chi_q(j)(e^2/r_{ij})\, d\tau_j\,, \tag{11.49}$$

where z_q is the charge on the atom q in the core, that is, the number of electrons that atom q contributes to the π-electron system. In hydrocarbons, and, in general, in molecules in which all the π centers are one-electron contributors, all the z_q's are, of course, unity.

The core Hamiltonian differs from molecule to molecule, and at different atoms in the same molecule, owing to the potential field due to distant atoms. If we substitute expression (11.48) into (11.45) and (11.46), we have

$$\begin{aligned}\alpha_p^{\mathrm{C}} = {} & \langle \chi_p(i)|\, \mathbf{T}(i) + \mathbf{V}_p(i)\, |\chi_p(i)\rangle + \sum_{q \neq p} \langle \chi_p(i)|\, \mathbf{V}_q(i)\, |\chi_p(i)\rangle \\ & + \sum_v \langle \chi_p(i)|\, \mathbf{V}_v^*(i)\, |\chi_p(i)\rangle;\end{aligned} \tag{11.50}$$

$$\begin{aligned}\beta_{pq}^{\mathrm{C}} = {} & \langle \chi_p(i)|\, \mathbf{T}(i) + \mathbf{V}_p(i) + \mathbf{V}_q(i)\, |\chi_q(i)\rangle + \sum_{t \neq p,q} \langle \chi_p(i)|\, \mathbf{V}_t(i)\, |\chi_q(i)\rangle \\ & + \sum_v \langle \chi_p(i)|\, \mathbf{V}_v^*(i)\, |\chi_q(i)\rangle.\end{aligned} \tag{11.51}$$

The first term in α_p^{C} can be taken as a constant for a given AO. If the AO χ_p is an eigenfunction of the isolated atom having the Hamiltonian $\mathbf{T} + \mathbf{V}_p$ so that

$$[\mathbf{T}(i) + \mathbf{V}_p(i)]\chi_p(i) = W_p\chi_p(i), \tag{11.52}$$

then this first term is just the energy of an electron in the orbital χ_p, W_p, which may be equated to the negative of the atomic ionization energy I_p.

By the use of expression (11.49), we can write

$$\langle \chi_p(i) | \mathbf{V}_q(i) | \chi_p(i) \rangle = -(q:pp) - z_q(pp \mid qq), \tag{11.53}$$

where $(pp \mid qq)$ is the two-center Coulomb repulsion integral [compare expression (11.30)] and $(q:pp)$ is an integral defined as

$$(q:pp) \equiv -\langle \chi_p(i) | \mathbf{V}_q{}^*(i) | \chi_p(i) \rangle. \tag{11.54}$$

The integral $(q:pp)$ is called the Coulomb penetration integral between χ_p and neutral atom q, and its negative represents the energy of interaction between the electron density $\chi_p{}^2$ and the neutral atom q. Thus expression (11.50) can be rewritten as

$$\alpha_p{}^{\mathrm{C}} = -I_p - \sum_{q \neq p} z_q(pp \mid qq) - \left[\sum_{q \neq p} (q:pp) + \sum_{v} (v:pp) \right]. \tag{11.55}$$

This equation is the fundamental formula for determination of the values of the core Coulomb integrals $\alpha_p{}^{\mathrm{C}}$.

In many cases, only differences of penetration integrals $(p:qq) - (q:pp)$ are needed in calculations of energies and intensities of electronic transitions. Pariser and Parr[5] assumed that the penetration integrals $(q:pp)$ were zero for $R_{pq} \geq 2.80$ Å, and gave formulas for determination of the differences $(p:qq) - (q:pp)$ for $R_{pq} \leq 2.80$ Å. However, later work has usually taken all the penetration integrals to be zero, which is more consistent with the zero-overlap approximation. At least in hydrocarbons, the penetration integrals may be assumed to make a small, constant contribution to each π center. That is, if we write

$$\alpha_p{}^{\mathrm{C}} = -I_p - \sum_{q \neq p} z_q(pp \mid qq) - C_p\,, \tag{11.56}$$

where C_p is the sum of penetration integrals in expression (11.55), the C_p may be taken to be a small constant and is usually neglected.

Now we turn to the core resonance integrals β_{pq}^{C}. The first term in expression (11.51) can be taken as a constant for given AO's and a given interatomic distance. In the zero-differential-overlap approximation all the remaining terms in β_{pq}^{C} are taken to be zero, since, for example,

$$\langle \chi_p(i) | \mathbf{V}_t(i) | \chi_q(i) \rangle = -(t:pq) - z_t(pq \mid tt). \tag{11.57}$$

This means that the core resonance integral can be taken to be a constant for given AO's and a given interatomic distance, being substantially independent of neighboring bonds or atoms. As in the similar approximation in the simple LCAO MO method, β_{pq}^{C} is usually taken to be zero except for atoms p and q directly bonded together, and β_{pq}^{C} for directly bonded atoms is treated as a basic empirical parameter which is a constant for a given type of bond.

Pariser and Parr[6] chose β^{C} values of -2.92 and -2.39 ev for the ethylenic bond (assumed to have a length of 1.35 Å) and a C—C bond in benzene (assumed to have a length of 1.39 Å), respectively. (Afterwards, Pariser[13] used a constant β^{C} value of -2.371 ev for all the C—C bonds in acenes, from benzene through pentacene.) Furthermore, they assumed the following relation between the core resonance integral β_{pq}^{C} and the bond length R_{pq} for C—C $2p\pi$–$2p\pi$ bonds, which fits the above values for ethylene and benzene:

$$\beta_{pq}^{\mathrm{C}} = -6442 \exp(-5.6864 R_{pq}) \text{ ev}. \tag{11.58}$$

On the basis of this assumption, a value of -1.68 ev was assigned to β^{C} for the central "single" bond in butadiene, which was assumed to have a length of 1.46 Å. As in the simple LCAO MO method, we may write

$$\beta_{pq}^{\mathrm{C}} = k_{pq}^{\mathrm{C}} \beta^{\mathrm{C}}, \tag{11.59}$$

where β^{C} is the core resonance integral for a standard C—C bond, for example, a C—C bond in benzene, and k_{pq}^{C} is a parameter which is a constant for a given bond and is taken to be zero if p and q are not directly bonded together.

Substitution of expressions (11.59) and (11.56) with neglect of penetration integrals into expression (11.47) leads to the following expression:

$$h_{[kr]}^{\mathrm{C}} = -\sum_{p} c_{k,p} c_{r,p} \Big[I_p + \sum_{q \neq p} z_q (pp \mid qq) \Big] + \sum_{p} \sum_{q \to p} c_{k,p} c_{r,q} k_{pq}^{\mathrm{C}} \beta^{\mathrm{C}}, \tag{11.60}$$

where the symbol $q \to p$ indicates that atom q is directly bonded to atom p. If r is set equal to k, we obtain an expression for the core energy of ψ_k, $\varepsilon_k{}^{\mathrm{C}}$ $(= h_{[kk]}^{\mathrm{C}})$.

Thus, when we have chosen a set of starting MO's and evaluated a relatively small number of Coulomb repulsion integrals over AO's of types $(pp \mid pp)$ and $(pp \mid qq)$ and the core resonance integrals for neighboring AO's, we can evaluate all the required matrix elements of the Hamiltonian, and when we have decided on the extent of configuration interaction to be taken into account, we can determine the approximate functions and energies of electronic states.

11.3.4. Formulation of F Matrix Elements

As is seen in expressions (11.23)–(11.25), matrix elements of **H** are conveniently expressed in terms of electronic repulsion integrals over MO's and **F** matrix elements between MO's. Matrix elements of $\mathbf{h}^{\mathrm{C}}$ between MO's appear in expressions for **F** matrix elements [(11.16) and (11.17)], but they do not appear explicitly in the expressions for matrix elements of **H**. Therefore, it seems to be convenient to formulate here the **F** matrix elements between MO's in terms of integrals over AO's.

The **F** matrix elements between MO's are expressed as follows [compare expression (11.16)]:

$$F_{[kr]} = h^{\mathrm{C}}_{[kr]} + \sum_{f}^{\mathrm{occ}} \{2[kr \mid ff] - [kf \mid rf]\}, \tag{11.61}$$

where the symbol $\sum^{\mathrm{occ}}$ represents the summation over all the MO's occupied in the ground electron configuration. When we expand the MO's by expression (11.28), we obtain

$$\begin{aligned} F_{[kr]} = \sum_{p}\sum_{q} c_{k,p}c_{r,q}F_{(pq)} &= \sum_{p}\sum_{q} \{c_{k,p}c_{r,q}h^{\mathrm{C}}_{(pq)} \\ &+ \sum_{f}^{\mathrm{occ}}(2c^{2}_{f,q}c_{k,p}c_{r,p} - c_{f,p}c_{f,q}c_{k,p}c_{r,q})(pp \mid qq)\}, \end{aligned} \tag{11.62}$$

where $F_{(pq)}$ is a quantity called the matrix element of **F** between AO's χ_p and χ_q. We now introduce the π-electron density P_{qq} and the π-bond order P_{pq}:

$$P_{qq} = 2\sum_{f}^{\mathrm{occ}} c^{2}_{f,q}\,; \qquad P_{pq} = 2\sum_{f}^{\mathrm{occ}} c_{f,p}c_{f,q}\,. \tag{11.63}$$

Expression (11.62) then becomes

$$\begin{aligned} \sum_{p}\sum_{q} c_{k,p}c_{r,q}F_{(pq)} = \sum_{p}\sum_{q} &\{c_{k,p}c_{r,q}h^{\mathrm{C}}_{(pq)} \\ &+ [c_{k,p}c_{r,p}P_{qq} - \tfrac{1}{2}c_{k,p}c_{r,q}P_{pq}](pp \mid qq)\}. \end{aligned} \tag{11.64}$$

Equating like terms on both sides of this expression leads to the following expressions:

$$F_{(pq)} = \beta^{\mathrm{C}}_{pq} - \tfrac{1}{2}P_{pq}(pp \mid qq), \qquad \text{for } p \neq q; \tag{11.65}$$

$$F_{(pp)} = \alpha_{p}{}^{\mathrm{C}} + \tfrac{1}{2}P_{pp}(pp \mid pp) + \sum_{q \neq p} P_{qq}(pp \mid qq). \tag{11.66}$$

Substitution for $\alpha_{p}{}^{\mathrm{C}}$ in expression (11.66) by expression (11.56) with neglect of C_p gives

$$F_{(pp)} = -I_p + \tfrac{1}{2}P_{pp}(pp \mid pp) + \sum_{q \neq p}(P_{qq} - z_q)(pp \mid qq). \tag{11.67}$$

Expressions (11.65) and (11.67) have been given by Pople.[7] The **F** matrix elements between MO's are expressed in terms of the **F** matrix elements between AO's as follows:

$$F_{[kr]} = \sum_p c_{k,p} c_{r,p} F_{(pp)} + \sum_p \sum_{q \neq p} c_{k,p} c_{r,q} F_{(pq)} . \qquad (11.68)$$

When r is set equal to k in this expression, we obtain an expression for the diagonal elements of the **F** matrix for MO's, that is, the zeroth-order SCF orbital energies, $\varepsilon_k{}^{\mathrm{S}}$ $(= F_{[kk]})$.

The steps of a calculation by the Pariser–Parr method may be summarized as follows:

(1) Calculate the starting MO's by the simple LCAO MO method.

(2) Calculate the bond-order matrix elements (that is, the π-electron densities P_{qq} and the π-bond orders P_{pq}) [compare expressions (11.63)].

(3) Determine the valence state ionization energies I_p or the valence state electron affinities A_p, the core resonance integrals β_{pq}^{C} [compare expression (11.58)], and the Coulomb repulsion integrals $(pp \mid pp)$ and $(pp \mid qq)$ [compare expressions (11.39)–(11.41)].

(4) Calculate the **F** matrix elements between AO's, $F_{(pq)}$ [compare expressions (11.65) and (11.67)].

(5) Calculate the **F** matrix elements between MO's, $F_{[kr]}$ [compare expression (11.68)], and the electronic repulsion integrals over MO's, $[kl \mid rs]$ [compare expressions (11.34)–(11.36)], needed for the Hamiltonian matrix elements [compare expressions (11.23)–(11.25)].

(6) Diagonalize the Hamiltonian matrix.

11.4. The Semiempirical SCF LCAO MO Method (The Pople Method)

In the SCF MO method the wavefunctions of electronic states are written as antisymmetrized products of MO's as in the ASMO method, but the MO's are determined so as to give the best possible total energy. In this method the Hamiltonian for a π electron in a π-electron system whose ground electron configuration is a closed-shell configuration is written as follows:[3]

$$\mathbf{h}^{\mathrm{SCF}} = \mathbf{h}^{\mathrm{C}} + \sum_f^{\mathrm{occ}} (2\mathbf{J}_f - \mathbf{K}_f) \equiv \mathbf{F} , \qquad (11.69)$$

where $\mathbf{J}_f$ and $\mathbf{K}_f$ are operators called the Coulomb and exchange operators, respectively, and defined by

$$\mathbf{J}_f \psi_k(i) = \left[\int \psi_f{}^2(j)(e^2/r_{ij})\, d\tau_j \right] \psi_k(i); \qquad (11.70)$$

$$\mathbf{K}_f \psi_k(i) = \left[\int \psi_f(j) \psi_k(j)(e^2/r_{ij})\, d\tau_j \right] \psi_f(i). \qquad (11.71)$$

The matrix elements of $\mathbf{J}_f$ and $\mathbf{K}_f$ between MO's, say ψ_k and ψ_r, are given by

$$\langle\psi_k| \mathbf{J}_f |\psi_r\rangle = [kr \mid ff]; \tag{11.72}$$

$$\langle\psi_k| \mathbf{K}_f |\psi_r\rangle = [kf \mid rf]. \tag{11.73}$$

The SCF MO's are to be the eigenfunctions of the SCF Hamiltonian, $\mathbf{h}^{\text{SCF}}$ or $\mathbf{F}$, that is, must satisfy the following relation:

$$\mathbf{F}\psi_k(i) = \varepsilon_k\psi_k(i). \tag{11.74}$$

The foregoing $\mathbf{F}$ matrix elements, given by expression (11.61), are just the matrix elements of the SCF Hamiltonian $\mathbf{F}$. If the SCF MO's are to be expressed as linear combinations of AO's as in expression (11.28), the matrix elements can be expanded in terms of the $\mathbf{F}$ matrix elements between AO's as in expression (11.68).

The SCF MO's for the closed-shell ground electron configuration can be defined as those which give the best single determinant wavefunction for the ground state, not needing to be improved by configuration interaction with any singly-excited configurations. If this is so, the matrix elements of the total electronic Hamiltonian $\mathbf{H}$ such as given by expression (11.23) must be zero. This implies that the $\mathbf{F}$ matrix elements between different MO's $F_{[kr]}$ must be zero, and this amounts to saying that the SCF MO's must be eigenfunctions of the SCF Hamiltonian $\mathbf{F}$.

If both sides of the SCF equation (11.74) are multiplied by $\psi_k(i)$ and integrated, there results

$$F_{[kk]} = \varepsilon_k S_{[kk]}, \tag{11.75}$$

where $S_{[kk]}$ is the overlap integral $\langle\psi_k \mid \psi_k\rangle$. Substitution for the MO by the LCAO expansion gives

$$\sum_p \sum_q c_{k,p}c_{k,q}F_{(pq)} = \varepsilon_k \sum_p \sum_q c_{k,p}c_{k,q}S_{(pq)}. \tag{11.76}$$

By the use of the variation principle (compare Section 2.2), and assuming that the $F_{(pq)}$'s were constants, we obtain a set of equations of the following type:

$$\sum_q c_{k,q}(F_{(pq)} - \varepsilon_k S_{(pq)}) = 0, \qquad \text{for each} \quad p. \tag{11.77}$$

This would be the familiar set of secular equations if the $\mathbf{F}$ matrix elements between AO's, $F_{(pq)}$, did not also depend upon the $c_{k,p}$'s. An iterative procedure must be used for the solution of these equations. Starting with a set of approximate MO's, say those obtained by the simple LCAO MO method, one calculates the bond-order matrix elements P_{pq}, and hence the $\mathbf{F}$ matrix elements between AO's, $F_{(pq)}$, by means of Eqs. (11.65) and (11.67). One can then calculate the energies of a new set of MO's by solving the secular determinantal equation

$$|F_{(pq)} - \varepsilon S_{(pq)}| = 0, \tag{11.78}$$

that is, by diagonalizing the **F** matrix for AO's, and hence one can determine the coefficients of AO's in the MO's in the usual manner (compare Subsection 2.3.2). If the π-electron system has $2m$ π electrons, the $2m$ π electrons are allotted into the m lowest of these new MO's in the ground electron configuration. These m MO's are used to recalculate the bond-order matrix elements and hence the **F** matrix elements between AO's. This process is repeated until self-consistency is reached, that is, until a new set of MO's is essentially the same as its precursor.

The m lowest of the SCF MO's are used in building up the ground state wavefunction. In addition to these m MO's, a number of higher energy MO's are determined at the same time, which are called "virtual" MO's. The virtual MO's can be used to build up excited configurations. If the MO's are completely self-consistent, the ground state function is, by definition, the correct function, and cannot be improved by configuration interaction with singly-excited configurations, that is, all the **H** matrix elements between the ground and singly-excited configurations are zero. However, if the MO's are not completely self-consistent, the ground state function can be improved by mixing it with singly-excited configurations, although this configuration interaction is smaller than with less satisfactory MO's. The closer the MO's are to the correct SCF MO's, the less configuration interaction will have to be invoked. The best single determinant wavefunction for the ground state can still be improved by mixing it with doubly-excited configurations.

Even when the MO's are completely self-consistent for the ground state, the Hamiltonian matrix elements between excited configurations can be nonzero [compare expressions (11.24) and (11.25)]. Therefore, nearly correct wavefunctions of excited state must be obtained by including configuration interaction between excited configurations. However, the use of the SCF MO's will likely somewhat reduce the importance of the configuration interaction.

So far only the SCF method for closed-shell configurations has been dealt with. The problem of obtaining SCF orbitals for an open-shell configuration is rather more difficult than for a closed-shell configuration. This problem has been discussed by Pople and Nesbet,[16] by Longuet-Higgins and Pople,[17] by Lefebvre,[18] and by Roothaan.[19]

[16] J. A. Pople and R. K. Nesbet, *J. Chem. Phys.* **22,** 571 (1954).

[17] H. C. Longuet-Higgins and J. A. Pople, *Proc. Phys. Soc.* (*London*) **A68,** 591 (1955).

[18] R. Lefebvre, *J. Chim. Phys.* **54,** 168 (1957).

[19] C. C. J. Roothaan, *Rev. Modern Phys.* **32,** 179 (1960).

11.5. Application of the Pariser–Parr Method to Even Alternant Hydrocarbons

11.5.1. PRELIMINARY CONSIDERATIONS

For even alternant hydrocarbons, owing to the similarity of all the π centers in a molecule and also to the pairing properties of the π MO's, the formulation of the **H** matrix elements between electron configurations is markedly simplified.

Now we take as the starting MO's the orthonormal MO's derived through the simple LCAO MO method. The MO's are given by expression (11.28), and their energies are expressed as follows:

$$\varepsilon_k = \alpha - w_k \beta. \tag{11.79}$$

The orthonormality of the MO's is expressed as:

$$\sum_p c_{k,p} c_{r,p} = \delta_{kr}. \tag{11.80}$$

As mentioned in Section 7.2, in an even alternant hydrocarbon, it is possible to divide the carbon π centers into two classes, "starred" and "unstarred." The number of starred π centers and that of unstarred π centers are equal to each other, and the bonding MO's ψ_{+k}, which have negative w values and are occupied in the ground configuration, and the antibonding MO's ψ_{-k}, which have positive w values and are normally unoccupied, are related by pairs, that is, the orbital energies are related by

$$w_{+k} = -w_{-k}, \tag{11.81}$$

and, with regard to the coefficients of AO's in these MO's, there are the following relations:

$$c_{+k,p} = c_{-k,p}, \quad \text{if } p \text{ is a starred } \pi \text{ center;} \tag{11.82}$$

$$c_{+k,p} = -c_{-k,p}, \quad \text{if } p \text{ is an unstarred } \pi \text{ center.} \tag{11.83}$$

In addition,

$$P_{pp}\left(= 2 \sum_f c_{+f,p}^2\right) = 1, \tag{11.84}$$

in which p may be starred or not.

Symbols (s) and (d) are introduced: p, q(s) indicates that p and q belong to the same class, that is, p and q are both starred or both unstarred; p, q(d) indicates that p and q belong to different classes, that is, one of p and q is starred and the other unstarred. Then, Theorem 8 in Section 7.2 can be expressed as follows:

$$P_{pq}\left(=2\sum_f c_{+f,p}c_{+f,q}\right)=0, \qquad \text{if} \quad p, q(\text{s}) \quad \text{and} \quad p \neq q. \tag{11.85}$$

Equations (11.81)–(11.85) summarize the Coulson–Rushbrooke theorems which are most useful for the present purpose. As has been shown by Pople,[7] and as will be verified later, not only the set of MO's derived by the simple LCAO MO method, but also a set of MO's obtained at each stage of the iterative procedure of the Pople SCF LCAO MO method has similar pairing properties. Therefore, even when such a set of MO's is used as the starting set of MO's in place of the set of MO's obtained by the simple LCAO MO method, the following discussions still are valid.

By the use of the symbols (s) and (d), the MO electronic repulsion integrals can be expressed as follows:

$$[kr \mid ls] = [kr \mid ls](\text{s}) + [kr \mid ls](\text{d}), \tag{11.86}$$

where $[kr \mid ls](\text{s})$ represents the sum of the terms with the AO integrals $(pp \mid qq)$ for $p, q(\text{s})$, that is, $\sum_p \sum_q c_{k,p}c_{r,p}c_{l,q}c_{s,q}(pp \mid qq)$ for $p, q(\text{s})$, and $[kr \mid ls](\text{d})$ represents the sum of the terms with the AO integrals $(pp \mid qq)$ for $p, q(\text{d})$. Similarly, we can write

$$J_{kr} = J_{kr}(\text{s}) + J_{kr}(\text{d}); \tag{11.87}$$

$$K_{kr} = K_{kr}(\text{s}) + K_{kr}(\text{d}). \tag{11.88}$$

The following relations are derived from the pairing properties of the AO coefficients of the MO's [compare Eqs. (11.82) and (11.83)]:

$$J_{+k,+r} = J_{+k,-r} = J_{-k,+r} = J_{-k,-r}\,; \tag{11.89}$$

$$K_{+k,-r} = K_{-k,+r}\,, \qquad K_{+k,+r} = K_{-k,-r}\,; \tag{11.90}$$

$$\begin{aligned}[][+k, -r \mid +l, -s] &= [+k, -r \mid -l, +s] = [-k, +r \mid +l, -s] \\ &= [-k, +r \mid -l, +s], \end{aligned} \tag{11.91}$$

$$[+k, +l \mid -r, -s] = [-k, -l \mid -r, -s] = [+k, +l \mid +r, +s];$$

$$K_{f,-k}(\text{s}) = K_{f,+k}(\text{s}), \qquad K_{f,-k}(\text{d}) = -K_{f,+k}(\text{d}); \tag{11.92}$$

$$\begin{aligned}[][f, -r \mid g, -s](\text{s}) &= [f, +r \mid g, +s](\text{s}), \\ [f, -r \mid g, -s](\text{d}) &= -[f, +r \mid g, +s](\text{d}); \end{aligned} \tag{11.93}$$

$$[f, -k \mid f, +k](\text{d}) = -[f, +k \mid f, -k](\text{d}) = 0. \tag{11.94}$$

For a hydrocarbon, since all the π centers are carbon atoms, the atomic ionization energy I_p, the atomic electron affinity A_p, and the one-center Coulomb repulsion integral $(pp \mid pp)$ can be taken to be constants for all the π centers. These constants are denoted by I_C, A_C, and γ_C, respectively.

From relations (11.80), (11.84), and (11.85) the following relations are derived:

$$\sum_f [+f, +f \mid k, r] = \tfrac{1}{2} \sum_p \sum_q P_{qq} c_{k,p} c_{r,p} (pp \mid qq)$$
$$= \tfrac{1}{2} \sum_p \sum_q c_{k,p} c_{r,p} (pp \mid qq)$$
$$= \frac{1}{2}\Big[\delta_{kr}\gamma_C + \sum_p \sum_{q \neq p} c_{k,p} c_{r,p} (pp \mid qq)\Big]; \tag{11.95}$$

$$\sum_f [+f, k \mid +f, r](\text{s}) = \tfrac{1}{2} \sum_p \sum_q P_{pq} c_{k,p} c_{r,q} (pp \mid qq)(\text{s})$$
$$= \tfrac{1}{2} \sum_p c_{k,p} c_{r,p} (pp \mid pp) = \tfrac{1}{2}\, \delta_{kr}\gamma_C; \tag{11.96}$$

$$\sum_f [+f, k \mid +f, r](\text{d}) = \tfrac{1}{2} \sum_p \sum_q P_{pq} c_{k,p} c_{r,q} (pp \mid qq)(\text{d}). \tag{11.97}$$

When $r = k$, these expressions are expressions for $\sum_f J_{+f,k}$, $\sum_f K_{+f,k}(\text{s})$, and $\sum_f K_{+f,k}(\text{d})$, respectively.

Since the z_q is unity for every carbon π center, if the sum of penetration integrals C_p is neglected, for hydrocarbons, expression (11.56) can be rewritten as follows:

$$\alpha_p{}^{C} = -I_C - \sum_{q \neq p} (pp \mid qq) = -A_C - \gamma_C - \sum_{q \neq p} (pp \mid qq)$$
$$= -A_C - \sum_q (pp \mid qq). \tag{11.98}$$

Even when C_p is not neglected, if it is taken to be the same for every carbon π center, it can be considered as having been absorbed in the constants I_C and A_C, since the values of these constants do not appear in the electronic excitation energies, just as α does not appear in the excitation energies in the simple LCAO MO approximation.

Since the MO's ψ are orthonormal eigenfunctions of an effective one-electron Hamiltonian $\mathbf{h}$, the following relation exists:

$$\langle \psi_k | \, \mathbf{h} \, | \psi_r \rangle = \delta_{kr} \varepsilon_k \,. \tag{11.99}$$

By expanding the MO's in this expression, we obtain

$$\sum_p c_{k,p} c_{r,p} \alpha + \sum_p \sum_{q \neq p} c_{k,p} c_{r,q} k_{pq} \beta = \delta_{kr} \varepsilon_k \,. \tag{11.100}$$

Therefore, by the use of relations (11.79) and (11.80), the following relation is derived:

$$\sum_{p}\sum_{q\neq p} c_{k,p}c_{r,q}k_{pq} = -\delta_{kr}w_k. \tag{11.101}$$

For simplification, parameter k^{C}_{pq} in expression (11.59) is set equal to parameter k_{pq} used in the simple LCAO MO method. Then, by the use of relations (11.80), (11.95), and (11.101), we can reduce the expression (11.60) for the MO core integrals to the following expression:

$$\begin{aligned} h^{\mathrm{C}}_{[kr]} &= -\delta_{kr}I_{\mathrm{C}} - \sum_{p}\sum_{q\neq p} c_{k,p}c_{r,p}(pp \mid qq) + \sum_{p}\sum_{q\neq p} c_{k,p}c_{r,q}k^{\mathrm{C}}_{pq}\beta^{\mathrm{C}} \\ &= -\delta_{kr}A_{\mathrm{C}} - 2\sum_{f}[+f, +f \mid k, r] - \delta_{kr}w_k\beta^{\mathrm{C}}. \end{aligned} \tag{11.102}$$

When $r = k$, we have

$$h^{\mathrm{C}}_{[kk]} = \varepsilon_k{}^{\mathrm{C}} = -A_{\mathrm{C}} - 2\sum_{f} J_{+f,k} - w_k\beta^{\mathrm{C}}. \tag{11.103}$$

If as usual the core resonance integrals β^{C}_{pq} and hence the core resonance parameters k^{C}_{pq} are taken to be nonzero only when p and q are directly bonded together, evidently the **F** matrix elements between AO's [compare expressions (11.65) and (11.67)] for even alternant hydrocarbons can be expressed as follows:

$$F_{(pq)} = k^{\mathrm{C}}_{pq}\beta^{\mathrm{C}} - \tfrac{1}{2}P_{pq}(pp \mid qq), \qquad \text{for} \quad p, q(\mathrm{d}); \tag{11.104}$$

$$F_{(pq)} = \delta_{pq}F_{(pp)}, \qquad \text{for} \quad p, q(\mathrm{s}); \tag{11.105}$$

$$F_{(pp)} = -I_{\mathrm{C}} + \tfrac{1}{2}\gamma_{\mathrm{C}} = -A_{\mathrm{C}} - \tfrac{1}{2}\gamma_{\mathrm{C}} \equiv F_{\mathrm{C}}. \tag{11.106}$$

The last equation shows that in an even alternant hydrocarbon $F_{(pp)}$ is the same for each carbon π center. This constant, F_{C}, does not appear in the expressions for the electronic excitation energies, just as α in the simple LCAO MO approximation.

As usual, the zero-overlap approximation (that is, $S_{(pq)} = \delta_{pq}$) is adopted. Then, by the use of the above expressions for the **F** matrix elements between AO's, the SCF secular equations (11.77) can be rewritten as

$$c^{\mathrm{S}}_{k,p}(F_{\mathrm{C}} - \varepsilon_k{}^{\mathrm{S}}) + \sum_{q\neq p} c^{\mathrm{S}}_{k,q}F_{(pq)} = 0 \qquad \text{for each} \quad p, \tag{11.107}$$

in which p and q belong to different classes, and the superscript S indicates quantities associated with the MO's to be determined by solving these equations. Now we take the following expressions for $F_{(pq)}$ and $\varepsilon_k{}^{\mathrm{S}}$:

$$F_{(pq)} = k^{\mathrm{S}}_{pq}F_{\mathrm{CC}}; \tag{11.108}$$

$$\varepsilon_k{}^{\mathrm{S}} = F_{\mathrm{C}} - w_k{}^{\mathrm{S}}F_{\mathrm{CC}}. \tag{11.109}$$

F_{CC} is a constant representing the $F_{(pq)}$ value for a standard C—C bond, and k_{pq}^{S} is a parameter, which is a constant for a given pair of p and q belonging to different classes, but which is zero for a pair of p and q belonging to the same class. After being divided through by F_{CC}, the secular equations (11.107) reduce to

$$c_{k,p}^{S} w_k^{S} + \sum_{q \neq p} c_{k,q}^{S} k_{pq}^{S} = 0 \qquad \text{for each} \quad p, \tag{11.110}$$

in which p and q belong to different classes. The form of these secular equations resembles closely that of the secular equations in the simple LCAO MO method [compare Eqs. (7.76)]. If w_{+k}^{S}, $c_{+k,p}^{S}$, and $c_{+k,q}^{S}$ define one solution of these secular equations, the set of $-w_{+k}^{S}$ $(= w_{-k}^{S})$, $c_{+k,p}^{S}$ $(= c_{-k,p}^{S})$, and $-c_{+k,q}^{S}$ $(= c_{-k,q}^{S})$ is also a solution. If p is starred, q's are unstarred. It has been thus verified that in an even alternant hydrocarbon the MO's to be obtained by solving the SCF secular equations have the pairing properties similar to those of the MO's derived by the simple LCAO MO method.

Substituting for $F_{(pp)}$ and $F_{(pq)}$ in expression (11.68) by (11.104)–(11.106), we obtain the following expression for the **F** matrix elements between MO's:

$$\begin{aligned} F_{[kr]} &= \sum_p c_{k,p} c_{r,p} F_C + \sum_p \sum_{q \neq p} c_{k,p} c_{r,q} [k_{pq}^{C} \beta^{C} - \tfrac{1}{2} P_{pq} (pp \mid qq)(\mathrm{d})] \\ &= \delta_{kr} (F_C - w_k \beta^{C}) - \sum_f [+f, k \mid +f, r](\mathrm{d}). \end{aligned} \tag{11.111}$$

According to this equation, the zeroth-order SCF orbital energies of paired MO's can be expressed as follows:

$$\varepsilon_{-k}^{S} = F_{[-k,-k]} = F_C + \kappa_k \,; \qquad \varepsilon_{+k}^{S} = F_{[+k,+k]} = F_C - \kappa_k , \tag{11.112}$$

where

$$\kappa_k \equiv w_{-k}(-\beta^{C}) + \sum_f K_{+f,+k}(\mathrm{d}). \tag{11.113}$$

11.5.2. The H Matrix Elements Involving the Ground and Singly-Excited Electron Configurations of an Even Alternant Hydrocarbon

Now let us formulate the **H** matrix elements between electron configurations of an even alternant hydrocarbon, on the basis of the preliminary considerations made in the preceding subsection.

The energy associated with the ground configuration is given by expression (11.18):

$$E(V_0) = \sum_k (\varepsilon_{+k}^{C} + \varepsilon_{+k}^{S}). \tag{11.114}$$

The core orbital energies $\varepsilon^{\mathrm{C}}_{+k}$ and the zeroth-order SCF orbital energies $\varepsilon^{\mathrm{S}}_{+k}$ are given by expressions (11.103) and (11.112), respectively, and F_{C} is given by expression (11.106). Suppose that the molecule has $2m$ carbon π centers and hence $2m$ π electrons. Then, needless to say, the molecule has m bonding and m antibonding π MO's. Evidently, we have the following relations:

$$\begin{aligned} 2\sum_k \sum_f J_{+f,+k} &= \sum_k \sum_p \sum_q c^2_{+k,p}(pp \mid qq) \\ &= \tfrac{1}{2} \sum_p \sum_q (pp \mid qq) \\ &= m\gamma_{\mathrm{C}} + \tfrac{1}{2} \sum_p \sum_{q \neq p} (pp \mid qq); \end{aligned} \tag{11.115}$$

$$\begin{aligned} \sum_k \sum_f K_{+f,+k}(\mathrm{d}) &= \sum_k \sum_f \sum_p \sum_q (c_{+f,p} c_{+f,q})(c_{+k,p} c_{+k,q})(pp \mid qq)(\mathrm{d}) \\ &= \tfrac{1}{4} \sum_p \sum_q P^2_{pq}(pp \mid qq)(\mathrm{d}). \end{aligned} \tag{11.116}$$

Therefore, $E(V_0)$ can be expressed as follows:

$$\begin{aligned} E(V_0) = &- m[A_{\mathrm{C}} + \gamma_{\mathrm{C}}] - \tfrac{1}{2} \sum_p \sum_{q \neq p} (pp \mid qq) - \sum_k w_{+k} \beta^{\mathrm{C}} \\ &- m[A_{\mathrm{C}} + \tfrac{1}{2}\gamma_{\mathrm{C}}] - \tfrac{1}{4} \sum_p \sum_q P^2_{pq}(pp \mid qq)(\mathrm{d}) - \sum_k w_{+k} \beta^{\mathrm{C}} \\ = &-m[2A_{\mathrm{C}} + \tfrac{3}{2}\gamma_{\mathrm{C}}] - \tfrac{1}{2} \sum_p \sum_{q \neq p} (pp \mid qq) \\ &- \tfrac{1}{4} \sum_p \sum_q P^2_{pq}(pp \mid qq)(\mathrm{d}) - 2 \sum_k w_{+k} \beta^{\mathrm{C}}. \end{aligned} \tag{11.117}$$

In building up the secular determinant for determination of state functions as linear combinations of configuration functions, the energy of the ground configuration, $E(V_0)$, is usually taken as the reference zero energy, that is, the energies of excited configurations are expressed as the values relative to $E(V_0)$. Of course, $E(V_0)$ does not appear in the expressions for off-diagonal matrix elements. Therefore, $E(V_0)$ does not appear in the secular determinant.

Only the ground and singly-excited configurations are taken into account. As usual, the antisymmetrized functions of the singly-excited configurations arising from the ground configuration by promoting one electron from an occupied orbital, say ψ_{+k}, to an unoccupied orbital, say ψ_{-r} (that is, $V_1^{+k,-r}$ and $T_1^{+k,-r}$) are simply expressed as V_1^{kr} and T_1^{kr}, respectively. Then, by substituting the appropriate expression (11.111) for $F_{[kr]}$'s in the general expressions (11.23)–(11.25), and by using the pairing properties of the MO's, we can express the matrix elements of $\mathbf{H}$ between the ground and singly-excited configurations in the following compact form:

$$\langle V_1^{kr}|\,\mathbf{H}\,|V_0\rangle = -\langle V_1^{rk}|\,\mathbf{H}\,|V_0\rangle = \sqrt{2}\sum_f [+f, -k \mid +f, +r](\mathrm{d}); \tag{11.118}$$

$$\begin{aligned}\langle V_1^{kr}|\,\mathbf{H}\,|V_1^{ls}\rangle - \delta_{+k,+l}\,\delta_{-r,-s}E(V_0) &= \delta_{+k,+l}\Big\{\delta_{-r,-s}(F_{\mathrm{C}} - w_{-r}\beta^{\mathrm{C}}) - \sum_f [+f, -r \mid +f, -s](\mathrm{d})\Big\} \\ &\quad - \delta_{-r,-s}\Big\{\delta_{+k,+l}(F_{\mathrm{C}} - w_{+k}\beta^{\mathrm{C}}) - \sum_f [+f, +k \mid +f, +l](\mathrm{d})\Big\} \\ &\quad - [+k, +l \mid -r, -s] + 2[+k, -r \mid +l, -s];\end{aligned} \tag{11.119}$$

$$\begin{aligned}\langle T_1^{kr}|\,\mathbf{H}\,|T_1^{ls}\rangle - \delta_{+k,+l}\,\delta_{-r,-s}E(V_0) &= \delta_{+k,+l}\Big\{\delta_{-r,-s}(F_{\mathrm{C}} - w_{-r}\beta^{\mathrm{C}}) - \sum_f [+f, -r \mid +f, -s](\mathrm{d})\Big\} \\ &\quad - \delta_{-r,-s}\Big\{\delta_{+k,+l}(F_{\mathrm{C}} - w_{+k}\beta^{\mathrm{C}}) - \sum_f [+f, +k \mid +f, +l](\mathrm{d})\Big\} \\ &\quad - [+k, +l \mid -r, -s].\end{aligned} \tag{11.120}$$

For a special case where $r = k$, owing to relation (11.94), expression (11.118) reduces to

$$\langle V_1^{kk}|\,\mathbf{H}\,|V_0\rangle = \sqrt{2}\sum_f [+f, -k \mid +f, +k](\mathrm{d}) = 0. \tag{11.121}$$

Similarly, for special cases, expressions (11.119) and (11.120) reduce to the following simple expressions:

For $k = l \neq r = s$,

$$\begin{aligned}E(V_1^{kr}) - E(V_0) = E(V_1^{rk}) - E(V_0) &= \varepsilon_{-r}^{\mathrm{S}} - \varepsilon_{+k}^{\mathrm{S}} - J_{+k,-r} + 2K_{+k,-r} \\ &= \kappa_r + \kappa_k - J_{+k,+r} + 2[K_{+k,+r}(\mathrm{s}) - K_{+k,+r}(\mathrm{d})];\end{aligned} \tag{11.122}$$

$$E(T_1^{kr}) - E(V_0) = E(T_1^{rk}) - E(V_0) = \kappa_r + \kappa_k - J_{+k,+r}, \tag{11.123}$$

where κ's are given by expression (11.113).

For $k = l = r = s$,

$$\begin{aligned}E(V_1^{kk}) - E(V_0) &= \varepsilon_{-k}^{\mathrm{S}} - \varepsilon_{+k}^{\mathrm{S}} - J_{+k,-k} + 2K_{+k,-k} \\ &= 2\kappa_k + J_{+k,+k}(\mathrm{s}) - 3J_{+k,+k}(\mathrm{d});\end{aligned} \tag{11.124}$$

$$E(T_1^{kk}) - E(V_0) = 2\kappa_k - J_{+k,+k}\,. \tag{11.125}$$

For $k \neq l \neq r \neq s$,

$$\begin{aligned}\langle V_1^{kr}| \mathbf{H} |V_1^{ls}\rangle &= \langle V_1^{rk}| \mathbf{H} |V_1^{sl}\rangle \\ &= -[+k, +l \mid -r, -s] + 2[+k, -r \mid +l, -s] \\ &= -[+k, +l \mid +r, +s] \\ &\quad + 2\{[+k, +r \mid +l, +s](\mathrm{s}) - [+k, +r \mid +l, +s](\mathrm{d})\};\end{aligned} \tag{11.126}$$

$$\langle T_1^{kr}| \mathbf{H} |T_1^{ls}\rangle = \langle T_1^{rk}| \mathbf{H} |T_1^{sl}\rangle = -[+k, +l \mid +r, +s]. \tag{11.127}$$

For $k \neq r \neq l = s$,

$$\begin{aligned}\langle V_1^{kr}| \mathbf{H} |V_1^{ll}\rangle &= \langle V_1^{rk}| \mathbf{H} |V_1^{ll}\rangle \\ &= -[+k, +l \mid -r, -l] + 2[+k, -r \mid +l, -l] \\ &= -[+k, +l \mid +r, +l] \\ &\quad + 2\{[+k, +r \mid +l, +l](\mathrm{s}) - [+k, +r \mid +l, +l](\mathrm{d})\};\end{aligned} \tag{11.128}$$

$$\langle T_1^{kr}| \mathbf{H} |T_1^{ll}\rangle = \langle T_1^{rk}| \mathbf{H} |T_1^{ll}] = -[+k, +l \mid +r, +l]. \tag{11.129}$$

For $k = r \neq l = s$,

$$\begin{aligned}\langle V_1^{kk}| \mathbf{H} |V_1^{ll}\rangle &= -[+k, +l \mid -k, -l] + 2[+k, -k \mid +l, -l] \\ &= -K_{+k,+l} + 2[J_{+k,+l}(\mathrm{s}) - J_{+k,+l}(\mathrm{d})];\end{aligned} \tag{11.130}$$

$$\langle T_1^{kk}| \mathbf{H} |T_1^{ll}\rangle = -K_{+k,+l}\,. \tag{11.131}$$

For $k = l \neq r \neq s$ or $k \neq l \neq r = s$,

$$\begin{aligned}\langle V_1^{kr}| \mathbf{H} |V_1^{ks}\rangle &= [V_1^{rk}| \mathbf{H} |V_1^{sk}\rangle \\ &= -\sum_f [+f, -r \mid +f, -s](\mathrm{d}) \\ &\quad - [+k, +k \mid -r, -s] + 2[+k, -r \mid +k, -s] \\ &= \sum [+f, +r \mid +f, +s](\mathrm{d}) - [+k, +k \mid +r, +s] \\ &\quad + 2\{[+k, +r \mid +k, +s](\mathrm{s}) - [+k, +r \mid +k, +s](\mathrm{d})\};\end{aligned} \tag{11.132}$$

$$\begin{aligned}\langle T_1^{kr}| \mathbf{H} |T_1^{ks}\rangle &= \langle T_1^{rk}| \mathbf{H} |T_1^{sk}\rangle \\ &= \sum_f [+f, +r \mid +f, +s](\mathrm{d}) - [+k, +k \mid +r, +s].\end{aligned} \tag{11.133}$$

For $k = l = s \neq r$ or $k = r = s \neq l$,

$$\begin{aligned}\langle V_1^{kr}|\,\mathbf{H}\,|V_1^{kk}\rangle &= \langle V_1^{rk}|\,\mathbf{H}\,|V_1^{kk}\rangle \\ &= \sum_f [+f, +k \mid +f, +r](\mathrm{d}) \\ &\quad - [+k, +k \mid +k, +r] + 2[+k, -r \mid +k, -k] \\ &= \sum_f [+f, +k \mid +f, +r](\mathrm{d}) \\ &\quad + [+k, +k \mid +k, +r](\mathrm{s}) - 3[+k, +k \mid +k, +r](\mathrm{d});\end{aligned} \tag{11.134}$$

$$\begin{aligned}\langle T_1^{kr}|\,\mathbf{H}\,|T_1^{kk}\rangle &= \langle T_1^{rk}|\,\mathbf{H}\,|T_1^{kk}\rangle \\ &= \sum_f [+f, +k \mid +f, +r](\mathrm{d}) - [+k, +k \mid +k, +r].\end{aligned} \tag{11.135}$$

For $k = s \neq l = r$,

$$\langle V_1^{kr}|\,\mathbf{H}\,|V_1^{rk}\rangle = -K_{+k,+r} + 2K_{+k,-r} = K_{+k,+r}(\mathrm{s}) - 3K_{+k,+r}(\mathrm{d}); \tag{11.136}$$

$$\langle T_1^{kr}|\,\mathbf{H}\,|T_1^{rk}\rangle = -K_{+k,+r}. \tag{11.137}$$

11.5.3. Plus and Minus Configurations

Singly-excited configurations fall into two types: the type of V_1^{kk} and T_1^{kk} and the type of V_1^{kr} and T_1^{kr} $(k \neq r)$. Because of the pairing properties of the MO's, configurations of type V_1^{kk} and T_1^{kk} as well as the ground configuration have uniform electron densities: a π-electron density of unity at each π center. Configurations of type V_1^{kk} and T_1^{kk} may be called symmetrically-excited configurations. These configurations are usually nondegenerate. The $\mathbf{H}$ matrix element between a symmetrically-excited configuration and the ground configuration is invariably zero [compare Eq. (11.121)].

As is evident from Eqs. (11.122) and (11.123), V_1^{kr} and V_1^{rk} form a degenerate pair, and T_1^{kr} and T_1^{rk} also form a degenerate pair, even when the electronic interaction has been included. The $\mathbf{H}$ matrix elements representing the interactions of such paired configurations with a symmetrically-excited configuration are invariably equal to each other [compare Eqs. (11.128), (11.129), (11.134), and (11.135)], and the $\mathbf{H}$ matrix elements representing the interactions of such paired configurations with the ground configuration invariably have the same absolute magnitude and opposed signs [compare Eq. (11.118)]. In addition, the $\mathbf{H}$ matrix element between a configuration of a degenerate pair, say V_1^{kr}, and a configuration of a different degenerate pair, say V_1^{ls}, is always equal to the H matrix element between the other configurations of the two degenerate pairs, that is, V_1^{rk} and V_1^{sl} [compare Eqs. (11.126), (11.127), (11.132), and (11.133)].

We construct the symmetry combinations of paired configurations:

$$2^{-1/2}(V_1^{kr} \pm V_1^{rk}) \equiv {}^{\pm}V_1^{kr}; \qquad (11.138)$$

$$2^{-1/2}(T_1^{kr} \pm T_1^{rk}) \equiv {}^{\pm}T_1^{kr}. \qquad (11.139)$$

${}^{+}V_1^{kr}$ and ${}^{+}T_1^{kr}$ are called plus configurations, and ${}^{-}V_1^{kr}$ and ${}^{-}T_1^{kr}$ minus configurations. As is evident from the above considerations, a plus configuration can interact only with other plus configurations and symmetrically-excited configurations; a minus configuration can interact only with other minus configurations and, if it is a singlet, with the ground configuration. Thus we can factorize each of the secular determinants for singlet and triplet configurations into two determinants: one involving only plus and symmetrically-excited configurations and another involving only minus configurations (and the ground configuration).

The energies of V_1^{kr} and V_1^{rk} and those of T_1^{kr} and T_1^{rk} are given by Eqs. (11.122) and (11.123), respectively, and the H matrix elements between V_1^{kr} and V_1^{rk} and between T_1^{kr} and T_1^{rk} are given by (11.136) and (11.137), respectively. The energies of plus and minus configurations are therefore given by

$$E({}^{+}V_1^{kr}) - E(V_0) = [E(T_1^{kr}) - E(V_0)] + 3K_{+k,+r}(\text{s}) - 5K_{+k,+r}(\text{d}); \qquad (11.140)$$

$$E({}^{-}V_1^{kr}) - E(V_0) = E({}^{-}T_1^{kr}) - E(V_0) = [E(T_1^{kr}) - E(V_0)] + K_{+k,+r}; \qquad (11.141)$$

$$E({}^{+}T_1^{kr}) - E(V_0) = [E(T_1^{kr}) - E(V_0)] - K_{+k,+r}. \qquad (11.142)$$

In general, the value of $K_{+k,+r}$ is positive, that is, the value of the H matrix element between T_1^{kr} and T_1^{rk} is negative. Therefore, a triplet plus configuration is invariably lower in energy than the corresponding triplet minus configuration. In most cases, especially in aromatic hydrocarbons, the H matrix element between V_1^{12} and V_1^{21} is positive, and hence the minus configuration ${}^{-}V_1^{12}$ is lower than the corresponding plus configuration ${}^{+}V_1^{12}$. Even in such cases some of the H matrix elements between singlet degenerate configurations are negative and hence some singlet plus configurations are lower than the corresponding singlet minus configurations. In a few molecules, such as *s-trans*-butadiene, the matrix element between V_1^{12} and V_1^{21} is negative, and hence the singlet plus configuration ${}^{+}V_1^{12}$ is lower than the corresponding minus configuration ${}^{-}V_1^{12}$. The energy difference between ${}^{+}V_1^{kr}$ and ${}^{+}T_1^{kr}$ is equal to $4K_{+k,-r}$ and is positive. This means that a triplet plus configuration is invariably lower than the corresponding singlet plus configuration.

The off-diagonal elements of the **H** matrix involving plus and minus configurations are given by the following equations:

$$\langle {}^{-}V_1^{kr} | \mathbf{H} | V_0 \rangle = \sqrt{2}\, \langle V_1^{kr} | \mathbf{H} | V_0 \rangle = 2 \sum_f [+f, -k \mid +f, +r](\mathrm{d}) \quad \text{[compare (11.118)];} \qquad (11.143)$$

$$\langle {}^{+}V_1^{kr} | \mathbf{H} | V_1^{ll} \rangle = \sqrt{2}\, \langle V_1^{kr} | \mathbf{H} | V_1^{ll} \rangle \quad \text{[compare (11.128) and (11.134)];} \qquad (11.144)$$

$$\langle {}^{+}T_1^{kr} | \mathbf{H} | T_1^{ll} \rangle = \sqrt{2}\, \langle T_1^{kr} | \mathbf{H} | T_1^{ll} \rangle \quad \text{[compare (11.129) and (11.135)];} \qquad (11.145)$$

$$\langle {}^{\pm}V_1^{kr} | \mathbf{H} | {}^{\pm}V_1^{ls} \rangle = \langle V_1^{kr} | \mathbf{H} | V_1^{ls} \rangle \pm \langle V_1^{kr} | \mathbf{H} | V_1^{sl} \rangle \quad \text{[compare (11.126) and (11.132)];} \qquad (11.146)$$

$$\langle {}^{\pm}T_1^{kr} | \mathbf{H} | {}^{\pm}T_1^{ls} \rangle = \langle T_1^{kr} | \mathbf{H} | T_1^{ls}] \pm \langle T_1^{kr} | \mathbf{H} | T_1^{sl} \rangle \quad \text{[compare (11.127) and (11.133)].} \qquad (11.147)$$

Needless to say, all the matrix elements between a plus configuration and the ground configuration, between a plus configuration and a minus configuration, and between a minus configuration and a symmetrically-excited configuration are zero.

Equations (11.146) and (11.147) can be expanded as follows:

$$\langle {}^{+}V_1^{kr} | \mathbf{H} | {}^{+}V_1^{ls}] = -[+k, +l \mid +r, +s] - [+k, +s \mid +r, +l] + 4\{[+k, +r \mid +l, +s](\mathrm{s}) - [+k, +r \mid +l, +s](\mathrm{d})\}; \qquad (11.148)$$

$$\langle {}^{-}V_1^{kr} | \mathbf{H} | {}^{-}V_1^{ls} \rangle = \langle {}^{-}T_1^{kr} | \mathbf{H} | {}^{-}T_1^{ls}] = -[+k, +l \mid +r, +s] + [+k, +s \mid +r, +l]; \qquad (11.149)$$

$$\langle {}^{+}T_1^{kr} | \mathbf{H} | {}^{+}T_1^{ls} \rangle = -[+k, +l \mid +r, +s] - [+k, +s \mid +r, +l]. \qquad (11.150)$$

For the special cases where $l = k$, these expressions must be replaced by the following expressions:

$$\langle {}^{+}V_1^{kr} | \mathbf{H} | {}^{+}V_1^{ks} \rangle = \langle T_1^{kr} | \mathbf{H} | T_1^{ks} \rangle \text{ [compare (11.133)]} + 3[+k, +r \mid +k, +s](\mathrm{s}) - 5[+k, +r \mid +k, +s](\mathrm{d}); \qquad (11.151)$$

$$\langle {}^{-}V_1^{kr} | \mathbf{H} | {}^{-}V_1^{ks} \rangle = \langle {}^{-}T_1^{kr} | \mathbf{H} | {}^{-}T_1^{ks} \rangle = \langle T_1^{kr} | \mathbf{H} | T_1^{ks} \rangle + [+k, +r \mid +k, +s]; \qquad (11.152)$$

$$\langle {}^{+}T_1^{kr} | \mathbf{H} | {}^{+}T_1^{ks} \rangle = \langle T_1^{kr} | \mathbf{H} | T_1^{ks} \rangle - [+k, +r \mid +k, +s]. \qquad (11.153)$$

A singlet minus configuration can interact only with the ground configuration and other singlet minus configurations, and a triplet minus configuration can interact only with other triplet minus configurations. As is evident from Eqs. (11.141), (11.149), and (11.152), the energy of a singlet minus configuration is equal to the energy of the corresponding triplet minus configuration, and the **H** matrix element between two singlet minus configurations is equal to the **H** matrix element between the corresponding two triplet minus configurations. Therefore, provided that small interactions of singlet minus configurations with the ground configuration are neglected, it can be said that the energies of corresponding singlet and triplet minus states will be the same after as well as before inclusion of configuration interaction.

Both plus and minus configurations have uniform electron densities, just as the ground and symmetrically-excited configurations. Therefore, in alternant hydrocarbons, any excited states have also uniform electron densities.

The ground configuration behaves like a minus configuration, interacting only with singlet minus configurations. Therefore, the ground configuration may be considered as a minus configuration, and V_0 may be written as ${}^{(-)}V_0$ or ${}^{-}V_0$. On the other hand, symmetrically-excited configurations behave like plus configurations, interacting only with other symmetrically-excited configurations and plus configurations. Therefore, symmetrically-excited configurations may be considered as plus configurations: V_1^{kk} may be written as ${}^{(+)}V_1^{kk}$ or ${}^{+}V_1^{kk}$; T_1^{kk} as ${}^{(+)}T_1^{kk}$ or ${}^{+}T_1^{kk}$. Electronic states constructed from plus configurations are called plus states, and those constructed from minus configurations are called minus states.

The electron densities $\psi_{+k}\psi_{-r}$ and $\psi_{+r}\psi_{-k}$ are identical because of the pairing properties of the MO's. This means that the transition densities of V_1^{kr} and V_1^{rk} are identical or, in other words, the moments of the transitions from V_0 to V_1^{kr} and V_1^{rk} are equal in magnitude and act in the same direction. It follows that in the transition from V_0 to ${}^{-}V_1^{kr}$ these transition moments exactly cancel one another: this transition is forbidden. On the other hand, the transition from V_0 to ${}^{+}V_1^{kr}$ can be allowed. In general, the transition moments between any two plus states and between any two minus states are zero: only transitions between plus and minus states can be allowed.

11.5.4. Application of the Pariser–Parr Method to Benzene

To illustrate the procedure of calculation by the Pariser–Parr method, the method is applied here to benzene. The real functions of π MO's given in Table 3.5 are used as starting MO's, and only the singly-excited configurations arising from one-electron transitions from the highest bonding degenerate MO's, ψ_{+1} and ψ_{+2}, to the lowest antibonding degenerate MO's, ψ_{-1} and ψ_{-2}, that is, V_1^{11}, V_1^{22}, V_1^{12}, V_1^{21}, and their triplet counterparts, are taken into account.

Now, for simplification, the one-center Coulomb repulsion integral, $(pp \mid pp)_{\mathrm{CC}}$ or γ_{C}, is represented by γ_1, and the two-center integrals $(pp \mid qq)_{\mathrm{CC}}$ for p and q at *ortho*, *meta*, and *para* positions are represented by γ_2, γ_3, and γ_4, respectively. The values shown in Table 11.1 are used for these integrals, and a value of -2.39 ev is used for the core resonance integral β^{C}.

From the values of the AO coefficients of the MO's, we obtain

$$\sum_f K_{+f,+1}(\mathrm{d}) = \sum_f K_{+f,+2}(\mathrm{d}) = \tfrac{1}{3}\gamma_2 + \tfrac{1}{6}\gamma_4 = 3.25 \text{ ev}.$$

Consequently, since $w_{-1} = w_{-2} = 1$,

$$\kappa_1 = \kappa_2 = -\beta^{\mathrm{C}} + \sum_f K_{+f,+1}(\mathrm{d}) = 2.39 + 3.25 = 5.64 \text{ ev}$$

[compare (11.113)].

Therefore,

$$(\varepsilon_{-1}^{\mathrm{S}} \text{ or } \varepsilon_{-2}^{\mathrm{S}}) - (\varepsilon_{+1}^{\mathrm{S}} \text{ or } \varepsilon_{+2}^{\mathrm{S}}) = 2\kappa_1 = 11.28 \text{ ev}.$$

The required electronic repulsion integrals over MO's can be calculated as follows:

$$J_{+1,-1} = J_{+2,-2} = \tfrac{1}{4}(\gamma_1 + \gamma_2 + \gamma_3 + \gamma_4) = +7.0475 \text{ ev};$$

$$K_{+1,-1} = K_{+2,-2} = \tfrac{1}{4}(\gamma_1 - \gamma_2 + \gamma_3 - \gamma_4) = +0.9475 \text{ ev};$$

$$J_{+1,-2} = J_{+2,-1} = \tfrac{1}{12}(\gamma_1 + 5\gamma_2 + 5\gamma_3 + \gamma_4) = +6.6025 \text{ ev};$$

$$K_{+1,-2} = K_{+2,-1} = [+1, -2 \mid +2, -1]$$

$$= \tfrac{1}{12}(\gamma_1 + \gamma_2 - \gamma_3 - \gamma_4) = +0.6225 \text{ ev};$$

$$[+1, +2 \mid -1, -2] = \tfrac{1}{12}(\gamma_1 - \gamma_2 - \gamma_3 + \gamma_4) = +0.2225 \text{ ev};$$

$$[+1, -1 \mid +2, -2] = \tfrac{1}{12}(\gamma_1 - 5\gamma_2 + 5\gamma_3 - \gamma_4) = -0.2975 \text{ ev}.$$

By the use of the formulas given in Subsection 11.5.2, the **H** matrix elements are calculated as follows:

Diagonal Elements

$$E(T_1^{11}) - E(V_0) = E(T_1^{22}) - E(V_0) = 4.2325 \text{ ev};$$

$$E(T_1^{12}) - E(V_0) = E(T_1^{21}) - E(V_0) = 4.6775 \text{ ev};$$

$$E(V_1^{11}) - E(V_0) = E(V_1^{22}) - E(V_0) = 6.1275 \text{ ev};$$

$$E(V_1^{12}) - E(V_0) = E(V_1^{21}) - E(V_0) = 5.9225 \text{ ev};$$

Off-Diagonal Elements

$$\langle T_1^{11}|\,\mathbf{H}\,|T_1^{22}\rangle = -0.2225 \text{ ev}; \qquad \langle T_1^{12}|\,\mathbf{H}\,|T_1^{21}\rangle = -0.2225 \text{ ev};$$

$$\langle V_1^{11}|\,\mathbf{H}\,|V_1^{22}\rangle = -0.8175 \text{ ev}; \qquad \langle V_1^{12}|\,\mathbf{H}\,|V_1^{21}\rangle = +1.0225 \text{ ev}.$$

All the other off-diagonal elements vanish.

As a result of configuration interaction, the functions and energies of excited states of benzene are obtained as shown in Table 3.7. All the

TABLE 11.2

COMPARISON OF THEORETICAL AND EXPERIMENTAL ENERGY VALUES (E, IN ev) OF π-ELECTRONIC STATES OF BENZENE

	Experimental				Theoretical			
Band	E_{onset}	$E_{\max}$	f	State	E^a	E^b	E^c	E^d
α	4.71	4.89	0.002	$^1B_{2u}$	4.90	4.710	5.0	5.9
p	5.96	6.17	0.10	$^1B_{1u}$	5.31	5.960	5.8	7.3
β, β'	6.76	6.98	0.69	$^1E_{1u}$	6.95	6.548*	8.0	9.8
t_p	3.59	3.8	—	$^3B_{1u}$	4.01	3.590	1.5	3.1
$t_\beta, t_{\beta'}$	—	—	—	$^3E_{1u}$	4.45	4.149	2.2	4.4
t_α	—	—	—	$^3B_{2u}$	4.90	4.710	3.0	5.8

[a] R. Pariser and R. G. Parr (1953), Reference 5: the Pariser–Parr method. The parameter values used are as follows: the effective nuclear charge (Z') for Slater carbon $2p\pi$ AO's = 3.25; $\beta^{\text{C}} = -2.39$ ev; $(pp\,|\,pp) = 10.53$ ev; $(pp\,|\,qq) = 7.30$ ev (*ortho*, $R_{pq} = 1.39$ Å), 5.46 ev (*meta*, $R_{pq} = 2.41$ Å), 4.90 ev (*para*, $R_{pq} = 2.78$ Å).

[b] R. Pariser (1956), Reference 13: the Pariser–Parr method. The parameter values used are as follows: $Z' = 3.18$; $\beta^{\text{C}} = -2.371$ ev; $(pp\,|\,pp) = 10.959$ ev; $(pp\,|\,qq) = 6.895$ ev (*ortho*, $R_{pq} = 1.390$ Å), 5.682 ev (*meta*, $R_{pq} = 2.407$ Å), 4.978 ev (*para*, $R_{pq} = 2.780$ Å). * $f = 2.215$.

[c] M. Goeppert-Mayer and A. L. Sklar (1938), Reference 1: LCAO ASMO (nonempirical).

[d] R. G. Parr, D. P. Craig, and I. G. Ross, *J. Chem. Phys.* **18**, 1561 (1950): LCAO ASMO + CI (nonempirical).

transitions from the ground state to these excited states except the β and β' states (that is, $^+[V]_1^{12}$ and $^{(+)}[V]_1^{22}$) have zero transition moments. The transitions to the β and β' states, which are degenerate, have nonzero transition moments in the directions of the x and y axes, respectively (for the coordinate system assigned to the benzene molecule, see Fig. 3.2), and the magnitudes of the transition moments are calculated to be $+R$ and $-R$, respectively, in which R represents the length of a C—C bond in benzene.

The values of energies of excited states of benzene thus calculated by the Pariser–Parr method are in fairly good agreement with the values obtained experimentally from the electronic absorption spectrum. When the values used for the core resonance integral and Coulomb repulsion integrals are varied, naturally the results will be varied. In Table 11.2, the values of energies of excited states of benzene calculated by the Pariser–Parr method are compared with the spectroscopically observed values and the values calculated by more rigorously theoretical methods.

Chapter 12

Relations between Electronic Absorption Spectra and Geometry of Molecules. Biphenyl and Related Compounds

12.1. Introduction

The interaction of π electrons in a π system is maximal when the skeleton of the system is planar. If for any reason the system deviates from the planar geometry, the interaction of π electrons will be reduced to the corresponding extent. Consequently, the π-electronic states, and hence the electronic spectrum, will be affected by the change in the geometry of the system. In the latter half of this book, that is, in this chapter and succeeding chapters, relations between electronic absorption spectra and geometry are discussed for various types of π system.

In a linear conjugated system, steric hindrance to the planarity of the system is considered to a first approximation to be relieved only by twist of essential single bond or bonds. In general, twist of a bond does not affect the σ-bond energy. Since an essential single bond has a relatively low π-bond order, the decrease in the delocalization energy of π electrons caused by twist of such a bond will be relatively small [compare Subsection 7.1.6, especially Eq. (7.68)]. Furthermore, in most cases it is expected that twist of such a bond will rapidly reduce the steric repulsion between atoms not directly bonded together.

The resonance parameter k for a twisted bond is related to the angle of twist θ by Eq. (9.34). Therefore, if it is assumed that, as suggested in Chapter 10, the position and intensity of certain types of band of related systems can be correlated well with the simple LCAO MO calculation, it may be expected that the position and intensity of bands can be related to the angle of twist by calculating the π MO's and their energies as functions of the parameter k for the twisted bond. From this point of view, the author has tried to elucidate the relations between electronic absorption spectra and geometry of some typical conjugated systems by means of the simple LCAO MO method and obtained fairly satisfactory results. In this chapter and the succeeding four chapters, the relations in linear conjugated hydrocarbons are discussed mainly on the basis of the author's work.

In correlating the observed spectra with the MO calculations, the author has assumed as follows: the molecules having the most probable nuclear configuration in the ground electronic state make the most important contribution to absorption, and the absorption maximum of each band corresponds to the vertical transition in such molecules. From the discussions made in Chapter 5 it will be evident that this assumption is quite reasonable. Since a vertical transition does not involve any changes in the nuclear configuration and in the momentum of vibrational motion of the nuclei, during the most probable vibronic transition in an electronic transition from the ground electronic state to an excited π-electronic state there must be no changes in energies other than the π-electronic energy, such as σ-bond energy, steric repulsion energy, and kinetic energy. Therefore, the energy of the most probable vibronic transition must be equal to the difference in π-electronic energy between the excited and ground electronic states at the most probable nuclear configuration in the ground electronic state. This means that the wave number at the absorption maximum of each band should correspond to the energy difference between the two π MO's or the two electronic states calculated with the most probable or equilibrium nuclear configuration in the ground electronic state. Analogously, it can also be justified to correlate the observed intensity of each band with the oscillator strength calculated with the ground state equilibrium nuclear configuration.

12.2. Biphenyl*

12.2.1. Steric Effects in the Spectra of Biphenyl and Related Compounds

The spectra of biphenyl in solutions show a structureless intense band at about 247 mμ (λ_{max} 247.4 mμ, ϵ_{max} 16700 in n-heptane; λ_{max} 247.7 mμ, ϵ_{max} 16600 in ethanol). This band is provisionally called the A band. The

* Subsections 12.2.1–12.2.4 are, so to speak, a revised version of Reference 1.

position of this band is not markedly different in various saturated hydrocarbons and alcohols. However, it considerably varies in other solvents. For example, the values of λ_{max} in chloroform and in carbon tetrachloride are 249.5 and 256.5 mμ, respectively.[1]

The A band of biphenyl is progressively shifted toward shorter wavelengths with concurrent decreases in the intensity as one, and then two methyl groups are introduced into the *o*- and *o*′-positions of the molecule. This hypsochromic and hypochromic shift is considered to be due to the deviation of the most probable geometry of the molecule from planarity caused by the steric interference of the substituents. The steric interference is considered to be relieved principally by twist of the coannular (1—1′) bond.

When the twist of a bond is large enough to eliminate almost completely the π–π interaction across the bond, the spectrum is similar to the sum of the spectra of the component parts of the molecule on either side of the bond. Thus, for example, the spectrum of bimesityl does not exhibit the band characteristic of biphenyl-type compounds (that is, the A band), but is closely similar to the sum of the spectra of two mesitylene or two isodurene molecules.[2] It is inferred that the two benzene rings in bimesityl are almost perpendicularly oriented. Similar situation appears to occur in the spectrum of 1-phenylnaphthalene. In this compound, the coplanarity of the benzene plane and the naphthalene plane is probably hindered by interference between hydrogen atoms of the phenyl and naphthyl groups. Some similarity between the spectrum of this compound and the sum of the spectra of benzene and naphthalene has been pointed out.[3] Furthermore, the spectrum of 1,1′-binaphthyl is very similar to the spectrum of naphthalene, and the spectrum of 9,10-diphenylanthracene is very similar to the spectrum of 9,10-dimethylanthracene.[4] These facts are also attributed to the nonplanarity of the molecules caused by twist of the coannular bonds.

12.2.2. The Simple LCAO MO Calculation on Biphenyl

With regard to the numbering and coordinate axes for the carbon skeleton of the π system of biphenyl, the conventions illustrated in Fig. 12.1 are adopted. According to X-ray crystal analysis, the biphenyl molecule has a

[1] H. Suzuki, *Bull. Chem. Soc. Japan* **32,** 1340 (1959).

[2] L. W. Pickett, G. F. Walter, and H. France, *J. Am. Chem. Soc.* **58,** 2296 (1936); M. T. O'Shaughnessy, and W. H. Rodebush, *ibid.* **62,** 2906 (1940); E. Marcus, W. M. Lauer, and R. T. Arnold, *ibid.* **80,** 3742 (1958).

[3] R. N. Jones, *Chem. Rev.* **32,** 1 (1943).

[4] R. N. Jones, *J. Am. Chem. Soc.* **67,** 2127 (1945).

planar or nearly planar geometry in the crystalline state. The planar model of biphenyl belongs to point group D_{2h}. A twisted model [compare Fig. 12.2] with the twist angle except 90° belongs to D_2. When the twist angle is 90° the system belongs to D_{2h}. In the simple LCAO MO approximation we can treat the biphenyl π system with the twist angle varying from 0 to 90° as belonging to point group D_2.

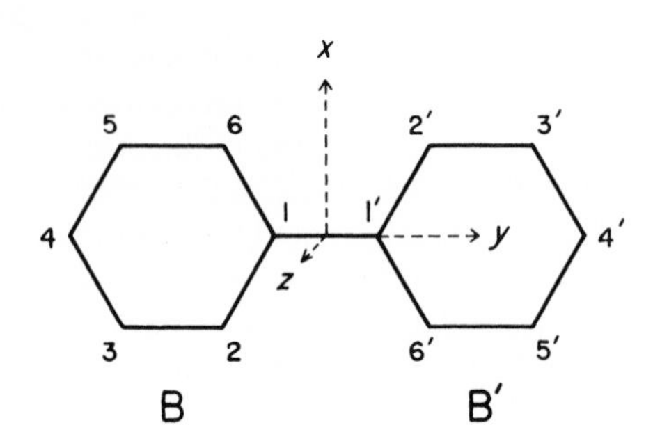

FIG. 12.1. The skeleton of biphenyl in the planar form.

The π–π resonance integral for the 1—1′ bond, $\beta_{11'}$, is expressed as $k_{11'}\beta$, in which β represents the standard resonance integral, that is, the π–π resonance integral for a C—C bond in benzene. The angle of twist of the 1—1′ bond, that is, the angle between the two benzene ring planes, is

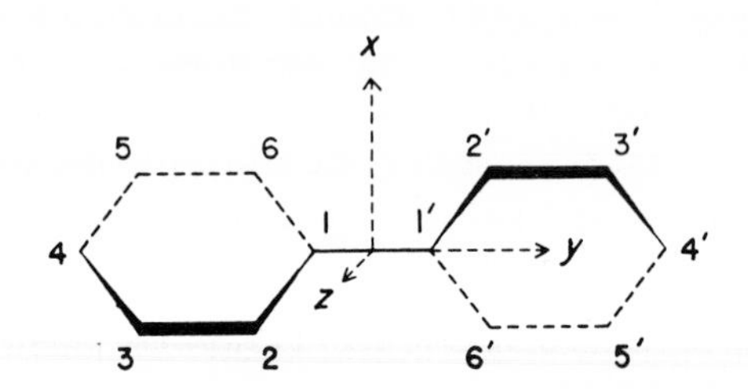

FIG. 12.2. The skeleton of biphenyl in a twisted form.

denoted by $\theta_{11'}$, and the length of the 1—1′ bond by $R_{11'}$. Subsequently, for simplification, the subscript 11′ in these symbols is omitted.

It is assumed that the relations among k, θ, and R are given by Eqs. (9.34) and (9.33) or (9.35). According to the X-ray crystal analysis by Dhar,[5] in the crystalline biphenyl R is 1.48 Å and θ is 0°. When R_0 and R_{90} are

[5] J. Dhar, *Indian J. Phys.* **7,** 43 (1932); *Proc. Natl. Inst. Sci. India, P.A.* **15,** 11 (1949).

assumed to be 1.48 and 1.54 Å, respectively, values of $k(R_0, 0°)$ and $k(R_{90}, 0°)$ are 0.858 and 0.771, respectively, and the relation between k and θ is given by Eq. (9.39). In this equation the terms with $\cos^2 \theta$ and $\cos \theta \cdot (1 - \cos \theta)$ arise from the possible variation in R associated with the variation in θ, and are seen to be of subsidiary importance. As discussed in Subsection 9.3.4, the choice of a value of 1.54 Å for R_{90} is somewhat questionable. However, the results of the calculation are not sensitive to the value of R_{90}. According to the more recent X-ray crystal analysis by Robertson,[6] the biphenyl molecule in the crystalline state is not strictly planar, and the value of R in the molecule is 1.4966 ± 0.0025 Å after corrected for libration. The value of k for $R = 1.497$ Å and $\theta = 0°$ is 0.834. When it is assumed that the value of R remains unchanged by twist, k is related to θ by

$$k = 0.834 \cos \theta. \tag{12.1}$$

However, the use of this equation in place of Eq. (9.39) results in increases in the values of θ calculated for biphenyl in solution and in vapor phase and *o*-substituted biphenyls only by at most 1–2°, as will be shown later. For the present Eq. (9.39) is adopted.

The 12-row secular determinant for determination of π MO's of biphenyl as linear combination of 12 carbon $2p\pi$ AO's can be factorized into two 4-row and two 2-row determinants. Molecular orbitals ψ_{+2} and ψ_{-3} belong to symmetry species a_1 (or a_{1u}) of point group D_2 (or D_{2h}), ψ_{+3} and ψ_{-2} to b_2 (or b_{2g}), ψ_{+6}, ψ_{+4}, ψ_{-1}, and ψ_{-5} to b_1 (or b_{1u}), and ψ_{+5}, ψ_{+1}, ψ_{-4}, and ψ_{-6} to b_3 (or b_{3g}).

Symmetry species a_1 (a_{1u}) and b_2 (b_{2g}) are symmetric with respect to rotation C_2 about the lengthwise axis (y axis) of the conjugated system. The MO's belonging to these symmetry species must have no contribution of the AO's on the axis, that is, χ_4, χ_1, $\chi_{1'}$, and $\chi_{4'}$. This means that these MO's must be nonbonding with respect to the 1—1′ bond and must be not affected by any variation in the k value for this bond. Actually, the form and energy of these MO's can be expressed as follows, independently of the value of k:

$$\begin{aligned} a_1 &: \psi_{-3} = \phi_{-2(+)}; & \psi_{+2} &= \phi_{+2(+)}; \\ b_2 &: \psi_{-2} = \phi_{-2(-)}; & \psi_{+3} &= \phi_{+2(-)}; \\ & w_{-3} = w_{-2} = +1; & w_{+2} &= w_{+3} = -1. \end{aligned} \tag{12.2}$$

In these expressions, $\phi_{-2(+)}$ and $\phi_{-2(-)}$, for example, represent the following symmetry combinations of the benzene π MO's:

$$\phi_{-2(\pm)} \equiv 2^{-1/2}(\phi_{-2} \pm \phi_{-2'}), \tag{12.3}$$

[6] G. B. Robertson, *Nature* **191**, 593 (1961).

where ϕ_{-2} is the π MO of one separate benzene ring B and $\phi_{-2'}$ is the corresponding π MO of another benzene ring B′. The notation of the π MO's of benzene is according to Table 3.5. As usual, the w's are orbital energies in units of $-\beta$ relative to α. Obviously, ψ_{-2} and ψ_{-3} form a degenerate pair, and ψ_{+2} and ψ_{+3} form another degenerate pair.

TABLE 12.1

RESULTS OF THE SIMPLE LCAO MO CALCULATION ON BIPHENYL

Characteristics of the $\psi_{+1} \rightarrow \psi_{-1}$ Transition as Functions of the Twist Angle of the Coannular "Single" Bond

k	1.1	1.0	0.9	0.858	0.8	0.7	0.6
θ, deg.	—	—	—	0	20.2	33.9	44.0
R, Å	—	—	—	1.480	1.484	1.490	1.497
ΔE_{11}, $-\beta$	1.361	1.409	1.460	1.481	1.512	1.567	1.623
D_{11}, Å^2	—	—	—	2.678	2.550	2.329	2.112
k	0.5	0.4	0.3	0.2	0.1	0	
θ, deg.	52.6	60.6	68.1	75.4	82.7	90	
R, Å	1.504	1.511	1.518	1.525	1.532	1.540	
ΔE_{11}, $-\beta$	1.682	1.742	1.805	1.869	1.934	2	
D_{11}, Å^2	1.899	1.693	1.494	1.307	1.130	0.966	

The four MO's belonging to b_1 (or b_{1u}) and the four MO's belonging to b_3 (or b_{3g}) vary with k. When the value of k becomes zero these MO's reduce to the symmetry combinations of the MO's of the two benzene rings as follows:

$$b_3 : \psi_{-6} = \phi_{-3(-)}, \quad \psi_{-4} = \phi_{-1(-)}, \quad \psi_{+1} = \phi_{+1(-)}, \quad \psi_{+5} = \phi_{+3(-)};$$
$$b_1 : \psi_{-5} = \phi_{-3(+)}, \quad \psi_{-1} = \phi_{-1(+)}, \quad \psi_{+4} = \phi_{+1(+)}, \quad \psi_{+6} = \phi_{+3(+)}. \tag{12.4}$$

In the simple MO approximation the A band is interpreted as being due to the S–S one-electron transition from the highest occupied MO, ψ_{+1}, to the lowest vacant MO, ψ_{-1}. This interpretation is supported by the calculation including the electronic interaction, as will be shown later. The transition from ψ_{+1} to ψ_{-1} is allowed and polarized along the y axis. The transition energy, ΔE_{11} (in $-\beta$) $\equiv \Delta w_{11} \equiv w_{-1} - w_{+1} = 2w_{-1}$, and the dipole strength of the transition, $D_{11} = M_{11}^2$, depend on the value of k and hence on the value of θ. Table 12.1 shows the values of ΔE_{11} and D_{11} for various values of k. As the value of k decreases, the value of ΔE_{11} increases, and the value of

D_{11} decreases. This means that as the angle of twist increases the A band will shift toward shorter wavelengths and decrease in intensity. By the way, the partial π-bond order* of the 1—1′ bond for ψ_{+1}, $p_{11'}^{(+1)}$ $(= -p_{11'}^{(-1)})$, is negative, and its absolute value increases gradually as the value of k decreases, from 0.131 for $k = 0.858$ to 0.167 for $k = 0$. The extra-resonance energy [0.285 $(-\beta)$ for $k = 0.858$] is almost completely proportional to k^2, and the total π-bond order* of the 1—1′ bond (0.323 for $k = 0.858$) is roughly proportional to k.

12.2.3. Correlation of the Calculated Transition Energy with the Observed Position of the A Band

As mentioned in Section 12.1, the wave number ν at the absorption maximum of a band should correspond to the calculated transition energy ΔE for the ground state equilibrium conformation, which is characterized by particular values of θ and R. In order to correlate the observed wave number with the calculated transition energy, two compounds having the extreme conformations with $\theta = 0°$ and $\theta = 90°$ are chosen as references, and it is assumed that ν and ΔE are linearly related between the two extreme points fixed by these two references. When such compounds are not available, if there are two compounds with considerably different known values of θ, these compounds can be used as the references. The reference with the smaller value of θ is called the small-angle reference, and ν and ΔE for this reference are denoted by ν_{S} and ΔE_{S}, respectively. On the other hand, the reference with the larger value of θ is called the large-angle reference, and ν and ΔE for this reference are denoted by ν_{L} and ΔE_{L}, respectively.† Then, since for the A band of the biphenyl system the values of ν and ΔE should be the larger for the larger value of θ, the value of ΔE for a partially twisted form is obtained from the observed value of ν by the following linear interpolation formula:

$$\Delta E = \Delta E_{\mathrm{L}} - (\Delta E_{\mathrm{L}} - \Delta E_{\mathrm{S}})(\nu_{\mathrm{L}} - \nu)/(\nu_{\mathrm{L}} - \nu_{\mathrm{S}}). \tag{12.5}$$

From the value of ΔE the corresponding value of k can be obtained, and

* The partial π-bond order and the (total) π-bond order are quantities defined by expressions (7.58) and (7.59), respectively. For a twisted bond it seems to be more reasonable to take the products of these quantities and $\cos \theta_{ij}$ as the "partial π-bond order" and the "(total) π-bond order," respectively.

† In the original paper (Ref. 1), the small- and large-angle references were called the longer and shorter wavelength references, respectively, and the symbols for ν and ΔE for these references were reverse.

then, from the value of k the values of θ, R, and some relevant quantities for the most probable conformation of the molecule can readily be interpolated. The choice of numerical values for ΔE_L, ν_L, ΔE_S, and ν_S remains the only difficult problem.

When the value of k is reduced to zero the MO's of the biphenyl system are reduced to the MO's of two isolated benzenes. Therefore, benzene is chosen as the large-angle reference, and the center of gravity of singlets of benzene (see Subsection 10.3.3), 48,000 cm^{-1}, is taken as ν_L. The value of ΔE_L is $2(-\beta)$. As will be mentioned later, the calculation including the electronic interaction shows that the biphenyl A band is reduced to the benzene p band when the twist angle becomes 90°. The wave number of the benzene p band is almost completely identical with the center of gravity of singlets of benzene. Therefore, the choice of 48,000 cm^{-1} for ν_L corresponding to the ΔE_L value of 2 seems to be adequate.

Since the crystalline biphenyl is known to be planar or nearly planar, it is chosen as the small-angle reference. If the value of k for this reference is taken to be 0.858, the value of ΔE_S is $1.482\,(-\beta)$. It is almost certain that biphenyl in solution, as well as in vapor, does not have the same molecular geometry as in the crystalline state. The biphenyl spectrum in the crystalline state measured by the pressed KCl disk technique shows an intense, structureless band with a maximum at 253 mμ,[7] which is to be compared with the wavelength at the maximum of the A band in the solution spectrum, about 247 mμ in n-heptane. In order to compare the solid-state spectrum with the solution spectrum, it is necessary to allow for the solid-state effect or the so-called normal red-shift (see Subsection 6.4.3). An absorption maximum observed in the spectrum of a sample prepared by the pressed disk technique almost always is located at longer wavelength than the corresponding maximum in the spectrum of solution in saturated hydrocarbon solvents, even with "rigid" compounds, such as naphthalene and anthracene, which cannot at least approximately assume other conformations in solution than in the crystalline state. This red-shift may be supposed to be due to the effect of interactions between molecules in the crystal. By reference to the magnitudes of the red-shifts of various bands of naphthalene and anthracene and of the first S–S bands (A bands) of *trans*-stilbene and related compounds (see Subsection 6.4.3), the magnitude of the normal red-shift of the A band of biphenyl is estimated at about 2.0 to 3.5 mμ. After being corrected for the normal red-shift by this value, the wavelength of the absorption maximum of the A band of the "isolated" planar biphenyl molecule is assumed to

[7] J. Dale, *Acta Chem. Scand.* **11**, 650 (1957).

be 251–249.5 mμ, and the corresponding wave number, 39,841–40,080 cm^{-1}, is taken as ν_S .

12.2.4. Estimation of the Twist Angle in Biphenyl in Solution and in Vapor

The difference between the above assumed position of the A band of the "isolated" planar biphenyl molecule, 249.5–251.0 mμ, and the position of the band of biphenyl in solution in n-heptane, 247 mμ, is assumed to be due to the difference in the most probable conformation of the molecule between in the crystal and in the solution. From comparison of the results of measurements of the infrared absorption spectra of biphenyl in solutions and in the pressed KBr disk, Dale[8] has inferred that biphenyl in solution is nonplanar. By the use of the above assumed values of ν and ΔE for the references, from the value of λ_{max} of the A band in the spectrum of the solution in n-heptane the value of θ in the most preferred conformation of biphenyl in solution is estimated at 19–23°.

The absorption maximum of the biphenyl A band in the vapor phase occurs at considerably shorter wavelength than in the solution, presumably indicating that the deviation from coplanarity of the two benzene rings is larger in the vapor than in the solution (compare Fig. 12.3). The biphenyl spectra in the vapor state at various temperatures from 170 to 520°C were measured by Almasy and Laemmel.[9] The spectrum is not largely affected by temperature at least over this range. Whereas the molar extinction coefficient at the absorption maximum of the A band decreases gradually from 12,050 at 170°C to 9400 at 520°C when the temperature is raised, the oscillator strength of the band is maintained almost constant independently of the temperature: at 170°C it is 0.316 and the average values at 170, 260, 360, and 520°C are 0.311 $\pm$ 0.0035. It is noteworthy that these f values for the vapor spectra are considerably smaller than the corresponding value for the solution spectrum, 0.411. This difference is almost certainly significant and is thought to be mainly due to the difference in the most probable conformation of the molecule between the two states (see Subsection 6.2.3). The position of the absorption maximum is gradually shifted toward longer wavelengths from 42,000 cm^{-1} (238.1 mμ) at 170°C to 41,750 cm^{-1} (239.5 mμ) at 520°C, as the temperature is raised. From these results the position of the maximum at 20°C was inferred, by extrapolation, to be 42,100 cm^{-1} (237.5 mμ).

[8] J. Dale, *Acta Chem. Scand.* **11,** 640 (1957).
[9] F. Almasy and H. Laemmel, *Helv. Chim. Acta* **33,** 2092 (1950).

In order to compare the solution spectra with the vapor spectra, it is necessary to allow for the solvent effect. After being corrected for the solvent effect by means of Bayliss' equation [Eq. (6.1)], the wave number of the maximum of the biphenyl A band in the vapor state, which is to be directly compared with the corresponding value in the solution spectrum, is evaluated

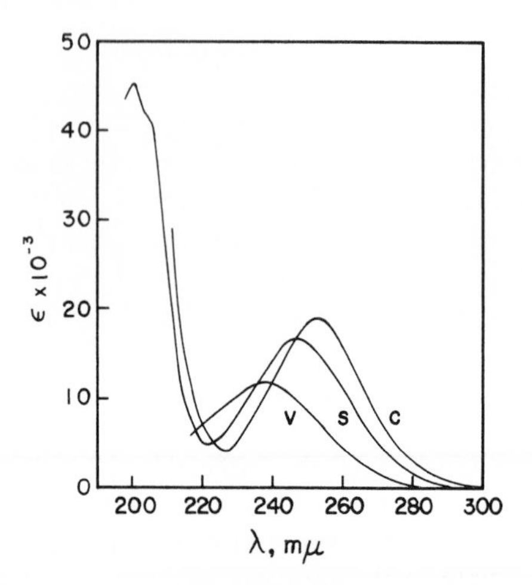

FIG. 12.3. Absorption spectra of biphenyl in various states. C, the crystal spectrum measured by Dale (Ref. 7) by the pressed KCl-disk technique (the extinction scale is arbitrary because the reliable absolute extinction value could not be obtained); S, the solution spectrum (solvent: heptane); V, the vapor spectrum measured by Almasy and Laemmel (Ref. 9) at 170°C.

to be 41,870 cm^{-1} (at 170°C) or 41,970 cm^{-1} (at 20°C). On the basis of these values for ν, the values of θ for vapor biphenyl at 170 and at 20°C are estimated at 40–42 and 41–43°, respectively.

In this way, the value of θ in the most preferred conformation of biphenyl in solution has been estimated at about 20°, and the value for the vapor has been estimated at about 41–43°. These values coincide fairly well with values estimated by other workers on different grounds (see below).

From electron diffraction study Bastiansen[10] deduced a value of 45 $\pm$ 10° for the angle θ in the vapor state. More recently, Almenningen and Bastiansen[11] found a value of 41.6 $\pm$ 2.0° for this angle (R = 1.489 Å) by

[10] O. Bastiansen, *Acta Chem. Scand.* **4,** 926 (1950).

[11] A. Almenningen and O. Bastiansen, *Det Kgl Norske Videnskabers Selskabs Skrifter* **1958,** No. 4.

more accurate electron diffraction study. This value agrees almost completely with the value estimated above from the electronic absorption spectrum.

Of course, the nonplanar conformation characterized by this value of θ must be by no means rigid, but it should be considered to be the most preferred or equilibrium conformation. Probably, the force constant with respect to the twisting of the coannular bond is comparatively small, energy separations between the vibrational energy levels of the torsional mode are comparatively small, and hence the probability of finding any angle in the neighborhood of the equilibrium position is rather large. This is considered to be responsible for the fact that the biphenyl A band is structureless (see Section 5.1). According to Almenningen and Bastiansen,[11] to obtain a probability one half the probability of the equilibrium form, one has to rotate away from the equilibrium an angle of approximately 17° in either direction. As the temperature is raised, the probability distribution curve must become broader, and the probability of the equilibrium conformation must become smaller. This probably explains the foregoing fact that the intensity of the absorption maximum of the biphenyl vapor A band decreases gradually as the temperature is raised, while the oscillator strength and position of the band are maintained almost constant.

Almenningen and Bastiansen[11] found that 4,4′-bipyridyl, an analog of biphenyl, has also a nonplanar equilibrium conformation in the gaseous state. They gave values of 37.2° and 1.465 Å to the twist angle and length of the coannular bond, respectively, in the equilibrium conformation.

Samoilov and Dyatkina[12] inferred the twist angle of the coannular bond of biphenyl to be about 30° from calculations of resonance energy and steric repulsion energy between the nearest hydrogen atoms. Adrian[13] concluded from similar calculations that the potential energy curve with respect to the twist angle would have a shallow minimum between 20 and 30°. In addition, calculations by Coulson and Longuet-Higgins indicate that the decrease in repulsion energy between the *ortho* hydrogen atoms and the decrease in resonance energy resulting from nonplanarity will be practically balanced for twist angles up to about 20°. All these values are compatible with the values estimated above for the solution state.

12.2.5. Assignment of Absorption Bands of Biphenyl

The absorption spectrum of biphenyl in *n*-heptane shows two intense, structureless bands in the near ultraviolet region, one with $\lambda_{max} = 247.4$ mμ

[12] S. Samoilov and M. Dyatkina, *Zhur. Fiz. Khim.* **22,** 1294 (1948).
[13] F. J. Adrian, *J. Chem. Phys.* **28,** 608 (1958).

(5.01 ev) and $\epsilon_{max} = 16{,}700$, and one with $\lambda_{max} = 200.5$ mμ (6.18 ev) and $\epsilon_{max} = 45{,}100$. The former is the so-called A band. The latter is subsequently referred to as the C band. The spectrum shows an inflection at about $\lambda = 205.0$ mμ (6.05 ev) and $\epsilon = 41{,}400$. The vestigial band corresponding to this inflection is referred to as the B band.

The biphenyl spectrum in the vapor phase shows two bands in the vacuum ultraviolet region: a band with a maximum at about 194 mμ (6.4 ev) and a broad band from about 163 mμ (7.6 ev) to 170 mμ (7.3 ev).[14] The former is probably the C band.

These various biphenyl bands cannot be thoroughly interpreted by the simple MO method, and several workers have tried to interpret the biphenyl spectrum by various advanced methods. London[15] and Davydov[16] calculated the energies of excited electronic states of biphenyl by regarding the electronic states of biphenyl as arising from interaction of the electronic states of the two benzene rings. According to their theories, the two nondegenerate singlet excited states (α and p states) and one degenerate singlet state (β and β' states) of benzene split into eight excited singlet states in biphenyl, four of which give allowed transitions from the ground state and four of which do not. They ascribed the A band to the lowest-energy allowed transition to the $^1B_{3u}$ state (the energy relative to the ground state: London, 5.2 ev; Davydov, 4.78 ev) originating from the benzene α state (the energy calculated by Goeppert-Mayer and Sklar,[17] 5.0 ev; the observed energy, 4.6 ev), the B band to the allowed transition to the $^1B_{2u}$ state (London, 6.6 ev; Davydov, 6.27 ev) originating from the benzene p state (calculated, 5.8 ev; observed, 6.1 ev), and the C band to the allowed transition to the $^1B_{2u}$ state (London, 8.1 ev; Davydov, 6.44 ev) originating from the benzene β' state (calculated, 8.0 ev; observed, 6.9 ev). If these assignments were correct, the A band would be polarized in the direction of the short axis of the molecule (the x axis), and the B and C bands would be polarized in the direction of the long axis of the molecule (the y axis).

On the other hand, Platt[18] has assumed that the A band corresponds to the p band of benzene and that a weak band corresponding to the benzene α band is also present in the biphenyl spectrum, but is hidden under the much stronger A band. He placed the suggested hidden band at about 244 mμ.

[14] E. P. Carr and H. Stücklen, *J. Chem. Phys.* **4,** 760 (1936).
[15] A. London, *J. Chem. Phys.* **13,** 396 (1945).
[16] A. S. Davydov, "Theory of Molecular Excitons," translated by M. Kasha and M. Oppenheimer, Jr., Sect. 39. McGraw-Hill, New York, 1962.
[17] M. Goeppert-Mayer and A. L. Sklar, *J. Chem. Phys.* **6,** 645 (1938).
[18] J. R. Platt, *J. Chem. Phys.* **19,** 101 (1951).

Afterwards, Wenzel[19] has inferred that the hidden band may be located at about 275 mμ or about 271 mμ, and that the intensity of this band may be of the same order as the benzene and naphthalene α bands.

The existence of the hidden biphenyl band has experimentally been corroborated by Dale.[7] As mentioned already, he measured the biphenyl spectrum in the crystalline state by the pressed KCl disk technique, and found that the spectrum is very similar to the solution spectrum, showing only an intense, structureless band with a maximum at 253 mμ. This band is almost certainly the A band. On the other hand, the spectra of thin films of solidified melts or sublimed crystals of biphenyl were quite different from the KCl-disk and solution spectra, and revealed a band showing vibrational fine structure with a maximum at about 275 mμ. The molar extinction coefficient of the maximum was very roughly estimated at about 1000–2000. In addition, whereas the KCl-disk and solution spectra of *p*-terphenyl and of *p,p'*-quaterphenyl showed intense, structureless bands at considerably longer wavelengths than the corresponding biphenyl band (A band), the spectra of solidified melts and sublimed crystals of these compounds, similarly to the corresponding biphenyl spectra, showed weak bands with structure at about 275 mμ. These weak bands at about 275 mμ are probably due to transitions polarized in the direction of the short axis (x axis) of the molecule, which are normally hidden under the intense A bands. This revelation of the hidden bands in the spectra of solidified melts and of sublimed crystals is interpreted as follows: since the incident light is almost parallel to the long axis (y axis) of the molecules in the thin films of crystals prepared by solidification of melts or by sublimation, owing to the orientation of molecules in these samples, the A band, which is probably due to the transition polarized in the direction of the long axis of the molecule, is so reduced in intensity that the weak hidden transition is revealed. The hidden band at about 275 mμ is subsequently referred to as the H band.

The H band also appears in solution spectra of sterically hindered biphenyl derivatives. According to Everitt and his co-workers,[20] the spectra of *o,o'*-dialkylbiphenyls in solution exhibit weak bands having vibrational fine structure with maxima at about 263.5 mμ and at about 270–271 mμ (ϵ_{max} < 1000), in addition to shorter wavelength bands. The solution spectrum of *o*-methylbiphenyl also exhibits a weak band at about 270 mμ ($\epsilon_{max} = 900$). These weak bands are believed to correspond to the biphenyl H band, and this revelation is probably due to the marked hypsochromic shift and reduction in intensity of the A band caused by larger deviation from planarity of

[19] A. Wenzel, *J. Chem. Phys.* **21,** 403 (1953).

[20] P. M. Everitt, D. M. Hall, and E. E. Turner, *J. Chem. Soc.* **1956,** 2286.

the conformation of the conjugated system in these sterically hindered biphenyl derivatives as compared with biphenyl itself.

The H band of the biphenyl system also appears at about 260–280 mμ in the spectra of derivatives of biphenyl bridged at the *o* and *o'* positions by three-, four-, and five-membered aliphatic chains.[21] As will be mentioned in a later section, as the size of the bridge ring becomes larger, the A band shifts toward shorter wavelength and reduces in intensity, and the appearance of the H band becomes clearer. Especially, in the spectra of nine-membered-ring bridged biphenyls (that is, biphenyl derivatives bridged at the *o* and *o'* positions by a five-membered chain), the A band appears only as an inflection at about 231 mμ ($\epsilon = 5550$), and the vibrational fine structure of the H band is clearly resolved.

Thus the existence of the H band at about 270 mμ (4.6 ev) can be considered to have been established, and it is almost certain that this band corresponds to the benzene α band. Therefore, the foregoing assignments by London and by Davydov are incorrect, and the A band is probably due to the longitudinally polarized transition to the $^1B_{2u}$ state originating from the benzene *p* state.

Longuet-Higgins and Murrell[22] calculated the energies of electronic states of biphenyl by the method of composite molecule (see Chapter 20). In their calculation, eight locally-excited (abbreviated as LE) electron configurations corresponding to the α, p, β, and β' states of the two separate benzene rings and eight charge-transfer (abbreviated as CT) singlet electron configurations arising from one-electron transfer from the doubly degenerate highest bonding π MO's of each benzene ring to the doubly degenerate lowest antibonding π MO's of another benzene ring are taken into account, and singlet excited states of biphenyl are constructed as linear combinations of these sixteen electron configurations.

According to the results of the calculation, the energies (relative to the ground state) of the four lowest singlet excited states of a planar biphenyl molecule (D_{2h}) are as follows: 4.69 ev ($^1B_{3u}$), 4.70 ev ($^1B_{1g}$), 5.22 ev ($^1B_{2u}$), and 6.04 ev ($^1B_{2u}$). Transitions from the ground state ($^1A_{1g}$) to A_{1g} and B_{1g} states are forbidden. On the other hand, transitions to $^1B_{3u}$ states can have nonzero transition moments in the direction of the short axis (x axis) of the molecule, and transitions to $^1B_{2u}$ states can have nonzero transition moments in the direction of the long axis (y axis) of the molecule. The two lowest excited states, $^1B_{3u}$ at 4.69 ev and $^1B_{1g}$ at 4.70 ev, are the antisymmetric and symmetric combinations of the LE configurations corresponding to the α states of the two benzene rings, respectively, with minor contributions of

[21] K. Mislow, S. Hyden, and H. Schaefer, *J. Am. Chem. Soc.* **84,** 1449 (1962).
[22] H. C. Longuet-Higgins and J. N. Murrell, *Proc. Phys. Soc.* (*London*) **A68,** 601 (1955).

four CT configurations arising from one-electron transfers $\phi_{+1} \rightarrow \phi_{-2'}$; $\phi_{+1'} \rightarrow \phi_{-2}$; $\phi_{+2} \rightarrow \phi_{-1'}$; and $\phi_{+2'} \rightarrow \phi_{-1}$. The transitions to these states are probably responsible for the H band. The ${}^1B_{2u}$ state at 5.22 ev is the antisymmetric combination of the LE configurations corresponding to the p states of the two benzene rings (${}^1LE_p^{(-)}$) with a considerably large contribution of the antisymmetric combination of the LE configurations corresponding to the β' states of the two benzene rings (${}^1LE_{\beta'}^{(-)}$). The ${}^1B_{2u}$ state at 6.04 ev is the combination ${}^1LE_{\beta'}^{(-)}$ with some contribution of the combination ${}^1LE_p^{(-)}$. Both these two ${}^1B_{2u}$ states also have some contributions of the CT configurations arising from one-electron transfers $\phi_{+1} \rightarrow \phi_{-1'}$; $\phi_{+1'} \rightarrow \phi_{-1}$; $\phi_{+2} \rightarrow \phi_{-2'}$; and $\phi_{+2'} \rightarrow \phi_{-2}$. The transitions to these states are strongly allowed and are polarized along the long molecular axis. The A and B bands are attributed to these transitions.

Similar results have been obtained by the author by the use of the Pariser–Parr method. The calculation and its results will be outlined in the succeeding subsection.

12.2.6. Application of the Pariser–Parr Method to Biphenyl

Using the MO's obtained by the simple LCAO MO method as the starting MO's, the author applied the Pariser–Parr method (see Chapter 11) to biphenyl. Of spectroscopic interest are, probably, the four highest bonding MO's, ψ_{+1}–ψ_{+4}, and the four lowest antibonding MO's, ψ_{-1}–ψ_{-4}. The ground electron configuration (V_0) and the sixteen singlet and sixteen triplet singly-excited electron configurations arising from one-electron transitions from these four bonding MO's to these four antibonding MO's were taken into account. Evidently, the singlet configurations belong to symmetry species of point group D_2 (or D_{2h}) as follows: A_1 (A_{1g}): V_0, V_1^{14}, V_1^{41}, V_1^{23}, V_1^{32}; B_1 (B_{1g}): V_1^{12}, V_1^{21}, V_1^{34}, V_1^{43}; B_3 (B_{3u}): V_1^{13}, V_1^{31}, V_1^{24}, V_1^{42}; B_2 (B_{2u}): V_1^{11}, V_1^{22}, V_1^{33}, V_1^{44}. The symmetry species of the triplet configurations are, of course, the same as those of the corresponding singlet configurations. Each of the two secular determinants for singlet and triplet configurations can be factorized into four secular determinants by the symmetry. Furthermore, each of the determinants for the symmetry species A_1 (A_{1g}), B_1 (B_{1g}), and B_3 (B_{3u}) can be factorized into two determinants for plus and minus configurations. By the way, the symbols for electron configurations are based on the usual notation: for example, V_1^{14} represents the singlet electron configuration arising from the one-electron transition from ψ_{+1} to ψ_{-4}.

Values of AO Coulomb repulsion integrals were used which were calculated by the use of the equations of Pariser and Parr [Eqs. (11.40) and (11.41)] from

TABLE 12.2

RESULTS OF THE CALCULATION ON BIPHENYL BY THE PARISER–PARR METHOD (A) ENERGIES (E, IN UNITS OF EV) OF STATES RELATIVE TO THE GROUND ELECTRON CONFIGURATION

State			$k = 0.858$ $\theta = 0°$ D_{2h}	$k = 0.5$ $\theta = 52.6°$ D_2	$k = 0^*$ $\theta = 90°$ D_{2d}	$k = 0^{**}$ $\theta = 90°$ D_{2d}	
$^1\Phi_7$	$^+[V]_1^{12}$	$^1B_{1(g)}$	6.38	6.56*	6.71*	6.94*	$^1LE_\beta^{(+)}$
$^1\Phi_6$	$^+[V]_1^{13}$	$^1B_{3(u)}$	6.38*	6.56*	6.71*	6.94*	$^1LE_\beta^{(-)}$
$^1\Phi_5$	$[V]_1^{22+33}$	$^1B_{2(u)}$	6.15*	6.18*	6.35*	6.94*	$^1LE_{\beta'}^{(-)}$
$^1\Phi_4$	$[V]_1^{11}$	$^1B_{2(u)}$	4.71*	5.04*	5.27*	5.30	$^1LE_p^{(-)}$
$^3\Phi_8$	$^+[T]_1^{14}$	$^3A_{1(g)}$	4.60	4.48	4.45	4.46	$^3LE_{\beta'}^{(+)}$
$^1\Phi_3$	$^-[V]_1^{12}$	$^1B_{1(g)}$	4.55	4.61	4.67	4.90	$^1LE_\alpha^{(+)}$
$^1\Phi_2$	$^-[V]_1^{13}$	$^1B_{3(u)}$	4.55	4.61	4.67	4.90	$^1LE_\alpha^{(-)}$
$^3\Phi_7$	$^-[T]_1^{12}$	$^3B_{1(g)}$	4.55	4.61	4.67	4.90	$^3LE_\alpha^{(+)}$
$^3\Phi_6$	$^-[T]_1^{13}$	$^3B_{3(u)}$	4.55	4.61	4.67	4.90	$^3LE_\alpha^{(-)}$
$^1\Phi_1$	$^+[V]_1^{23}$	$^1A_{1(g)}$	4.51	4.41	4.38	5.30	$^1LE_p^{(+)}$
$^3\Phi_5$	$[T]_1^{22+33}$	$^3B_{2(u)}$	4.36	4.39	4.45	4.46	$^3LE_{\beta'}^{(-)}$
$^3\Phi_4$	$^+[T]_1^{23}$	$^3A_{1(g)}$	4.22	4.10	4.01	4.02	$^3LE_p^{(+)}$
$^3\Phi_3$	$^+[T]_1^{12}$	$^3B_{1(g)}$	4.12	4.18	4.23	4.46	$^3LE_\beta^{(+)}$
$^3\Phi_2$	$^+[T]_1^{13}$	$^3B_{3(u)}$	4.12	4.18	4.23	4.46	$^3LE_\beta^{(-)}$
$^3\Phi_1$	$[T]_1^{11}$	$^3B_{2(u)}$	3.46	3.82	4.01	4.02	$^3LE_p^{(-)}$
$^1\Phi_0$	$[V]_0$	$^1A_{1(g)}$	−0.02	−0.01	0	0	G

the interatomic distances calculated by allowing for the twist angle θ and the length R of the 1—1′ bond corresponding to a given value of k. A value of −2.39 ev was used for the core resonance integral for a C—C bond in benzene rings, β^C, and the core resonance integral for the coannular bond, $\beta_{11'}^C$, was assumed to be given by $k\beta^C$.

In Table 12.2 are summarized the values of the energies (relative to the energy of the ground configuration) of the lowest fifteen excited states and of the ground state calculated for the k values of 0.858, 0.5, and 0, which correspond to the θ values of 0, 52.6, and 90°, respectively. The symbols for states have been chosen on the basis of the following principle: if the electron configuration that makes the largest contribution to a state is, for example, V_1^{kr}, the state is denoted by $[V]_1^{kr}$. By the way, V_1^{22+33} and T_1^{22+33} represent, respectively, the symmetric combinations of V_1^{22} and V_1^{33} and of T_1^{22} and T_1^{33}:

$$V_1^{22\pm33} \equiv 2^{-1/2}(V_1^{22} \pm V_1^{33}); \qquad T_1^{22\pm33} \equiv 2^{-1/2}(T_1^{22} \pm T_1^{33}). \qquad (12.6)$$

For $k = 0$ the calculation was made in two ways. In one case (denoted by

$k = 0^*$), all the AO Coulomb repulsion integrals were taken into account, as for $k = 0.858$ and $k = 0.5$. In another case (denoted by $k = 0^{**}$), only the AO Coulomb repulsion integrals within each benzene ring were taken into account, those over the two benzene rings being neglected. The case $k = 0^{**}$ is the case where there is no interaction between the two benzene rings. In the last column of Table 12.2 the wavefunctions of the electronic states for $k = 0^{**}$ are shown by the use of the following notation. For example, the singlet LE configuration corresponding to the p state ($^1B_{1u}$ state) of the benzene ring B is denoted by $^1LE_{p(\mathrm{B})}$, and the one corresponding to the p state of the benzene ring B′ is denoted by $^1LE_{p(\mathrm{B'})}$. These two LE configurations have, of course, the same energy. The symmetric and antisymmetric combinations of these configurations are denoted by $^1LE_p^{(+)}$ and $^1LE_p^{(-)}$, respectively:

$$2^{-1/2}(^1LE_{p(\mathrm{B})} + {}^1LE_{p(\mathrm{B'})}) \equiv {}^1LE_p^{(+)}; \tag{12.7}$$

$$2^{-1/2}(^1LE_{p(\mathrm{B})} - {}^1LE_{p(\mathrm{B'})}) \equiv {}^1LE_p^{(-)}. \tag{12.8}$$

Similarly, the triplet LE configuration corresponding to the t_p state ($^3B_{1u}$ state) of the benzene ring B is denoted by $^3LE_{p(\mathrm{B})}$, the one corresponding to the t_p state of the benzene ring B′ is denoted by $^3LE_{p(\mathrm{B'})}$, and the symmetric and antisymmetric combinations of these two configurations are denoted by $^3LE_p^{(+)}$ and $^3LE_p^{(-)}$, respectively. Other LE configurations and their symmetry combinations are analogously denoted. The ground configuration is denoted by G. The table shows that when the interaction between the two benzene rings vanishes completely the state $^+[V]_1^{11}$, for example, reduces to the antisymmetric combination of the LE configurations corresponding to the p states of the two benzene rings.

When the interaction between the two benzene rings vanishes completely, the nine highest singlet and eight highest triplet states except for $^+[V]_1^{14}$ reduce to the symmetry combinations of CT configurations and their energies converge to a common value of 11.28 ev. The state $^+[V]_1^{14}$ reduces to $^1LE_{\beta'}^{(+)}$.

In Table 12.2, the asterisk suffixed to the values of state energies indicates that the transitions from the ground state to those states are allowed. The calculated values of the dipole strength D of the transitions from the ground state to excited states $^+[V]_1^{12}$, $^+[V]_1^{13}$, $^+[V]_1^{22+33}$, and $^+[V]_1^{11}$ are shown in Table 12.3. The letters in parentheses following the D values indicate the direction of the transition moment. The transitions to all the other states listed in Table 12.2 are forbidden. The wavefunctions of the five lowest singlet excited states and of the lowest triplet state for k values of 0.858, 0.5, and 0^{**} are shown in Table 12.4.

The H band at about 270 mμ is attributed to the forbidden transitions to degenerate minus states $^-[V]_1^{12}$ and $^-[V]_1^{13}$, which correspond to the benzene

TABLE 12.3

RESULTS OF THE CALCULATION ON BIPHENYL BY THE PARISER–PARR METHOD (B) TRANSITION DIPOLE STRENGTHS (D, IN UNITS OF Å^2)

State			$k = 0.858$ $\theta = 0°$	$k = 0.5$ $\theta = 52.6°$	$k = 0^*$ $\theta = 90°$	$k = 0^{**}$ $\theta = 90°$
${}^1\Phi_7$	${}^+[V]_1^{12}$	${}^1B_{1(g)}$	0	0.702 (z)	1.932 (z)	1.932 (z)
${}^1\Phi_6$	${}^+[V]_1^{13}$	${}^1B_{3(u)}$	3.312 (x)	2.873 (x)	1.932 (x)	1.932 (x)
${}^1\Phi_5$	$[V]_1^{22+33}$	${}^1B_{2(u)}$	3.942 (y)	3.866 (y)	3.853 (y)	3.864 (y)
${}^1\Phi_4$	$[V]_1^{11}$	${}^1B_{2(u)}$	0.833 (y)	0.284 (y)	0.012 (y)	0

TABLE 12.4

RESULTS OF THE CALCULATION ON BIPHENYL BY THE PARISER–PARR METHOD (C) COEFFICIENTS OF ELECTRON CONFIGURATIONS IN STATE FUNCTIONS[a]

State			Coefficient		
$[V]_1^{22+33}$	k	E, ev	V_1^{11}	V_1^{22+33}	V_1^{44}
	0.858	6.15	+0.442	−0.896	+0.042
	0.5	6.18	+0.519	−0.846	+0.124
	0**	6.94	+0.5	$-2^{-1/2}$	+0.5
$[V]_1^{11}$	k	E, ev	V_1^{11}	V_1^{22+33}	V_1^{44}
	0.858	4.71	+0.867	+0.434	+0.246
	0.5	5.04	+0.778	+0.533	+0.332
	0**	5.30	+0.5	$+2^{-1/2}$	+0.5
${}^-[V]_1^{12}$ (${}^-[V]_1^{13}$)	k	E, ev	${}^-V_1^{12}$ (${}^-V_1^{13}$)	${}^-V_1^{34}$ (${}^-V_1^{24}$)	
	0.858	4.55	+0.867	−0.497	
	0.5	4.61	+0.811	−0.585	
	0**	4.90	$+2^{-1/2}$	$-2^{-1/2}$	
${}^+[V]_1^{23}$	k	E, ev	${}^+V_1^{23}$	${}^+V_1^{14}$	
	0.858	4.51	+0.728	+0.686	
	0.5	4.41	+0.721	+0.693	
	0**	5.30	$+2^{-1/2}$	$+2^{-1/2}$	
$[T]_1^{11}$	k	E, ev	T_1^{11}	T_1^{22+33}	T_1^{44}
	0.858	3.46	+0.880	+0.357	+0.313
	0.5	3.82	+0.749	+0.528	+0.401
	0**	4.02	+0.5	$+2^{-1/2}$	+0.5

[a] The coefficients are listed under the corresponding electron configurations.

α state. Similarly to the transition to the benzene α state, the transitions to these states are probably made allowed, with polarization along the short molecular axis (x axis), by coupling with vibration. The data in Table 12.2 show that the energy of these transitions is comparatively insentive to variation in the twist angle, in agreement with observations.

The A band is attributed to the allowed transition to $^+[V]_1^{11}$, and the B band to the allowed transition to $^+[V]_1^{22+33}$. These states correspond to the benzene p and β' states, respectively, and belong to symmetry species B_2 (or B_{2u}). The transitions to these states are polarized along the long molecular axis (y axis). It is noteworthy that the major contributor to state $^+[V]_1^{11}$ is configuration $^+V_1^{11}$. This means that the transition to this state may be considered as the one-electron transition from ψ_{+1} to ψ_{-1} in the simple MO approximation. The energy of the transition to this state increases gradually and the dipole strength of this transition decreases rather rapidly with increasing angle of twist. This trend is quite similar to that found by the simple LCAO MO calculation (see Subsection 12.2.2). The C band is probably due to the transition to $^+[V]_1^{13}$ and $^+[V]_1^{12}$, which correspond to the benzene β state, if this band belongs to a band system different from that of the B band.

Besides the H band at about 275 mμ, a very feeble band with fine structure has been found at about 290 mμ in the spectra of liquid and solid films of biphenyl.[23] Probably the same band has also been found in the spectrum of a single crystal of biphenyl at low temperature measured with polarized light[24,25] and in the spectrum of a solution of biphenyl at low temperature.[26] This feeble band is probably due to the forbidden transition to the lowest singlet excited state, $^+[V]_1^{23}$. In addition, biphenyl appears to have very feeble band or bands in the region of wavelengths longer than 300 mμ, which are probably due to transitions to triplet excited states.

12.3. *o*-Substituted Biphenyls

12.3.1. *o*-ALKYL- AND *o,o'*-DIALKYLBIPHENYLS*

The steric interference of the two benzene rings in biphenyl is greatly increased when some or all of the hydrogen atoms at the 2, 2′, 6, and 6′

* This subsection is a revised version of Reference 27.

[23] A. R. Deb, *Indian J. Phys.* **27,** 305 (1953).

[24] R. Coffman and D. S. McClure, *Can. J. Chem.* **36,** 48 (1958); D. S. McClure, *ibid.* **36,** 59 (1958).

[25] V. L. Broude, E. A. Izrailevich, A. L. Liberman, M. I. Onoprienko, O. S. Pakhomova, A. F. Prikhot'ko, and A. I. Shatenshtein, *Optika i Spektroskopiya* **5,** 113 (1958).

[26] Y. Kanda (Kyushu University, Japan), personal communication.

positions of biphenyl are replaced by other atoms or groups.[27] Thus, for example, the A band of *o*-methylbiphenyl is at considerably shorter wavelength and of considerably lower intensity than that of biphenyl. This hypsochromic and hypochromic shift of the A band, caused by introduction of an *o*-methyl substituent, is believed to indicate that the twist angle θ in the most probable conformation of *o*-methylbiphenyl is considerably larger than that of biphenyl. Furthermore, the spectrum of *o*,*o*′-dimethylbiphenyl exhibits

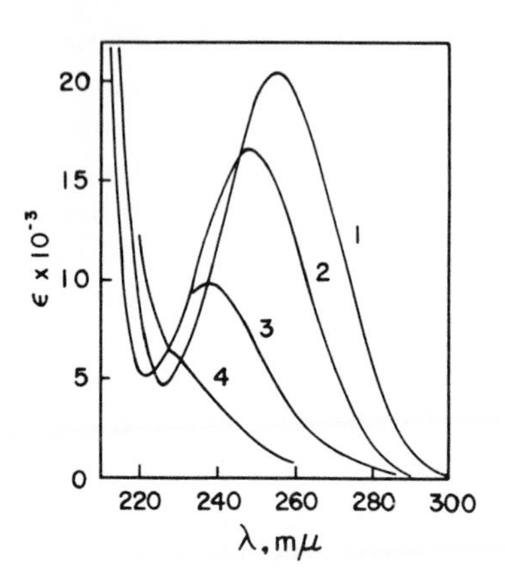

FIG. 12.4. Absorption spectra of biphenyls in ethanol. 1, *p*,*p*′-dimethylbiphenyl; 2, biphenyl; 3, *o*-methylbiphenyl; 4, *o*,*o*′-dimethylbiphenyl.

no distinct maximum of the A band, but exhibits only an inflection at further shorter wavelength than the position of the A band of *o*-methylbiphenyl (see Fig. 12.4).

For a detailed analysis of the steric effects of progressively introducing *ortho* substituents into the biphenyl chromophore, alkyl groups have the advantage of causing the least electronic perturbation of the parent spectrum, so that the identification of the A band is not complicated by appearance of other band systems arising from the electronic interaction of the substituents with the aromatic rings. The data on the A bands of some alkylbiphenyls are summarized in Table 12.5, in which that of biphenyl is included for comparison. While the *o*-methyl substitution gives rise to a distinct hypsochromic and hypochromic shift of the A band, the *p*-methyl substitution gives rise to a distinct bathochromic and hyperchromic shift of the band. The introduction of a second methyl group into the *p*′-position roughly doubles the magnitude

[27] H. Suzuki, *Bull. Chem. Soc. Japan* **32,** 1350 (1959).

TABLE 12.5

THE A BANDS OF ALKYLATED BIPHENYLS[a]

Biphenyl	Solvent[b]	λ_{max}, mμ	ϵ_{max}
Unsubstituted	E	247.7	16600
	H	247.4	16700
4-Methyl-	E	253	19000
3-Methyl-[c]	P	249	16300
2-Methyl-[d]	E	235	10500
2-Ethyl-[d]	E	233	10500
2-*n*-Propyl-[d]	E	233	10000
2-Isopropyl-[d]	E	233	11000
2-*n*-Butyl-[d]	E	233	10500
4,4′-Dimethyl-	E	255.6	20500
4,4′-Diethyl-[e]	E	256.5	22500
4,4′-Diisoproyl-[e]	E	256.5	23500
3,3′-Dimethyl-[c]	P	250.5	16100
2,2′-Dimethyl-	E	227*	6800
2,2′-Diethyl-[e]	E	227*	6000
2,2′-Diisopropyl-[e]	E	227*	5500
2,6-Dimethyl-[c]	P	231*	5600
2,6,2′-Trimethyl-[c]	P	230**	4000

[a] The asterisk on wavelength values denotes inflections; the double askterisk denotes a very faint inflection.

[b] Solvent: E = ethanol; H = heptane; P = light petroleum, b.p. 100–120°.

[c] Reference 28.

[d] E. A. Braude and W. F. Forbes, *J. Chem. Soc.* **1955,** 3776.

[e] Reference 20.

of the shift. The *m*-methyl substitution exerts almost no distinct effect on the A band: both the spectra of *m*-methyl- and *m,m′*-dimethylbiphenyl resemble the biphenyl spectrum.

The twist angle of the coannular bond in *o*-alkyl- and *o,o′*-dialkylbiphenyls can be calculated from the position of the A band by extending the method developed in Subsections 12.2.2–12.2.4 for biphenyl. It is to be noted here that the effect of the *ortho* substituents upon the A band must include not only the steric hypsochromic effect but also the electronic bathochromic effect. In the application of the method to alkylbiphenyls, new MO calculations are avoided by assuming that the relation between k and ΔE is the same as in biphenyl itself, and the possible electronic bathochromic effect of the *o*-alkyl substituents is allowed for by choosing appropriate values of ΔE and ν for the two references for correlation of the observed ν with the calculated ΔE. For example, in application of the method to *o*-methylbiphenyl, in order to allow for the possible bathochromic effect of the *o*-methyl group in the absence of steric effects, it is assumed that this is equivalent to the effect of a *p*-methyl group, and from the observed value of λ_{max} of *p*-methylbiphenyl,

TABLE 12.6

RESULTS OF THE CALCULATION OF THE TWIST ANGLE IN BIPHENYLS[a]

Biphenyl	λ_{max} mμ	Results from Eq. (9.39) ΔE_{11}	k	θ	R, Å	Results from Eq. (12.1) ΔE_{11}	k	θ
Unsubstituted								
(solution)	247	1.523	0.780	23°	1.49	1.535	0.758	25°
(vapor)	238	1.617	0.611	43	1.50	1.627	0.595	44.5
2-Methyl-	235	1.719	0.438	58	1.51	1.726	0.428	59
2-Ethyl-	233	1.743	0.399	60.7	1.51	1.749	0.392	62
2,2′-Dimethyl-	227*	1.826	0.267	70	1.52	1.830	0.262	72
2,2′-Diethyl-	227*	1.829	0.262	70.3	1.52	1.833	0.257	72

[a] The asterisk on wavelength values denotes inflections.

253 mμ, the corresponding wave number, 39,526 cm^{-1}, is taken as ν_S. Since the observed spectrum of *p*-methylbiphenyl is the spectrum of ethanol solution, the value of ΔE found applicable to biphenyl in solution, 1.523 $(-\beta)$ for $k = 0.780$, is taken as ΔE_S. Toluene is taken as the large-angle reference for which $k = 0$ and $\Delta E_L = 2\ (-\beta)$. Each of bands of toluene is located at longer wavelength by about 5 mμ than the corresponding band of benzene (see Subsection 10.6.3). The "center of gravity of singlets" of toluene, which is estimated at 46,882 cm^{-1}, is taken as ν_L. It is then possible to calculate ΔE, k, θ, and R for *o*-methylbiphenyl. Similar treatment allows calculations for other *ortho*-substituted biphenyls. The results are shown in Table 12.6, in which the results for biphenyl in various states obtained in Subsection 12.2.4 are also included. In this table, besides the values calculated on the basis of the assumption that k is related to θ by Eq. (9.39), those calculated by the use of Eq. (12.1) are also shown. Evidently the differences between the values of θ calculated in the two ways are quite small.

The value of θ for 2-methylbiphenyl is in fairly good agreement with the value of 62° deduced from a scale model in which an *ortho* hydrogen atom just touches the *ortho* methyl group. The values of θ for 2-ethyl-, 2-*n*-propyl-, 2-isopropyl-, and 2-*n*-butylbiphenyl are substantially identical with one another, and they are merely slightly larger than the value for the methyl analog. This probably indicates that methyl or methylene groups attached to the methylene group which is directly attached to the 2-position can adopt conformations which cause little or no additional interference over the one methyl group.

The spectrum of 2,6-dimethylbiphenyl is closely similar to that of 2,2'-dimethylbiphenyl, exhibiting no distinct maximum of the A band but an inflection at about 230 mμ,[28] in contrast with the spectrum of 2-methylbiphenyl. 2,6-Dimethylbiphenyl involves the steric interferences between *o*-methyl groups and *o'*-hydrogen atoms on both sides of the molecule, while 2-methylbiphenyl involves the steric interference between an *o*-methyl group and an *o'*-hydrogen atom on one side of the molecule. The steric hindrance to the coplanarity of the two benzene rings will, therefore, be severer in the former compound than in the latter compound. When 2,2'-dimethylbiphenyl assumes a conformation that has the two *o*-methyl groups in a *trans*-like disposition, the steric situation in this molecule will be similar to that in the molecule of 2,6-dimethylbiphenyl. The spectral similarity between 2,2'- and 2,6-dimethylbiphenyl presumably indicates that the inflection at about 227 mμ in the spectrum of 2,2'-dimethylbiphenyl is due to the vestigial A band of this compound in a *trans*-like conformation and that the value of θ calculated from the position of this inflection is for this *trans*-like conformation. The probability distribution curve of this compound with respect to the twist angle probably has two maxima, one at a *trans*-like conformation and another at a *cis*-like conformation. The angle of twist in the *cis*-like conformation will be larger than that in the *trans*-like conformation, and the molecules having the *cis*-like conformation will absorb at shorter wavelength than do the molecules having the *trans*-like conformation. According to an X-ray crystal analysis, the 4,4'-diamino derivative of 2,2'-dimethylbiphenyl, that is, *m*-tolidine or 2,2'-dimethylbenzidine, assumes in the crystalline state a *cis*-like conformation in which the twist angle of the coannular bond is 86° and the length of the bond is 1.55 Å.[29]

The spectra of 2,6,2'-trimethyl-, 2,6,2',6'-tetramethyl-, and 2,4,6,2',4',6'-hexamethylbiphenyl show no sign of any contribution from the A band, and

[28] G. H. Beaven, *in* "Steric Effects in Conjugated Systems" (G. W. Gray, ed.), p. 22. Butterworths, London and Washington, D.C., 1958.

[29] F. Fowweather, *Acta Cryst.* **5**, 820 (1952).

are thus distinctly different from the spectra of 2,6- and 2,2′-dimethylbiphenyl.[28] These highly methylated biphenyls involve necessarily the steric interference between *o*-methyl and *o*′-methyl groups in a *cis*-like disposition. The spectrum of 2,2′-di-*tert*-butylbiphenyl also shows no sign of any contribution from the A band.[28] The spectra of these highly hindered biphenyls are rather similar to, but not completely quantitatively equivalent to, the additive absorption of two independent alkylbenzene chromophores. Probably the twist angle of the coannular bond is nearly 90° in these compounds.

12.3.2. Biphenyl Derivatives with *Ortho* Substituents other than Alkyl Groups

The method applied to alkylbiphenyls in the preceding subsection is not applicable to biphenyl derivatives with *ortho* substituents other than alkyl groups, because in these derivatives the perturbation of the biphenyl chromophore by electronic effects of the substituents is too strong. However, steric effects closely similar to those of *o*-alkyl substituents are observed with the *o*-substituents other than the alkyl groups.

TABLE 12.7

THE A BANDS AND INTERPLANAR ANGLES OF 2,2′-DIHALOGENATED BIPHENYLS[a]

Biphenyl	The A band[b] (solvent: ethanol)		Interplanar angle (θ)	
	λ_{max}, mμ	ϵ_{max}	(electron diffraction), deg.	(dipole moment), deg.
2,2′-Difluoro-	233.5	13800	60 ± 5	77
2,2′-Dichloro-	230*	6600	74 ± 5	81
2,2′-Dibromo-	228**	12000	75 ± 5	85
2,2′-Diiodo-	—	—	79 ± 5	84

[a] Data taken from Reference 28.

[b] The asterisk on wavelength values denotes an inflection; the double asterisk denotes a very faint inflection.

The spectra of 2,2′-dihalobiphenyls show a progressive reduction in intensity and increasing hypsochromic shift of the A band with the increasing size of the halogen atom in the sequence F, Cl, Br. The spectrum of 2,2′-diiodobiphenyl qualitatively resembles that of *o*-iodotoluene. In Table 12.7 the data on the A bands of 2,2′-dihalobiphenyls are shown, together with the values of θ (taken as 0° for the planar *cis* conformation) estimated from electron diffraction and dipole moment studies. The values of θ shown in this table correspond nearly to *cis*-like conformations in which the two *ortho* halogen atoms are in van der Waals contact. The 4,4'-diamino derivative of 2,2'-dichlorobiphenyl, that is, 2,2'-dichlorobenzidine, has been found by X-ray crystal analysis to assume in the crystalline state a *cis*-like conformation in which the twist angle θ is about 72°.[30]

Biphenyl derivatives substituted at *ortho* positions by bulky substituents, such as amino, nitro, and carboxyl groups, show spectra closely resembling those of the two halves of the molecule,[31] suggesting the near-orthogonality of the substituted phenyl moieties. For example, the spectrum of 6,6'-dinitrobiphenyl-2,2'-dicarboxylic acid (dinitrodiphenic acid) closely resembles that of *m*-nitrobenzoic acid, the spectra of 2-amino-, 2,2'-diamino-, and 2,2'-diamino-6,6'-dimethylbiphenyl resemble the spectrum of aniline, and the spectra of the protonated amines resemble the spectrum of anilinium ion. Dipole moment measurements have yielded the following values of θ (taken as 0° for the planar *cis* conformation): 2,2'-diaminobiphenyl, 79°; 2,2'-diamino-6,6'-dimethylbiphenyl, 67°; 2,2'-bisdimethylaminobiphenyl, 120°. The spectrum of 2,2'-dinitrobiphenyl resembles that of nitrobenzene. From dipole moment measurements, it has been concluded that the dinitrobiphenyl assumes a *cis*-like equilibrium conformation.

12.4. *o,o'*-Bridged Biphenyls*

The A band of biphenyls bridged at the *o*- and *o'*-positions by an aliphatic chain suffers a monotonic hypsochromic shift and decrease in intensity as the size of the bridge chain increases[32]: for compounds of types I*b*, I*c*, I*d*, and I*e* the A band assumes values of λ_{max} of 264 mμ (log ϵ_{max}, 4.23–4.26), 248–250 mμ (log ϵ_{max}, 4.18–4.23), 235–239 mμ (log ϵ_{max}, 4.0–4.1), and 231 mμ (inflection, log ϵ 3.75), respectively.[21] The spectrum of a nine-membered-ring bridged (or five-atom-bridged) biphenyl, IV, shows the A band only as an inflection at 231 mμ and is nearly superposable on that of the open-chain analog, 2,2'-diethylbiphenyl. The spectra of some *o,o'*-bridged biphenyls are compared in Fig. 12.5. The weak bands observed at about 260–280 mμ are probably the H bands.

* This section is a revised version of Reference 32.

[30] D. L. Smare, *Acta Cryst.* **1,** 150 (1948).

[31] K. Mislow, M. A. W. Glass, R. E. O'Brien, P. Rutkin, D. H. Steinberg, J. Weiss, and C. Djerassi, *J. Am. Chem. Soc.* **84,** 1455 (1962).

[32] H. Suzuki, *Bull. Chem. Soc. Japan* **32,** 1357 (1959).

I

I*a* : $n = 5$

I*b* : $n = 6$

I*c* : $n = 7$

I*d* : $n = 8$

I*e* : $n = 9$

II

III

IV

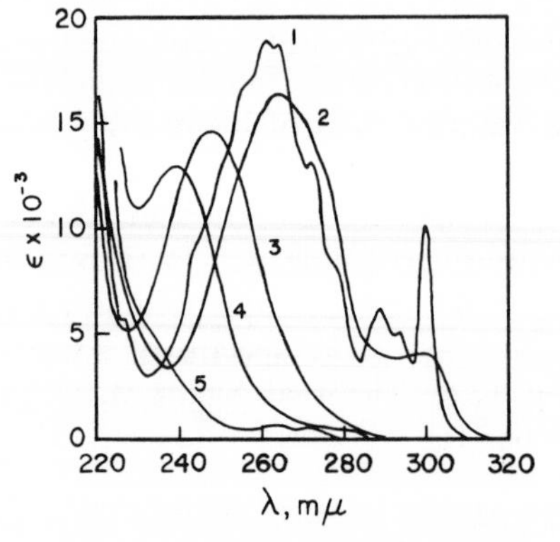

FIG. 12.5. Absorption spectra of alicyclic ring 2,2′-bridged biphenyls in 95% ethanol. 1, I*a* (solvent: heptane); 2, I*b*; 3, II, R = H, B = CHCOOH; 4, III, $R_1 = R_2 = COOC_2H_5$; 5, IV. Curve 1, from Reference 32; curve 2, from References 21 and 4; curves 3, 4, and 5, from Reference 21.

The first member of the series of *o,o′*-bridged biphenyls, fluorene (I*a*), has a spectrum considerably similar to that of the second member of the series, 9,10-dihydrophenanthrene (I*b*), but the spectrum of fluorene shows much more well-developed vibrational structure, presumably because of the strained near-planar geometry of the molecule. In the spectra of solutions in ethanol, the A band of fluorene has the maximum at about 260 mμ with ϵ of 19,000, and the A band of 9,10-dihydrophenanthrene has the maximum at about 264 mμ with ϵ of about 17,000.

The A band of fluorene is at considerably longer wavelength than that of biphenyl. This may be due partly to the forced coplanarity or near-coplanarity of the two benzene rings of fluorene, and partly to the hyperconjugative effect of the methylene group. The interplanar angle between the two benzene rings in 9,10-dihydrophenanthrene has been estimated to be about 20° from a model in which the saturated carbon atoms maintain their tetrahedral valence angle. Accordingly, the bathochromic shift of the A band of this compound relative to that of biphenyl may be attributed mainly to the normal electronic effect of alkyl groups. The maximum of the A band of 4,5-methylene-9,10-dihydrophenanthrene ($\lambda_{max} = 272$ mμ, $\epsilon_{max} =$ 18,500, in ethanol) is at longer wavelength than those of fluorene and 9,10-dihydrophenanthrene. In this compound the conflicting steric requirement of the two bridges seems to be most favorably compromised in an approximately planar geometry. Accordingly, the bathochromic shift of the A band of this compound relative to those of fluorene and 9,10-dihydrophenanthrene may be due partly to the planarity or near-planarity of the geometry, and partly to the electronic effect of the additional dimethylene or methylene bridge.

In applying our method to the series of *o,o′*-bridged biphenyls, it seems to be most adequate to take 9,10-dihydrophenanthrene as the small-angle reference. A justification for taking this compound rather than fluorene as the reference is that this makes a pertinent allowance for the electronic bathochromic effect of two methylene groups attached at the *o*- and *o′*-positions of biphenyl, and further justification is found in the fact that the molecule of fluorene is presumed to contain a considerably large strain: in fact, the long biphenyl axis in fluorene has been found to be nonlinear.[33]

Assuming the value of θ for the small-angle reference to be 20°, we obtain $\Delta E_S = 1.5115(-\beta)$. The wave number at the maximum of the A band of 9,10-dihydrophenanthrene, 37,880 cm^{-1} (264 mμ), is taken as ν_S. Similarly to

[33] G. M. Brown and M. H. Bartner, *Acta Cryst.* **7,** 139 (1954); D. M. Burns and J. Iball, *Nature* **173,** 635 (1954).

the previous treatment of *o*-alkyl- and *o,o'*-dialkylbiphenyls, toluene is taken as the large-angle reference. That is, the center of gravity of singlets of toluene, 46,882 cm^{-1} (213.3 mμ), is taken as ν_L, and this value of ν_L is assumed to correspond to the ΔE_L value of $2(-\beta)$. On the basis of these reference values, the interplanar angles θ between two benzene rings in *o,o'*-bridged biphenyls can be calculated by the usual procedure from the position of the A band. The results are summarized in Table 12.8.

The interplanar angle in 2,2'-three-atom-bridged biphenyls has been estimated at about 50° from a scale model. The angle in higher homologs cannot be accurately estimated from models, since these compounds must have more than one conformation corresponding to the minimum valence-angle strain. Thus the angle in 2,2'-four-atom-bridged biphenyls has been inferred to assume about 60–65° or alternatively 75–80°. The values of θ

TABLE 12.8

CHARACTERISTICS OF 2,2'-BRIDGED BIPHENYLS

Compound	The A band	Results of calculation		
	λ_{max}, mμ	ΔE_{11}, $-\beta$	k	θ, deg.
I*a*	261.5	—	—	0
I*b*	264	1.5115	0.801	20
Compounds of type II (R = H)	248–250	1.644–1.627	0.565–0.594	47.3–44.5
Compounds of type III	236.5–239.5	1.751–1.722	0.387–0.434	61.5–58.0
IV	231	1.805	0.300	68
II, B = CH_2, R = CH_3	240	1.805	0.300	68

calculated from the position of the A band agree fairly well with the values estimated from models.

The A band of 2,2'-three-atom-bridged biphenyls ($\lambda_{max} = 249 \pm 2$ mμ, log $\epsilon_{max} \doteqdot 4.2$) suffers both a hypsochromic shift and a decrease in intensity to $\lambda_{max} = 242 \pm 2$ mμ and log $\epsilon_{max} \doteqdot 4.0$ on substitution at the 6- and 6'-positions by either methyl or chlorine.[31] The two *ortho* substituents must assume a *cis*-like disposition and must increase the interplanar angle in the parent bridged compounds. In application of our method to 6,6'-dimethyl-2,2'-trimethylenebiphenyl (II with B = CH_2 and R = CH_3), allowance for the electronic bathochromic effect of the methyl groups is made by adding about 9–10 mμ to the wavelengths of both the references or by subtracting the same amount from the observed wavelength of the maximum of the A band of this compound. In this way the interplanar angle in this compound (λ_{max} = 240 mμ, log ϵ_{max} = 4.06, in hexane) is calculated to be about 68°. This value coincides fairly well with the corresponding value for 2,2'-dimethylbiphenyl (see Table 12.6). It should be noted that the A band of this bridged compound shows a distinct maximum of considerably large intensity, in contrast with the A band of 2,2'-dimethylbiphenyl, which, as mentioned previously, shows no distinct maximum and only shows a weak inflection. Owing to the existence of the bridge, the probability distribution curve of this bridged compound with respect to the angle θ must be much steeper than that of 2,2'-dimethylbiphenyl, and it probably has a maximum corresponding to a *cis*-like conformation with a θ value which is smaller than the value for the preferred *cis*-like conformation of 2,2'-dimethylbiphenyl, and is incidentally comparable to the value for the preferred *trans*-like conformation of that compound.

In suitably *ortho* substituted biphenyls and compounds in which the *o*- and *o'*-positions of biphenyl are bridged by a more-than-three-membered aliphatic chain, the two benzene rings are not coplanar, and with such compounds optical isomerism can occur owing to the restricted rotation around the coannular bond. Indeed, many compounds of such types have been obtained in optically active form. The π-electronic transitions in such a twisted biphenyl system are generally optically active and are accompanied with a Cotton effect and with a circular dichroism.

Mislow and his co-workers[31] have gathered optical rotatory dispersion and ultraviolet absorption data for a large body of dissymmetric biaryls of known absolute configuration. According to them, the optical rotatory dispersion curves of 6,6'-dimethyl and 6,6'-dichloro derivatives of dissymmetric 2,2'-three-atom-bridged biphenyls indicate a Cotton effect of appreciable amplitude centered near 245 mμ (nearly the position of the A band) and that of very low amplitude near 280 mμ (nearly the position of the H band). The absolute configuration of the biphenyl skeleton uniquely determines the sign of the Cotton effect at the position of the A band: the sign of this Cotton effect is positive for the (*R*)-configuration, and the effect of the substituents on this Cotton effect is small. On the other hand, the sign of the Cotton effect at the position of the H band depends on the chemical nature of the substituents. By the way, recently, Grinter and Mason[34] attempted to calculate the dihedral angle between the planes of the anthracene nuclei in some optically-active derivatives (2,2'-dimethoxycarbonyl and 2,2'-dimethylene derivatives) of 1,1'-bianthryl by analyzing theoretically the electronic absorption and circular dichroism spectra of these compounds.

[34] R. Grinter and S. F. Mason, *Trans. Faraday Soc.* **60**, 274 (1964).

12.5. Polyphenyls*

In the series of *p*-polyphenyls, the first intense band, that is, the A band, shifts toward longer wavelengths and increases in intensity with the increasing number of benzene rings.[35] Data listed in Table 12.9 show that the magnitude of the wavelength shift of the band associated with introduction of one additional phenyl group decreases progressively with the increasing number of benzene rings in the *p*-polyphenyl. The wavelength of the maximum of

TABLE 12.9

CHARACTERISTICS OF *p*-POLYPHENYLS

		The A band[b]			Results of calculation		
Compound	n^a	λ_{max} mμ	ϵ_{max}	ν_{max} cm^{-1}	ΔE_{11} $-\beta$	D_{11} Å^2	$c_{1,p}{}^e$
(Benzene)	1	203[c]	7,400[c]	49,260[c]	2.000	0.966	0.5774
Biphenyl	2	247.4	16,700	40,420	1.4814	2.678	0.3983
p-Terphenyl	3	276.5	31,800	36,170	1.2819	4.275	0.2935
p-Quaterphenyl	4	292[d]	55,000[d]	34,250	1.1820	5.595	0.2275
p-Quinquiphenyl	5	—	—	—	1.1241	6.621	0.1828
p-Sexiphenyl	6	308[d]	—	32,470	1.0876	7.414	0.1508

[a] n represents the number of benzene rings in the molecule.
[b] Solvent, hexane or heptane.
[c] Data for the p band.
[d] Data taken from Reference 37.
[e] $c_{1,p}$ represents the coefficients of the AO's on the *para* positions of the outer benzene rings in MO ψ_{+1} as well as ψ_{-1}.

the band, λ_{max}, appears to approach asymptotically a limited value as the length of the conjugated system increases. A similar trend has been observed with the first intense band in the series of conjugated polyenes (see Subsection 10.4.2).

In the simple LCAO MO approximation the A band of a *p*-polyphenyl can be considered to be due to the one-electron transition from the highest bonding π MO (ψ_{+1}) to the lowest antibonding π MO (ψ_{-1}). This transition has a nonzero transition moment in the direction of the long molecular axis. Of course, w_{+1} is equal to $-w_{-1}$, and the transition energy, ΔE_{11}, is equal to $2w_{-1}(-\beta)$. Biphenyl, *p*-terphenyl, and *p,p'*-quaterphenyl have been reported

* This entire section is a revised version of Reference 35.

[35] H. Suzuki, *Bull. Chem. Soc. Japan* **33,** 109 (1960).

to have planar conformations in the crystalline state, in which the length of the coannular bonds is approximately 1.48 Å.[36] Values of the energy (ΔE_{11}) and dipole strength ($D_{11} = M_{11}^2$) of the one-electron transition from ψ_{+1} to ψ_{-1} in *p*-polyphenyls and some relevant data calculated by the use of values of 0.858 and 1.48 Å for the resonance parameter and length of the coannular bonds are also shown in Table 12.9. There is an approximately linear relation between the wave number of the maximum of the A band, ν_{max}, and the calculated transition energy, ΔE_{11}. Furthermore, when ϵ_{max} is plotted against D_{11}, a smooth correlation curve is obtained.

In contrast with the *all-para*-polyphenyl series, in the *all-meta*-polyphenyl series there is virtually no change in the position of the first intense band ($\lambda_{max} = 253 \pm 2$ mμ, in chloroform) after biphenyl up to the highest member of the series yet prepared, in which the number of benzene rings, *n*, is sixteen, while the intensity of the band increases as the series ascends.[37] In *all-meta*-polyphenyls, conjugative interaction between any two adjacent benzene rings is possible and effective, but conjugative interaction between more than two rings is unimportant. Therefore, the *all-meta*-polyphenyl containing *n* benzene rings may be considered to act as an absorbing entity made up of ($n - 1$) isolated biphenyl units. Actually, the molar extinction coefficient of the maximum of the A band is roughly proportional to ($n - 1$).

In *o*-polyphenyls there should be severe steric interference between *ortho*-related benzene rings: the most probable conformations of *o*-polyphenyls cannot be planar. Therefore, an *o*-polyphenyl (for example, *o*-terphenyl) is expected to show a spectral feature appreciably different from that of its *para* isomer. Actually, this has been shown to be the case.[7]

Now it is attempted to infer the most probable geometry of *o*-terphenyl from the electronic absorption spectrum by means of the simple LCAO MO calculation. In this case, *p*-terphenyl in the crystalline state is taken as the small-angle reference, and benzene is taken as the large-angle reference. The spectra of *p*- and *o*-terphenyl in the crystalline state measured by the pressed KCl-disk technique do not significantly differ from the respectively corresponding spectra of hexane solutions.[7] Only a normal, or slightly larger than normal, red-shift is observed for each compound. This fact is considered to indicate that the twist angle of the coannular bonds, θ, in each compound in solution is not much larger than in the crystalline state. Thus, the A band of *p*-terphenyl is located at 276 mμ ($\epsilon_{max} = 32{,}100$) in the solution spectrum, and is located at 284 mμ in the KCl-disk spectrum. The magnitude

[36] L. W. Pickett, *J. Am. Chem. Soc.* **58,** 2299 (1936).
[37] A. E. Gillam and D. H. Hey, *J. Chem. Soc.* **1939,** 1170.

of the normal red-shift is estimated to be about 6 mμ at 284 mμ on the basis of the regular relation between the position of the band and the magnitude of the corresponding normal red-shift determined by Dale (see Subsection 6.4.3). Accordingly, a value of 35,971 cm^{-1} (284 − 6 = 278 mμ) is taken as ν_S. The value of ΔE_{11} calculated for the planar *p*-terphenyl molecule by the use of a value of 0.858 for the resonance parameter for the coannular bonds is 1.282($-\beta$), as shown in Table 12.9. This value is taken as ΔE_S. For the large-angle reference, as usual, values of 2($-\beta$) and 48,000 cm^{-1} are taken as ΔE_L and ν_L, respectively.

The spectrum of *o*-terphenyl in hexane solution exhibits an absorption maximum at 232 mμ (ϵ_{max} = 26,900), and the KCl-disk spectrum exhibits the corresponding maximum at 237 mμ. In addition, the solution spectrum shows a weaker band at about 252 mμ (ϵ_{max} = ca. 11,400), and the KCl-disk spectrum shows an inflection at about 250–260 mμ. This diffuse or vestigial band is regarded as the A band. From the position of the A band, 252 mμ, appropriate values of ΔE_{11}, k, θ, and R for *o*-terphenyl in solution are determined by the usual procedure as follows: $\Delta E_{11} = 1.504(-\beta)$, $k = 0.611$, $\theta = 43°$, and $R = 1.496$ Å. Thus the most probable geometry of *o*-terphenyl in solution is inferred to be the one in which the two outer phenyl groups are twisted about 43° out of the plane of the parent benzene ring.

The interplanar angle θ in *o*-terphenyl in the crystalline state has been determined by X-ray crystal analysis to be 50° or less, probably 45–50°.[38] In view of the close similarity between the solution spectrum and KCl-disk spectrum of this compound, the most probable geometry of this compound in solution is considered to be similar to that in the crystalline state. Therefore, it may be said that the value of θ, estimated above from the position of the A band, is quite adequate. By the way, the interplanar angle in *o*-terphenyl in the vapor state has been very roughly determined by electron diffraction method to be about 90°.[39] The outer phenyl groups of hexaphenylbenzene in the vapor state have also been found by electron diffraction studies to be nearly orthogonal to the central benzene ring, and probably oscillate with no appreciable restriction in a rather limited angle interval of approximately ±10° from the orthogonal form.[40]

[38] C. J. B. Clews and K. Lonsdale, *Proc. Roy. Soc.* (*London*) **A161,** 493 (1937).

[39] I. L. Karle and L. O. Brockway, *J. Am. Chem. Soc.* **66,** 1974 (1944).

[40] A. Almenningen, O. Bastiansen, and P. N. Skancke, *Acta Chem. Scand.* **12,** 1215 (1958).

Chapter 13

Styrene and Related Compounds*

13.1. Styrene and Its Alkylated Derivatives

13.1.1. Spectra

Data on the electronic absorption spectra of styrene and alkylstyrenes in solution in saturated hydrocarbon or alcohol are summarized in Table 13.1, and the spectra of some styrenes are shown in Fig. 13.1.[1]

The intense band of styrene at about 248 mμ is the so-called conjugation band, which is considered to be due to one-electron transition from the highest occupied π MO, ψ_{+1}, to the lowest vacant π MO, ψ_{-1}, in the simple MO approximation. This band is hereafter called the A band. It is evident that introduction of one or more alkyl substituents into the sterically hindered positions, that is, the *o*-, α-, and *cis*-β-positions, exerts an hypsochromic and hypochromic effect on the A band, while introduction of an alkyl substituent into the *p*- or the *trans*-β-position exerts a bathochromic and hyperchromic effect. It is worthwhile noting that the A bands of the sterically hindered derivatives are almost completely structureless, in contrast with the styrene A band, which has vibrational fine structure. The weak band appearing with fine structure in the region between 270 and 290 mμ corresponds probably to the benzene α band, similarly to the biphenyl H band.

13.1.2. The Simple LCAO MO Calculation

The skeleton of the styrene π system is illustrated in Fig. 13.2. In application of the simple LCAO MO method to this system, the length of the ethylenic bond (1′—2′ bond) is assumed to be 1.34 Å, and a value of 1.08 is used for the parameter k for this bond. Unfortunately, no reliable data on

* This chapter is a revised version of Reference 1.

[1] H. Suzuki, *Bull. Chem. Soc. Japan*, **33**, 619 (1960).

TABLE 13.1

THE A BANDS OF STYRENES[a]

No.	Styrene	Solvent[b]	λ_{max}, mμ	ϵ_{max}
1	Unsubstituted	C[c]	248	—
		E[d]	248	14620
		H[e]	247.7	14760
			242.7	14600
2	*p*-Methyl-	C[c]	253	—
3	*trans*-β-Methyl-	E[f]	250	17300
4	*trans*-β-Ethyl-	E[d]	251	17640
5	*o*-Methyl-	C[c]	246	—
6	*o,p*-Dimethyl-	C[c]	251	—
7	α-Methyl-	H	241.5	9780
		E[d]	243	11400
8	α,β-Dimethyl-	E[g]	244	8700
9	α-Ethyl-	E[d]	240	10100
10	α-*n*-Propyl-	E[d]	239	9390
11	α-Isopropyl-	E[d]	234	7800
12	*cis*-β-Methyl-	E[f]	240.6	13800
13	β,β-Dimethyl-	H	247.2*	6690
			239.4	9060
14	*o,o'*-Dimethyl-	C[c]	238	—
15	*o,p,o'*-Trimethyl-	O[h]	245	7000
16	α,β,β-Trimethyl-	H	245.3*	6210
			237.8	8440
17	α,*o*-Dimethyl-	E[i]	No maximum	

[a] The asterisk on wavelength values denotes inflections.

[b] Solvent: C, cyclohexane; E, ethanol; H, heptane; O, octane.

[c] L. H. Schwartzman and B. B. Corson, *J. Am. Chem. Soc.* **78,** 322 (1956).

[d] C. G. Overberger and D. Tanner, *J. Am. Chem. Soc.* **77,** 369 (1955).

[e] D. F. Evans, *J. Chem. Phys.* **23,** 1429 (1955).

[f] R. Y. Mixer, R. F. Hech, S. Winstein, and W. G. Young, *J. Am. Chem. Soc.* **75,** 4094 (1953).

[g] Y. Hirschberg, *J. Am. Chem. Soc.* **71,** 3241 (1949).

[h] K. C. Bryant, G. T. Kennedy, and E. M. Tanner, *J. Chem. Soc.* **1949,** 2389.

[i] M. J. Murray and W. S. Gallaway, *J. Am. Chem. Soc.* **70,** 3867 (1948).

the molecular geometry of styrene appear to exist. From an electron diffraction investigation on β-bromostyrene this molecule has been found to have a planar conformation in which the length of the essential single C—C bond, that is, the 1—1′ bond, is 1.48 ± 0.03 Å.[2] When it is assumed that in the most probable conformation of styrene the interplanar angle θ between the benzene ring and the ethylenic bond is 0° and the length R of the

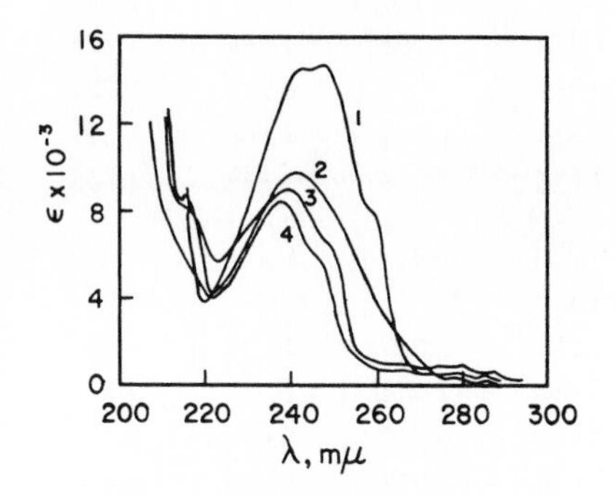

FIG. 13.1. Absorption spectra of styrenes in heptane. 1, styrene; 2, α-methylstyrene; 3, β,β-dimethylstyrene; 4, α,β,β-trimethylstyrene.

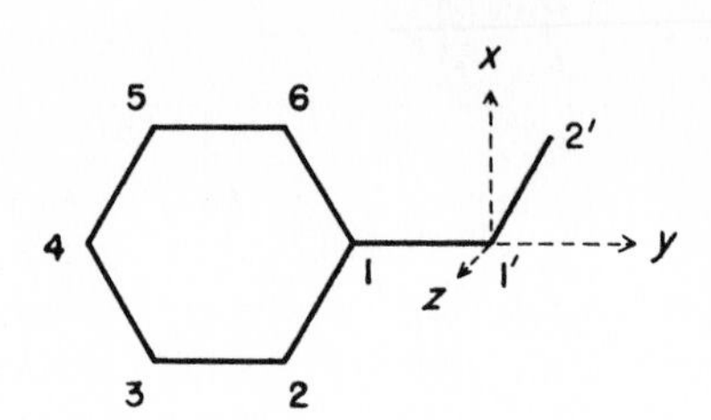

FIG. 13.2. The skeleton of styrene.

1—1′ bond is 1.48 Å, the value of the parameter k for the 1—1′ bond is 0.858. As in the case of biphenyl, the relations between θ, R, and k for the 1—1′ bond are assumed to be given by Eqs. (9.38) and (9.39).

As already mentioned, the A band can be considered as being due to the S–S one-electron transition from ψ_{+1} to ψ_{-1}. The calculated values of the energy of this transition, $\Delta E_{11}(-\beta) \equiv \Delta w_{11} \equiv w_{-1} - w_{+1} = 2w_{-1}$, and of the dipole strength of this transition, $D_{11} = M_{11}^2$, as functions of the k parameter for the 1—1′ bond are shown in Table 13.2. In the calculation of the dipole strength of the transition, all the valence angles of each carbon atom have been assumed to be 120°. The plane of the ethylenic bond has been taken as the xy plane, and the direction of the 1—1′ bond as the direction of the y axis. The z component of the transition moment M_{11} is invariably zero.

The value of ΔE_{11} increases and the value of D_{11} decreases with the decreasing value of k. This means that the A band will shift toward shorter wavelengths and decrease in intensity with the increasing twist angle of

[2] M. Igarashi, S. Cho, and K. Someno, *Nippon Kagaku Zasshi* **81**, 23 (1960).

TABLE 13.2

RESULTS OF THE SIMPLE LCAO MO CALCULATION ON STYRENE

Characteristics of the $\psi_{+1} \rightarrow \psi_{-1}$ Transition as Functions of the Twist Angle of the C—C "Single" Bond

k	1.0	0.9	0.858	0.8	0.7	0.6
θ, deg.	—	—	0	20.2	33.9	44.0
R, Å	—	—	1.480	1.484	1.490	1.497
ΔE_{11}, $-\beta$	1.396	1.451	1.474	1.508	1.568	1.631
$D_{11,x}$, Å^2	—	—	0.262	0.250	0.228	0.204
$D_{11,y}$, Å^2	—	—	1.857	1.764	1.606	1.452
D_{11}, Å^2	—	—	2.120	2.014	1.834	1.656
k	0.5	0.4	0.3	0.2	0.1	0
θ, deg.	52.6	60.6	68.1	75.4	82.7	90
R, Å	1.504	1.511	1.518	1.525	1.532	1.540
ΔE_{11}, $-\beta$	1.697	1.765	1.834	1.904	1.967	2
$D_{11,x}$, Å^2	0.178	0.148	0.113	0.069	0.018	0
$D_{11,y}$, Å^2	1.305	1.168	1.046	0.954	0.924	0.966
D_{11}, Å^2	1.483	1.316	1.159	1.024	0.943	0.966

the 1—1′ bond, in agreement with observation. To a first approximation, ψ_{+1} can be considered as arising from a mixture of ϕ_{+1} of benzene and ϕ_{+1} of ethylene, and ψ_{-1} can be considered as arising from a mixture of ϕ_{-1} of benzene and ϕ_{-1} of ethylene. When the interplanar angle θ becomes 90°, ψ_{+1} and ψ_{-1} of styrene reduce to ϕ_{+1} and ϕ_{-1} of benzene, respectively, and, similarly to the biphenyl A band, the styrene A band reduces to the benzene p band.

13.1.3. Estimation of the Twist Angle of the Essential Single Bond

In order to estimate the twist angle of the 1—1′ bond in sterically hindered alkylstyrenes from the position of the A band, the electronic bathochromic effect of the alkyl groups should be allowed for. This allowance can be made by choosing adequately the small-angle and large-angle references in the correlation of the observed wave number of the maximum of the A band with the calculated transition energy, in a manner similar to the treatment of alkylbiphenyls in the preceding chapter.

As usual, the calculated transition energy ΔE for the planar conformation is denoted by ΔE_S, and the corresponding wave number ν is denoted by ν_S. The value of ΔE_{11} for the k value of 0.858, $1.4744(-\beta)$, is taken as the value of ΔE_S. In principle, the wave number at which the molecule under consideration would exhibit the maximum of the A band if the molecule were planar is taken as ν_S. For example, for treatment of *o*-methylstyrene, the wave number of the maximum of the A band of *p*-methylstyrene, 39,526 cm^{-1} (253 mμ), is taken as the value of ν_S, since the electronic bathochromic effect of a methyl group can be considered to be approximately equivalent in the *o*- and *p*-positions.

From the comparison of the spectrum of *p*-methylstyrene with that of styrene and the comparison of the spectrum of *o,p*-dimethylstyrene with that of *o*-methylstyrene, it may be presumed that the magnitude of the bathochromic shift of the A band per a methyl substituent at the *p*- or the *o*-position of styrene is about 5 mμ. If the shift is assumed to be approximately additive, it is inferred that *o,p*-dimethylstyrene would have the maximum of the A band at about 258 mμ ($= 248 + 5 \times 2$ mμ) if it were not for the steric hypsochromic effect. Accordingly, for treatment of this compound, a value of 38,760 cm^{-1} (258 mμ) is taken as the value of ν_S. In a similar manner, the magnitude of the electronic bathochromic shift of the A band per a methyl substituent at the α-position (that is, the 1′-position) or at the β-position (that is, the 2′-position) is assumed to be 2 mμ and to be additive. Then, for example, the value of ν_S for treatment of α,β,β-trimethylstyrene becomes 39,370 cm^{-1} (254 mμ $= 248 + 2 \times 3$ mμ). For treatment of α-ethyl-, α-*n*-propyl-, and α-isopropylstyrene, the ν value of *trans*-β-ethylstyrene, 39,841 cm^{-1} (251 mμ), is taken as the value of ν_S.

The value of ΔE_{11} for $k = 0$, $2(-\beta)$, is denoted by ΔE_L, and the corresponding wave number is denoted by ν_L. For treatment of the styrenes which have no substituent on the benzene ring, the center of gravity of singlets of benzene, 48,000 cm^{-1} (208.3 mμ), is taken as ν_L. Absorption bands of toluene, *m*-xylene, and mesitylene are at longer wavelengths by about 5, 9, and 11 mμ, respectively, than the corresponding bands of benzene (see Subsection 10.6.3). Accordingly, the values for these methylated benzenes corresponding to the center of gravity of singlets of benzene may be taken to be approximately 46,882 cm^{-1} (213.3 mμ $= 208.3 + 5$ mμ), 46,019 cm^{-1} (217.3 mμ $= 208.3 + 9$ mμ), and 45,600 cm^{-1} (219.3 mμ $=$ $208.3 + 11$ mμ), respectively. These values are taken as ν_L for treatment of styrene derivatives which are correspondingly substituted on the benzene ring. For example, for treatment of *o,p,o*′-trimethylstyrene, the value of 45,600 cm^{-1} is taken as ν_L.

When the two references have been determined for each styrene derivative, the value of ΔE corresponding to the observed ν value of the compound can be obtained by the use of the postulated linear relation of ΔE to ν between the two references [Eq. (12.5)]. Then, from the value of ΔE the corresponding value of k for the 1—1′ bond can be interpolated by reference to Table 13.2, and from the value of k values of θ and R can be obtained

TABLE 13.3

RESULTS OF THE CALCULATION OF THE INTERPLANAR ANGLE IN STYRENES

Group	No.	Styrene	λ, mμ	λ_S, mμ	λ_L, mμ	ΔE, $-\beta$	k	θ, deg.	R, Å
I	1	Unsubstituted	248	—	—	1.475	0.858	0	1.480
	2	*p*-Methyl-	253	—	—	—	—	0	—
	3	*trans*-β-Methyl-	250	—	—	—	—	0	—
	4	*trans*-β-Ethyl-	251	—	—	—	—	0	—
II	5	*o*-Methyl-	246	253	213.3	1.555	0.722	31	1.489
	6	*o,p*-Dimethyl-	251	258	217.3	1.553	0.725	30.7	1.489
	7	α-Methyl-	241.5	250	208.3	1.567	0.702	33	1.490
	8	α,β-Dimethyl-	244	252	208.3	1.557	0.719	31.5	1.489
II′	9	α-Ethyl-	240	251	208.3	1.592	0.662	38	1.493
	10	α-*n*-Propyl-	239	251	208.3	1.604	0.642	40	1.494
	11	α-Isopropyl-	234	251	208.3	1.661	0.555	48	1.500
III	12	*cis*-β-Methyl-	240.6	250	208.3	1.578	0.685	35.2	1.491
	13	β,β-Dimethyl-	239.4	252	208.3	1.607	0.639	40	1.494
IV	14	*o,o*′-Dimethyl-	238	258	217.3	1.711	0.480	54	1.505
	15	*o,p,o*′-Trimethyl-	245	263	219.3	1.669	0.543	49	1.500
	16	α,β,β-Trimethyl-	237.8	254	208.3	1.638	0.590	45	1.498
V	17	α,*o*-Dimethyl-	—	—	—	—	—	(90)	—

by the use of Eqs. (9.39) and (9.38). The results of the calculation are summarized in Table 13.3, in which λ is the wavelength of the maximum of the A band of the compound concerned, and λ_S and λ_L are the wavelengths corresponding to ν_S and ν_L, respectively.

Although much reliance may not be placed on the individual numerical values of θ, the results of the calculation appear to lead to the conclusion that these styrenes can be classified into the following groups with respect to the type or degree of the steric interaction involved in the molecule.

Group I. The compounds which have no substituent at hindering positions, that is, *o*-, α-, and *cis*-β-positions. The compounds of the entry numbers 1–4 in the table belong to this group. The most probable conformations of these compounds are probably planar or nearly planar.

Group II. The compounds which have one hindering methyl group either at an *o*-position or at the α-position. These compounds are considered to involve the steric interference either between the *o*-methyl group and the α-hydrogen atom, or between the α-methyl group and an *o*-hydrogen atom. To this group belong the compounds of the entry numbers 5–8. The β-methyl group in "α,β-dimethylstyrene" (the entry number 8) is probably *trans* to the phenyl group. The calculated interplanar angles for these compounds range from 31 to 33°.

Group II′. The compounds which have an ethyl group or a higher alkyl group at the α-position. The compounds of the entry numbers 9–11 belong to this group. As the size of the alkyl group increases, the A band shifts to shorter wavelength and decreases in intensity, and the calculated interplanar angle becomes larger.

Group III. The compounds having one hindering methyl group at the *cis*-β-position. These compounds involve the steric interference between the *cis*-β-methyl group and an *o*-hydrogen atom. The compounds of the entry numbers 12 and 13 belong to this group. The calculated interplanar angles for these compounds are 35 and 40°, respectively, and are slightly larger than those for the compounds belonging to Group II.

Group IV. The compounds having two hindering methyl groups either at both the *o*-positions or at the α- and the *cis*-β-position. These compounds involve the steric interferences between a methyl group and a hydrogen atom on both sides of the benzene nucleus. The compounds of the entry numbers 14–16 belong to this group. The calculated interplanar angles for these compounds range from 45 to 54°, and are considerably larger than those for the compounds belonging to Group II and those for the compounds belonging to Group III.

Group V. α,*o*-Dimethylstyrene (the entry number 17) belongs to this group. If the most preferred conformation of this compound were a conformation involving the steric interferences between the α-methyl group and the *o*-hydrogen atom and between the *o*-methyl group and the *cis*-β-hydrogen atom, this compound would belong to Group IV. This compound has, however, been reported to show no absorption maximum in the region between 210 and 300 mμ (see Table 13.1), in contrast with the compounds belonging to Group IV. If this is true, it is inferred that the foregoing conformation is not the most preferred conformation. Probably, the most preferred conformation of this compound is an *s-cis* conformation in which the two methyl groups just contact with each other.

13.1.4. THE SPECTRUM OF PHENYLACETYLENE

It seems convenient to refer here to the spectrum of phenylacetylene, an analog of styrene. On replacement of the ethylenic bond of styrene by an acetylenic bond, each absorption band shifts to shorter wavelength. In Table 13.4 the spectral data of benzene, styrene, and phenylacetylene are compared.

TABLE 13.4

COMPARISON OF THE SPECTRA OF BENZENE, PHENYLACETYLENE, AND STYRENE IN ETHANOL

Compound	The α band		The p band	
	λ, mμ	log ϵ	λ_{max}, mμ	log ϵ_{max}
Benzene	243–263	2.2–2.4	204	3.8
Phenylacetylene	260–280	2.1–2.6	245	4.2
Styrene	273–291	2.7–2.9	248	4.2

The short length of the triple bond must lead to a relatively high value of the resonance parameter k for the bond: when the bond length is assumed to be 1.19 Å, the value of k for the bond is estimated at 1.378. In addition, the acetylenic carbon atom is more electronegative than the ethylenic carbon atom. For these reasons, the π electrons will be more tightly held in the triple bond than in the double bond, and hence the delocalization of the π electrons will be less effective in phenylacetylene than in styrene. A similar situation is encountered in comparison of the spectrum of tolan with that of *trans*-stilbene (see Subsection 14.1.4).

13.2. 1-Phenylcyclohexene and Its Derivatives

1-Phenylcyclohexene has the same conjugated system as that of styrene, and the spectrum of this compound in methanol solution shows the A band at 247 mμ, nearly the same position as that of the styrene A band. However, in contrast with that of styrene, the most preferred conformation of the conjugated system of this compound is probably nonplanar. If the methylene groups at the 3- and the 6-position are regarded as being equivalent to methyl groups, the geometry of the relevant part of this compound is considered to be similar to that of α,β-dimethylstyrene with the β-methyl group in the position *trans* to the phenyl group, which shows the A band at 244 mμ and has been inferred to have the interplanar angle of about 32° (see Table 13.3). Accordingly, the position of the A band of 1-phenylcyclohexene is considered to be the result of the steric hypsochromic effect and electronic bathochromic effect of the tetramethylene bridge upon the position of the styrene A band. Introduction of substituents into the *o*-, 2-, and 6-positions of this compound will appreciably increase the dihedral angle between the benzene ring plane and the ethylenic bond plane.

Ultraviolet absorption and proton magnetic resonance spectra of 1-phenylcyclohexene and its derivatives have been reported by Garbisch.[3] He has arbitrarily assigned 1-phenylcyclohexene to an approximately planar conformation with the dihedral angle of $0 \pm 20°$, and has estimated the dihedral angles in 6- and *o*-substituted derivatives from the olefinic proton chemical shifts, using theoretical value of Johnson and Bovey[4] for shieldings experienced by a proton nucleus in the neighborhood of a benzene ring. The data on the A band and the dihedral angles estimated in this way are summarized in Table 13.5. According to Garbisch, these values of the dihedral angle θ represent the time-averaged values which may not coincide with the equilibrium dihedral angle values, and which may be reliable to not better than $\pm 15°$.

The method used to estimate the θ in alkylstyrenes from the position of the A band can also be applied to these 1-phenylcyclohexenes. The small-angle reference compound is, of course, the parent compound, 1-phenylcyclohexene: $\lambda_S = 247$ mμ, $\nu_S = 40{,}486$ cm^{-1}. As usual, benzene is taken as the large-angle reference: $\Delta E_L = 2(-\beta)$, and $\nu_L = 48{,}000$ cm^{-1}. The choice of a numerical value for ΔE_S is the problem, because of the uncertainty of the equilibrium or most preferred conformation of the conjugated system of 1-phenylcyclohexene. If the dihedral angle θ in this compound, θ_S, is assumed to be 0°, we obtain $k_S = 0.858$ and $\Delta E_S = 1.4744(-\beta)$; if the angle θ_S is assumed to be 20°, we obtain $k_S = 0.801$ and $\Delta E_S = 1.5076(-\beta)$. The values of θ calculated by this method on the basis of these two values of ΔE_S are shown in the last two columns of Table 13.5. For compounds 9 and 17, allowance for the electronic bathochromic effect of the *o*- or the *p*-methyl group has been made by subtracting 5 mμ from the observed λ values.

[3] E. W. Garbisch, Jr., *J. Am. Chem. Soc.* **85,** 927 (1963).

[4] C. E. Johnson, Jr. and F. A. Bovey, *J. Chem. Phys.* **29,** 1012 (1958).

TABLE 13.5

THE A BANDS AND INTERPLANAR ANGLES (θ) OF 1-PHENYLCYCLOHEXENES[a]

No.	1-Phenylcyclohexene	The A band[b,c]		Interplanar angle θ, deg.		
		λ_{max}, mμ	log ϵ_{max}	from n.m.r.[b]	calcd. from $\theta_S = 0°$	calcd. from $\theta_S = 20°$
1	Unsubstituted	247	4.09	0	0	20
2	6-Methyl-	242	4.05	43	27	33
3	6-Isopropyl-	239.5	4.02	43	33	38
4	6-*tert*-Butyl-	239.5	4.00	48	33	38
5	6,6-Dimethyl-	227*	3.65	90	56	58
6	6,6-Diethyl-	227*	3.68	66	56	58
7	6-Phenyl-	246.5	4.05	0	9	22
8	6-Methyl-6-phenyl-	240	3.99	48	32	37
9	*o*-Methyl-	228*	3.75	69, 100	62	64
10	*o,o'*-Dimethyl-	No A band		90	90	90
11	6-(2-Hydroxy-2-propyl)-	239.5	4.02	42	33	38
12	4,4-Dimethyl-6-(2-hydroxy-2-propyl)-	236	4.01	61	41	45
13	6-Acetyl-	244	4.05	36	21	28
14	6-Acetyl-4,4-dimethyl-	244	4.09	38	21	28
15	6-Acetyl-6-ethyl-	242	3.94	42	27	33
16	6-Nitro-	238.5	4.05	38	35	39
17	4-*tert*-Butyl-6-nitro-*p*-methyl- (*trans*)	246	4.13	36	29	35
18	4-Methyl-6-nitro- (*trans*)	239	4.05	35	34	39
19	4-Methyl-6-nitro- (*cis*)	236	4.05	62	41	45
20	4,4-Dimethyl-6-nitro-	236.5	4.05	60	40	44
21	6-Bromo-6-nitro-	225*	3.96	70	59	61

[a] The asterisk on wavelength values denotes inflections.

[b] Reference 3.

[c] The solvent is methanol for all the compounds except compound 20, for which the solvent is isooctane.

13.3. 1,1-Diphenylethylene and Its Methylated Derivatives

In 1,1-diphenylethylene the two phenyl groups cannot be simultaneously coplanar with the ethylenic bond. The most preferred conformation of this compound is presumed to be a conformation in which both the phenyl groups are rotated out to the same extent as each other from the plane of the ethylenic bond, or alternatively, a conformation in which one of the phenyl groups is almost perpendicularly rotated out of the plane of the ethylenic bond, leaving the other phenyl group coplanar with the ethylenic bond.

TABLE 13.6

THE A BANDS OF 1,1-DIPHENYLETHYLENES IN HEPTANE

1,1-Diphenylethylene	λ_{max}, mμ	ϵ_{max}
Unsubstituted	248	14,270
	235	13,730
2-Methyl-[a]	ca. 248	ca. 13,000
2,2-Dimethyl-	248	11,700
(Styrene)[b]	248	14,760

[a] Ramart-Lucas and P. Amagat, *Bull. soc. chim. France* [4] **51,** 108 (1932).
[b] D. F. Evans, *J. Chem. Phys.* **23,** 1429 (1955).

The data on the ultraviolet absorption spectra of 1,1-diphenylethylene and its methylated derivatives are presented in Table 13.6, in which the data on the spectrum of styrene are also included for comparison. In addition, the spectra of 1,1-diphenylethylene and its 2,2-dimethyl derivative are shown in Fig. 13.3. It is especially of interest that the absorption maxima of the principal bands (that is, the A bands) of the four compounds listed in Table 13.6 appear at the almost identical wavelength.

The close similarity between the spectrum of 1,1-diphenylethylene and that of styrene would be considered as supporting the second of the two hypotheses mentioned above concerning the most preferred conformation of 1,1-diphenylethylene. However, the close similarity between the spectrum of 1,1-diphenylethylene and the spectra of its methylated derivatives seems to suggest that the first hypothesis is more plausible, especially because the dimethyl derivative cannot assume the conformation of the second hypothesis owing to the steric interferences between the phenyl groups and the methyl groups at the *cis*-β-positions relative to the phenyl groups.

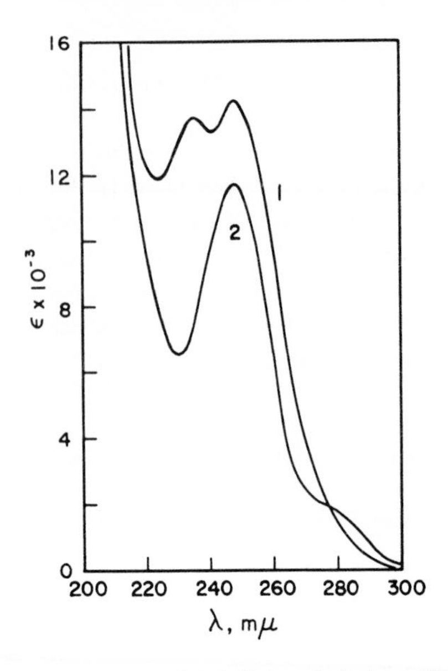

Fig. 13.3. The absorption spectra of 1,1-diphenylethylene (1) and 1,1-diphenyl-2,2-dimethylethylene (2) in heptane.

The first hypothesis is adopted for the present, and the resonance parameter for each essential single C—C bond connecting a phenyl group to an ethylenic carbon atom is denoted by k'. When the k' is replaced by $2^{-1/2}k$, the secular determinant for determination of the π MO's as linear combinations of the carbon $2p\pi$ AO's can be factorized into one determinant of identical form as the one for styrene, in which k represents the resonance parameter for the 1—1′ bond, and one determinant of identical form as the one for benzene. Therefore, the calculation similar to that for styrenes can be made for the 1,1-diphenylethylenes: from the position of the A band,

TABLE 13.7

Results of the Calculation of the Twist Angle in 1,1-Diphenylethylenes

1,1-Diphenyl-ethylene	λ, mμ	λ_S, mμ	λ_L, mμ	ΔE, $-\beta$	k	k'	θ, deg.	R, Å
Unsubstituted	248	248	208.3	1.475	0.858	0.607	43.5	1.496
2-Methyl-	248	250	208.3	1.496	0.821	0.581	46	1.498
2,2-Dimethyl-	248	252	208.3	1.515	0.788	0.557	48	1.500

the values of ΔE, k, k', θ, and R can, in turn, be estimated. The results of the calculation are summarized in Table 13.7, in which the symbols have the same significance as in Table 13.3. In 1,1-diphenylpropene the positions of the two phenyl groups are not completely equivalent, and therefore, their twist angles must differ, though probably slightly, from each other. The values shown in this table for this compound are the results obtained on the basis of the assumption that the twist angles of the two phenyl groups are equal.

It is noteworthy that the introduction of methyl groups into the 2-position of 1,1-diphenylethylene affects only slightly the calculated angle θ. The existence of the additional steric interference by the methyl groups in spite of the apparent absence of a wavelength shift of the A band is evident from the fact that the intensity of the band decreases with the increasing number of methyl substituents. The apparent absence of a wavelength displacement of the A band is considered to be the result of the compensation of the hypsochromic shift due to the steric interference by the bathochromic shift due to the electronic (probably hyperconjugation) effect of the methyl groups. The calculated values of θ for these compounds, especially for the dimethyl compound, are very similar to those for the styrene derivatives belonging to Group IV (see Table 13.3).

Chapter 14

Stilbene and Related Compounds

14.1. Stilbene*

14.1.1. THE SPECTRA OF *TRANS*- AND *CIS*-STILBENE

The spectrum of *trans*-stilbene in *n*-heptane solution exhibits three main band systems in the near ultraviolet region. These band systems are provisionally called the A, B, and C bands in the order of decreasing wavelength. The A band, which appears near 294 mμ, and the B band, which appears near 229 mμ, have well-resolved vibrational fine structures. The vibrational subbands of the A band at 320.5, 306.9, 294.1, and 283 mμ are termed the α, β, γ, and δ subbands, respectively, of which the γ subband is the most intense one. The spectrum of this compound in ethanol solution is very similar to that of the *n*-heptane solution.[1]

The spectrum of *cis*-stilbene differs considerably from that of the *trans* isomer. The longest wavelength band, the A band, is structureless and occurs at considerably shorter wavelength and with lower intensity than the corresponding band of the *trans* isomer. The B band is also structureless and occurs at slightly shorter wavelength and with higher intensity than the corresponding band of the *trans* isomer. The spectra of *trans*- and *cis*-stilbene are shown in Fig. 14.1, and the data on the spectra are given in Table 14.1.

The marked difference between the spectra of the two geometrical isomers is conceivably due, at least in part, to the difference in the most preferred conformation of the two isomers. While the most preferred conformation of the *trans* isomer is probably planar or nearly planar, that of the *cis* isomer must be nonplanar owing to the steric interference between the two phenyl groups.

* Subsections 14.1.1–14.1.3 are a revised version of Reference 1.

[1] H. Suzuki, *Bull. Chem. Soc. Japan* **33,** 379 (1960).

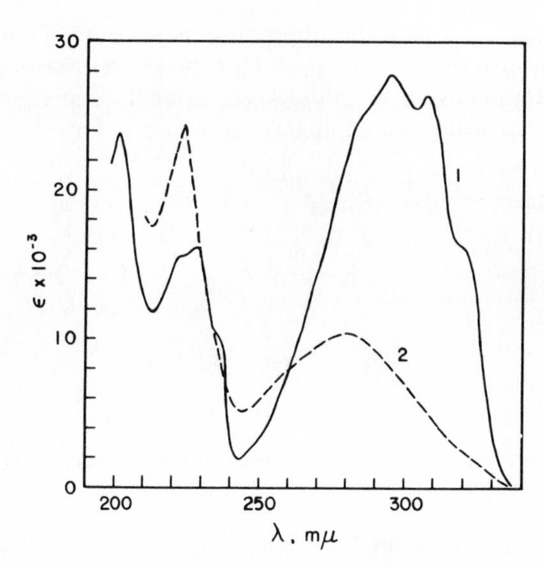

FIG. 14.1. The absorption spectra of *trans*- and *cis*-stilbene. 1, *trans*-stilbene in heptane; 2, *cis*-stilbene in ethanol.

TABLE 14.1

THE SPECTRA OF *TRANS*- AND *CIS*-STILBENE[a]

Compound	A band			B band		C band	
	λ_{max} mμ	ϵ_{max}	f	λ_{max} mμ	ϵ_{max}	λ_{max} mμ	ϵ_{max}
trans-Stilbene[b]	320.5*	16000	0.739	236*	10400	201.5	23900
	306.9	26500		228.5	16200		
	294.1	27950		222*	15500		
	283*	24500					
cis-Stilbene[c]	280	10450	0.323	224	24400	—	—

[a] The asterisk on wavelength values denotes inflections.
[b] In heptane; Reference 1.
[c] In ethanol; Reference 21.

14.1.2. THE SIMPLE LCAO MO CALCULATION ON STILBENE

Planar *trans*, *cis*, and abstract linear forms of the skeleton of the conjugated system of stilbene are shown in Figs. 14.2a, b, and c, respectively. Symbols B and B′ represent the two benzene rings, and symbol E represents the ethylenic bond. Symmetrically twisted models are shown in Figs. 14.3a, b, and c.

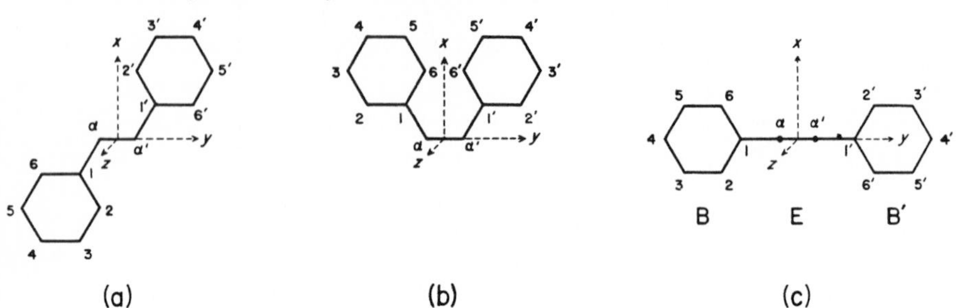

FIG. 14.2. Planar forms of the skeleton of stilbene. (a) The *trans* form. (b) The *cis* form. (c) The abstract linear form.

In these models the dihedral angles between each benzene ring plane and the ethylenic bond plane, that is, the angle of twist of the α—1 bond ($\theta_{\alpha 1}$) and that of the α'—1′ bond ($\theta_{\alpha' 1'}$), are assumed to be equal to each other, and are denoted by a common symbol θ. The *trans* form has symmetry element $C_2(z)$,

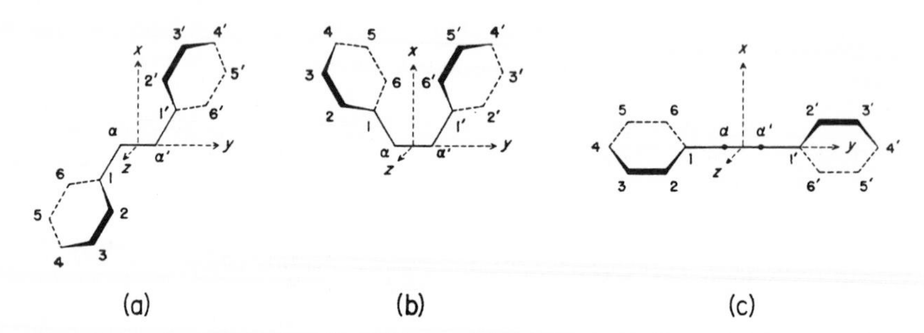

FIG. 14.3. Symmetrically twisted forms of the skeleton of stilbene. (a) The *trans* form. (b) The *cis* form. (c) The abstract linear form.

and belongs to point group C_{2h}, C_2, or C_{2h} according as $\theta = 0°, 0° < \theta < 90°$, or $\theta = 90°$. The *cis* form has symmetry element $C_2(x)$, and belongs to point group $C_{2v}(x)$, $C_2(x)$, or $C_{2v}(x)$ according as $\theta = 0°$, $0° < \theta < 90°$, or $\theta = 90°$. The abstract linear model belongs to point group D_{2h}, D_2, or D_{2h} according as $\theta = 0°$, $0° < \theta < 90°$, or $\theta = 90°$. With the simple LCAO MO method, which neglects interactions between nonneighboring atoms, the conjugated system of stilbene can be treated conveniently on the basis of an abstract linear model belonging to point group D_2.

According to the X-ray crystal analysis of *trans*-stilbene by Robertson and Woodward,[2] the molecule in the crystalline state has a planar or nearly planar structure in which the length of the essential double bond (that is, the α—α' bond) is 1.33 ± 0.04 Å and the length of each of the essential single bonds (that is, the α—1 and α'—1′ bonds) is 1.44 ± 0.02 or 1.45 ± 0.02 Å. As mentioned in Subsection 9.3.4, the length of the double bond in ethylene has been found to be about 1.333–1.339 Å, and the length of the essential double bonds in 1.3-butadiene to be 1.337 ± 0.005 Å. It may therefore be presumed that the length of the α—α' bond in the stilbene system does not significantly vary with the extent of conjugation. A round value of 1.34 Å is taken as the length of this bond, and a value of 1.080 is assigned to the resonance parameter for this bond ($k_{\alpha\alpha'}$). As mentioned above, the angles $\theta_{\alpha 1}$ and $\theta_{\alpha' 1'}$ are assumed to be equal, and are denoted simply by θ. Similarly, the length of the α—1 bond ($R_{\alpha 1}$) and that of the α'—1′ bond ($R_{\alpha' 1'}$) are denoted by R, and the resonance parameters for these bonds ($k_{\alpha 1}$ and $k_{\alpha' 1'}$) are denoted by k. The value of R is assumed to vary from 1.445 Å (R_0) for $\theta = 0°$ to 1.54 Å (R_{90}) for $\theta = 90°$, according to expression (9.35). The value of k for $\theta = 0°$ and $R = 1.445$ Å, that is, $k(R_0, 0°)$, is estimated at 0.9084. The value of k is assumed to vary from 0.9084 for $\theta = 0°$ to 0 for $\theta = 90°$ according to the following expression [see expression (9.37)]:

$$\begin{aligned} k &= 0.7711 \cos\theta + 0.1373 \cos^2\theta \\ &= 0.9084 \cos\theta - 0.1373 \cos\theta\cdot(1 - \cos\theta). \end{aligned} \tag{14.1}$$

Then the wavefunctions and energies of the π MO's and related quantities can be calculated as functions of k, and hence of θ.

As usual, the π MO's are designated according to the two-directional notation: for example, the highest bonding MO is denoted by ψ_{+1}, and the lowest antibonding MO is denoted by ψ_{-1}. The symmetry properties of the π MO's in various models are summarized in Table 14.2.

Molecular orbitals ψ_{-2} and ψ_{-3} form a degenerate pair ($w_{-2} = w_{-3} = +1$), and ψ_{+2} and ψ_{+3} form another degenerate pair ($w_{+2} = w_{+3} = -1$). When the π MO's of one benzene ring B are denoted by ϕ_m and those of another benzene ring B′ by $\phi_{m'}$, these four MO's of the stilbene system are given by expressions (12.2), similarly to the corresponding MO's of the biphenyl system. The remaining 10 π MO's vary with the value of k.[3] The

[2] J. M. Robertson and I. Woodward, *Proc. Roy. Soc.* (*London*) **A162,** 568 (1937).

[3] H. Suzuki, *Bull. Chem. Soc. Japan.* **25,** 145 (1952).

orbital energies for $k = 0.9084$ are as follows:

$$w_{-1} = -w_{+1} = 0.5711; \qquad w_{-4} = -w_{+4} = 1.1256;$$
$$w_{-5} = -w_{+5} = 1.5016; \qquad w_{-6} = -w_{+6} = 2.0501; \qquad (14.2)$$
$$w_{-7} = -w_{+7} = 2.1831.$$

TABLE 14.2

THE SYMMETRY PROPERTIES OF THE π MO'S OF THE STILBENE SYSTEM IN VARIOUS MODELS

	Linear model		*Trans* form		*Cis* form	
	Planar	Twisted	Planar	Twisted	Planar	Twisted
MO	D_{2h}	D_2	$C_{2h}(z)$	$C_2(z)$	$C_{2v}(x)$	$C_2(x)$
−3, +2	a_{1u}	a_1	a_u	a	b_1	b
−6, −4, +1, +5, +7	b_{1u}	b_1	a_u	a	b_1	b
−2, +3	b_{2g}	b_2	b_g	b	a_2	a
−7, −5, −1, +4, +6	b_{3g}	b_3	b_g	b	a_2	a

When the value of k becomes zero, ψ_{-5} and ψ_{+5} reduce to the antibonding and bonding π MO's of the ethylenic bond, respectively:

$$b_3 : \psi_{-5} = \phi_{-1(\mathrm{E})}, \qquad w_{-5} = +1.080;$$
$$b_1 : \psi_{+5} = \phi_{+1(\mathrm{E})}, \qquad w_{+5} = -1.080. \qquad (14.3)$$

At the same time, the remaining 8 MO's reduce to the symmetry combinations of the benzene orbitals:

$$b_3 : \psi_{-7} = \phi_{-3(-)}, \qquad \psi_{-1} = \phi_{-1(-)}, \qquad \psi_{+4} = \phi_{+1(-)}, \qquad \psi_{+6} = \phi_{+3(-)};$$
$$b_1 : \psi_{-6} = \phi_{-3(+)}, \qquad \psi_{-4} = \phi_{-1(+)}, \qquad \psi_{+1} = \phi_{+1(+)}, \qquad \psi_{+7} = \phi_{+3(+)};$$
$$w_{-7} = w_{-6} = +2, \quad w_{-4} = w_{-1} = +1, \quad w_{+1} = w_{+4} = -1,$$
$$w_{+6} = w_{+7} = -2. \qquad (14.4)$$

The A band of the stilbene system can be approximated as being due to the one-electron transition from ψ_{+1} to ψ_{-1}, and is considered to reduce to the p band of benzene when the value of θ becomes 90°. When $\theta \neq 90°$, this transition in the *cis* form has a nonzero transition moment in the direction of the y axis, and the transition moment of this transition in the *trans* form has a nonzero x component, in addition to a nonzero y component of the same magnitude as the transition moment in the corresponding *cis* form.

Analogously to the case of butadiene (see Subsection 8.2.2), the x components of the two partial transition moments are in the opposite directions and cancel each other in the *cis* form, while in the *trans* form they are in the same direction and result in the sum. Consequently, the transition moment in the *trans* form is invariably larger than that in the corresponding *cis* form.

TABLE 14.3

RESULTS OF THE SIMPLE LCAO MO CALCULATION ON STILBENE

Characteristics of the $\psi_{+1} \rightarrow \psi_{-1}$ Transition as Functions of the Twist Angle of the C—C "Single" Bonds

k	1.1	1.0	0.9084	0.9	0.8	0.7	0.6
θ, deg.	—	—	0	7.2	26.5	37.7	46.2
R, Å	—	—	1.4450	1.446	1.455	1.464	1.474
ΔE_{11}, $-\beta$	0.992	1.068	1.1421	1.149	1.235	1.326	1.422
D_{11} (*cis*), Å^2	—	—	2.2062	2.183	1.918	1.665	1.429
D_{11} (*trans*), Å^2	—	—	4.2250	4.174	3.593	3.079	2.599
k	0.5	0.4	0.3	0.2	0.1	0	
θ, deg.	54.0	61.4	68.6	75.6	82.7	90	
R, Å	1.484	1.495	1.505	1.516	1.528	1.540	
ΔE_{11}, $-\beta$	1.521	1.625	1.731	1.838	1.940	2	
D_{11} (*cis*), Å^2	1.210	1.006	0.816	0.631	0.427	0.242	
D_{11} (*trans*), Å^2	2.168	1.785	1.455	1.178	0.977	0.966	

The calculated values of the energy of this transition, $\Delta E_{11}(\text{in } -\beta) \equiv \Delta w_{11} \equiv w_{-1} - w_{+1} = 2w_{-1}$, and of the dipole strength of this transition, $D_{11} = M_{11}^2$ (in Å^2), as functions of k are summarized in Table 14.3. As the value of k decreases the transition energy increases and the transition dipole strength decreases. This means that the A band will shift to shorter wavelength and reduce in intensity as the twist angle θ increases.

14.1.3. GEOMETRY OF *TRANS*- AND *CIS*-STILBENE

By the use of the postulated linear relation (12.5) between the calculated transition energy, ΔE, and the observed wave number of the absorption maximum, ν, the appropriate value of ΔE for a sterically hindered stilbene can be determined from the wave number of the maximum of the A band, and from the value of ΔE the values of θ and some other quantities can be estimated by reference to Table 14.3. It is necessary, again, to choose the values of ΔE and ν for the small-angle and the large-angle reference. As usual, ΔE_L and ν_L are taken to be $2(-\beta)$ (the value of ΔE_{11} for $\theta = 90°$) and 48,000 cm^{-1} (the center of gravity of singlets of benzene, or the wave number of the *p* band of benzene), respectively. The value of ΔE_{11} for $\theta = 0°$, 1.1421 $(-\beta)$, is taken as ΔE_S. Then, ν_S should be the wave number of the maximum of the A band of a planar stilbene.

The solid-state spectrum of *trans*-stilbene measured by the pressed KCl disk technique differs considerably in the shape of the absorption curve from the solution spectrum. Thus, the most intense peak of the A band in the KCl-disk spectrum is the δ subband, while that in the *n*-heptane-solution spectrum is the γ subband. In addition, the subbands of the A band in the KCl-disk spectrum are at longer wavelengths than the corresponding subbands in the *n*-heptane-solution spectrum. The magnitude of this bathochromic shift is about 4.5 mμ for the γ subband and about 5.5 or 6 mμ for the β subband. However, these values are considerably smaller than those (about 8 mμ for the γ subband and about 9 mμ for the β subband) expected from the regular relation between the magnitude of the shift and the position of the band determined by Dale for rigid molecules (see Subsection 6.4.3). This fact may be considered to indicate that the most probable conformation of *trans*-stilbene in solution does not significantly differ from the one in the crystalline state, and is therefore planar or nearly planar. Accordingly, *trans*-stilbene in solution can safely be taken as the small-angle reference in which $\theta = 0°$. In the present treatment, the wave number of the most intense peak (that is, the γ subband) of the A band in the *n*-heptane-solution spectrum of *trans*-stilbene, 34,002 cm^{-1} (294.1 mμ), is taken as ν_S.

From the position of the A band of *cis*-stilbene in solution, 280 mμ, the values of k and θ for the most probable conformation of this compound are determined to be 0.787 and 28°, respectively. This value of θ is in fairly good agreement with the value of 25° estimated from a scale model. The extra-resonance energy, that is, the π-electronic energy relative to the sum of the π-electronic energies of separate two benzene rings and one ethylenic bond, is calculated to be 0.703 $(-\beta)$ for *trans*-stilbene ($k = 0.9084$) and to be 0.530 $(-\beta)$ for *cis*-stilbene ($k = 0.787$). The difference of 0.173 $(-\beta)$ corresponds to the energy difference between the two isomers of 3 kcal./mole estimated from the equilibrium of the thermal isomerization.[4]

The A band in the vapor spectrum of *trans*-stilbene appears to be at considerably shorter wavelength than in the solution spectrum. Thus the vapor spectrum shows a discontinuous band in the region from 284.4 to

[4] G. B. Kistiakowsky and W. R. Smith, *J. Am. Chem. Soc.* **56**, 638 (1934); T. W. J. Taylor and A. R. Murray, *J. Chem. Soc.* **1938**, 2078.

265.6 mμ with the most intense peaks at 284.4 and 283.4 mμ.[5] This fact probably indicates that the most preferred conformation of *trans*-stilbene in the vapor state is nonplanar.

A compound which has a structure formed by connecting the two *para* positions of the phenyl groups of *cis*-stilbene by a dimethylene bridge shows no distinct absorption bands in the region of wavelengths longer than 230 mμ,[6] in contrast with *cis*-stilbene, indicating further hindrance of conjugation in the already hindered molecule.

14.1.4. The Spectrum of Tolan

The replacement of the ethylenic bond of stilbene by an acetylenic bond to give tolan (diphenylacetylene) results in a hypsochromic shift of the A band. The A band of tolan has well-developed vibrational fine structure, of which the third subband, that is, the γ subband, is the most intense peak (λ_{max} = 280 mμ, log ϵ_{max} = 4.50, in *n*-heptane).[7] This hypsochromic shift may be attributed mainly to the short bond length of the acetylenic bond as compared with the ethylenic bond (see Subsection 13.1.4). According to the X-ray crystal analysis by Robertson and Woodward,[8] tolan in the crystalline state has a planar structure in which the length of the essential triple bond and that of each essential single bond are 1.19 ± 0.02 and 1.40 ± 0.02 Å, respectively. From these values of the bond lengths the values of the resonance parameters for these bonds are calculated to be 1.378 and 0.983, respectively, and the energy of the transition from ψ_{+1} to ψ_{-1}, ΔE_{11}, is calculated to be 1.270 $(-\beta)$. This value of ΔE_{11} is to be compared with the corresponding value for *trans*-stilbene, 1.142 $(-\beta)$.

14.1.5. Application of the Pariser–Parr Method to Stilbene

The author calculated the π-electronic states of the stilbene system by the Pariser–Parr method (see Chapter 11), using as the starting MO's the MO's obtained by the simple LCAO MO method in the manner described in Subsection 14.1.2, and taking account of, in addition to the ground electron configuration (V_0), the 25 singlet and the 25 triplet singly-excited electron configurations which arise from one-electron transitions from the highest 5 bonding MO's, ψ_{+1}–ψ_{+5}, to the lowest 5 antibonding MO's, ψ_{-1}–ψ_{-5} (that is,

[5] Y. Kanda, *Mem. Faculty Sci. Kyushu Univ. Ser. C, Chem.* **1,** 189 (1950).

[6] K. C. Dewhirst and D. J. Cram, *J. Am. Chem. Soc.* **80,** 3115 (1958).

[7] H. Suzuki, *Bull. Chem. Soc. Japan* **33,** 389 (1960).

[8] J. M. Robertson and I. Woodward, *Proc. Roy. Soc.* (*London*) **A164,** 436 (1938).

V_1^{kr}'s and T_1^{kr}'s in which $k = 1, 2, \ldots, 5$; $r = 1, 2, \ldots, 5$). As in the similar calculation on the biphenyl system (see Subsection 12.2.6), the value of the core resonance integral for a C—C bond in the benzene rings, β^{C}, was taken to be -2.39 ev (the value used originally by Pariser and Parr in the calculation

TABLE 14.4

RESULTS OF THE CALCULATION ON STILBENE BY THE PARISER–PARR METHOD (A) ENERGIES (E) OF STATES RELATIVE TO THE GROUND CONFIGURATION[a]

Trans form				*Cis* form			
		$\theta = 0°$	$\theta = 30°$			$\theta = 0°$	$\theta = 30°$
State	Sym.	E, ev	E, ev	State	Sym.	E, ev	E, ev
$^-[V]_1^{1}$	$A_{(g)}$	6.339	6.541	$^+[V]_1^{13}$	$A_{(1)}$	6.182*	6.349*
$^-[T]_1^{14}$	$A_{(g)}$	6.261	6.482	$^-[V]_1^{14}$	$A_{(1)}$	5.930	6.166
$^+[V]_1^{12}$	$B_{(u)}$	6.260*	6.412*	$^-[T]_1^{14}$	$A_{(1)}$	5.878	6.122
$^+[V]_1^{13}$	$A_{(g)}$	6.112	6.343*	$^+[V]_1^{12}$	$B_{(2)}$	5.836*	6.126*
$[V]_1^{22(+33)}$	$B_{(u)}$	5.570*	5.605*	$[V]_1^{33(+22)}$	$B_{(2)}$	5.394*	5.542*
$^+[V]_1^{14}$	$A_{(g)}$	5.144	5.169*	$^+[V]_1^{14}$	$A_{(1)}$	5.287*	5.178*
$^-[V]_1^{13}$	$A_{(g)}$	4.621	4.706	$^-[V]_1^{12}$	$B_{(2)}$	4.624	4.691
$^-[T]_1^{13}$	$A_{(g)}$	4.621	4.706	$^-[T]_1^{12}$	$B_{(2)}$	4.624	4.691
$^+[T]_1^{15}$	$B_{(u)}$	4.620	4.535	$^+[T]_1^{15}$	$B_{(2)}$	4.615	4.533
$^-[V]_1^{12}$	$B_{(u)}$	4.614	4.703	$^-[V]_1^{13}$	$A_{(1)}$	4.574	4.680
$^-[T]_1^{12}$	$B_{(u)}$	4.614	4.703	$^-[T]_1^{13}$	$A_{(1)}$	4.573	4.680
$^+[T]_1^{23}$	$A_{(g)}$	4.347	4.368	$^+[T]_1^{23}$	$A_{(1)}$	4.348	4.368
$[T]_1^{22(+33)}$	$B_{(u)}$	4.297	4.259	$[T]_1^{33(+22)}$	$B_{(2)}$	4.305	4.243
$^+[T]_1^{13}$	$A_{(g)}$	4.219	4.291	$^+[T]_1^{13}$	$A_{(1)}$	4.212	4.278
$^+[T]_1^{12}$	$B_{(u)}$	4.207	4.304	$^+[T]_1^{12}$	$B_{(2)}$	4.183	4.305
$[V]_1^{11}$	$B_{(u)}$	4.136*	4.425*	$[V]_1^{11}$	$B_{(2)}$	3.889*	4.286*
$^+[T]_1^{14}$	$A_{(g)}$	3.863	3.895	$^+[T]_1^{14}$	$A_{(1)}$	3.860	3.896
$[T]_1^{11}$	$B_{(u)}$	2.539	2.858	$[T]_1^{11}$	$B_{(2)}$	2.526	2.843
$[V]_0$	$A_{(g)}$	−0.080	−0.061	$[V]_0$	$A_{(1)}$	−0.064	−0.053

[a] The asterisks suffixed to the values of state energies indicate that the transitions from the ground state to those states are allowed.

on benzene), and the core resonance integral for a C—C bond, say the i—j bond, was assumed to be $k_{ij}\beta^{\mathrm{C}}$, in which k_{ij} is the resonance parameter for the i—j bond used in the simple LCAO MO calculation.

A part of the results of the calculation is shown in Tables 14.4 and 14.5. In Table 14.4 are shown the calculated energies (in electron volts, relative to the energy of the ground electron configuration) of the ground state and of the lowest eighteen excited states. The symbols for states have been chosen on the basis of the principle mentioned in Subsection 12.2.6. In the present case, differently from the case of biphenyl, V_1^{22} and V_1^{33} are not degenerate, and T_1^{22} and T_1^{33} are also not degenerate. However, V_1^{22} and V_1^{33}, as well as T_1^{22}

TABLE 14.5

RESULTS OF THE CALCULATION ON STILBENE BY THE PARISER–PARR METHOD (B) ENERGIES (ΔE, IN ELECTION VOLTS UNITS) AND DIPOLE STRENGTHS (D, IN Å^2 UNITS) OF TRANSITIONS FROM THE GROUND STATE

Trans form

State	$\theta = 0°$, $C_{2h}(z)$		$\theta = 30°$, $C_2(z)$		$\theta = 90°$, $C_{2h}(z)$		
	ΔE	D	ΔE	D	ΔE	D	State
$^+[V]_1^{12}$	6.340	2.432(xy)	6.473	1.996(xy)	6.94	0	$^1LE_\beta^{(-)}$
$^+[V]_1^{13}$	6.192	0	6.404	0.603(z)	6.94	3.864(z)	$^1LE_\beta^{(+)}$
$[V]_1^{22(+33)}$	5.650	0.941(xy)	5.666	1.241(xy)	6.94	3.864(xy)	$^1LE_{\beta'}^{(-)}$
$^+[V]_1^{14}$	5.224	0	5.230	0.000(z)	5.30	0	$^1LE_p^{(+)}$
$[V]_1^{11}$	4.216	3.399(xy)	4.486	2.731(xy)	5.30	0	$^1LE_p^{(-)}$

Cis form

State	$\theta = 0°$, $C_{2v}(x)$		$\theta = 30°$, $C_2(x)$		$\theta = 90°$, $C_{2v}(x)$		
	ΔE	D	ΔE	D	ΔE	D	State
$^+[V]_1^{13}$	6.246	0.517(x)	6.402	0.457(x)	6.94	0	$^1LE_\beta^{(+)}$
$^+[V]_1^{12}$	5.900	2.976(y)	6.179	2.708(yz)	6.94	3.864(z)	$^1LE_\beta^{(-)}$
$[V]_1^{33(+22)}$	5.458	0.085(y)	5.595	0.199(yz)	6.94	0.966(y)	$^1LE_{\beta'}^{(-)}$
$^+[V]_1^{14}$	5.351	0.064(x)	5.231	0.048(x)	5.30	0	$^1LE_p^{(+)}$
$[V]_1^{11}$	3.953	1.069(y)	4.339	1.182(yz)	5.30	0	$^1LE_p^{(-)}$

and T_1^{33}, are nearly degenerate and make nearly equal contributions to each state. $^+[V]_1^{22(+33)}$, for example, represents the state to which V_1^{22} makes the most important contribution and to which V_1^{33} makes a nearly equal, though slightly smaller, contribution with the same sign.

Transitions from the ground state to excited states $^+[V]_1^{11}$, $^+[V]_1^{14}$, $^+[V]_1^{22(+33)}$ (or $^+[V]_1^{33(+22)}$), $^+[V]_1^{13}$, and $^+[V]_1^{12}$ can be allowed (indicated by the asterisks suffixed to the values of state energies in Table 14.4). The calculated dipole strengths (D, in Å^2) and energies (ΔE, in electron volts) of these transitions are shown in Table 14.5. The transitions to the other states listed in Table 14.4 are all forbidden. In Table 14.5 the letters in parentheses following the

D values indicate the directions in which the transition moment has nonzero components. In the last column of this table are shown the wavefunctions of the states when the interactions between the fragments B, B′, and E are completely eliminated ($\theta = 90°$). The symbols for the wavefunctions have the same meaning as in Table 12.2. For example, when the interactions completely vanish the lowest singlet excited state, $^{+}[V]_1^{11}$, reduces to $^1LE_p^{(-)}$, that is, the antisymmetric combination of the LE electron configurations corresponding to the *p* states of the two separate benzene rings ($^1LE_{p(B)}$ and $^1LE_{p(B')}$). At the same time, $^{-}[V]_1^{12}$ and $^{-}[V]_1^{13}$ reduce to $^1LE_\alpha^{(-)}$ and $^1LE_\alpha^{(+)}$ (the calculated energy: 4.90 ev), respectively, and the lowest triplet state, $^{+}[T]_1^{11}$, reduces to $^3LE_p^{(-)}$ (the calculated energy: 4.02 ev).

In the planar *trans* form, the magnitudes of the *x* and *y* components of the electric transition moment of the lowest-energy allowed transition, $^{-}[V]_0 \rightarrow {}^{+}[V]_1^{11}$, are calculated to be $+1.335$ and $+1.271$ Å, respectively. From these values the angle (φ) that the transition moment makes with the *y* axis is calculated to be $+46.6°$: the transition moment may be said to be almost along the long axis of the molecule. The calculated values of φ of transitions $^{-}[V]_0 \rightarrow {}^{+}[V]_1^{22(+33)}$ and $^{-}[V]_0 \rightarrow {}^{+}[V]_1^{12}$ are $+61.8$ and $-4.6°$, respectively. By the way, the φ values of one-electron transitions $\psi_{+1} \rightarrow \psi_{-1}$, $\psi_{+2} \rightarrow \psi_{-2}$ (as well as $\psi_{+3} \rightarrow \psi_{-3}$), and $\psi_{+1} \rightarrow \psi_{-2}$ (as well as $\psi_{+2} \rightarrow \psi_{-1}$) in the simple LCAO approximation are $+43.7$, $+60.0$, and $-30.0°$, respectively.

The A band is almost certainly due to the transition to the lowest singlet excited state, $^{+}[V]_1^{11}$. This state has, needless to say, V_1^{11} as the most important contributor. The wavefunction of this state for the planar *trans* form can be expressed as follows:

$$\begin{aligned} {}^{+}[V]_1^{11}\ (\theta = 0°,\ \textit{trans}) \\ &= 0.946\ V_1^{11} + 0.097\ V_1^{22} + 0.099\ V_1^{33} + 0.252\ V_1^{44} + 0.119\ V_1^{55} \\ &\quad - 0.073\ {}^{+}V_1^{12} - 0.032\ {}^{+}V_1^{15} - 0.038\ {}^{+}V_1^{34} - 0.002\ {}^{+}V_1^{25}. \qquad (14.5) \end{aligned}$$

As the twist angle θ becomes larger, the contribution of V_1^{11} decreases gradually, and the contributions of V_1^{22}, V_1^{33}, and V_1^{44} increase gradually. Evidently, as the value of θ increases, the energy of the transition to this state increases, and the dipole strength of the transition decreases.

The bands due to the forbidden transitions to $^{-}[V]_1^{12}$ and $^{-}[V]_1^{13}$ correspond to the benzene α band, and probably are hidden under the A band. The energies of these transitions are comparatively insensitive to variation in the twist angle.

In the *trans* form, the B band is probably due to the transition to $^{+}[V]_1^{22(+33)}$, and the C band is probably due to the transition to $^{+}[V]_1^{12}$ (and $^{+}[V]_1^{13}$). As the twist angle becomes larger the energies of these transitions increase. In the *cis* form, the B band is perhaps mainly due to the transition to $^{+}[V]_1^{12}$.

14.2. Sterically Hindered Stilbene Derivatives*

14.2.1. The Spectra

The spectra of a series of α-alkyl and α,α′-dialkyl derivatives of *trans*- and *cis*-stilbene are shown in Fig. 14.4.[9–11] The spectra of the *n*-heptane solutions do not significantly differ from the spectra of the ethanol solutions. With regard to these spectra the following generalizations can be deduced: (1) the A band

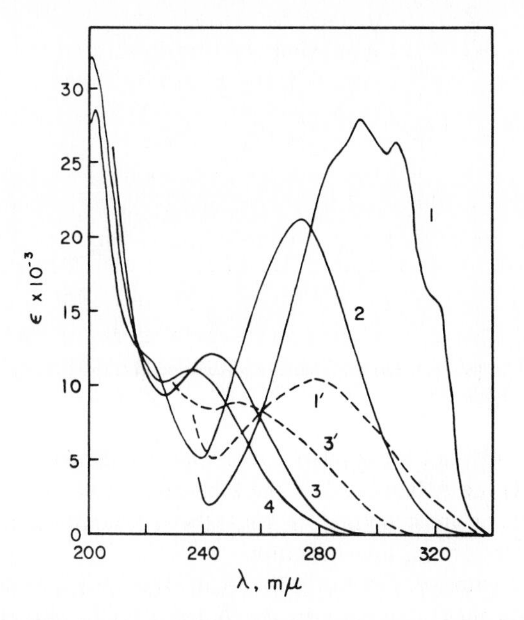

Fig. 14.4. Absorption spectra of stilbenes in heptane. 1, *trans*-stilbene; 2, *trans*-α-methylstilbene; 3, *trans*-α,α′-dimethylstilbene; 4, *trans*-α,α′-diethylstilbene; 1′, *cis*-stilbene (in ethanol); 3′, *cis*-α,α′-dimethylstilbene.

* Subsections 14.2.1 and 2 are abstracts of References 9 and 10, and Subsection 14.2.3 is an abstract of Reference 11.

[9] H. Suzuki, *Bull. Chem. Soc. Japan* **33,** 396 (1960).

[10] H. Suzuki, *Bull. Chem. Soc. Japan* **33,** 406 (1960).

[11] H. Suzuki, *Bull. Chem. Soc. Japan* **33,** 410 (1960).

of *trans*-stilbene as well as that of *cis*-stilbene undergoes increasing hypsochromic and hypochromic shifts with increasing number and size of substituents; (2) the A band of a *trans* isomer is usually more intense than that of its *cis* isomer; (3) the A band of the α-methyl derivative occurs at longer wavelength in the *trans* compound than in the *cis* compound as in the case of the

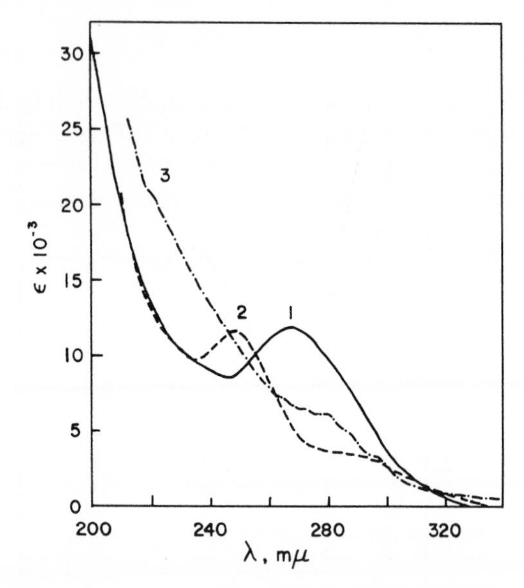

FIG. 14.5. Absorption spectra of *trans*-α,α′-dihalostilbenes in heptane. 1, dichloro-; 2, dibromo-; 3, diiodo-.

parent compounds, but the reverse is true with α,α′-dimethyl and α,α′-diethyl derivatives; (4) the A bands of all the hindered derivatives as well as the A band of *cis*-stilbene are structureless, while the A band of *trans*-stilbene has the characteristic vibrational fine structure.

The spectra of *trans*-α,α′-dihalostilbenes in *n*-heptane solution are shown in Fig. 14.5. With these compounds, the A band is also structureless, and is shifted toward shorter wavelengths as the substituents become larger. The diiodo compound in *n*-heptane dissociates gradually into tolan and iodine, and consequently the absorption due to tolan is detected in the spectrum. The inflection at near 220.5 mμ probably indicates the vestigial A band of the diiodo compound.

The spectra of some *trans*-stilbene derivatives bearing methyl groups on the benzene nuclei are shown in Fig. 14.6. The spectra of the 4-methyl and

4,4′-dimethyl derivatives resemble the spectrum of the parent compound, *trans*-stilbene, showing characteristic vibrational fine structure. However, the A bands of these *p*-methyl derivatives are at longer wavelengths and have higher intensities than the A band of the parent compound. In contrast, the

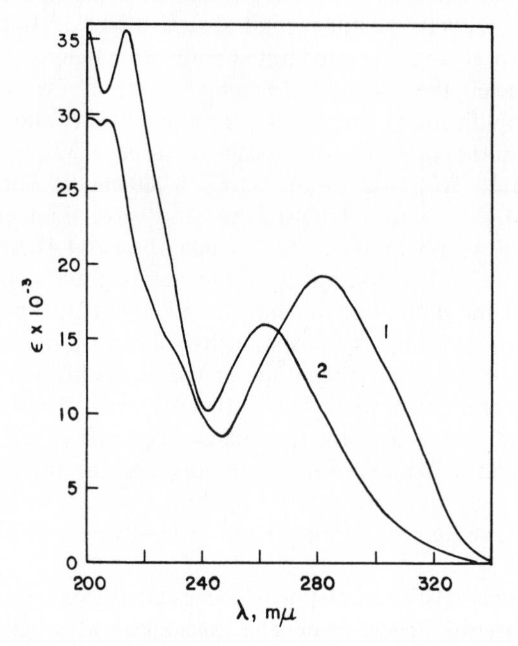

FIG. 14.6. The absorption spectra of 2,4,6-trimethyl-*trans*-stilbene (1) and 2,4,6,2′,4′,6′-hexamethyl-*trans*-stilbene (2) in heptane.

A bands of the derivatives having methyl groups at the *o*-positions are structureless, and these are at shorter wavelengths and have lower intensities than the A band of the parent compound.

The loss of the vibrational fine structure, hypsochromic shift, and hypochromic shift of the A band observed with the α- and the *o*-substituted stilbenes as well as *cis*-stilbene are undoubtedly due to the steric hindrance which prevents the conjugated system from assuming a planar conformation. The loss of the vibrational fine structure of the first intense band (the so-called conjugation band) in the spectra of sterically hindered compounds is also observed with many other conjugated systems. The relation between the shape of the conjugation band and the steric hindrance to the planarity of the conjugated system will be discussed in Chapter 15.

14.2.2. Calculation of the Twist Angle

The dihedral angles between each benzene ring plane and the ethylenic bond plane, that is, the twist angles of the α—1 and α′—1′ bonds, θ, in sterically hindered derivatives of stilbene can be calculated by the method used to calculate the θ in *cis*-stilbene in Subsection 14.1.3. In the application of the method to α- and α,α′-substituted stilbenes having no substituent on the benzene nuclei, the possible electronic effects of the substituents are ignored for simplification, and the same values of ΔE and ν for the two references as in the case of *cis*-stilbene are used: $\Delta E_S = 1.1421\ (-\beta)$; $\nu_S = 34{,}002\ \text{cm}^{-1}$; $\Delta E_L = 2\ (-\beta)$; $\nu_L = 48{,}000\ \text{cm}^{-1}$. For treatment of *trans*- and *cis*-4,4′,α,α′-tetramethylstilbene, the wave number of the most intense peak (the γ subband) of the A band of *trans*-4,4′-dimethylstilbene, 33,190 cm^{-1} (301.3 mμ), is taken as ν_S; the center of gravity of singlets or the wave number of the *p* band of toluene, 46,882 cm^{-1} (213.3 mμ), is taken as ν_L. For treatment of 2,4,6,2′,4′,6′-hexamethylstilbene(1,2-dimesitylethylene), it is assumed that the mean magnitude of the bathochromic shift of the A band per a *para* or an *ortho* methyl substituent is 3 mμ. That is, it is assumed that the A band of the hypothetical planar molecule of dimesitylethylene would be at about 312 mμ (294 + 3 × 6 mμ), and the corresponding wave number, 32,051 cm^{-1}, is taken as ν_S. In this case, the center of gravity of singlets or the wave number of the *p* band of mesitylene, 45,600 cm^{-1} (219.3 mμ), is taken as ν_L.

Since 2,4,6-trimethylstilbene (1-phenyl-2-mesitylethylene) is unsymmetrical, the twist angle of the Ph—C bond and that of the Mes—C bond must be different. Therefore, our method cannot be applied to this compound. By the way, on the basis of the above assumption that the bathochromic shift per a *para* or an *ortho* methyl substituent is about 3 mμ, it is presumed that the A band of *trans*-2,4,6-trimethylstilbene would be at about 273.7 mμ (282.7 − 3 × 3 mμ) if it were not for the bathochromic shift by the electronic effect of the methyl substituents. This presumed position coincides almost completely with the observed position of the A band of *trans*-α-methylstilbene (273.5 mμ). Similarly, the A band of *trans*-1,2-dimesitylethylene is presumed to be at about 244.7 mμ (262.7 − 3 × 6 mμ) if it were not for the bathochromic shift by the electronic effect of the methyl substituents. This presumed position coincides fairly well with the observed position of the A band of *trans*-α,α′-dimethylstilbene (243.3 mμ).

The results of the calculation are summarized in Table 14.6. In α-methylstilbene as well as α-methyl-α′-ethylstilbene, the positions of the two phenyl groups are not equivalent, and therefore, their angles of twist must differ

TABLE 14.6

THE A BANDS AND RESULTS OF CALCULATION OF TWIST ANGLES OF STILBENES

	The A band			Results of the calculation				
Stilbene	Solvent[a]	λ_{max}, mμ	ϵ_{max}	ΔE, $-\beta$	k	θ, deg.	R, Å	RE, $-\beta$
trans-	H	294.1	27950	1.142	0.908	0	1.445	0.704
cis-	E	280	10450	1.247	0.787	28	1.456	0.530
trans-α-Methyl-	H	273.5	21100	1.299	0.730	34.5	1.462	0.457
cis-α-Methyl-	E	267	9340	1.354	0.671	40	1.467	0.387
trans-α,α′-Dimethyl-	H	243.3	12270	1.577	0.446	58	1.490	0.173
cis-α,α′-Dimethyl-	H	252	8880	1.490	0.531	51.5	1.481	0.243
trans-α-Methyl-α′-ethyl-	E	240	11950	1.612	0.412	60.5	1.493	0.146
trans-α,α′-Diethyl-	H	236.6	11090	1.649	0.378	63	1.497	0.124
cis-α,α′-Diethyl-	E	244	7740	1.570	0.453	57.5	1.489	0.178
trans-α,α′-Dichloro-	H	268	12000	1.345	0.680	39	1.467	0.397
trans-α,α′-Dibromo-	H	248.5	11570	1.525	0.497	54.5	1.485	0.212
trans-α,α′-Diiodo-	H	220.5[b]	20800	1.838	0.200	75.7	1.517	0.034
trans-4,4′,α,α′-Tetramethyl-	E	245	15000	1.620	0.404	61	1.494	0.140
cis-4,4′,α,α′-Tetramethyl-	E	258	10000	1.491	0.530	51.7	1.481	0.242
trans-2,4,6,2′,4′,6′-Hexamethyl-	H	262.7	16000	1.523	0.498	54.3	1.485	0.212

[a] Solvent: H, heptane; E, ethanol.
[b] The wavelength at an inflection.

from each other. In spite of this, the results for these compounds, obtained on the basis of the assumption that the twist angles of the two phenyl groups were equal, are also shown for comparison. It is to be noted that the calculated values of the angle of twist for α- and α,α′-substituted compounds, especially for α,α′-dihalo compounds, are possibly rather smaller than the actual values because of the neglect of the possible electronic bathochromic effect of the substituents in the calculation.

The calculated value of θ for *trans*-α,α′-dimethylstilbene, 58°, is in good agreement with the value of 56–60° estimated from a scale model in which the interference radius of a methyl group is taken to be 2.0 Å. The calculated value of θ for *trans*-α,α′-dibromostilbene, 54.5°, is also in fairly good agreement with the value of 48–52° estimated from a scale model in which the interference radius of a bromine atom is 1.91 Å. The calculated value of θ for the chloro analog, 39°, is appreciably smaller than the value of 43–48° estimated from a scale model in which the interference radius of a chlorine atom is 1.80 Å.

The change of the α- and α′-substituents from methyl groups to ethyl groups affects only slightly the position of the A band and hence the value of θ. A similar phenomenon was seen in the comparison between *o*-methyl- and *o*-ethylbiphenyl (see Subsection 12.3.1). On the other hand, replacement of the methyl group in β-methylcinnamic acid by an ethyl group has a greater steric consequence. The λ_{max} (in mμ) and ϵ_{max} (in parentheses) of the first intense π–π band (the conjugation band or A band) of cinnamic acid and its β-alkyl derivatives in 94% ethanol have been reported as follows:[12] cinnamic acid, 273 (22,000); β-methylcinnamic acid, 260 (15,000); β-ethylcinnamic acid, 244 (10,000).

It is noteworthy that the A bands of *trans*-α,α′-dialkylstilbenes are at shorter wavelengths than those of the corresponding *cis* isomers, contrary to a generally accepted view that *trans* isomers of conjugated compounds should exhibit the conjugation band at longer wavelength than do the corresponding *cis* isomers. This fact is considered to indicate that with these compounds the *trans* isomers are less conjugated than the corresponding cis isomers. Indeed, the calculated values of θ for the *trans* isomers are larger than those for the corresponding *cis* isomers, and the calculated values of *RE* (the extra-resonance energy) of the *trans* isomers are smaller than those of the corresponding *cis* isomers. This conclusion seems to be supported by some pieces of information on the properties of these compounds such as the exaltations of molecular refraction and the nuclear magnetic resonance spectra.

The exaltation of molecular refraction can be considered as a measure of the extent of conjugation. The exaltation is much greater in *trans*-stilbene than in the *cis* isomer. On the other hand, in contrast with the case of stilbene, the exaltation is, though slightly, smaller in *trans*-α,α′-dimethylstilbene than in the *cis* isomer.[13] This reversal is similar to what happens with the electronic absorption spectra.

The α-hydrogens of *cis*-stilbene give rise to the proton magnetic resonance at a higher magnetic field than do those of the *trans* isomer,[14] indicating that the α-hydrogens of the *cis* isomer are more shielded than those of the *trans* isomer. In contrast with the α-hydrogens of stilbenes, the hydrogens of the

[12] M. Takahasi, *Bull. Chem. Soc. Japan* **31,** 756 (1958).

[13] K. v. Auwers, *Ber.* **62,** 693 (1929); *Ann.* **499,** 123 (1932).

[14] D. Y. Curtin, H. Gruen, and B. A. Shoulders, *Chem. & Ind.* (*London*) **1958,** 1205.

α-methyl groups of *cis*-α,α′-dimethylstilbene and of the α-ethyl groups of *cis*-α,α′-diethylstilbene give rise to resonance at lower fields than do the corresponding hydrogens of the *trans* isomers.[15] That is, the hydrogens are less shielded in the *cis* isomers than in the corresponding *trans* isomers.

As mentioned already, *trans*-stilbene is thermodynamically much more stable than the *cis* isomer, and the energy difference has been determined to be about 3 kcal./mole from the equilibrium of the thermal *cis-trans* isomerization. Similarly, the *trans* isomers of *p*-nitrostilbene and *p,p*′-dinitrostilbene are more stable than the corresponding *cis* isomers by 7.0 ± 0.4 and 4.6 ± 0.3 kcal./mole, respectively.[16] The difference of stability between the isomers is by far smaller in the case of α,α′-dimethylstilbene than in the case of stilbene. The equilibrium mixture yielded by heating either *trans*- or *cis*-α,α′-dimethylstilbene with a small amount of concentrated sulfuric acid for 20 hr at 210°C contained the *trans* and the *cis* isomer in the ratio of the amounts 55:45–60:40.[17] The fact that the *trans* isomer is still, though slightly, also more stable in this case than the *cis* isomer would appear to be rather surprising, being contrary to the expectation from the absorption spectra. However, this fact is understandable if one considers that the steric repulsion between the two methyl groups in the *cis* isomer, which cannot be reduced by the rotation of the phenyl groups, contributes to destabilization of the molecule.

14.2.3. *p,p*′-Diphenylstilbene and Its α,α′-Dimethyl Derivative

The spectra of *trans-p,p*′-diphenylstilbene and its α,α′-dimethyl derivative in *n*-heptane are shown in Fig. 14.7. The first intense band (that is, the A band) of the diphenylstilbene has indistinct vibrational structure. On the other hand, the A band of the dimethyl derivative is almost structureless, and is at considerably shorter wavelength and of lower intensity than that of the parent compound. This situation is very similar to that observed in the comparison between *trans*-stilbene and its α,α′-dimethyl derivative, and is undoubtedly explained in terms of the steric effect of the methyl substituents.

Now the simple LCAO MO method is applied to this conjugated system. Similarly to the treatment in the preceding subsections, the value of the resonance parameter for the ethylenic bond (the α—α′ bond) is assumed to be 1.080. The value of the resonance parameter for the essential single bond

[15] M. Katayama, S. Fujiwara, H. Suzuki, Y. Nagai, and O. Simamura, *J. Mol. Spectroscopy* **5**, 85 (1960).

[16] C. M. Anderson, L. G. Cole, and E. C. Gilbert, *J. Am. Chem. Soc.* **72**, 1263 (1950).

[17] O. Simamura and H. Suzuki, *Bull. Chem. Soc. Japan* **27**, 231 (1954).

between the two benzene rings in each biphenyl part is assumed to be 0.780, a value which has been found to be appropriate for the corresponding parameter in biphenyl in solution (see Subsection 12.2.4). The resonance parameter for each essential single bond in the stilbene part of the molecule is taken as the variable. This parameter is denoted simply by k.

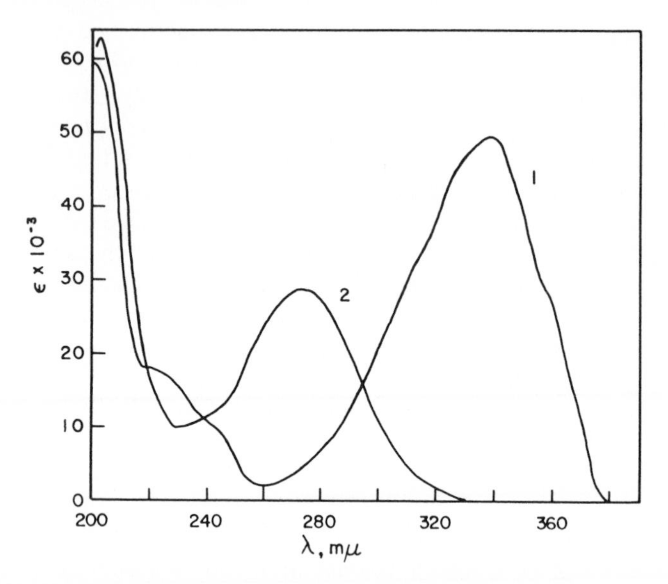

FIG. 14.7. The absorption spectra of *trans-p,p'*-diphenylstilbene (1) and its α,α'-dimethyl derivative (2) in heptane.

When as the value of k is taken the corresponding value for planar *trans*-stilbene, 0.9084, the energy of the allowed transition from the highest occupied π orbital to the lowest vacant π orbital, ΔE_{11}, is calculated to be 1.018 $(-\beta)$. This value is taken as ΔE_S, and the wave number of the maximum of the A band of *trans-p,p'*-diphenylstilbene, 29,638 cm^{-1} (337.4 mμ), is taken as ν_S. When k is zero the value of ΔE_{11} is 1.523 $(-\beta)$. This value is taken as ΔE_L, and the wave number of the maximum of the A band of biphenyl in n-heptane, 40,486 cm^{-1} (247 mμ), is taken as ν_L.

On the basis of the above reference values and of the postulated linear relation between ΔE and ν, the appropriate value of ΔE_{11} for the α,α'-dimethyl derivative is calculated to be 1.342 $(-\beta)$ from the wave number of the maximum of the A band of this compound in n-heptane, 36,590 cm^{-1} (273.3 mμ), and then the following values are obtained for this compound: $k = 0.461$;

$\theta = 57°$; $R = 1.488$ Å. R represents the length of each essential single bond in the stilbene part, and θ represents the angle of twist of this bond. The value of θ for this α,α'-dimethyl compound, 57°, coincides almost completely with the value for *trans*-α,α'-dimethylstilbene, 58° (see Table 14.6).

14.3. Tetraphenylethylene and Related Compounds*

14.3.1. The Spectra

The spectra of tetraphenylethylene (α,α'-diphenylstilbene), triphenylethylene (α-phenylstilbene), and 1,1,2-triphenylpropene (α-phenyl-α'-methylstilbene) in n-heptane are compared with the spectrum of *trans*-stilbene in Fig. 14.8.

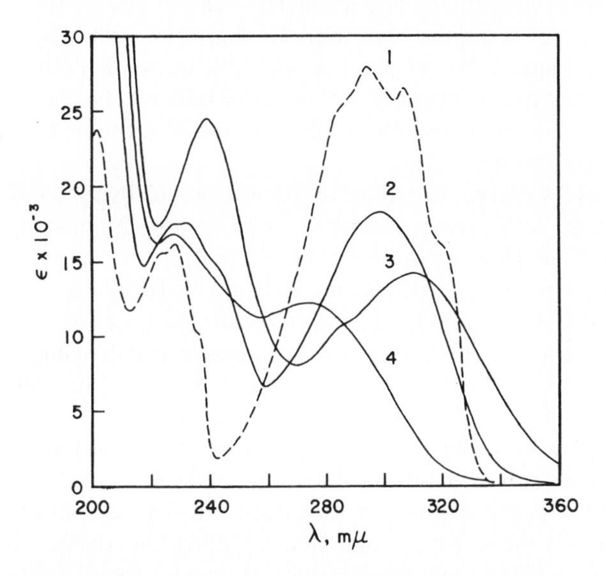

Fig. 14.8. Absorption spectra of triphenylethylene, tetraphenylethylene, and related compounds in heptane. 1, *trans*-stilbene; 2, triphenylethylene; 3, tetraphenylethylene; 4, 1,1,2-triphenylpropene.

Unlike the parent compound, *trans*-stilbene, the phenylated stilbenes show no vibrational structure in the A band. In the series of *trans*-stilbene and its α-phenylated derivatives, as the number of phenyl groups increases, both the A and the B band shift to longer wavelengths. At the same time, the

* This section is based mainly on Reference 7.

intensity of the A band decreases, while the intensities of the B and C bands increase. It is of interest that the ratio of ϵ_{max} of the C bands of *trans*-stilbene, triphenylethylene, and tetraphenylethylene are roughly 1:2:3.

14.3.2. The Molecular Geometry

According to the results of a measurement of the spectra of tri- and tetraphenylethylene in chloroform, the spectra appear to differ only slightly from the spectrum of *trans*-stilbene.[18] Jones[19] explained this apparent spectral similarity as follows. "The three phenyl groups of triphenylethylene and the four phenyl groups of tetraphenylethylene cannot be accommodated in a strainless planar structure; the strain can be eased either by a small rotation of all of the phenyl groups along their 1,4-axes, or by a larger displacement of one of the phenyl groups of triphenylethylene or two of the *trans*-related phenyl groups of tetraphenylethylene, leaving two remaining *trans*-phenyl groups in the plane of the ethylenic bond. This second hypothesis leaves the molecule of tetraphenylethylene with a *trans*-stilbene chromophoric system together with two insulated phenyl groups, the absorption of which is too feeble to affect the spectrum."

When it is assumed, according to Jones' second hypothesis, that tetraphenylethylene and triphenylethylene have conformations in which the two *trans*-related phenyl groups are coplanar with the ethylenic bond and in which the other phenyl groups are insulated, evidently the MO calculation on the stilbene system described in Subsection 14.1.2 is also applicable to these compounds. In this case, since the A bands of these compounds are at longer wavelengths than the A band of *trans*-stilbene, the values of k for these compounds should be larger than the value for *trans*-stilbene. As long as Jones' second hypothesis is adopted, since the most probable conformation of *trans*-stilbene is already planar, the large values of k for these compounds as compared with the value for *trans*-stilbene must be explained only by the assumption that the essential single bonds linking the coplanar *trans*-related phenyl groups to the ethylenic carbon atoms in these compounds were shorter than the corresponding bonds in *trans*-stilbene. This is undoubtedly implausible. This inference seems to be supported by the fact that the intensity of the A band decreases markedly in the order *trans*-stilbene > triphenylethylene > tetraphenylethylene. If all these compounds had a *trans*-stilbene chromophore as in Jones' second hypothesis, the A bands of these compounds would have approximately the same intensity.

[18] B. Arends, *Ber.* **64,** 1936 (1931).
[19] R. N. Jones, *J. Am. Chem. Soc.* **65,** 1818 (1943).

Now, it is assumed, according to Jones' first hypothesis, that all of the phenyl groups in tetraphenylethylene are rotated out of the ethylenic bond plane by the same angle θ. In this model the values of the resonance parameters for the four essential single bonds connecting the phenyl groups to the ethylenic carbon atoms are equal to one another. These resonance parameters are denoted by a common symbol k'. When k' is replaced by $2^{-1/2}k$, the secular determinant for determination of the π MO's of this system as linear combinations of the carbon $2p\pi$ AO's can be factorized into one determinant identical in form with the secular determinant for the stilbene system and two determinants identical with the secular determinant for benzene. The result of the calculation on the stilbene system can therefore be used for this compound. When the two references for correlation of ΔE_{11} and ν_A are taken as the same as before, the appropriate value of ΔE_{11} corresponding to the observed position of the A band ($\lambda_{max} = 308.7$ mμ) is determined by extrapolation to be 1.044 ($-\beta$), and then the following values are determined by the usual procedure: $k = 1.032$; $k' = 0.730$; $\theta = 34.5°$; $R = 1.462$ Å; $RE = 0.906$ ($-\beta$). The one-electron transition from ψ_{+1} to ψ_{-1} is allowed and is polarized in the direction of the central ethylenic bond axis (that is, the y axis). The dipole strength of this transition, $D_{11} = M_{11}^2$, is calculated to be 2.566 Å^2. This value is to be compared with the corresponding value of 4.221 Å^2 for *trans*-stilbene. The low calculated value of D_{11} for tetraphenylethylene as compared with the value for *trans*-stilbene explains the low intensity of the A band of the former compound ($\epsilon_{max} = 14{,}200$, $f = 0.393$) as compared with the A band of the latter compound ($\epsilon_{max} = 27{,}950$, $f = 0.739$). It is concluded that the most preferred conformation of tetraphenylethylene is a conformation in which all the phenyl groups are rotated out of the plane of the ethylenic bond. Similarly, it is inferred that the most preferred conformation of triphenylethylene is a conformation in which at least the two phenyl groups attached to the same ethylenic carbon atom are rotated out of the plane of the ethylenic bond.

The A band of methyltriphenylethylene ($\lambda_{max} = 274.5$ mμ, $\epsilon_{max} = 12{,}200$) is at considerably shorter wavelength than that of triphenylethylene ($\lambda_{max} = 298.5$ mμ, $\epsilon_{max} = 18{,}220$) and coincides in position almost completely with the A band of *trans*-α-methylstilbene ($\lambda_{max} = 273.5$ mμ, $\epsilon_{max} = 21{,}100$) (compare Figs. 14.8 and 14.4). From this fact it would be inferred that the most preferred conformation of methyltriphenylethylene might be a conformation in which the phenyl group at the *trans*-position to the methyl group was completely rotated out of the ethylenic bond plane, leaving the two remaining *trans*-related phenyl groups in a conformation similar to the most preferred conformation of *trans*-α-methylstilbene. However, the fact that the intensity of the A band of methyltriphenylethylene is much smaller than that of *trans*-α-methylstilbene suggests that this inference is unlikely. This situation is very similar to that encountered in the comparison between the spectra of triphenylethylene and of *trans*-stilbene.

14.4. The Influence of Environment on the Spectra of Stilbene and of Related Compounds

14.4.1. THE SOLVENT EFFECT OF BENZENE*

While the spectra of *n*-heptane, methanol, and ethanol solutions of *trans*-stilbene closely resemble one another, chloroform, carbon tetrachloride, and benzene as the solvent exert considerably large bathochromic effects on the A band, and more or less blur the vibrational fine structure of the band. Data on the spectra of *trans*-stilbene in various solvents are shown in Table 14.7.

TABLE 14.7

THE A BAND OF *TRANS*-STILBENE IN VARIOUS SOLVENTS

Solvent	The β subband		The γ subband		
	λ_{max}, mμ	ϵ_{max}	λ_{max}, mμ	ϵ_{max}	$\epsilon_\beta/\epsilon_\gamma$
Heptane	306.9	26,500	294.1	27,950	0.949
Methanol	306.8	27,350	294.0	28,050	0.975
Ethanol (95%)	307.3	27,200	294.5	27,850	0.977
Chloroform	309.9	27,150	297.9	27,450	0.989
Carbon tetrachloride	311.8	24,750	299.4	25,050	0.988
Benzene	311.2	25,600	298.8	25,340	1.010
Mixed alcohols (−130°C)[a]	311.2	38,500	297.5	22,500	1.711

[a] Reference 21.

It is to be noted that the distribution of intensity among the vibrational subbands of the A band varies considerably with the solvent. The ratio of the intensity of the β subband to that of the γ subband for each solvent is listed in the last column of Table 14.7. It is seen that the ratio increases in the order *n*-heptane $<$ methanol $\approx$ ethanol $<$ chloroform $\approx$ carbon tetrachloride $<$ benzene: while in solvents other than benzene the γ subband is the most intense subband of the A band, in benzene the β subband is the most intense

* This subsection is based on References 1 and 11.

subband. The A band is much more blurred in benzene than in the other solvents. The special solvent effect of benzene is also seen with the spectra of stilbene derivatives such as *p*-methoxy- and *p*-nitro-*trans*-stilbene. These facts seem to suggest the existence of a special interaction between the π orbitals of the solute molecule and those of benzene.

The bathochromic shift of the A band by changing the solvent from *n*-heptane to benzene is also observed with other stilbene derivatives, such as α,α'-diphenylstilbene (tetraphenylethylene) and p,p'-diphenylstilbene (1,2-dibiphenylylethylene). The magnitude of the shift appears to be related to the most preferred conformation of the molecule. Thus it seems that the following generalization can be deduced: whereas the shifts for planar molecules are about 5 mμ, the shifts for probably nonplanar molecules are about 3 mμ and are considerably smaller than the shifts for planar molecules.

14.4.2. The KCl-Disk Spectra of Stilbenes

The features of the solid-state spectra of stilbenes measured by the pressed KCl disk technique as compared with those of the spectra of *n*-heptane solutions also appear to reflect the most preferred conformations of the molecules.[20]

trans-Stilbene, *trans*-*p*-phenylstilbene, *trans*-1-phenyl-2-(2-fluorenyl)-ethylene, and *trans*-p,p'-diphenylstilbene are *trans*-stilbene-type compounds which have no sterically hindering substituent. The most preferred conformations of these compounds are considered to be planar or nearly planar. It is a common distinctive feature of the spectra of these compounds that the A band exhibits more or less well-resolved vibrational fine structure. With these compounds, when the state is changed from solution to solid, a marked redistribution of intensity among the vibrational subbands of the A band occurs: the subband which appears as the most intense peak in the solution spectrum appears in the KCl-disk spectrum as an inflection at longer wavelength than the most intense peak, and the subband which appears in the solution spectrum as an inflection at shorter wavelength than the most intense peak becomes the most intense peak in the KCl-disk spectrum.

A typical example is *trans*-stilbene. With this compound, the most intense subband of the A band in the solution spectrum is the γ subband, and that in the KCl-disk spectrum is the δ subband. In this connection, it has been known that the temperature affects markedly the relative intensities of the vibrational subbands. In the spectrum of *trans*-stilbene in mixed alcohols

[20] This subsection is an abstract of H. Suzuki, *Bull. Chem. Soc. Japan* **33**, 944 (1960).

at −130°C measured by Beale and Roe,[21] the vibrational fine structure is much more sharply resolved, the intensities of the α and β subbands are much greater, and the intensity of the γ subband is much smaller than in the solution spectrum measured at room temperature. In this low-temperature

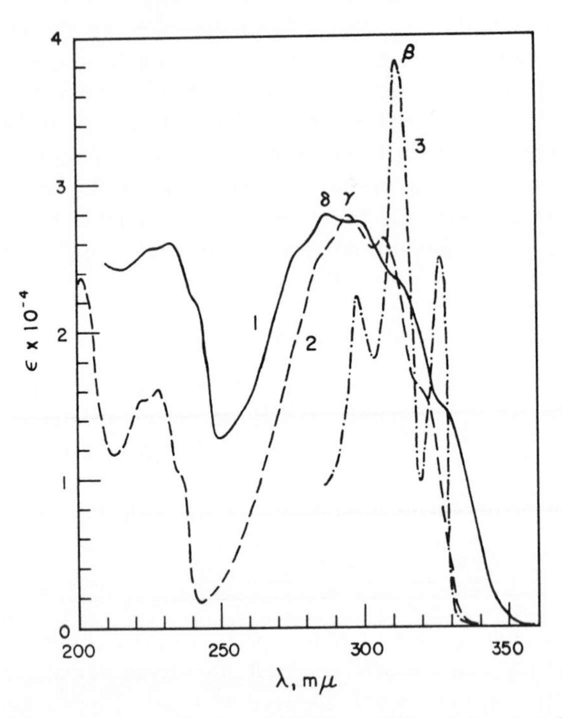

FIG. 14.9. Absorption spectra of *trans*-stilbene under various conditions. 1, the KCl-disk spectrum; 2, the spectrum of the heptane solution; 3, the low-temperature spectrum (in mixed alcohols at −130°C, redrawn from Beale and Roe's paper, Reference 21).

spectrum the most intense subband is the β subband. The three spectra of *trans*-stilbene are shown in Fig. 14.9.

The wavelength difference (in mμ) between a band in the KCl-disk spectrum and the corresponding band in the *n*-heptane-solution spectrum is denoted by Δλ. For the vibrational subbands of the A bands of the probably planar *trans*-stilbene-type compounds, the value of Δλ is positive, and increases fairly regularly with the wavelength of the subband in the solution

[21] R. N. Beale and E. M. F. Roe, *J. Chem. Soc.* **1953**, 2755.

spectrum, λ, from about 3 mμ at 270 mμ to about 5 mμ at 300 mμ, and to about 9.5 mμ at 350 mμ. A similar regular relation between $\Delta\lambda$ and λ has been found by Dale for the spectra of naphthalene and anthracene (see Subsection 6.4.3). According to him, $\Delta\lambda$ for these compounds, that is, the "normal red-shift," increases regularly with the wavelength of the band, from about 2 mμ at 220 mμ to about 8 mμ at 300 mμ, and to about 19 mμ at 400 mμ. Thus it may be said that the *trans*-stilbene-type compounds show the red-shifts of normal magnitudes in themselves, which are smaller than the magnitudes of the normal red-shifts in naphthalene and anthracene. The latter compounds are, to use Dale's words, "rigid" compounds which obviously cannot assume other conformations in solution than in the crystalline state. Therefore, the above-mentioned fact may be considered as evidence indicating that the most preferred conformations of these stilbenes in solution are not significantly different from those in the solid state and are probably planar.

While the probably planar compounds show the red-shifts of normal magnitudes, some compounds which are sterically hindered and evidently nonplanar show the shifts of more or less abnormal magnitudes, as will be mentioned below. The A bands of these sterically hindered compounds have no distinct vibrational structure, in contrast with the foregoing sterically unhindered compounds, and as a matter of fact, the shapes of the KCl-disk spectra of these sterically hindered compounds are apparently not markedly different from those of the solution spectra.

The value of $\Delta\lambda$ for the A band ($\lambda = 298.5$ mμ) of triphenylethylene is about 5.3 mμ. This value is, though very slightly, greater than the value expected from the foregoing correlation between $\Delta\lambda$ and λ. The value of $\Delta\lambda$ for the A band ($\lambda = 308.7$ mμ) of tetraphenylethylene is about 14.2 mμ. This value is much greater than the expected value. The extraordinarily large shift especially in the case of tetraphenylethylene probably indicates that the most probable conformation of this compound in the solid state is considerably different from that in solution. It is inferred that the interplanar angles between the phenyl groups and the ethylenic bond are probably smaller in the most probable conformation in the solid state than in that in solution.

The value of $\Delta\lambda$ for the A band ($\lambda = 262.7$ mμ) of *trans*-1,2-dimesitylethylene is about 1.7 mμ, which is slightly smaller than the expected value. The values of $\Delta\lambda$ for the A bands of *trans*- and *cis*-α,α'-dimethylstilbene and *trans*-α,α'-diethylstilbene are much smaller than the expected values. The most probable conformations of these compounds are undoubtedly nonplanar. It seems reasonable to correlate the smallness of the shifts with the nonplanarity of the molecular structures: with these alkyl-substituted, nonplanar compounds the intermolecular interactions in the crystals are probably smaller than in the case of the planar compounds.

14.5. Vinylogs of Stilbene and of Tetraphenylethylene

14.5.1. 1,4-Diphenylbutadiene and 1,1,4,4-Tetraphenylbutadiene

Data on the spectra of 1,4-diphenyl-1,3-butadienes and of some related compounds are shown in Table 14.8.

The spectrum of *all-trans*-1,4-diphenylbutadiene is similar in shape to that of *trans*-stilbene. The first band (that is, the A band) has vibrational

TABLE 14.8

THE SPECTRA OF 1,4-DIPHENYL-1,3-BUTADIENES AND RELATED COMPOUNDS

		The A band		The B band	
Compound	Solvent	λ_{max}, mμ	ϵ_{max}	λ_{max}, mμ	ϵ_{max}
1,4-Diphenyl-1,3-butadiene					
trans-trans-[a]	Hexane	345	34,000	232	12,000
		328	52,700		
		315	46,400		
cis-trans-[b]	Hexane	313	30,300	239	17,500
cis-cis-[b]	Hexane	300	29,400	—	—
1,1,4,4-Tetraphenyl-1,3-butadiene[c]	Ethanol	343	37,250	249	22,800
1,2-Bisdiphenylmethylene-cyclobutane[c]	Ethanol	351	21,400	258	23,000

[a] E. R. Blout and V. W. Eager, *J. Am. Chem. Soc.* **67**, 1315 (1945).
[b] J. H. Pinckard, B. Wille, and L. Zechmeister, *J. Am. Chem. Soc.* **70**, 1938 (1948).
[c] K. B. Alberman, R. N. Haszeldine, and F. B. Kipping, *J. Chem. Soc.* **1952**, 3284.

fine structure with well-resolved three peaks, and the second band (the B band) shows rudimentary vibrational structure. When one or two ethylenic bonds are converted to the *cis* configuration, the A band undergoes a marked hypsochromic and hypochromic shift and loses the vibrational fine structure. The situation is quite similar to that with stilbene. Evidently these effects must be largely due to the steric inhibition of planarity of the molecular geometry in the *cis* isomers.

The relation of 1,1,4,4-tetraphenylbutadiene to *trans-trans*-1,4-diphenylbutadiene is very similar to that of tetraphenylethylene to *trans*-stilbene. Thus, on passing from the 1,4- to the 1,1,4,4-substituted butadiene, the A band undergoes a considerable decrease in intensity, a small bathochromic shift, and loss of vibrational structure, and the B band undergoes a marked increase in intensity, a small bathochromic shift, and loss of vibrational structure. Evidently, as in tetraphenylethylene, in the tetraphenylbutadiene two phenyl groups attached to the same carbon atom cannot be coplanar with each other.

When a dimethylene bridge is introduced into the 2- and 3-positions of 1,1,4,4-tetraphenyl-1,3-butadiene to give 1,2-bisdiphenylmethylenecyclobutane, the A band markedly decreases in intensity and slightly shifts to longer wavelength, and the B band very slightly increases in intensity and slightly shifts to longer wavelength. These effects are probably related to the forced planar *s-cis* conformation of the butadiene system in the bridged molecule.

14.5.2. α,ω-Diphenylpolyenes and $\alpha,\alpha,\omega,\omega$-Tetraphenylpolyenes

The longest wavelength intense band, that is, the A band, of any α,ω-diphenylpolyene C_6H_5—$(CH{=}CH)_n$—C_6H_5 shows distinct vibrational fine structure, just as the A band of *trans*-stilbene, and, as mentioned already in Subsection 10.4.2, the A band shifts to longer wavelength with the increasing length of the conjugated polyene chain.

In the series of $\alpha,\alpha,\omega,\omega$-tetraphenylpolyenes,

$$(C_6H_5)_2C{=}CH{-}(CH{=}CH)_{n-2}{-}CH{=}C(C_6H_5)_2,$$

the longest wavelength intense band (the A band) also shifts to longer wavelength as the length of the conjugated polyene chain becomes longer: the A band of an $\alpha,\alpha,\omega,\omega$-tetraphenylpolyene is at about 20 mμ longer wavelength than that of the corresponding α,ω-diphenylpolyene.[22]

A prominent characteristic of the A bands of the tetraphenylpolyenes as compared with the A bands of the diphenylpolyenes is loss of the vibrational fine structure. Thus the A bands of lower members ($n \leq 3$) of this series are almost completely structureless. In the A band of the fourth member of this series, tetraphenyloctatetraene, a rudimentary inflection appears at longer wavelength than the maximum, and in the A bands of the fifth and sixth members of this series two broad peaks and a considerably distinct inflection at shorter wavelength than the peaks appear.

[22] G. Kortüm and G. Dreesen, *Chem. Ber.* **84,** 182 (1951).

14.6. Biphenylene Derivatives of Ethylene, Butadiene, and Hexatriene

The spectra of the following compounds have been reported by Bergmann and Hirshberg[23] and by Magoon and Zechmeister:[24]

Ia 1-Biphenylene-2-phenylethylene $[(C_6H_4)_2]C{=}CHC_6H_5$

Ib 1-Biphenylene-2,2-diphenylethylene $[(C_6H_4)_2]C{=}C(C_6H_5)_2$

Ic Dibiphenylene-ethylene $[(C_6H_4)_2]C{=}C[(C_6H_4)_2]$

II*a* 1-Biphenylene-4-phenyl-1,3-butadiene

$$[(C_6H_4)_2]\overset{1}{C}{=}\overset{2}{C}H{-}\overset{3}{C}H{=}\overset{4}{C}HC_6H_5$$

II*b* 1-Biphenylene-4,4-diphenyl-1,3-butadiene

$$[(C_6H_4)_2]C{=}CH{-}CH{=}C(C_6H_5)_2$$

II*c* 1,4-Dibiphenylene-1,3-butadiene

$$[(C_6H_4)_2]C{=}CH{-}CH{=}C[(C_6H_4)_2]$$

III*a* 1-Biphenylene-6-phenyl-1,3,5-hexatriene

$$[(C_6H_4)_2]\overset{1}{C}{=}\overset{2}{C}H{-}\overset{3}{C}H{=}\overset{4}{C}H{-}\overset{5}{C}H{=}\overset{6}{C}HC_6H_5$$

III*c* 1,6-Dibiphenylene-1,3,5-hexatriene

$$[(C_6H_4)_2]C{=}CH{-}CH{=}CH{-}CH{=}C[(C_6H_4)_2].$$

In the above structural formulas $[(C_6H_4)_2]$ represents a 2,2′-biphenylene group. The Roman numericals in the symbols for compounds, I, II, and III, signify the numbers of double bonds in the open chains, and the letters *a*, *b*, and *c* signify α-biphenylene-ω-phenyl, α-biphenylene-ω,ω-diphenyl, and α,ω-dibiphenylene compounds, respectively. The data on the spectra of these compounds are shown in Table 14.9, in which the data on the spectra of triphenylethylene (I*a*′), tetraphenylethylene (I*b*′), and 1,1,4,4-tetraphenyl-1,3-butadiene (II*b*′) are also shown for the purpose of comparison.

The spectra of these compounds are considerably similar to one another on the whole, showing three main band systems in the wavelength region longer than about 230 mμ. The longest wavelength bands (the A bands) of the sterically unhindered compounds, *trans*-II*a*, II*c*, etc., have vibrational fine structure: in most cases three distinct peaks can be detected and some vestigial subbands appear as shoulders or inflections. The vibrational fine structure of the A band is blurred or almost completely lost in the spectra of the sterically hindered compounds, such as I*a*, I*b*, I*c*, and *cis*-II*a*. The second band (the B band) is at considerably shorter wavelength than the A band, and in most cases it has two vibrational peaks. The third band (the

[23] E. D. Bergmann and Y. Hirshberg, *Bull. soc. chim. France* **1950,** 1091.

[24] E. F. Magoon and L. Zechmeister, *J. Am. Chem. Soc.* **77,** 5642 (1955).

TABLE 14.9

THE SPECTRA OF BIPHENYLENE DERIVATIVES OF ETHYLENE, BUTADIENE, AND HEXATRIENE[a]

Compound	Solvent[c]	A band[b] λ_{max} mμ	A band[b] ϵ_{max} ×10⁻⁴	B band λ_{max} mμ	B band ϵ_{max} ×10⁻⁴	B′ band λ_{max} mμ	B′ band ϵ_{max} ×10⁻⁴
I*a*′[d]	Hp	298.5	1.8	232	1.8	—	—
I*b*′	Hp	308.7	1.4	238.5	2.4	—	—
I*a*[e]	D	326	1.5	260	3.0	228	3.6
I*b*[e]	D	338	1.4	260	3.5	235	4.7
				252	3.4		
I*c*[e]	D	458	2.3	272	4.0	244	7.8
				258	4.0		
II*b*′[f]	E	343	3.7	249	2.3	—	—
trans-II*a*[g]	Hx	390*	3.2	273	2.4	240	4.2
		Δ373	4.7	261	3.1		
		360*	4.0				
cis-II*a*[g]	Hx	357.5	3.5	269	2.7	238	3.7
				260	3.1		
II*b*[e]	D	388	4.3	274	2.5	244	4.4
				264	3.5		
II*c*[e]	D	444	4.9	278	4.5	242	5.4
		Δ418	5.4	269	4.8		
3,5-*di*-*trans*-III*a*[g]	Hx	410	6.4	276	2.7	242	3.6
		Δ388	6.8	267	2.8		
		366*	4.2				
3-*cis*-5-*trans*-III*a*[g]	Hx	408	4.9	283*	1.7	242	3.6
		Δ385.5	5.7	276	2.7		
		365*	3.8	267	3.1		
3-*trans*-5-*cis*-III*a*[g]	Hx	385	5.3	276	2.7	242	3.6
				267	3.1		
3,5-*di*-*cis*-III*a*[g]	Hx	379	4.8	276	2.7	242	3.6
				267	3.1		
trans-III*c*[g]	CH	465	7.7	280	3.5	240	5.7
		Δ434	7.7	270	4.0		
		410	4.4				
cis-III*c*[g]	CH	458.5	4.8	278	3.2	240	5.7
		Δ428.5	5.4	270	4.0		
		403*	3.7				

[a] The asterisk on wavelength values denotes inflections.
[b] Symbol Δ denotes the most intense peak of fine structure of the A band.
[c] Solvent: CH, cyclohexane; D, dioxan; E, ethanol; Hp, heptane; Hx, hexane.
[d] Reference 7.
[e] Reference 23.
[f] K. B. Alberman, R. N. Haszeldine, and F. B. Kipping, *J. Chem. Soc.* **1952,** 3284.
[g] Reference 24.

B′ band) is at slightly shorter wavelength than the B band. Perhaps the B and B′ bands belong to the same band system.

The data listed in Table 14.9 are for solutions in various solvents. However, since all the compounds are hydrocarbons, the wavelengths of bands are considered to be only slightly dependent on the solvent: the data may be directly compared with one another, without making any special allowance for the difference in the solvent effect. Comparison of the data seems to lead to the following generalizations.

(1) As mentioned above, while the A bands of the sterically unhindered compounds (*trans*-II*a*, II*c*, 3,5-*ditrans*-III*a*, 3-*cis*-5-*trans*-III*a*, *trans*-III*c*, and *cis*-III*c*) have considerably well-resolved vibrational fine structure, the A bands of the sterically hindered compounds (I*a*′, I*b*′, I*a*, I*b*, I*c*, II*b*′, *cis*-II*a*, II*b*, 3-*trans*-5-*cis*-III*a*, and 3,5-*dicis*-III*a*) are almost completely structureless. In I*a*′ and the compounds of the *b* series (I*b*′, I*b*, II*b*′, and II*b*) two phenyl groups are attached to the same ethylenic carbon atom. Evidently, these two phenyl groups cannot simultaneously be coplanar with the ethylenic bond. In I*a* and I*b* there should be steric interference between the biphenylene group and the phenyl group or goups. In I*c* there should be severe steric interference between the two biphenylene groups. In *cis*-II*a*, 3-*trans*-5-*cis*-III*a*, and 3,5-*dicis*-III*a* the terminal ethylenic bond of the aliphatic polyene chain has the *cis* configuration: these compounds have a sterically hindered *cis* bond.

(2) On passing from a compound of the *a* series to the corresponding compound of the *b* series, the A band undergoes a slight bathochromic shift and a marked decrease in intensity (for example, I*a*′ → I*b*′, I*a* → I*b*, and *trans*-II*a* → II*b*). At the same time, the B and B′ bands undergo small bathochromic shifts and considerably large increases in intensity.

(3) On passing from a compound of the *b* series to the corresponding compound of the *c* series, that is, on replacement of the two phenyl groups attached to the same carbon atom by a 2,2′-biphenylene group, the A band undergoes a considerably large bathochromic shift (I*b* → I*c*, $\Delta\lambda = 120$ mμ; II*b* → II*c*, $\Delta\lambda = 30$ mμ; I*a*′ → I*a*, $\Delta\lambda = 27.5$ mμ; I*b*′ → I*b*, $\Delta\lambda = 29$ mμ; II*b*′ → II*b*, $\Delta\lambda = 45.5$ mμ). In the four examples other than I*b* → I*c*, the magnitudes of the shifts are nearly comparable to one another, and the changes in intensity are comparatively small. In I*b* → I*c* the shift and the increase in intensity are extraordinarily large: while I*b* is pale yellow, I*c* is dark red. This effect is considered to be due to the severe steric interference between the two biphenylene groups in I*c*. By the way, on the replacement of the two phenyl groups by a biphenylene group the B band moderately shifts to longer wavelength and moderately increases in intensity.

(4) Rearrangement of an *all-trans* isomer to a sterically unhindered *cis* isomer slightly blurs the vibrational structure of the A band, and causes an appreciable decrease in intensity and a small hypsochromic shift of the band (*trans*-III*a* → 3-*cis*-III*a*, $\Delta\lambda = -2.5$ mμ; *trans*-IIIc → *cis*-IIIc, $\Delta\lambda = -5.5$ mμ).

(5) On the other hand, rearrangement of an *all-trans* isomer to a sterically hindered *cis* isomer brings about an almost complete loss of the vibrational structure of the A band, a marked decrease in intensity, and a comparatively large hypsochromic shift of the band (*trans*-II*a* → *cis*-II*a*, $\Delta\lambda = -15.5$ mμ; *trans*-III*a* → 5-*cis*-III*a*, $\Delta\lambda = -3$ mμ; *trans*-III*a* → 3,5-*dicis*-III*a*, $\Delta\lambda = -9$ mμ).

(6) In contrast with the case of conjugated polyenes, in the spectra of *cis* isomers there appears no "*cis* peak" (see Subsection 10.4.3). This fact may be understandable since the *trans-cis* rearrangements do not significantly alter the nearly straight overall shape of the conjugated system in this type of compound.

(7) Both in the *a* series and in the *b* series, the A band shifts to longer wavelength and markedly increases in intensity as the length of the conjugated system is increased by introduction of an additional ethylenic bond (I*a* → *trans*-II*a*, $\Delta\lambda = 47$ mμ; *trans*-II*a* → *trans*-III*a*, $\Delta\lambda = 15$ mμ; I*b* → II*b*, $\Delta\lambda = 50$ mμ). The magnitude of the shift on the change I*a* → *trans*-II*a* as well as I*b* → II*b* is much larger than that on the change *trans*-II*a* → *trans*-III*a*. The extraordinarily large shifts on passing from compounds of type I to the corresponding compounds of type II may be attributed to the fact that the steric interference between the biphenylene group attached to one end of the ethylenic bond and the phenyl group or groups attached to another end in I*a* and I*b* is relieved by insertion of one additional ethylenic bond. The B band shifts similarly to the A band.

(8) In the *c* series the situation is quite different. The intensity of the A band increases as the length of the conjugated system increases, in the order I*c* < II*c* < *cis*-III*c* < *trans*-III*c*. However, the wavelength of the A band increases in the order II*c* < *cis*-III*c* < *trans*-III*c* (red-orange) < I*c* (dark red). Of course, MO treatment of planar models for these compounds shows that the wavelength of the first intense band (A band) should increase in the order I*c* < II*c* < III*c*. The extraordinarily long wavelength of the A band of I*c* is undoubtedly attributed to the distortion of the molecule from the planar structure by the severe steric interference between the two biphenylene groups.

If I*c* had a planar structure, the overcrowded carbon atoms (carbon atoms at the 3- and 3′-positions of each 2,2′-biphenylene group) would be only 2.5 Å and the hydrogen atoms attached to them only 0.7 Å apart. Nyburg[25] has shown by means of three-dimensional X-ray crystal analysis that the molecule, while centrosymmetrical, is not planar. The benzene rings in one biphenylene group are tilted up, and in the other down, out of the general molecular plane so as to provide better clearance between the over-crowded atoms. It is known that, while a twist of an essential single bond in an alternant hydrocarbon causes a hypsochromic shift of the first main band, a distortion of an aromatic hydrocarbon molecule on the whole causes in most cases a bathochromic shift of the first main band (the *p* band) of the molecule (see Chapter 18).

[25] S. C. Nyburg, *Acta Cryst.* **7**, 779 (1954).

Chapter 15

Relations of the Intensity and Shape of Conjugation Bands to the Geometry of Conjugated Systems*

15.1. Intensity of Conjugation Bands

15.1.1. The Transition Moment as a Function of the Twist Angle

In linear alternant hydrocarbon systems such as biphenyl, styrene, and stilbene, steric hindrance to planarity of the conjugated system exerts a hypochromic as well as a hypsochromic effect on the first intense π–π^* band, that is, the conjugation band or the A band. In the preceding chapters, the hypsochromic effect has been related to the increase in the energy of the one-electron transition from the highest bonding π MO (ψ_{+1}) to the lowest antibonding π MO (ψ_{-1}), ΔE_{11}, associated with the increase in the twist angle of the essential single bond, θ, in the most preferred conformation of the molecule in the ground electronic state. Conceivably, the steric hypochromic effect is to be related to the decrease in the dipole strength of the transition, D_{11} $(= M_{11}^2)$, associated with the increase in θ.[1]

The calculated values of D_{11} for the biphenyl, styrene, and stilbene systems for various values of θ were shown in Tables 12.1, 13.2, and 14.3, respectively. In Fig. 15.1 the values of D_{11} are plotted against θ. The oscillator strength, f, is related to the dipole strength of the transition by

$$f \fallingdotseq 1.085 \times 10^{-5} \nu D, \qquad (15.1)$$

where ν is the wave number at which the absorption due to the transition occurs and D is the dipole strength of the transition in units of Å^2. The wave number ν has been correlated with ΔE_{11}. Accordingly, the value of f

* This chapter is a revised version of Reference 1.

[1] H. Suzuki, *Bull. Chem. Soc. Japan* **35**, 1715 (1962).

can be calculated as a function of θ for each system. By the way, the ΔE and ν for a given value of θ are denoted by $\Delta E(\theta)$ and $\nu(\theta)$, respectively. The calculated value of f is designated as f_{theor}. The f_{theor} for the stilbene system is also graphed in Fig. 15.1. It can be clearly seen that the dipole strength and oscillator strength of the transition in each system become smaller as the value of θ becomes larger.

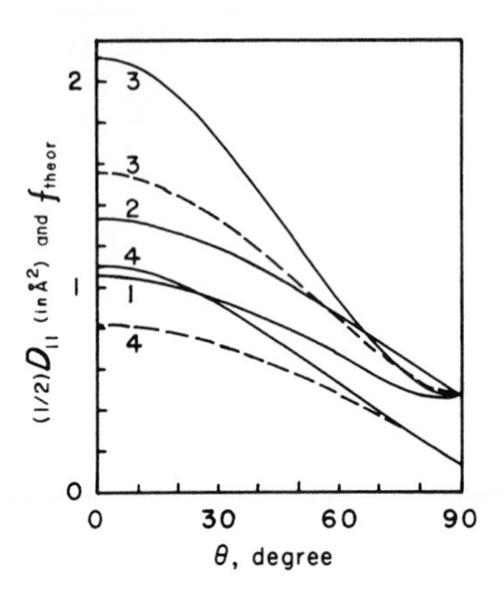

FIG. 15.1. The dipole strength, D_{11}, and theoretical oscillator strength, f_{theor}, of the A band as functions of the angle of twist of the essential single bond, θ. Solid lines are for $\frac{1}{2}D_{11}$ (in Å^2); broken lines, for f_{theor}. 1, styrene; 2, biphenyl; 3, *trans*-stilbene; 4, *cis*-stilbene.

15.1.2. COMPARISON OF THEORETICAL VALUES OF OSCILLATOR STRENGTH WITH EXPERIMENTAL VALUES

The oscillator strength of an absorption band can be estimated from the absorption curve by the use of

$$f \fallingdotseq 4.319 \times 10^{-9} \int \epsilon_\nu \, d\nu, \tag{15.2}$$

where ϵ_ν is the molar extinction coefficient at wave number ν, and the integral is the so-called integrated intensity of the band. The value of f obtained in this way is designated as f_{exp}.

The values of f_{theor} and f_{exp} for the A bands of some conjugated hydrocarbons are shown in Table 15.1, together with some relevant data. The values of θ shown in this table are those estimated in the preceding chapters from the wavelength of the absorpton maximum of the A band of each compound. The values of D_{11} and f_{theor} are the values calculated for these

TABLE 15.1

THE THEORETICAL AND EXPERIMENTAL VALUES OF THE OSCILLATOR STRENGTHS (f) OF THE A BANDS OF CONJUGATED HYDROCARBONS AND RELEVANT DATA

No.	Compound	λ_{max} mμ	$\epsilon_{max} \times 10^{-2}$	θ, deg.	D_{11} Å^2	f_{theor}	f_{exp}	f_{theor}/f_{exp}	$f_{exp} \times 10^7/\epsilon_{max}$	$\Delta\nu^a \times 10^{-1}$ cm^{-1}
1	Butadiene (in *n*-hexane)	217	209	0	2.334	1.167	0.53	2.20	254	—
2	Styrene	248	148	0	2.120	0.927	0.327	2.84	222	—
3	Biphenyl (KCl disk)	253	—	0	2.677	1.148	—	—	—	—
4	Biphenyl	247	167	23	2.503	1.100	0.411	2.68	246	254
5	Biphenyl (vapor, 170°C)	238	121	42	2.159	0.984	0.316	3.11	262	—
6	*o*-Methylbiphenyl	235	100	58	1.772	0.818	0.261	3.13	261	260
7	*trans*-Stilbene	294	280	0	4.221	1.557	0.739	2.11	264	307
8	*trans*-α,α′-Dimethylstilbene	243	123	58	1.951	0.870	0.350	2.49	285	323
9	*trans*-α,α′-Diethylstilbene	237	111	63	1.696	0.778	0.330	2.36	298	335
10	*cis*-Stilbene (in ethanol)	280	105	28	1.882	0.729	0.323	2.26	309	311
11	*cis*-α,α′-Dimethylstilbene	252	89	52	1.276	0.549	0.329	1.67	371	450
12	*cis*-α,α′-Diethylstilbene (in ethanol)	244	77	58	1.110	0.494	0.270	1.83	349	372
13	Tetraphenylethylene	309	142	35	2.566	0.902	0.393	2.30	278	266
14	Tolan	280	316	0	4.804	1.862	0.700	2.66	222	233

[a] $\Delta\nu$: the wave number interval between the absorption maximum and the point of half the maximum intensity at longer wavelength than the maximum.

θ values. All the f_{exp} values and other spectral data are for the solution in *n*-heptane at room temperature except where otherwise indicated. It is difficult to determine unambiguously the intrinsic band envelope and hence the integrated intensity of the A band as differentiated from neighboring bands. For each of compounds 4, 7,' 10, 13, and 14, the area under the absorption curve at wave numbers lower than the absorption minimum between the A band and the neighboring band lying at higher wave numbers was taken as an approximation of the integrated intensity of the A band, since with these compounds the A band is comparatively well resolved from its neighbor. With compounds 6, 8, 9, 11, and 12, the A band itself is situated at quite a high wave number, and has quite a low intensity; hence, the resolution of the band from the neighboring band is very unsatisfactory. For these compounds, twice the area under the absorption curve at wave numbers lower than the absorption maximum was taken as a rough approximation of the integrated intensity of the A band.

As mentioned already, the value of f of the first intense π–π^* band of a conjugated system calculated by the simple MO method is in general too high by a factor of about 2–3. This general tendency is due to the essential character of the method, which does not take account of interelectronic interactions and hence tends to overestimate the delocalization of the molecular orbitals. In this respect, the present calculation is not an exception. However, apart from this discrepancy in the absolute values, quite a good correspondence is found between the f_{theor} values and the f_{exp} values. Thus it may be concluded that the hypochromic as well as hypsochromic shift of the A band caused by steric hindrance to planarity of the conjugated system has been almost satisfactorily explained in terms of the change in the π-electronic states of the system associated with the change in the most probable conformation of the system.

15.2. Shape of Conjugation Bands

15.2.1. Potential Energy Curves for the Twisting Mode

It is a notable fact that the A band of *trans*-stilbene exhibits a well-resolved vibrational fine structure, even at room temperature, whereas the A bands of sterically-hindered related compounds are structureless. As mentioned in preceding chapters, a similar situation can be found with many other conjugated systems.

Now it is attempted to relate the shape of the A band to the steric hindrance to planarity of the conjugated system. With this end in view, the potential energy curves for the twisting of the essential single bonds are first examined.

In Fig. 15.2 the π-electronic energy ($E_\pi{}^G$) of the ground state (Ψ_0) of the stilbene system and that ($E_\pi{}^E$) of the excited state (Ψ_1^{11}) which arises from one-electron transition from ψ_{+1} to ψ_{-1}, calculated by the simple LCAO MO method, are plotted against θ, the curves being shown as broken lines.

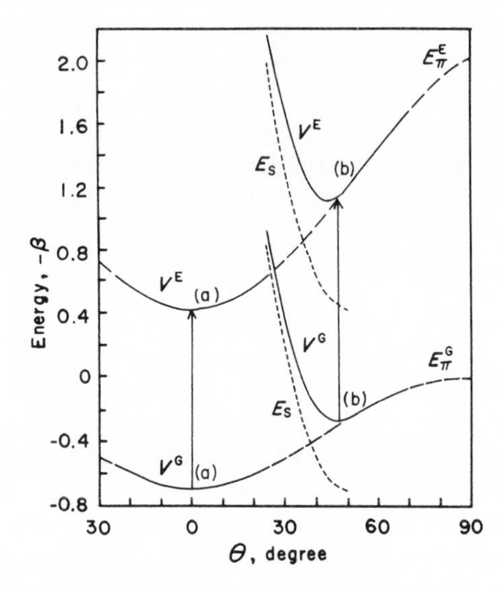

FIG. 15.2. Potential energy diagram of the stilbene system. The V curves (solid lines) represent potential energies as functions of the angle of twist of the essential single bonds, θ: (a) for the sterically unhindered compound; (b) for a sterically hindered compound. E_π, π-electronic energies (broken lines); E_S, steric repulsion energy (dotted lines).

In this figure, the energy of the ground state when θ is 90° is taken as the reference zero energy, and the unit of energy is $-\beta$. The superscripts G and E refer to the ground and the excited electronic state, respectively. The quantity $-E_\pi{}^G$ is the so-called extra-resonance energy and is approximately proportional to k^2 and hence to $\cos^2 \theta$. The difference between $E_\pi{}^E$ and $E_\pi{}^G$ is the transition energy, $\Delta E_{11} = \Delta w_{11}$, and it increases, roughly speaking, linearly with the decreasing value of k and hence of $\cos \theta$. The $E_\pi{}^E \sim \theta$ curve is, of course, steeper than the $E_\pi{}^G \sim \theta$ curve. The curves for the steric repulsion energy (E_S) as a function of θ for a hindered case are arbitrarily depicted with dotted lines. The curves for the resultant potential energy (V) as the sum of E_π and E_S for the unhindered and the hindered case are shown as solid lines.

In the unhindered case, the V-curves coincide with the E_π-curves. It follows that in this case, $\theta_{eq}^{G} = \theta_{eq}^{E} = 0°$, where θ_{eq} represents the equilibrium angle of twist, that is, θ at the minimum of the V-curve. On the other hand, in the hindered case the V of each electronic state is the sum of E_π of the state and E_S, which is considered to be identical in both the states. Since $E_\pi{}^{E}$ increases more rapidly than does $E_\pi{}^{G}$ as θ becomes larger, there must be in this case the relation $\theta_{eq}^{G} > \theta_{eq}^{E} > 0°$. The E_S-curve may be quite steep. The E_π-curves are steeper at a larger value of θ, except when θ is extremely large. Therefore, if the steric hindrance is not so slight, the V-curves in the sterically hindered case may be steeper than the corresponding ones in the unhindered case. At least, there seems to be no reason to believe that the latter should be distinctly steeper than the former.

15.2.2. Distribution of Conformations in the Ground State

The population ratio of the conformation characterized by a given value of θ in the ground electronic state is denoted by $P^{G}(\theta)$. As mentioned above, the V^{G}-curve in the sterically hindered case may be steeper than that in the unhindered case. Accordingly, the distribution curve of the population of conformations characterized by various values of θ in the ground electronic state, namely, the $P^{G}(\theta) \sim \theta$ curve, which can be considered to have its maximum approximately at θ_{eq}^{G}, may be broader in the unhindered case than in the hindered case. However, if $P^{G}(\theta)$ is plotted against $\Delta E(\theta)$ or $\nu(\theta)$, a distribution curve with quite different features will result. The conformation characterized by θ can be considered to be also characterized by $\Delta E(\theta)$ or $\nu(\theta)$. $\Delta E(\theta)$ and $\nu(\theta)$ vary roughly linearly with k and hence with $\cos\theta$. Therefore, if θ_{eq}^{G} is small, even though the distribution curve with respect to θ is comparatively broad, the distribution curve with respect to $\Delta E(\theta)$ or $\nu(\theta)$ will be quite steep. This means that, in this case, even quite a wide variation of θ will give rise to the transition in quite a narrow range of wave numbers. On the other hand, if θ_{eq}^{G} is fairly large, even if θ varies within a comparatively limited range around θ_{eq}^{G}, the variation of θ will give rise to the transition in a comparatively wide range of wave numbers. Thus, from the consideration of the distribution of conformations in the ground electronic state, it may be expected that the conjugation bands of sterically hindered compounds are broader than that of the unhindered parent compound.

By the way, since $-E_\pi{}^{G}$ is approximately proportional to $\cos^2\theta$, and since there is an approximately linear relationship between ΔE as well as ν and $\cos\theta$, $-E_\pi{}^{G}$ can be roughly related with ΔE and ν as follows:

$$-E_\pi{}^{G}(\theta) \propto [\Delta E(90°) - \Delta E(\theta)]^2 \propto [\nu(90°) - \nu(\theta)]^2. \qquad (15.3)$$

15.2.3. Intensity Distribution among Transitions of a Vibrational Progression

The total probability of the transition from a vibrational substate of the ground electronic state to an excited electronic state is distributed among transitions to vibrational substates of the excited electronic state in proportion to the square of the overlap integrals of the wavefunctions of the two vibrational substates involved (see Sections 4.2 and 5.1).

The vibrational quantum number is represented by v, and, as mentioned above, quantities associated with the ground and the excited electronic state are signified by superscripts G and E, respectively. The lowest vibrational substate of the ground electronic state, that is, the substate for which $v^{G} = 0$, is most densely populated, and the transition from this substate should make the most important contribution to the absorption band.

In general, in a substate with $v = 0$, there is a broad maximum of the probability density distribution approximately above the minimum of the potential energy curve. On the other hand, in a substate with $v \neq 0$, there is a broad maximum of the probability density distribution in the neighborhood of each of the points of intersection of the vibrational energy level and the potential energy curve, that is, the classical turning points of the vibrational motion. In a series of transitions from the substate with $v^{G} = 0$, the vibrational overlap integral has its maximum value for the transition to the v^{E} substate which has the broad maximum of its probability distribution vertically above the minimum of the potential energy curve of the ground electronic state. Therefore, the further the minima of the potential energy curves of the two electronic states are separated from each other, and the steeper the upper potential energy curve is, the higher will be the v^{E} value of the most probable transition from the substate with $v^{G} = 0$, and the greater will be the variety of the v^{E} substates, among transitions to which the total transition probability is distributed.

As mentioned already, in the unhindered case $\theta_{eq}^{G} = \theta_{eq}^{E} = 0°$. Therefore, with respect to the progression due to the twisting vibration, in this case the transition probability from the lowest vibrational substate of the ground electronic state will be concentrated mostly in the so-called 0–0 band. On the other hand, in the hindered case, $\theta_{eq}^{G} > \theta_{eq}^{E} > 0°$, and the upper potential energy curve may be steeper than that in the unhindered case. In this case, therefore, the transition probability from the lowest vibrational substate of the ground electronic state will be broadly distributed among transitions to a comparatively large number of vibrational substates of the excited electronic state.

15.2.4. Symmetry Properties of Vibrational Modes

When the electronic-vibrational coupling is taken into account, the symmetry selection rule should be considered on the basis of the symmetry of the initial and final vibronic states (see Subsections 4.3.2 and 5.1.3). Thus, the intensity-determining integral for a vibronic transition can be nonzero only if the direct product of the representations to which the initial and final vibronic state functions belong coincides with, or contains, the irreducible representation to which at least one of the coordinates belongs (see Subsection 3.4.3). The symmetry properties of a vibronic state can be represented by the direct product of the representation to which the electronic state function belongs and the representation to which the vibrational state function belongs. A change of vibrational quanta of a totally symmetric vibration does not change the symmetry of the state. Therefore, in principle, only excitations of totally symmetric vibrations can occur accompanying an allowed electronic transition. For example, the stretching vibration of the coannular bond (the 1—1′ bond) of biphenyl and that of the central ethylenic bond (the α—α′ bond) of stilbene are totally symmetric. Therefore, these vibrational modes can couple with the symmetry-allowed electronic transitions responsible for the A bands of these compounds.

Now let us consider the symmetry properties of the torsional vibrations of the essential single bonds of these compounds.

A planar model of biphenyl belongs to point group D_{2h}, and the torsional vibration of the coannular bond in this model belongs to representation A_{1u} of this group. Planar models of *trans*- and *cis*-stilbene belong to point groups $C_{2h}(z)$ and $C_{2v}(x)$, respectively (see Subsection 14.1.2). The symmetrical torsional vibrations of the essential single bonds in these compounds as illustrated in Fig. 14.3 belong to representation A_u of point group $C_{2h}(z)$ and representation A_2 of point group $C_{2v}(x)$, respectively. That is, in planar models, these vibrational modes are not totally symmetric, and hence, these cannot couple with symmetry-allowed electronic transitions such as the transitions responsible for the A bands of these compounds.

On the other hand, twisted or nonplanar models of biphenyl, *trans*- and *cis*-stilbene (see Figs. 12.2 and 14.3) belong to point groups D_2, $C_2(z)$, and $C_2(x)$, respectively, and in the twisted models the above-mentioned torsional vibrations of the essential single bonds belong to the respective totally symmetric representations (that is, A_1, A, and A) of these point groups. Therefore, excitations of these vibrations can occur accompanying symmetry-allowed electronic transitions.

Thus, we can say as follows: in the planar conformation of each conjugated system, any change of quanta of the torsional vibration of the essential single bonds is forbidden to occur accompanying the symmetry-allowed electronic transition responsible for the A band, and hence only the 0–0 subband with respect to this vibration can occur in the A band; on the other hand, in a twisted conformation, this forbiddenness is broken down owing to the lowering of the symmetry of the conjugated system, and various multiquantum excitations of the torsional vibration can occur accompanying the symmetry-allowed electronic transition.

15.2.5. Shape of the A Bands

From the above considerations it is expected that in the unhindered case the absorption will occur over a comparatively narrow range of wavelengths, and that as the steric hindrance becomes greater the absorption will occur over a wider range of wavelengths. Thus, as far as the twisting vibration about the essential single bonds is concerned, it is expected that the absorption band will be broadened by the steric hindrance.

This expected effect of the steric hindrance in broadening the absorption band may be of two kinds: (a) a broadening of the whole band system; (b) a broadening of individual absorption lines originating in the other kinds of vibrational mode, that is, a blurring of the fine structure of the band.

The half-band-width and the ratio of the integrated intensity or of the oscillator strength to the intensity of the absorption maximum may be taken as measures of the broadness of the band as a whole. Instead of the half-band-width, the wave number interval between the absorption maximum and the point of half the intensity of the maximum at longer wavelength than the maximum, which is referred to as $\Delta\nu$, is presented in Table 15.1, together with the f_{exp}/ϵ_{max} ratio for each compound. It seems that there is, in gross, a trend in conformity with the above expectations, although it is doubtful whether much significance can be attached to this apparent trend, because of the uncertainty involved in the estimation of these data. It may be safe to say that the effect of the steric hindrance of broadening the band as a whole is not clear. On the other hand, the effect of blurring the fine structure of the band and, eventually, of making the band structureless seems to be more convincing.

As was shown in Fig. 14.9, the A band of *trans*-stilbene exhibits quite a well-resolved fine structure. The separations of the peaks in the fine structure are about 1340 cm^{-1} at 20°C; these become 1480 cm^{-1} at −130°C. This fine structure is considered to originate in an ethylenic bond stretching vibration. As mentioned above, this stretching vibration is totally symmetric either in a planar or in a nonplanar conformation. In addition, the ethylenic bond undergoes the largest change in π-bond order on the $\psi_{+1} \rightarrow \psi_{-1}$ transition; hence, the minima of the potential energy curves for the ground and the excited electronic state with respect to the stretching vibration of this bond are probably at values of the bond length considerably different from each other. Therefore, it must be possible for transitions from the lowest vibrational substate of the ground electronic state to several vibrational substates

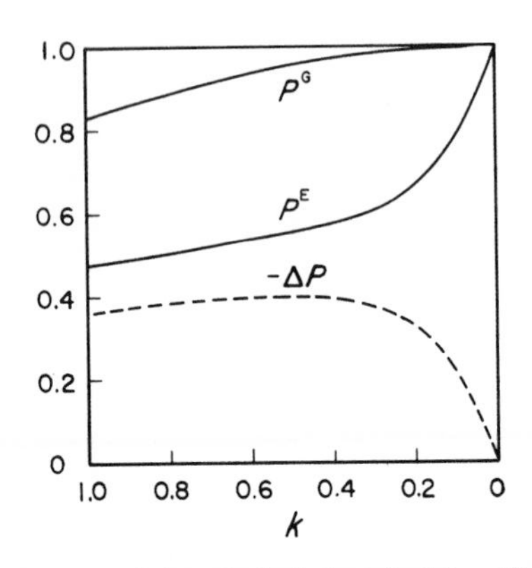

FIG. 15.3. The π-bond order of the ethylenic bond of the stilbene system as a function of the resonance parameter for the essential single bonds, k. P^G, the π-bond order in the ground state; P^E, the π-bond order in the first excited state $\Psi_1^{\prime 11}$. ΔP represents the difference $P^E - P^G$, that is, the change in the π-bond order on the one-electron transition from ψ_{+1} to ψ_{-1}.

of the excited electronic state to occur, and, since the energy separations of the vibrational substates must be comparatively large, these transitions will be revealed as the progression of the fine structure.

The values of the π-bond order of the ethylenic bond (the α—α' bond) of the stilbene system in the excited state $\Psi_1^{\prime 11}$ and in the ground state Ψ'_0, $P^E_{\alpha\alpha'}$ and $P^G_{\alpha\alpha'}$, and the difference between them, $\Delta P_{\alpha\alpha'} \equiv P^E_{\alpha\alpha'} - P^G_{\alpha\alpha'}$, are plotted against the resonance parameter for the essential single bonds, k, in Fig. 15.3. The value of $\Delta P_{\alpha\alpha'}$ is negative: the π-bond order of the ethylenic bond decreases on the excitation. Therefore, the equilibrium length of this bond in the excited state, $R^E_{\alpha\alpha',\text{eq}}$, is probably longer than that in the ground state, $R^G_{\alpha\alpha',\text{eq}}$. In the range where the value of k is larger than about 0.5, the absolute value of $\Delta P_{\alpha\alpha'}$ increases gradually as the value of k decreases. This probably means that the difference between $R^E_{\alpha\alpha',\text{eq}}$ and $R^G_{\alpha\alpha',\text{eq}}$ will increase with the increasing value of the twist angle, θ. It follows that the absorption intensity of the band will be distributed among a larger variety of vibronic transitions of the progression due to the stretching vibration in a hindered case than in the unhindered case. Perhaps this explains the foregoing apparent broadening of the band system as a whole by the steric hindrance.

As mentioned above, insofar as the twisting vibration about the essential single bonds is concerned, in the unhindered case the transition probability is considered to be concentrated in a considerably restricted wavelength region. Consequently, in this case the progression originating in the C=C stretching mode is not so blurred and will reveal itself as a comparatively well-resolved fine structure. On the other hand, in the case of hindered molecules, the transition probability will be distributed among a greater variety of vibronic transitions covering a wider wavelength region, and the possible fine structure will be thereby blurred. The same explanation will be applicable to other systems.

Chapter 16

Conjugated Dienes and Polyenes

16.1. Conjugated Dienes

16.1.1. The Simple LCAO MO Calculation on the Conjugated Diene System

The present chapter deals with the relation between the electronic absorption spectra and geometry of conjugated dienes and polyenes. The calculation on the π MO's of the conjugated diene system in the Hückel approximation was described in Section 8.2. In the present subsection, let us examine the variation of the π MO's brought about by twist of the central essential single bond, that is, the 1—1′ bond (see Fig. 8.1).

As mentioned already, the π MO's of the conjugated diene system and their energies in the simple LCAO MO approximation are given by expressions (8.15) and (8.14), respectively. According to an electron diffraction study,[1] 1,3-butadiene in the vapor state has probably almost exclusively the planar *s-trans* conformation in which the length of the essential single bond and that of each essential double bond are 1.483 $\pm$ 0.01 and 1.337 $\pm$ 0.005 Å, respectively. By assuming that the essential double bonds have a constant length of 1.34 Å, a constant value of 1.080 is assigned for the resonance parameters for these bonds, k_{12} and $k_{1'2'}$. The resonance parameter for the essential single bond, $k_{11'}$, is subsequently denoted by k, the subscript being omitted, as far as this abbreviation does not lead to confusion. Similarly, the length of this bond, $R_{11'}$, is denoted simply by R, and the dihedral angle between the two ethylenic bond planes, that is, the twist angle of the 1—1′ bond, $\theta_{11'}$, is simply denoted by θ. The value of R at $\theta = 0°$ is assumed to be 1.48 Å, and, as in the case of the biphenyl system, the relations of R and of k to θ

[1] A. Almenningen, O. Bastiansen, and M. Traetteberg, *Acta Chem. Scand.* **12,** 1221 (1958).

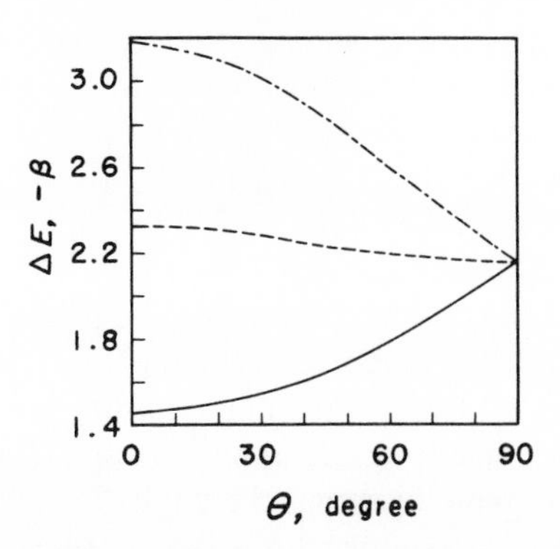

FIG. 16.1. The energies (ΔE, in units of $-\beta$) of electronic transitions in the conjugated diene system as functions of the interplanar angle between the two ethylenic bonds, θ. ———, $\psi_{+1} \rightarrow \psi_{-1}$; – – – –, $\psi_{+1} \rightarrow \psi_{-2}$, as well as $\psi_{+2} \rightarrow \psi_{-1}$; – - – - –, $\psi_{+2} \rightarrow \psi_{-2}$.

are assumed to be given by expressions (9.38) and (9.39), respectively. Part of the result of the simple LCAO MO calculation based on these assumptions is shown in Figs. 16.1 and 16.2, in which ΔE_{kr} and D_{kr} represent, respectively, the energy (in $-\beta$) and dipole strength (in Å^2) of the one-electron transition from ψ_{+k} to ψ_{-r}.

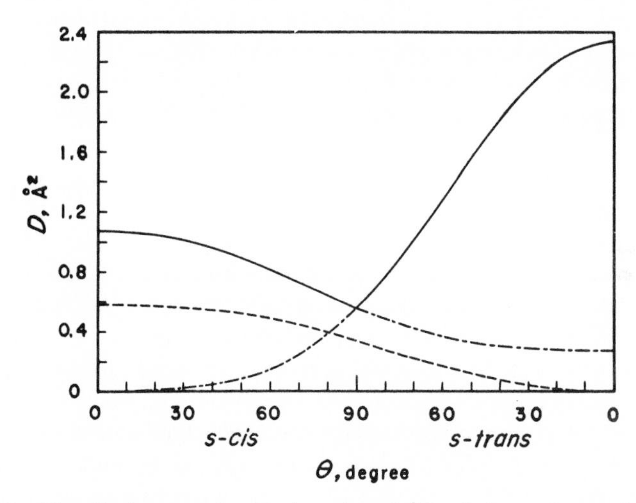

FIG. 16.2. The dipole strengths (D, in units of Å^2) of electronic transitions in the conjugated diene system as functions of the interplanar angle between the two ethylenic bonds, θ. ———, $\psi_{+1} \rightarrow \psi_{-1}$; – – – –, $\psi_{+1} \rightarrow \psi_{-2}$, as well as $\psi_{+2} \rightarrow \psi_{-1}$; – - – - –, $\psi_{+2} \rightarrow \psi_{-2}$.

The one-electron transition from ψ_{+1} to ψ_{-1} can be considered to be responsible for the first intense butadiene band (the A band) at about 217 mμ. As the twist angle θ increases, the energy of this transition, ΔE_{11}, increases, and the dipole strength of this transition, D_{11} $(= M_{11}^2)$, decreases. This means that the twist of the 1—1′ bond will exert a hypsochromic and hypochromic effect on the A band.

The value of θ is taken to be zero both for the planar *s-trans* and for the planar *s-cis* conformation. The direction of the 1—1′ bond is taken as the direction of the y axis of the rectangular coordinate system. For *s-trans* conformations the plane bisecting the angle θ is taken as the xy plane; for *s-cis* conformations the plane bisecting the angle θ is taken as the yz plane. Thus, for the planar *s-trans* conformation the coordinate system illustrated in Fig. 8.2 is assigned, and for the planar *s-cis* conformation the coordinate system obtained by interchanging the x and z axes in the system illustrated in Fig. 8.3 is assigned. The planar *s-trans* and *s-cis* conformations belong to point groups C_{2h} and C_{2v}, respectively, and nonplanar conformations belong to point group C_2: in all the conformations the principal symmetry axis is the z axis.

The transition moment M_{11} has a nonzero y component, and, in conformations other than the planar *s-cis* conformation, a nonzero x component. The magnitude of the y component is the same in *s-trans* and *s-cis* conformations having the same θ value, and decreases as the θ value increases. The magnitude of the x component is at its maximum in the planar *s-trans* conformation, and decreases rapidly as the conjugated system deviates from the planar *s-trans* conformation by rotation about the 1—1′ bond, to zero in the planar *s-cis* conformation. Consequently, the dipole strength D_{11} is invariably greater in an *s-trans* conformation than in the corresponding *s-cis* conformation.

The one-electron transitions from ψ_{+1} to ψ_{-2} and from ψ_{+2} to ψ_{-1} require the same energy and have the transition moments of the same magnitude and direction, owing to the pairing properties of the π MO's. The transition moments are in the direction of the z axis. Their magnitude is at its maximum in the planar *s-cis* conformation, and is zero in the planar *s-trans* conformation.

By the way, for planar conformations the extra-resonance energy and the π-bond order of the essential single bond are 0.328 $(-\beta)$ and 0.369, respectively. Both these quantities decrease as the twist angle θ increases, and vanish when $\theta = 90°$. On the other hand, the π-bond order of each essential double bond increases with θ, from 0.929 at 0° to 1 at 90°.

16.1.2. Application of the Pariser–Parr Method to the Conjugated Diene System

The effects of the configuration interaction in the conjugated diene system were outlined qualitatively in Subsection 10.3.2. In the present subsection the results of application of the Pariser–Parr method to this system are presented.

In the calculation, the simple LCAO MO's obtained as functions of k were used as the starting MO's, and the ground electron configuration (V_0) and the four singlet and four triplet singly-excited electron configurations (V_1^{kr}'s and T_1^{kr}'s) were taken into account. As usual, a value of -2.39 ev was assigned for the standard core resonance integral β^{C}, and the values of the core resonance integrals β_{ij}^{C} were assumed to be given by $k_{ij}\beta^{\mathrm{C}}$, in which k_{ij} is the resonance parameter used in the simple LCAO MO calculation.

The results of the calculation are summarized in Tables 16.1A, B, and C. In Table 16.1A, E represents the energy (in ev) of the state relative to the energy of the ground electron configuration. The energy level diagram for the planar *s-trans* conformation was shown already in Fig. 10.1. In Table 16.1B, D represents the dipole strength (in Å^2) of the transition from the ground state to each excited state, and letters in parentheses following the value of D indicate the directions of nonzero components of the transition moment. In Table 16.1C, ^1A, ^1B, and ^3B are the coefficients of electron configurations in the state functions [see expressions (10.17)]. Needless to say, coefficients ^1a, ^1b, and ^3b are given as the square roots of $1-(^1A)^2$, $1-(^1B)^2$, and $1-(^3B)^2$, respectively. The state functions for the conformation where $\theta = 90°$ are also shown in Table 16.1A. The symbols for these state functions have been chosen according to the principle described in Subsection 12.2.6. Thus, for example:

$$
\begin{aligned}
{}^1CT^{11'(\pm)} &= 2^{-1/2}({}^1CT^{11'} \pm {}^1CT^{1'1}); \\
{}^1LE^{11(\pm)} &= 2^{-1/2}({}^1LE^{11} \pm {}^1LE^{1'1'}),
\end{aligned}
\tag{16.1}
$$

where, for example, ${}^1CT^{11'}$ represents the singlet CT configuration arising from the one-electron transition from the bonding π orbital (ϕ_{+1}) of an ethylenic bond to the antibonding π orbital ($\phi_{-1'}$) of another ethylenic bond; ${}^1LE^{11}$ represents the singlet LE configuration arising from the one-electron transition from the bonding π orbital (ϕ_{+1}) of an ethylenic bond to the antibonding π orbital (ϕ_{-1}) of the same ethylenic bond.

TABLE 16.1

RESULTS OF THE CALCULATION ON THE CONJUGATED DIENE SYSTEM BY THE PARISER–PARR METHOD[a]

A. Energies (E, in electron volts units)

State	*s-trans* $\theta = 0°$ $k = 0.858$ C_{2h}		$\theta = 52.6°$ $k = 0.5$ C_2		$\theta = 90°$ $k = 0$ C_2			*s-cis* $\theta = 52.6°$ $k = 0.5$ C_2		$\theta = 0°$ $k = 0.858$ C_{2v}	
$^{+}[V]_1^{22}$	B_u	9.20*	B	8.11*	B	7.11	$^1CT^{11'(-)}$	B	7.94*	B_2	8.96*
$^{-}[V]_1^{12}$	A_g	7.19	A	7.14	A	7.11	$^1CT^{11'(+)}$	A	7.00	A_1	7.02
$^{+}[V]_1^{12}$	A_g	6.83	A	6.86*	A	6.99*	$^1LE^{11(+)}$	A	7.18*	A_1	7.34*
$^{+}[V]_1^{11}$	B_u	5.58*	B	5.96*	B	6.47*	$^1LE^{11(-)}$	B	5.66*	B_2	5.09*
$^{-}[V]_0$	A_g	−0.06	A	−0.02	A	0	G	A	−0.02	A_1	−0.04
$^{+}[T]_1^{22}$	B_u	8.19	B	7.52	B	7.11	$^3CT^{11'(-)}$	B	7.37	B_2	7.98
$^{-}[T]_1^{12}$	A_g	7.13	A	7.12	A	7.11	$^3CT^{11'(+)}$	A	6.98	A_1	6.98
$^{+}[T]_1^{12}$	A_g	4.07	A	3.78	A	3.60	$^3LE^{11(+)}$	A	3.78	A_1	4.14
$^{+}[T]_1^{11}$	B_u	2.57	B	3.19	B	3.60	$^3LE^{11(-)}$	B	3.19	B_2	2.63

B. Dipole strengths (D, in units of Å^2) of transitions from the ground state

State	*s-trans* $\theta = 0°$	$\theta = 52.6°$	$\theta = 90°$	*s-cis* $\theta = 52.6°$	$\theta = 0°$
$^{+}[V]_1^{22}$	0.182(x, y)	0.167(x, y)	0	0.038(x, y)	0.059(y)
$^{-}[V]_1^{12}$	0	0	0	0	0
$^{+}[V]_1^{12}$	0	0.251(z)	0.673(z)	1.024(z)	1.156(z)
$^{+}[V]_1^{11}$	2.406(x, y)	1.655(x, y)	1.122(x, y)	0.926(x, y)	1.013(y)

C. Coefficients of electron configurations in state functions

	s-trans $\theta = 0°$	$\theta = 52.6°$	$\theta = 90°$	*s-cis* $\theta = 52.6°$	$\theta = 0°$
^1A	0.996	0.999	1	0.998	0.997
^1B	0.991	0.986	$2^{-1/2}$	0.982	0.988
^3B	0.959	0.904	$2^{-1/2}$	0.905	0.961

[a] The asterisk indicates that the transition from the ground state to the excited state is allowed.

Since the calculation is based on many approximations and assumptions, too much reliance cannot be placed on the individual numerical values of the results. However, we can say that most of conclusions derived from the simple LCAO MO calculation have been supported by the results of the advanced MO calculation. For example, the results indicate that the absorption band due to the lowest energy S–S excitation will shift to shorter wavelength and decrease in intensity as the twist angle of the essential single bond increases, and that the intensity of the band will be higher in an *s-trans* conformation than in the corresponding *s-cis* conformation.

There are some new features that have been revealed by the advanced MO calculation. For example, while the results of the simple MO calculation indicate that the energy of the first excitation will be the same in the planar *s-trans* and *s-cis* conformations, the results of the advanced MO calculation indicate that the energy of the first S–S excitation will be lower in the planar *s-cis* conformation than in the planar *s-trans* conformation. It is actually known that the A bands of some compounds having the conjugated diene structure in the *s-cis* conformation are at considerably longer wavelength than the A bands of compounds having the conjugated diene structure in the *s-trans* conformation.

16.1.3. Spectra of *s-Trans* Conjugated Dienes

The position and intensity of the A band of the conjugated diene system have so far been related theoretically to the conformation of the system. However, the conformation of the conjugated system is not the only factor affecting the spectrum, but is only an important one of factors. For example, alkyl substitution has an appreciable electronic effect on the position and intensity of the A band. When one or two of the ethylenic bonds of the conjugated diene system constitutes a part of a ring system, the strain caused by the ring possibly affects the spectrum of the diene. It is to be noted that most dienes are comparatively labile, having a tendency to oxidize, polymerize, or isomerize. It is generally difficult to obtain the pure specimens, and the data on the spectra of dienes found in the literature are considered to contain inevitably uncertainty to a more or less extent, especially in the absorption intensity. One must therefore be careful in comparing the spectra of dienes.

The first intense band (the A band) of 1,3-butadiene in alcohol and saturated hydrocarbon solvents has a somewhat blurred vibrational fine structure ($\lambda_{max} = 217$ mμ, $\epsilon_{max} = 21 \times 10^3, f = 0.53$). Data on the A bands of this compound and of its methylated derivatives were already shown in

Tables 10.8 and 6.4. Table 10.8 shows that introduction of a methyl substituent into a sterically unhindered position (that is, the 1- or 4-position) of 1,3-butadiene causes a bathochromic shift of about 5 mμ and a small increase in intensity of the A band.

In 2-methyl-1,3-butadiene (isoprene) and 2,3-dimethyl-1,3-butadiene there should be slight steric interference between the methyl groups and the hydrogen atoms linked to the diene system. Although the effect of this steric interference is not clearly detected in the data listed in Table 10.8, it appears to reveal itself in the data listed in Table 6.4. Thus, the oscillator strength of the A band of isoprene is smaller than that of 1,3-butadiene, and that of *cis*-piperylene (*cis*-1,3-pentadiene) is smaller than that of the *trans* isomer. The smaller intensities of the A bands of isoprene and *cis*-piperylene as compared with the A bands of butadiene and *trans*-piperylene should be attributed to the steric effect.

When an ethylenic bond of the conjugated diene system forms part of a 5- or 6-membered alicyclic system, either exocyclic or endocyclic, the diene shows the A band at considerably longer wavelength than does the corresponding open-chain conjugated diene. Data on the A bands of some such compounds are shown in Table 16.2. Most conjugated dienes in which one of the two ethylenic bonds is exocyclic to a 6-membered ring show the A band at 236 $\pm$ 0.5 mμ. The A bands of conjugated dienes in which both the two ethylenic bonds are exocyclic to 6-membered rings are generally at 245–248 mμ. When one of the ethylenic bonds forms part of a 6-membered alicyclic system, while the other is in an open-chain system, the conjugated dienes show the A band at about 235 mμ.

The fact that the cyclic dienes show the A band at longer wavelengths than the A bands of the corresponding acyclic dienes may be explained in terms of the steric strain caused in the conjugated diene system of the former compounds by the ring formation. While a double bond exocyclic to a 6-membered ring is considered to have a tendency to distort the strainless staggered conformation of the chair form of the cyclohexane ring, the double bond is considered to be imposed a strain by the ring to distort the bond angle from the normal value of the sp^2 hybridization to a value near to the value of the sp^3 hybridization. A double bond endocyclic to a 6-membered ring is also considered to be imposed a strain by the ring. Actually, it is known that a double bond exocyclic to a 6-membered ring is less stable by about 3.5 kcal./mole than the corresponding endocyclic double bond, and, furthermore, it is known that the ionization potential of cyclopentene (about 9.0 ev) is appreciably smaller than that of the open-chain analog (about 9.5 ev) and that the ionization potential of cyclohexene is also appreciably smaller

TABLE 16.2

THE FIRST BANDS OF CYCLIC CONJUGATED DIENES (*s-TRANS*)

No.	Compound	Solvent	λ_{max}, mμ	ϵ_{max}
1	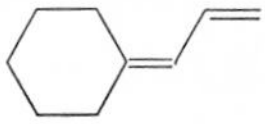	Ethanol	236.5	>7700[a]
2	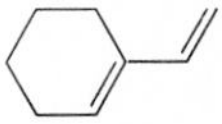	Ethanol	230	8500[a]
3	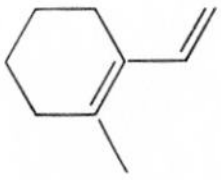	Ethanol	233	7200[a]
4	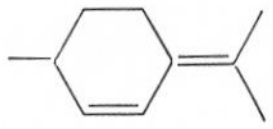 (2,4(8)-*p*-menthadiene)	Isooctane	244	15,800[b]
5	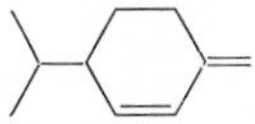 (β-phellandrene)	Ethanol	232	19,300[c]
6	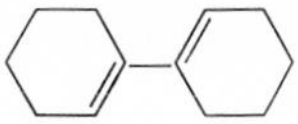	Hexane	238	18,200[d]

[a] H. Booker, L. K. Evans, and A. E. Gillam, *J. Chem. Soc.* **1940,** 1453.
[b] R. T. O'Connor and L. A. Goldblatt, *Anal. Chem.* **26,** 1726 (1954).
[c] J. B. Davenport, M. D. Sutherland, and T. F. West, *J. Applied Chem.* (*London*) **1,** 527 (1951).
[d] G. Laber, *Ann. Chem. Liebigs* **588,** 79 (1954).

than that of the open-chain analog. The mechanism by which such a strain or distortion affects the π-electronic states of the conjugated system is not clear. However, it seems to be reasonable to consider that such a strain or distortion will destabilize the ground electronic state by a greater amount than it does the excited electronic state.

16.1.4. SPECTRA OF *S-CIS* CONJUGATED DIENES

When both the two ethylenic bonds of a conjugated diene are in one ring, the compound is called a homoannular conjugated diene. In homoannular conjugated dienes the conjugated diene system assumes necessarily an *s-cis* conformation. Data on the A bands of 1,3-cyclohexadiene and related homoannular conjugated dienes are shown in Table 16.3.

TABLE 16.3

THE FIRST BANDS OF HOMOANNULAR CONJUGATED DIENES (*S-CIS*)

No.	Compound	Solvent	λ_{max}, mμ	ϵ_{max}
1		Hexane	256.5	8000[a]
2	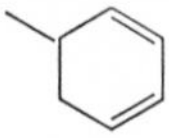	Ethanol	260	4910[a]
3	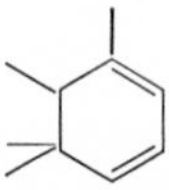	Hexane	263	4600[a]
	(α-pyronene)	Isooctane	263	5800[b]
4	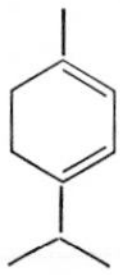 (α-terpinene)	Isooctane	265	6400[b]
5	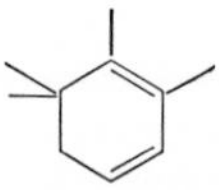 (β-pyronene)	Isooctane	264	5700[b]
6	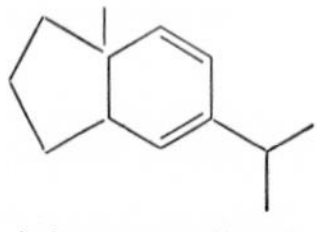 (picrotoxadiene)	Ethanol	258	4000[a]
7	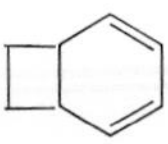	Cyclohexane	274	3340[a]
8	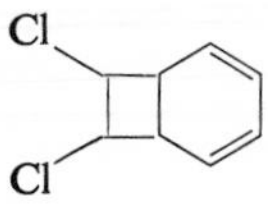 Cl, Cl	Cyclohexane	270	5640[a]
9	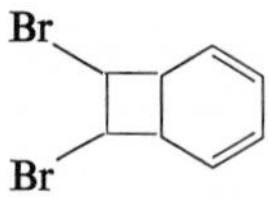 Br, Br	Cyclohexane	270	3640[a]

[a] J. J. Wren, *J. Chem. Soc.* **1956,** 2208.

[b] R. T. O'Connor and L. A. Goldblatt, *Anal. Chem.* **26,** 1726 (1954).

The A bands of these homoannular *s-cis* conjugated dienes are at markedly longer wavelengths and of markedly lower intensities than the A bands of *s-trans* conjugated dienes. The relatively low intensity of the A bands of *s-cis* conjugated dienes is in agreement with the theoretical prediction made from the MO calculation. The fact that the A bands of homoannular *s-cis* dienes are at much longer wavelengths than those of *s-trans* dienes may be due partly to the effect of the electronic repulsion, as revealed by the advanced MO calculation, and partly to the effect of the strain caused in the conjugated system by the ring.

The bathochromic effect of alkyl substitution at the 5- and 6-positions of 1,3-cyclohexadiene is normally small. However, introduction of a dimethylene bridge into the 5- and 6-positions causes a large bathochromic shift of 10 mμ or more. This effect may be attributed to the additional strain in the conjugated diene system brought about by the cyclobutane ring.

The bathochromic effect of the strain appears to be much more marked in polycyclic systems. In steroid dienes, compounds having the two ethylenic bonds of the conjugated diene system in one ring, especially in ring B, show the A band at about 280 mμ. For example, the λ_{max} and ϵ_{max} of the A band of 5,7-cholestadiene are 280 mμ and 11,100, respectively.[2]

A series of 1,3-cycloalkadienes (I) and a series of 1,2-dimethylenecycloalkanes (II) are two interesting series of compound which indicate clearly the relation between the conjugated diene absorption and the conformation of the conjugated system.

$(CH_2)_{n-4}$ $(CH_2)_{n-4}$

I II

Data on the A bands of a series of 1,3-cycloalkadienes (I) are shown in Table 16.4. The A bands of all these compounds have relatively low intensity characteristic of the *s-cis* conjugated diene system. It is seen that the position of the absorption maximum is greatly affected by the size of the ring, the smallest value of λ_{max} being reached at the C_9 compound. In the C_5 and the C_6 compound the conjugated diene system is probably maintained in the planar or nearly planar conformation, and, as mentioned already, especially in the C_6 compound there will be a considerable strain. The large values of λ_{max} in these compounds may be attributed to the forced planar or nearly

[2] L. Dorfman, *Chem. Rev.* **53**, 47 (1953).

TABLE 16.4

THE FIRST BANDS OF 1,3-CYCLOALKADIENES (I)[a]

n	Solvent	λ_{max}, mμ	ϵ_{max}
5	Hexane	238.5	3400
6	Hexane	256.5	8000
7	Isooctane	248	7500
8	Hexane	228	5600
9	Ethanol	219.5	2500
10	Ethanol	223	5000
11	Ethanol	225	6000
13	—	232	—

[a] A. E. Gillam and E. S. Stern, "An Introduction to Electronic Absorption Spectroscopy in Organic Chemistry," 2nd ed., p. 96. Arnold, London, 1957.

planar *s-cis* conformation of the conjugated system and to the strain in the conjugated system. As the size of the ring becomes larger, the conjugated diene system will deviate to a larger extent from the planar conformation by twist of the essential single bond, and at the same time the strain will be relieved, with a consequent hypsochromic shift of the A band. The deviation of the conformation of the conjugated diene system from the planarity will

TABLE 16.5

THE FIRST BANDS OF 1,2-DIMETHYLENECYCLOALKANES (II) IN HEXANE[a]

n	λ_{max}, mμ	ϵ_{max}
4[b]	255*	6460
	246	10,230
	237	9770
5[c]	248	10,470
6[d]	220	10,050
6[c]	222	4000–6600
		(log ϵ_{max} = 3.6–3.82)

[a] The asterisk on wavelength values denotes an inflection.

[b] J. J. Wren, *J. Chem. Soc.* **1956,** 2208.

[c] T. A. Blomquist and D. T. Longone, *J. Am. Chem. Soc.* **79,** 3916 (1957).

[d] W. J. Bailey and H. R. Golden, *J. Am. Chem. Soc.* **75,** 4780 (1953).

reach its maximum at the C_9 compound, and the conformation will get nearer again to the planarity as the size of the ring increases beyond the 9-membered ring.

Data on the A bands of a series of 1,2-dimethylenecycloalkanes (II) are shown in Table 16.5. 1,2-Dimethylenecyclobutane (II, $n = 4$) is, of course, highly strained, and 1,2-dimethylenecyclopentane (II, $n = 5$) contains probably some strain. In these compounds the conjugated diene system assumes probably nearly planar *s-cis* conformations. The nearly planar *s-cis* conformation and the strain must be responsible for the large values of λ_{max} of these compounds. The next higher homolog, 1,2-dimethylenecyclohexane (II, $n = 6$), shows the A band at markedly shorter wavelength. In this compound the two ethylenic bonds of the conjugated diene system are certainly not coplanar. Introduction of methyl substituents to the positions adjacent to the conjugated diene system of this compound causes an even further restriction to the planarity: 3,6-dimethyl-1,2-dimethylenecyclohexane does not exhibit the A band in the region longer than 220 mμ.[3]

16.2. Conjugated Polyenes

16.2.1. Linear Conjugated Polyenes

The longest-wavelength S–S band (the A band) of a linear conjugated polyene, which can be considered to be due to the one-electron transition from ψ_{+1} to ψ_{-1}, shifts to longer wavelength and increases in intensity as the length of the conjugated polyene system increases. This fact can be well explained on the basis of the simple LCAO MO calculation (see Subsection 10.4.2). The wavelength of the absorption maximum of the A band does not increase to infinity as predicted by the Hückel theory, but appears to approach an asymptotic value. As mentioned already in Subsection 10.4.2, this apparent convergence of the wavelength λ_{max} can be well explained by allowing for alternation of bond length of essential single and essential double bonds. A similar apparent convergence of λ_{max} of the A band to a limit value is also observed in other linear conjugated systems such as α,ω-diphenylpolyene and *p*-polyphenyl.

Introduction of an additional ethylenic bond to an *all-trans* C_{40} carotenoid brings about a bathochromic shift of the A band by approximately 20–22 mμ when the additional double bond is situated in the open chain, and by

[3] W. J. Bailey and R. L. Hudson, *J. Am. Chem. Soc.* **78,** 2806 (1956).

approximately 9–11 mμ when it is situated in the alicyclic part. The relatively small magnitude of the shift in partly alicyclic systems may be ascribed to the possible steric hindrance to coplanarity of the ethylenic bond in the alicyclic part and the linear chain polyene system.

The A bands of *all-trans* polyenes and α,ω-diphenylpolyenes have in general considerably well-resolved vibrational fine structure. When a double bond of a polyene or a double bond other than the terminal double bonds of the open chain in an α,ω-diphenylpolyene is rearranged from the *trans* to the *cis* configuration, this *trans-cis* rearrangement brings about a distinct decrease in intensity of the A band and an appearance of the so-called *cis* band (see Subsection 10.4.3), but it does not significantly affect the shape and position of the A band: it blurs only slightly the vibrational structure of the A band, and shifts the band only slightly to shorter wavelength (by approximately 3–4 mμ in C_{40} carotenoids). This fact suggests that the steric hindrance to the planarity of the conjugated system in such *cis* compounds is not so large.

When a terminal double bond of the open chain of an α,ω-diphenylpolyene undergoes *trans-cis* rearrangement, the situation is quite different. In this case the A band is shifted to considerably shorter wavelength and loses almost completely its vibrational fine structure.[4] In the resulting diphenylpolyene having a terminal (α—β) double bond in the *cis* configuration steric interference must exist between the hydrogen atom at the γ-position and an *ortho* hydrogen atom of the phenyl group at the α-position. The marked hypsochromic shift and loss of vibrational structure of the A band must be due to the steric hindrance to coplanarity of the benzene ring and the polyene system (see Subsection 14.5.2). Similar phenomena have been observed with some symmetrical enalazines such as cinnamalazines, $(C_6H_5—CH{=}CH—CH{=}N—)_2$, and phenylpentadienalazines, $(C_6H_5—CH{=}CH—CH{=}CH—CH{=}N—)_2$.[5]

16.2.2 Cyclooctatetraene

1,3,5,7-Cyclooctatetraene is known to exist in a nonplanar "tub" or "boat" form with alternating single and double bonds.[6] In this conformation the planes of adjacent double bonds are almost orthogonal and hence there should be little conjugative interaction between the bonds. Actually, this

[4] E. F. Magoon and L. Zechmeister, *J. Am. Chem. Soc.* **77**, 5642 (1955).

[5] J. Dale and L. Zechmeister, *J. Am. Chem. Soc.* **75**, 2379 (1953).

[6] W. B. Person, G. C. Pimentel, and K. S. Pitzer, *J. Am. Chem. Soc.* **74**, 3437 (1952); O. Bastiansen, L. Hedberg, and K. Hedberg, *J. Chem. Phys.* **27**, 1311 (1957).

compound behaves as a highly unsaturated hydrocarbon having localized ethylenic double bonds and has no aromatic character. However, contrary to the expectation from the almost complete localization of the double bonds in this nonplanar structure, the electronic absorption spectrum of this compound does not resemble the spectrum of ethylene or other compounds having an isolated double bond. Thus, this compound shows a relatively weak, broad absorption band at about 280 mμ (in the vapor state, $\lambda_{max} = 280$ mμ, $\log \epsilon_{max} = 2.45$; in cyclohexane solution, $\lambda_{max} = 283$ mμ, $\log \epsilon_{max} = 2.41$).[7] This band is, on one hand, at much longer wavelength and of much lower intensity than the ethylene π–π^* band, and on the other hand, it is at much shorter wavelength and of much lower intensity than the A band of the open-chain conjugated tetraene ($\lambda_{max} = 304$ mμ). The appearance of this band might be taken as indicating the existence of a weak conjugative interaction between adjacent double bonds. However, it may be more plausible to explain the appearance of this band in terms of the conjugative nonneighbor interaction (see Chapter 24) or in terms of the exciton delocalization (see Chapter 20) between alternating double bonds, that is, between the 1—2 and the 5—6 bond and between the 3—4 and the 7—8 bond.

[7] S. Miyakawa, I. Tanaka, and T. Uemura, *Bull. Chem. Soc. Japan* **24,** 136 (1951).

Chapter 17

Polymethine Dyes

17.1. Spectra of Odd-Membered Conjugated Systems

Linear conjugated systems are divided into two types, even-membered (type 1) and odd-membered (type 2). Conjugated polyenes and *p*-polyphenyls are typical examples of type 1, and polymethine dyes, which are dealt with in this chapter, belong to type 2.

The two types show considerably different features of light absorption. The first intense band (the A band) of a linear conjugated compound of either type shifts to longer wavelength as the length of the conjugated system increases. As mentioned already, in a series of compounds of type 1, the magnitude of the bathochromic shift of the A band per one additional mesomeric unit (for example, ethylenic bond or benzene nucleus) decreases as the length of the conjugated system increases, the wavelength of the maximum of the band, λ_{max}, being roughly proportional to the square root of the number of the conjugated π centers or of the mesomeric units in the system, and the wavelength λ_{max} appears to converge to a limit when the length of the system becomes infinite. In contrast, in a series of symmetrical conjugated compounds of type 2, for example, in a series of symmetrical cyanine dyes $(CH_3)_2N—(CH=CH)_n—CH=N^+(CH_3)_2$ (I) and in a series of oxopolyenolate ions $O=CH—(CH=CH)_n—O^-$ (II), the wavelength of the maximum of the first intense band increases by an almost constant amount as much as about 100 mμ for each additional ethylenic bond, and the successive increases appear to be nonconvergent. In both types of series the intensity of the first intense band is approximately proportional to the chain length.[1]

Unsymmetrical odd-membered conjugated compounds such as merocyanines $(CH_3)_2N—(CH=CH)_n—CH=O$ (III) are of an intermediate type

[1] L. G. S. Brooker, *Rev. Modern Phys.* **14**, 275 (1942).

in the light absorption properties, being rather similar to compounds of type 1. Thus, in a series of unsymmetrical odd-membered conjugated compounds, the magnitude of the bathochromic shift of the first intense band per one additional ethylenic bond is in general smaller than in a series of symmetrical odd-membered conjugated compounds, and the wavelength λ_{max} appears to converge. Data on the first intense bands in typical series of odd-membered conjugated systems, I, II, and III, are shown in Table 17.1.

TABLE 17.1

THE FIRST INTENSE BANDS OF TYPICAL ODD-MEMBERED CONJUGATED SYSTEMS[a]

n	System I		System II	System III
	λ_{max}, mμ	$\epsilon_{max} \times 10^{-2}$	λ_{max}, mμ	λ_{max}, mμ
0	224	145	—	—
1	312.5	645	267.5	283
2	416	1195	362.5	361.5
3	519	2070	455	421.5
4	625	2950	547.5	462.5
5	734.5	3530	(644)	491.5
6	848	(2200)	—	512.5

[a] The data were taken from Reference 5.

In conjugated polyenes there is alternation of essential double and essential single bonds. In *p*-polyphenyls the coannular bonds are essential single bonds, and are different in nature from the bonds in the benzene rings. As mentioned in Subsection 10.4.2, the apparent convergence of the wavelength λ_{max} in these series of compound can be explained by allowing for this bond alternation. In contrast, in symmetrical odd-membered conjugated systems, there is no reason to suppose that the conjugated chain should show bond alternation. Thus, for example, a symmetrical cyanine molecule I and an oxopolyenolate ion II can be represented as resonance hybrids of two equivalent structures as follows:

$$\underset{\text{I}a}{(CH_3)_2\ddot{N}{-}(CH{=}CH)_n{-}CH{=}N^+(CH_3)_2} \leftrightarrow \underset{\text{I}b}{(CH_3)_2N^+{=}CH{-}(CH{=}CH)_n{-}\ddot{N}(CH_3)_2}$$

$$\underset{\text{II}a}{\ddot{O}^-{-}(CH{=}CH)_n{-}CH{=}O} \leftrightarrow \underset{\text{II}b}{O{=}CH{-}(CH{=}CH)_n{-}\ddot{O}^-}$$

Therefore, the bonds in these molecules are neither essentially single nor essentially double. These symmetrical odd-membered conjugated compounds are isoconjugate with odd-membered polymethine anions

$CH_2^-—(CH{=}CH)_{n+1}—H$: the conjugated systems $[C—(C{=}C)_n]^-$ can be considered as prototypes of odd-membered conjugated dyes. According to the result of the calculation in the Hückel approximation on these conjugated systems shown in Table 10.6, the π-bond order varies from bond to bond even in these conjugated systems, but the variation of the π-bond order is not so large as in the conjugated polyene systems. The fact that the wavelength λ_{max} in a series of symmetrical odd-membered conjugated compounds is approximately proportional to the number of the conjugated atoms and is apparently nonconvergent is in agreement with the prediction from the Hückel theory based on the assumption that there is no bond alternation [see Eq. (10.25)].

The merocyanines III are typical examples of unsymmetrical odd-membered polymethine dyes. A merocyanine molecule may be represented as a resonance hybrid of the following two extreme structures:

$$\underset{\text{III}a}{(CH_3)_2\ddot{N}—(CH{=}CH)_n—CH{=}O} \leftrightarrow \underset{\text{III}b}{(CH_3)_2N^+{=}CH—(CH{=}CH)_n—\ddot{O}^-}$$

In this case the two extreme structures are not equivalent. At least in nonpolar solvents, the nonionic structure *a* will be lower in energy than the zwitter-ion structure *b*, so that the former contributes largely to the ground state and the latter predominates in an excited electronic state. Therefore, in such a molecule there will be distinct, though probably less distinct than in a conjugated polyene, bond alternation. Both the position and intensity of the first intense band of merocyanine dyes are sensitive to changes in the polarity of the solvent.[2,3] This is what will be expected if the ground and excited electronic states have very different electron distributions.

The location of the first intense band of an unsymmetrical odd-membered conjugated system of the merocyanine type, $A—(CH{=}CH)_n—CH{=}B$, depends not only on the length of the conjugated system and the polarity of the solvent but also on the electronegativities of the terminal groups A and B.[4] Thus, the first intense band is displaced to longer wavelength as the electron-attracting power of the terminal group B increases, in such an order as $=CH_2 < =NH < =O$. When the groups $=NH$ and $=O$ are replaced

[2] L. G. S. Brooker, G. H. Keyes, R. H. Sprague, R. H. VanDyke, E. VanLare, G. VanZandt, F. L. White, H. W. J. Cressman, and S. G. Dent, Jr., *J. Am. Chem. Soc.* **73,** 5332 (1951); L. G. S. Brooker and G. H. Keyes, *J. Am. Chem. Soc.* **73,** 5356 (1951); A. L. LeRosen and C. E. Reid, *J. Chem. Phys.* **20,** 233 (1952).

[3] E. G. McRae, *Spectrochim. Acta* **12,** 192 (1958).

[4] A. E. Gillam and E. S. Stern, "An Introduction to Electronic Absorption Spectroscopy in Organic Chemistry," 2nd ed., pp. 128–130. Arnold, London, 1957.

by the more strongly electron-attracting groups $=NH_2^+$ and $=OH^+$, respectively, the band is displaced to further longer wavelength. The band is also displaced to longer wavelength as the electron-repelling property of the terminal group A increases, in such an order as $H < OCH_3 < N(CH_3)_2$. As the electron-attracting property of B and the electron-repelling property of A increase, the zwitter-ionic resonance structure, $A^+{=}CH{-}(CH{=}CH)_n{-}B^-$, will be stabilized, and hence the excited electronic state will be stabilized, with a consequent bathochromic shift of the band.

As mentioned already, the intensity of the first intense band in either the conjugated polyene series or the symmetrical odd-membered polymethine series increases as the length of the conjugated system increases, both the oscillator strength and the maximum molar extinction coefficient being approximately proportional to the number of the π centers in the system. This implies that the half-band-width is constant or linearly varies in a series.

The first intense bands of symmetrical cyanines (for example, I) and those of conjugated polyenes have markedly different shapes. The bands of the cyanines are structureless and have a half-band-width of about 1000 cm^{-1},[5] while the bands of conjugated polyenes consist of a progression with three to five members in the upper state carbon–carbon stretching vibration, and have a comparatively large half-band-width of about 5000 cm^{-1}.[6] In conjugated polyenes, which have distinct bond alternation, the essential single and double bonds assume, respectively, a larger and a smaller double-bond character on the electronic excitation, and hence the equilibrium nuclear configuration in the excited state must considerably differ from that in the ground state. In this case the ground state equilibrium nuclear configuration corresponds to the turning points of higher vibrational levels of the upper state, so that a strong vibrational progression appears in the electronic absorption band (see Chapter 5). On the other hand, in the cyanines, the bond alternation is not distinct, and changes in the π-bond orders of the respective bonds and changes in the equilibrium nuclear configuration on the electronic excitation are expected to be comparatively small. In this case the absorption will occur predominantly at nearly the 0–0 band frequency, giving a band of small half-width.[7]

The intensity of the first intense band of merocyanines also increases as the length of the conjugated system increases. However, the maximum molar

[5] S. S. Malhotra and M. C. Whiting, *J. Chem. Soc.* **1960,** 3812.

[6] P. Nayler and M. C. Whiting, *J. Chem. Soc.* **1955,** 3037; F. Bohlmann and H. J. Mannhardt, *Chem. Ber.* **89,** 1307 (1956).

[7] J. R. Platt, *J. Chem. Phys.* **25,** 80 (1956).

extinction coefficient shows no proportionality to the number of the π centers in the system.

The maximum molar extinction coefficient and band-width of the first intense band of a merocyanine markedly depend on the polarity of the solvent, although the oscillator strength is approximately constant.[3] In nonpolar solvents, in which the nonpolar resonance structure, III*a*, contributes largely to the ground state and the zwitter-ionic resonance structure, III*b*, to the excited state, the conjugated system can be said to be similar to the conjugated polyene system, and for the same reason as in the case of the conjugated polyenes the first intense band has a comparatively large band width. The zwitter-ionic resonance structure is increasingly stabilized as the polarity of the solvent increases. In a suitable solvent in which the nonpolar and the zwitter-ionic structure have an equal energy, the two structures contribute equally to the ground and the excited state. Under such a condition, the bond alternation in the conjugated system is nearly lost, and hence, similarly to the case of the symmetrical cyanines, the intensity of the electronic band is concentrated on comparatively small members of the upper state vibrational progression, resulting in a high maximum molar extinction coefficient and a narrow band width. At the same time, under such a condition, the energy separation between the excited and the ground electronic state becomes minimum, and hence the wavelength λ_{max} reaches its maximum.[3]

17.2. Steric Effects in Spectra of Symmetrical Cyanines

17.2.1. Cyanines of Pyrrole Series

Brooker and his co-workers have studied the spectra of many cyanines, and the results are summarized in a review.[8] The data on the first intense bands of cyanines of pyrrole series taken from the review are shown in Table 17.2.

As mentioned in preceding chapters, when an even alternant linear conjugated system is forced out of planarity, the intensity of the first intense band (the A band) falls and the wavelength of the band shows a hypsochromic shift. The decrease in intensity seems to be quite general in other types of conjugated system, but the direction of the wavelength shift is not. In conjugated systems of the cyanine type, cases are known where the nonplanarity of the geometry leads to a bathochromic shift of the first intense band. Brunings and Corwin[9] first observed this effect in the case of dyes IV*a* (IV, R = H) and IV*b* (IV, R = CH_3). In the dye IV*a* the two pyrrole rings are

[8] L. G. S. Brooker, F. L. White, R. H. Sprague, S. G. Dent, Jr., and G. VanZandt, *Chem. Rev.* **41,** 325 (1947).

[9] K. J. Brunings and A. H. Corwin, *J. Am. Chem. Soc.* **64,** 593 (1942).

probably coplanar or nearly coplanar. The *o*-methyl groups make it impossible for the dye IV*b* to be planar. The first intense band (the A band) of

TABLE 17.2

THE FIRST BANDS OF CYANINES OF PYRROLE SERIES IN CHLOROFORM

Compound	λ_{max}, mμ	$\epsilon_{max} \times 10^{-4}$	$\Delta\lambda_{max}$, mμ	$\Delta\epsilon_{max} \times 10^{-4}$
IV*a* (IV, R = H)	473	13.5	—	—
IV*b* (IV, R = CH_3)	510	5.7	+37	−7.8
V*a* (V, R = H)	446	3.5	—	—
V*b* (V, R = CH_3)	479	1.25	+33	−2.25
VI*a* (VI, R = H)	595	17	—	—
VI*b* (VI, R = CH_3)	581	17	−14	0
VII*a* (VII, R = H)	536	6.3	—	—
VII*b* (VII, R = CH_3)	534	7.6	−2	+1.3

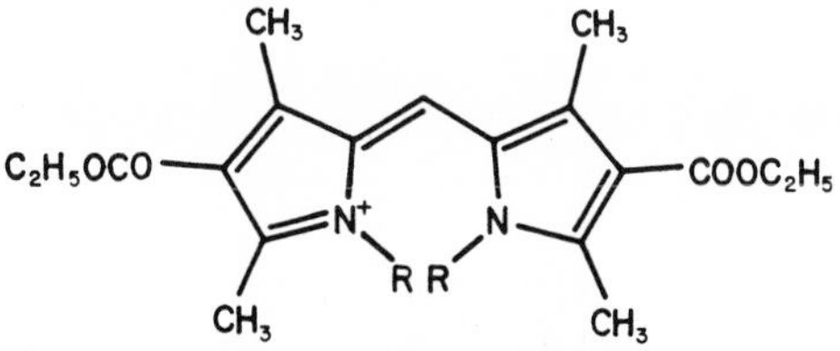

IV

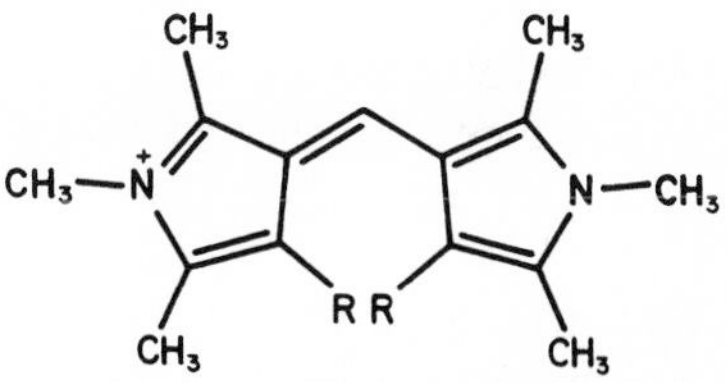

V

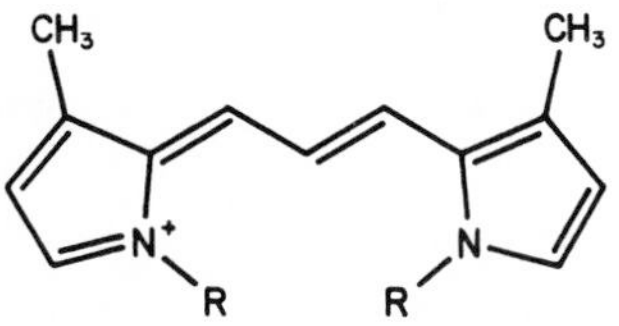

VI

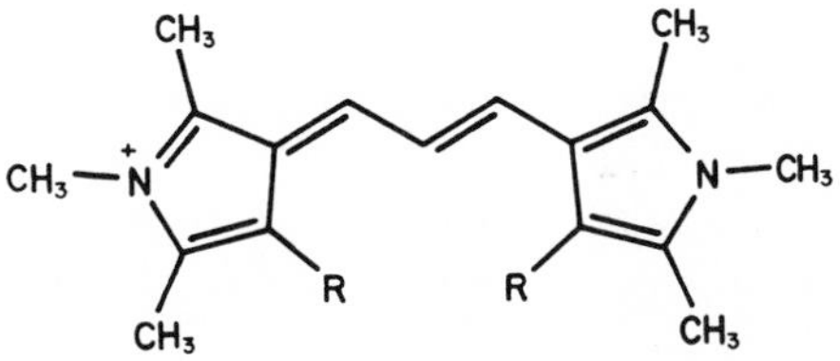

VII

this sterically hindered dye IV*b* is of lower intensity and at longer wavelength than that of the sterically unhindered dye IV*a* (see Table 17.2). Brooker and his co-workers[8] also observed a large bathochromic shift of the A band in passing from the planar dye V*a* to its nonplanar dimethyl derivative V*b*: the magnitude of the wavelength shift is almost identical with that for the isosteric pair IV*a*–IV*b*. In the sterically hindered dyes IV*b* and V*b* the steric interference between the *o*-methyl groups will be relieved mainly by twist of the two central bonds.

In higher vinylogs of these dyes the situation is quite different. The steric hindrance in dyes VI*b* (VI, R = CH_3) and VII*b* (VII, R = CH_3) is, if any, probably much smaller than in dyes IV*b* and V*b*. Actually, between VII*a* (VII, R = H) and VII*b* there is no such marked difference of the spectra as between the lower vinylogs pair, V*a* and V*b*. It is rather surprising that VI*b* absorbs at distinctly shorter wavelength than does VI*a* (VI, R = H) when the absorptions are compared in chloroform, although the positions of the absorption maxima are much closer in methanol. For this hypsochromic shift the electronic effect of the methyl groups attached directly to the nitrogen atoms may be responsible, at least in part if not wholly.

The interesting steric effect in the spectra of these cyanine dyes is explained in terms of the simple MO theory as follows. Dyes IV and V can be considered to be isoconjugate with odd-membered polymethine anions H—$(CH)_{11}^{-}$—H (IV′) and H—$(CH)_7^{-}$—H (V′), respectively.

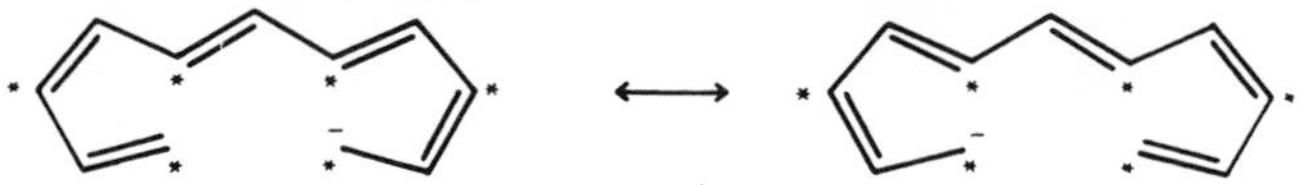

IV'

V'

Similarly, dyes VI and VII can be considered to be isoconjugate with odd-membered polymethine anions H—$(CH)_{13}^{-}$—H (VI′) and H—$(CH)_9^{-}$—H (VII′), respectively.

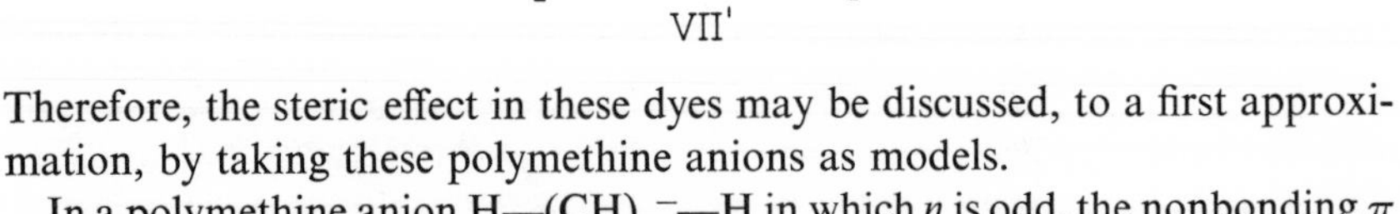

VI'

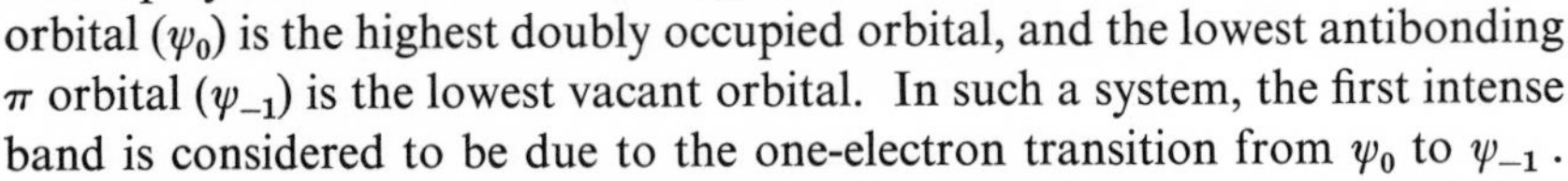

VII'

Therefore, the steric effect in these dyes may be discussed, to a first approximation, by taking these polymethine anions as models.

In a polymethine anion H—$(CH)_n^{-}$—H in which n is odd, the nonbonding π orbital (ψ_0) is the highest doubly occupied orbital, and the lowest antibonding π orbital (ψ_{-1}) is the lowest vacant orbital. In such a system, the first intense band is considered to be due to the one-electron transition from ψ_0 to ψ_{-1}. As explained in Subsection 10.4.4, when $(n + 1)/2$ is even as in IV′ and V′, the central atom (that is, the $[(n + 1)/2]$th atom) is unstarred, and the changes in the π-bond orders of the two central bonds, which connect this atom to the neighboring atoms, on the one-electron transition from ψ_0 to ψ_{-1} are negative. Therefore, twist of these bonds is expected to reduce the energy of this transition, with a consequent bathochromic shift of the band. This presumably is what happens in the sterically hindered dyes IV*b* and V*b*. On the other hand, when $(n + 1)/2$ is odd as in VI′ and VII′, the central atom is starred, and the partial π-bond orders of the two central bonds in orbitals ψ_{-1}, ψ_0, and ψ_{+1} are all zero. In this case, there is no changes in the π-bond orders of these bonds on the transition from ψ_0 to ψ_{-1}. Therefore, it is expected that, even if these bonds are twisted, the energy of the transition and hence the position of the band remain unchanged.

The above explanation can be put in another way as follows. Suppose that an n-membered polymethine chain in which n is odd is divided into three

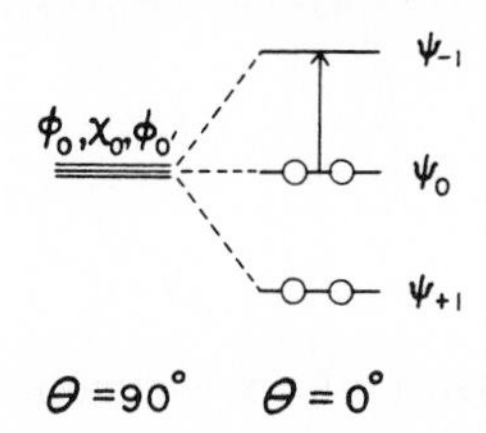

FIG. 17.1. Orbital energy levels of an n-membered polymethine chain in which $(n + 1)/2$ is even.

fragments, the central atom and two $(n - 1)/2$-membered polymethine chains, which are denoted by R and R′. When $(n + 1)/2$ is even as in IV′ and V′, $(n - 1)/2$ is odd: fragments R and R′ are odd alternant conjugated systems. The $2p\pi$ atomic orbital of the central atom is denoted by χ_0, and the nonbonding π orbitals of R and R′ are denoted by ϕ_0 and $\phi_{0'}$, respectively. When there are no conjugative interactions between these three fragments, that is, when R and R′ lie in planes perpendicular to the plane defined by the σ bonds at the central atom, the three orbitals χ_0, ϕ_0, and $\phi_{0'}$ have the same energy. When R and R′ are twisted toward the planar structure of the whole conjugated system, the three nonbonding orbitals will interact with one another and give three orbitals of different energies, as illustrated in Fig. 17.1. The resulting three orbitals can be considered to correspond approximately to ψ_{-1}, ψ_0, and ψ_{+1} in the whole conjugated system, and can be expressed as follows:

$$\psi_{-1} \doteqdot 2^{-1}(\phi_0 + \phi_{0'}) - 2^{-1/2}\chi_0;$$

$$\psi_0 \doteqdot 2^{-1/2}(\phi_0 - \phi_{0'});$$

$$\psi_{+1} \doteqdot 2^{-1}(\phi_0 + \phi_{0'}) + 2^{-1/2}\chi_0.$$

It is evident that the energy of the transition from ψ_0 to ψ_{-1} (as well as the energy of the transition from ψ_{+1} to ψ_0) will decrease when the conjugated system deviates from the planar structure through twist of the central two bonds.

When $(n + 1)/2$ is odd, fragments R and R′ are even alternant conjugated systems. The lowest antibonding and highest bonding orbitals of R are denoted by ϕ_{-1} and ϕ_{+1}, respectively, and those of R′ by $\phi_{-1'}$ and $\phi_{+1'}$, respectively. With increasing coplanarity of the three fragments, these orbitals interact more strongly with χ_0 and will split as illustrated in Fig. 17.2.

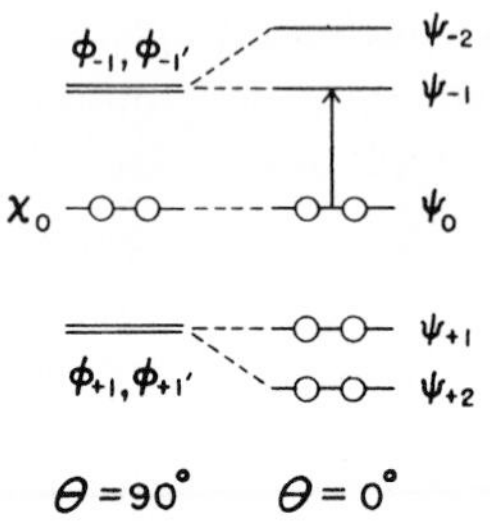

FIG. 17.2. Orbital energy levels of an n-membered polymethine chain in which $(n + 1)/2$ is odd.

The lowest antibonding, nonbonding, and highest bonding orbitals of the whole conjugated system can be expressed, to a first approximation, as follows:

$$\psi_{-1} \doteqdot 2^{-1/2}(\phi_{-1} - \phi_{-1'});$$

$$\psi_0 \doteqdot a\chi_0 + b(\phi_{-1} + \phi_{-1'}) - b(\phi_{+1} + \phi_{+1'});$$

$$\psi_{+1} \doteqdot 2^{-1/2}(\phi_{+1} - \phi_{+1'}).$$

The forms and energies of ψ_{-1} and ψ_{+1} are independent of the interplanar angles between the fragments. The energy of ψ_0 is also independent of the interplanar angles, but the form of this orbital varies with the angles in such a way that the coefficient a decreases from 1 and the coefficient b increases from 0 as the angles decrease from 90°. It is evident that any twist of the two central bonds gives no change in the energy of the transition from ψ_0 to ψ_{-1} (as well as in the energy of the transition from ψ_{+1} to ψ_0). It is also evident that the partial π-bond orders of the two central bonds in ψ_{-1}, ψ_0, and ψ_{+1} are zero.

In either case, when the planes of fragments R and R′ are perpendicular to the plane defined by the σ bonds at the central atom, that is, when the twist angles of the two central bonds are 90°, the orbitals of the fragments do not mix with one another, and hence the transition from ψ_0 to ψ_{-1}, as well as the transition from ψ_{+1} to ψ_0, is forbidden. The probabilities of these transitions will decrease as the twist angles increase.

The preceding idea may be also applicable to cations and radicals of odd-membered polymethine systems and other similar systems. For example, the steric effect on the energy levels of diarylcarbonium ions may be given by a picture similar to that shown in Fig. 17.2. Of course, in a cation, the highest occupied orbital is ψ_{+1}, and the lowest energy excitation is the one-electron transition from ψ_{+1} to ψ_0. In a radical, ψ_0 is single occupied, and the transitions from ψ_{+1} to ψ_0 and from ψ_0 to ψ_{-1} require the same energy in the simple MO approximation. The lowest energy excitation in a radical is considered to be the transition from the ground state to the lower of the two states arising from the interaction of the excited configurations of these two transitions.

17.2.2. Cyanines of Quinoline Series

The data on the first intense bands (the A bands) of some cyanines of quinoline series taken from the review of Brooker and his co-workers[8] are shown in Table 17.3.

In passing from the less hindered dyes VIII*a*, IX*a*, X*a*, and XI*a* to the more hindered corresponding dyes having additional two *o*-methyl groups, VIII*b*, IX*b*, X*b*, and XI*b*, respectively, the A band of each dye markedly decreases in intensity and shifts to longer wavelength. Notwithstanding that the steric hindrance in these hindered dyes of quinoline series is considered to be severer than in the pyrrole analogs, IV*b* and V*b*, owing to the different bond angles in the six- and five-membered rings, the magnitudes of the bathochromic shifts in the dyes of quinoline series are comparatively small.

With the higher vinylogs of VIII and IX, that is, with XII and XIII, in passing from the unhindered dyes XII*a* and XIII*a* to the derivatives having additional two *o*-methyl groups, XII*b* and XIII*b*, respectively, the A band of each dye undergoes also a bathochromic shift, but the magnitude of the shift is greater. Unlike the pyrrole analog VII*b*, the *o*-methyl derivatives XII*b* and XIII*b* are almost certainly sterically hindered, as indicated by the low intensities of the A bands of these derivatives as compared with those of the unhindered parent dyes XII*a* and XIII*a*. This difference is probably due to the different bond angles in the six- and five-membered rings.

TABLE 17.3

THE FIRST BANDS OF CYANINES OF QUINOLINE SERIES IN METHANOL

Compound	λ_{max}, mμ	$\epsilon_{max} \times 10^{-4}$	$\Delta\lambda_{max}$, mμ	$\Delta\epsilon_{max} \times 10^{-4}$
VIII*a* (VIII, $R_2 = H$, $R_3 = H$)	520	7.2	—	—
VIII*b* (VIII, $R_2 = CH_3$, $R_3 = H$)	527	4.1	+7	−3.1
VIII*c* (VIII, $R_2 = H$, $R_3 = CH_3$)	574.5	3.5	+54.5	−3.7
VIII*d* (VIII, $R_2 = H$, $R_3 = C_2H_5$)	576	3.1	+56	−4.1
IX*a* (IX, $R_2 = H$, $R_3 = H$)	523.5	7.6	—	—
IX*b* (IX, $R_2 = CH_3$, $R_3 = H$)	546	4.5	+22.5	−3.1
IX*c* (IX, $R_2 = H$, $R_3 = CH_3$)	584	2.1	+60.5	−5.5
IX*d* (IX, $R_2 = H$, $R_3 = C_2H_5$)	591	2.0	+67.5	−5.6
X*a* (X, $R_2 = H$)	588.5	8.6	—	—
X*b* (X, $R_2 = CH_3$)	614	5.0	+25.5	−3.6
XI*a* (XI, $R_2 = H$)	591	8.8	—	—
XI*b* (XI, $R_2 = CH_3$)	615	5.4	+24	−3.4
XII*a* (XII, $R_2 = H$)	604	18.5	—	—
XII*b* (XII, $R_2 = CH_3$)	640	9.7	+36	−8.8
XIII*a* (XIII, $R_2 = H$)	604	19	—	—
XIII*b* (XIII, $R_2 = CH_3$)	631	11.4	+27	−7.6

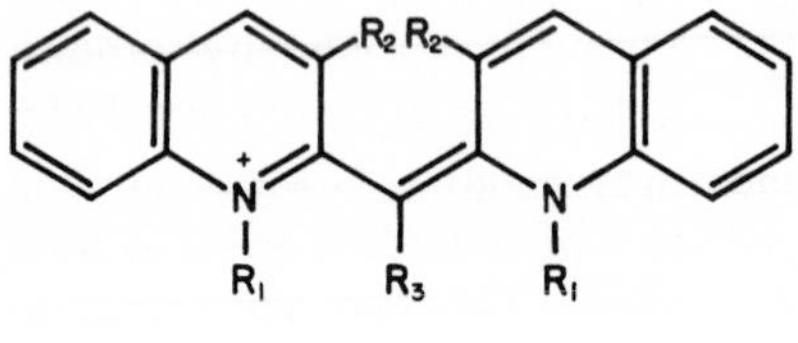

VIII : $R_1 = CH_3$
IX : $R_1 = C_2H_5$

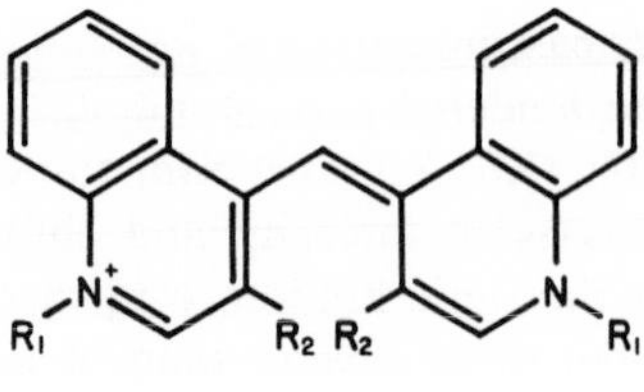

X : $R_1 = CH_3$
XI : $R_1 = C_2H_5$

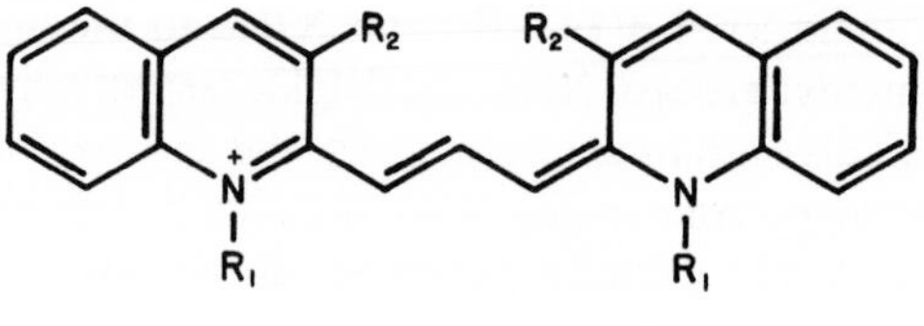

XII : $R_1 = CH_3$
XIII : $R_1 = C_2H_5$

The steric hindrance must be much severer in the dyes having a monomethine chain VIII*b* and IX*b* than in the higher vinylogs having a trimethine chain XII*b* and XIII*b*. It is noticeable that the magnitudes of the bathochromic shifts of the A bands in the lower vinylogs are smaller than in the higher vinylogs, in contrast with the case of the pyrrole analogs. This fact may be explained as follows.

In the odd alternant hydrocarbon anion that is isoconjugate with VIII and IX, and in the odd alternant hydrocarbon anion that is isoconjugate with X and XI, the central carbon atom of the methine chain is starred. When each of these odd-membered conjugated systems is divided into three fragments in the way described in the preceding subsection, the two outer fragments R and R′ are even alternant conjugated systems. Therefore, the feature of the steric effect on the energy levels of these systems can be represented by a picture similar to that depicted in Fig. 17.2. Thus, if the steric strain is relieved only by twisting the central methine bonds, the twist will lead, to a first approximation, to no wavelength shift of the A band. However, perhaps, the large steric strain in a severely hindered dye such as VIII*b*, IX*b*, X*b*, and XI*b* is not relieved only by twist of the central methine bonds, but it brings about distortion of the whole molecule, especially of the two aza-aromatic rings. As will be mentioned in the next chapter, such distortion should produce a general reduction of the absolute values of the relevant resonance integrals with a consequent bathochromic shift of bands. This could well account for the bathochromic shift of the A band observed in passing from VIII*a* to VIII*b*, from IX*a* to IX*b*, from X*a* to X*b*, and from XI*a* to XI*b*. This could also well account for the very large bathochromic shifts observed with the more highly hindered dyes VIII*c* and VIII*d* as compared with VIII*a* and with IX*c* and IX*d* as compared with IX*a*. Also with these highly hindered dyes, marked decreases in intensity occur.

In the odd alternant hydrocarbon anion that is isoconjugate with XII as well as XIII, the central atom of the trimethine chain is unstarred: in this system the two outer fragments R and R′ are odd alternant conjugated systems. Therefore, the feature of the steric effect on the energy levels of this system can be represented by a picture similar to that depicted in Fig. 17.1. Thus, if the steric strain is relieved largely by twist of the two central methine bonds, the shift will result in a bathochromic shift of the A band. In this conjugated system there are two additional open-chain bonds. Judging from the data on π-bond order changes on the lowest energy excitation in odd-membered polymethine chaines shown in Table 10.6, there seems to be a possibility that a twist of these bonds exerts a hypsochromic effect on the A band.

17.3. Steric Effects in Spectra of Highly Unsymmetrical Cyanines

As mentioned in the preceding section, with symmetrical cyanine dyes distortion from the planar structure of the conjugated system causes in almost all cases a bathochromic shift of the A band, although the extent of the shift varies from type to type. In contrast, with highly unsymmetrical cyanine dyes in which the terminal groups differ greatly in electron-donating property, the effect of the nonplanarity is hypsochromic. Sterically hindered dyes of both categories, however, behave alike in that they show a marked decrease in intensity of the A band compared with the related unhindered dyes. Data on the A bands of some representative unsymmetrical cyanine dyes, taken from the review of Brooker and his co-workers,[8] are shown in Table 17.4.

In the dyes with $R_2 = CH_3$, there should be a strong steric interference between the methyl group and a hydrogen atom on the dimethine (vinylene)

TABLE 17.4

THE FIRST BANDS OF UNSYMMETRICAL CYANINES IN METHANOL

Compound	λ_{max}, mμ	$\epsilon_{max} \times 10^{-4}$	$\Delta\lambda_{max}$, mμ	$\Delta\epsilon_{max} \times 10^{-4}$
XIV*a* (XIV, $R_2 = H$)	525	6.0	—	—
XIV*b* (XIV, $R_2 = CH_3$)	500	2.7	−25	−3.3
XV*a* (XV, $R_2 = H$)	525	5.8	—	—
XV*b* (XV, $R_2 = CH_3$)	493	2.2	−32	−3.6
XVI*a* (XVI, $R_2 = H$)	544	4.5	—	—
XVI*b* (XVI, $R_2 = CH_3$)	510	2.3	−34	−2.2
XVII*a* (XVII, $R_2 = H$)	546	4.6	—	—
XVII*b* (XVII, $R_2 = CH_3$)	511.5	2.4	−34.5	−2.2

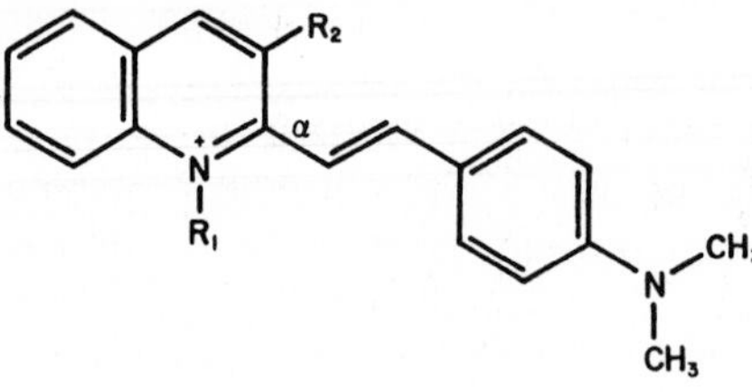

XIV : $R_1 = CH_3$
XV : $R_1 = C_2H_5$

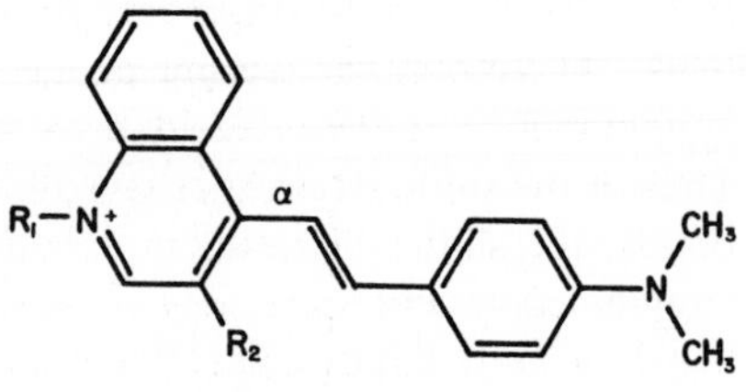

XVI : $R_1 = CH_3$
XVII : $R_1 = C_2H_5$

chain, and this steric interference will be relieved mainly by a twist of the linkage marked α. The A bands of these sterically hindered dyes are at shorter wavelengths by about 30 mμ and have lower intensities by a factor of about $\frac{1}{2}$ than those of the corresponding unhindered or less hindered dyes ($R_2 = H$).

As mentioned already, in an unsymmetrical odd-membered conjugated system, the two important extreme structures are not equivalent, and the conjugated system should therefore show a marked bond alternation. In such a system, steric hindrance to planarity of the conformation will be relieved mainly by twist of one or more essential single bonds, that is, bonds of lower π-bond order. Since the π-bond orders of essential single bonds must increase on the lowest-energy electronic excitation, twist of such bonds will produce a hypsochromic shift of the A band, as actually observed.

The conjugated system of dyes XIV and XV can be regarded as a resonance hybrid mainly of two extreme structures XIV′*a* and *b*, and the conjugated system of dyes XVI and XVII as a resonance hybrid mainly of two extreme structures XVI′*a* and *b*.

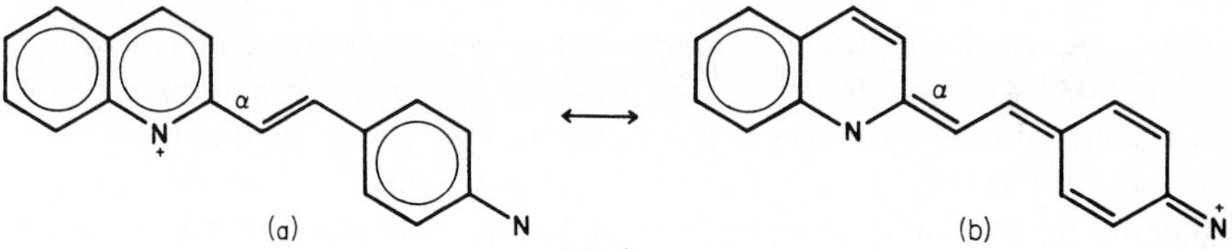

(a) ⟷ (b)

XIV'

(a) ⟷ (b)

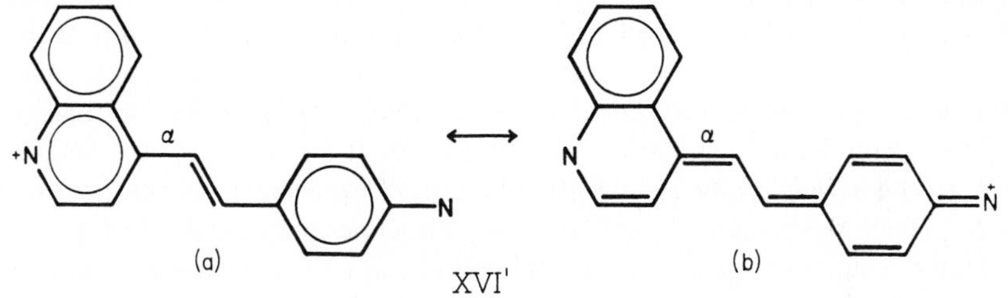

XVI'

In both the cases, structure *a*, in which both the benzene ring and the quinoline ring are benzenoid, is probably of lower energy than structure *b*, in which both the rings are quinonoid. Therefore, structure *a* will contribute mainly to the ground state, and structure *b* to the excited state. Accordingly, the linkage α will have a character closer to a single bond in the ground state and a character closer to a double bond in the excited state, and hence twist of this bond will produce a hypsochromic shift of the A band.

Chapter 18

Nonplanar Aromatic Systems

18.1. General

It has been found that nonbonded carbon atoms do not approach closer than 3.0 Å and that normal distances between nearest hydrogen atoms are 2.4–2.5 Å.[1] The adoptation of a planar geometry by some aromatic molecules would necessitate the compression of nonbonded atoms to within this forbidden range. Relief of such strain in overcrowded molecules could be brought about by changes in bond lengths and valence angles of the overcrowded atoms only, in such a way as to leave the remainder of the molecule coplanar, or by distortion of the molecule as a whole, distributing the strain over all the atoms instead of confining it to a small region of the molecule. The results of calculations by Coulson, Senent, and Herraéz[2] make it quite clear that distortion of the molecule as a whole requires much less energy than if distortion is confined to the overcrowded atoms only. Actually, many aromatic compounds have been found by means of X-ray crystal analysis to have nonplanar geometries, and in the small number of cases where a proper theoretical study has been made a good agreement is found between the distortions predicted and those found experimentally.[1] The present chapter deals with the effects of such distortions of aromatic ring systems on the electronic absorption spectra.

The most prominent effect of the distortion of aromatic rings on the electronic spectra seems to be to blur the vibrational fine structure. While the

[1] G. Ferguson and J. M. Robertson, *in* "Advances in Physical Organic Chemistry" (V. Gold, ed.), Vol. 1, p. 203. Academic Press, New York, 1963.

[2] C. A. Coulson and S. Senent, *J. Chem. Soc.* **1955**, 1813, 1819; C. A. Coulson, S. Senent, and M. A. Herraéz, *Anal. Fis. Quim.* **52B**, 515 (1956); S. Senent and M. A. Herraéz, *ibid.* **53B**, 257, 325 (1957).

spectra of nonovercrowded aromatic compounds such as benzene, naphtharene, and anthracene show bands having prominent vibrational fine structure, the vibrational structure is much more blurred in the spectra of overcrowded aromatic compounds such as benzo[*c*]phenanthrene (3,4-benzophenanthrene) (I), dibenzo[*c*,*g*]phenanthrene (3,4-5,6-dibenzophenanthrene, pentahelicene) (II), and phenanthro[3,4-*c*]phenanthrene (hexahelicene) (III). This effect may be attributed to the enhanced unseparability of the vibrational and electronic wavefunctions in distorted molecules.

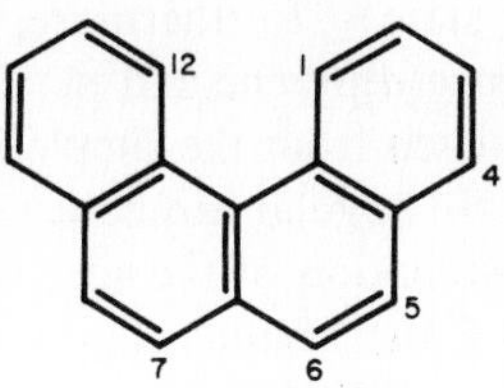

I

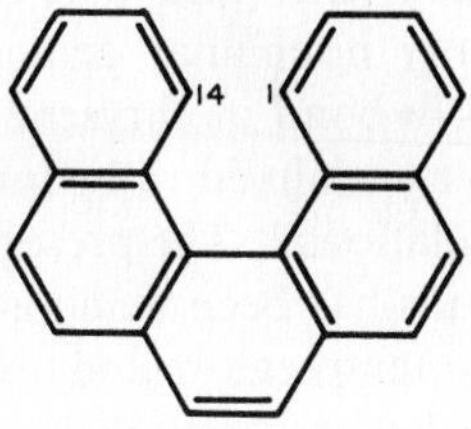

II

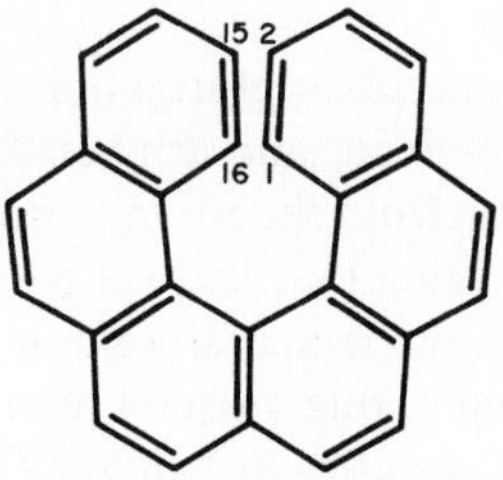

III

Introduction of alkyl substituents into sterically hindering positions further blurs the vibrational structure. A comparison of the spectra of chrysene (V), 5-methylchrysene, and 4,5-dimethylchrysene, for example, shows that the amount of fine structure in the spectra decreases markedly in that order.[3]

The blurred vibrational structure of the spectrum of benzophenanthrene I is further blurred by introduction of two methyl groups into the 1- and 12-positions.[4] On the other hand, introduction of a methylene bridge into the 1- and 12-positions of the benzophenanthrene, which is considered to constrain the aromatic ring system to assume a planar or nearly planar geometry, increases the amount of vibrational structure. Introduction of a dimethylene bridge into the 6- and 7-positions of the benzophenanthrene also increases the amount of vibrational structure. This is probably due to the

[3] R. A. Friedel, *Appl. Spectroscopy* **2**, 13 (1956).
[4] A. W. Johnson, *J. Org. Chem.* **24**, 833 (1959).

limitations imposed on the freedom of deformation of the aromatic ring system by the dimethylene bridge.

Even in phenanthrene (IV) there is the overcrowding of the two hydrogen atoms at positions 4 and 5, and this overcrowding is considered to be relieved mainly by displacement of the atoms out of the mean molecular plane and by bond-angle deformation, although no reliable experimental results on the geometry of phenanthrene are yet available. Some compounds containing angular ring systems similar to that of phenanthrene, for example, chrysene, triphenylene, and benzo[*a*]pyrene (1,2-benzopyrene), are known to have slightly nonplanar geometries in the crystalline state.[1] Furthermore, the 4a—4b bond in chrysene and similar bonds in benzo[*a*]pyrene and perylene (VI) are claimed to be longer than the values predicted from the simple MO calculations.[1] The presence of overcrowding in these angular aromatic compounds has been demonstrated by Clar[5] from examination of the spectra at room temperature and low temperature. At $-170°$C the β bands of nonovercrowded aromatic hydrocarbons such as benzene, naphthalene, anthracene, and pyrene become more distinct, showing much more fine structure, than at 18°C. This is explained by the cessation at low temperature of thermal collisions which produce molecular deformations. On the other hand, the β bands of phenanthrene, perylene, and triphenylene undergo only a small change in the shape on going from the room temperature to the low temperature. This is taken as indicating that there is permanent deformation of the molecular geometry at both the temperatures.

The distortion of aromatic ring systems causes also wavelength shifts and changes in intensity of absorption bands. The wavelength shifts are in most cases bathochromic but in some cases hypsochromic. The following sections concern mainly with the wavelength shifts. The changes in intensity appear to be hypochromic in most cases, but there appear to be some cases in which the changes are slightly hyperchromic. In general, the changes in intensity are small compared with the wavelength shifts.

18.2. Correlation of the Direction of the Wavelength Shift with the π-Bond Order of the Mainly Twisted Bond

Because of the pairing properties of the Hückel orbitals of alternant hydrocarbons, the partial π-bond orders of a bond in a bonding orbital and the corresponding antibonding orbital of an alternant hydrocarbon are equal in magnitude but have opposed signs: a bond which has a node in ψ_{+1}; for

[5] E. Clar, *Spectrochim. Acta* **4**, 116 (1950).

example, has no node in ψ_{-1}, and vice versa (see Section 7.2). It follows that any distortion of the molecule which raises ψ_{+1} will lower ψ_{-1}, and any which lowers ψ_{+1} will raise ψ_{-1}. In a linear conjugated even alternant hydrocarbon, every essential single bond has a node in ψ_{+1} and has no node in ψ_{-1}, and every essential double bond has no node in ψ_{+1} and has a node in ψ_{-1}. In such a system, a steric strain in the planar geometry is usually relieved mainly by a twist of one or more essential single bonds. Such a twist lowers ψ_{+1} and raises ψ_{-1}, and results in a hypsochromic shift of the first intense band (the A band), which is considered to be, to a first approximation, due to the one-electron transition from ψ_{+1} to ψ_{-1}.

In aromatic rings there is no bond alternation, and it is rather difficult to determine which bonds have a node in a particular orbital. As mentioned in the preceding section, the distortion due to the steric strain in an aromatic ring system is usually distributed over all the atoms instead of being confined to a small region of the molecule. Furthermore, the spectra of aromatic compounds are generally complicated, and, as mentioned in Chapter 10, any calculations based on a one-electron Hamiltonian cannot explain the overall features of the spectra. These facts make it considerably difficult to predict the directions of the wavelength shifts of bands brought about by the distortion of aromatic ring systems. However, the so-called *p* bands of aromatic compounds can be considered, to a first approximation, to arise from the one-electron transition from ψ_{+1} to ψ_{-1} (see Sections 10.5). Therefore, in the case where the steric strain is considered to be relieved mainly by a twist of a bond, according as the bond has or has not a node in ψ_{+1}, the *p* band is expected to be shifted hypsochromically or bathochromically by the distortion. From such a viewpoint, Cromartie and Murrell[6] attempted to predict the directions of the steric wavelength shifts of the *p* bands of some aromatic hydrocarbons. The forms of ψ_{+1} of the aromatic hydrocarbons are shown in Fig. 18.1, in which broken lines indicate the nodes in ψ_{+1}, and in Table 18.1 the directions of the wavelength shifts of the *p* bands of these compounds predicted from the sign of the partial π-bond order of the bond principally twisted in ψ_{+1} are compared with the observed shifts. It is seen that the predicted direction agrees with the observed one in all the cases examined.

In some cases the geometry of the strained molecule has not been determined by X-ray analysis, but Cromartie and Murrell inferred the probable mode of distortion from examination of molecular models and from the principle that twisting a bond of low π-bond order should lead to less destabilization of the π-electron system than twisting a bond of high π-bond

[6] R. I. T. Cromartie and J. N. Murrell, *J. Chem. Soc.* **1961,** 2063.

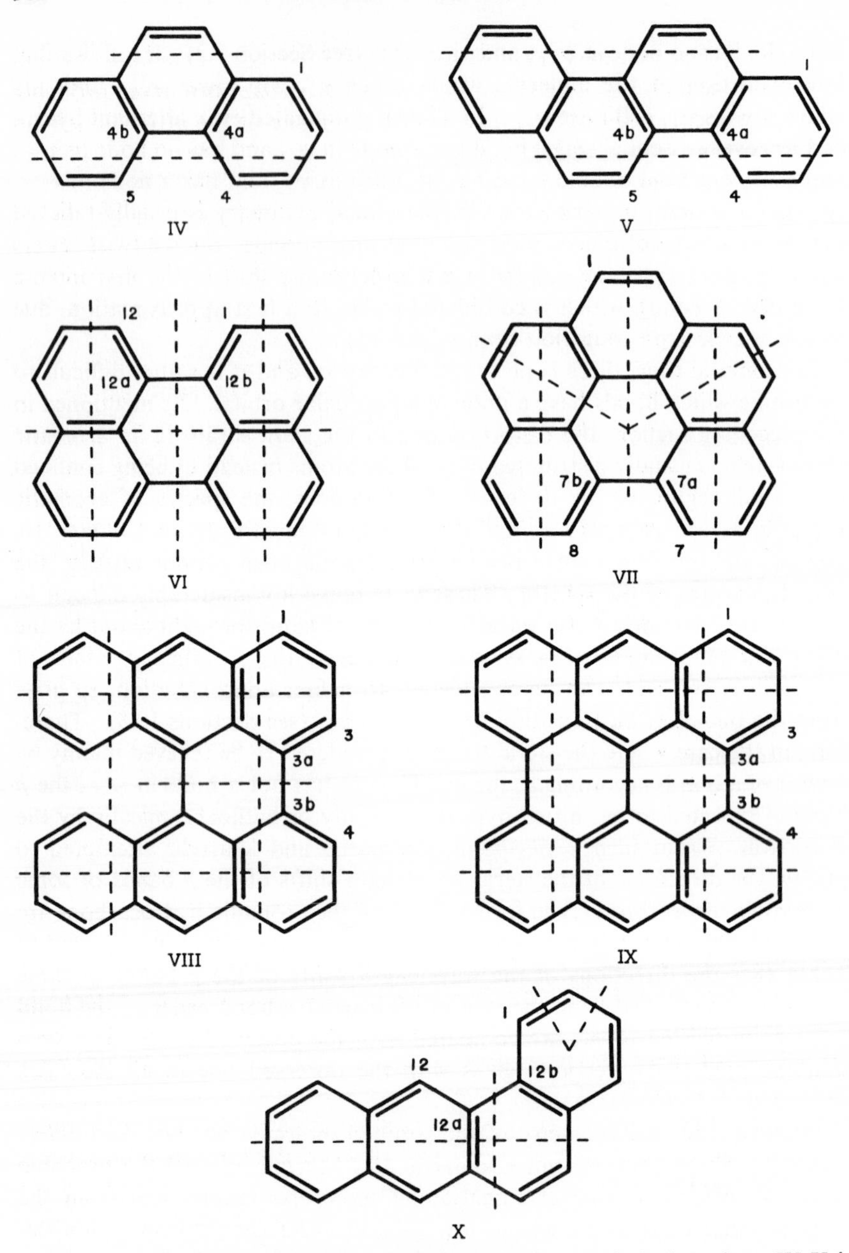

FIG. 18.1. The highest bonding π orbitals (ψ_{+1}) of aromatic hydrocarbons IV–X in Table 18.1. The broken lines indicate the nodes in ψ_{+1}.

order does. For example, examination of models shows that in 4,5-dimethylphenanthrene the steric interference of the methyl groups is relieved probably by displacement of the substituents out of the plane of the aromatic system and by twisting mainly about the 4a—4b bond, which has a low π-bond order. Similarly, in 4,5-dimethylchrysene the most twisted bond is inferred to be the 4a—4b bond. Since these bonds have no node in ψ_{+1}, such a distortion should

TABLE 18.1

SHIFTS OF THE *p* BANDS OF SOME AROMATIC HYDROCARBONS BY INTRODUCTION OF STERICALLY HINDERING SUBSTITUENTS[a]

Aromatic system	Substituent	Bond principally twisted	Frequency shift $\Delta\nu$ Predicted	Frequency shift $\Delta\nu$ Observed, cm^{-1}
Phenanthrene (IV)	2,5-dimethyl	4a—4b	−	−570
Phenanthrene (IV)	4,5-dimethyl	4a—4b	−	−2300
Chrysene (V)	4,5-dimethyl	4a—4b	−	−2300
Perylene (VI)	1,12-diacetoxy	12a—12b	+	+610
Benzo[*ghi*]perylene (VII)	7,8-di(hydroxymethyl)	7a—7b	+	+1110
Dibenzo[*a,o*]perylene (VIII) (helianthrene)	3,4-dimethyl	3a—3b	+	+850
Phenanthro[1,10,9,8-*opqra*]perylene (IX) (*meso*-naphthodianthrene)	3,4-dimethyl	3a—3b	+	+800
Benz[*a*]anthracene (X) (tetraphene)	1,12-dimethyl	12a—12b	+ (small)	0

[a] Reference 6.

exert a bathochromic effect on the *p* band. The observed shifts with these compounds are actually bathochromic, and the magnitudes of the shifts are much larger than expected from the possible electronic bathochromic effect of the substituents. In addition, as mentioned in the preceding section, the spectra of these sterically hindered compounds are much broader than those of the respective parent compounds. By the way, the nonplanarity of the geometry of these sterically hindered compounds has been confirmed by resolution of some simple derivatives of 4,5-dimethylphenanthrene into optical antipodes.[7]

[7] M. S. Newman and A. S. Hussey, *J. Am. Chem. Soc.* **62,** 3023 (1947). See also: M. S. Newman, *in* "Steric Effects in Organic Chemistry" (M. S. Newman, ed.), Chap. 10. Wiley, London and New York, 1956.

In 1,12-diacetoxyperylene the strain is believed to be relieved principally by twisting the two central (*peri*) bonds joining the two naphthalene units. These *peri*-bonds have a very low π-bond order and are known to have an almost complete single-bond length.[8] A similar type of distortion is assumed for derivatives of benzo[*ghi*]perylene (VII), helianthrene (VIII), and *meso*-naphthodianthrene (IX) bearing substituents at corresponding positions. In all these cases hypsochromic shifts are observed, in agreement with the prediction.

The benz[*a*]anthracene (X) system is a subtle case. This molecule may be compared with phenanthrene but unlike it the highest occupied orbital ψ_{+1}

TABLE 18.2

THE SPECTRA OF CHRYSENE AND ITS STERICALLY HINDERED METHYLATED DERIVATIVES IN 95% ETHANOL[a]

Chrysene (V)	λ_{max}, mμ	log ϵ_{max}	λ_{max}, mμ	log ϵ_{max}	λ_{max}, mμ	log ϵ_{max}
Unsubstituted	361	2.81	320	4.12	267	5.18
5-Methyl-	367	2.90	326	4.10	269	4.99
4,5-Dimethyl-	382	2.87	343	4.32	282	5.03

[a] The numerical data were estimated from the spectral curves presented in H. H. Jaffé and M. Orchin, "Theory and Applications of Ultraviolet Spectroscopy," pp. 446–447. Wiley, New York and London, 1962.

has a node across the 12a—12b bond. The absolute value of the partial π-bond order of this bond in ψ_{+1} is, however, rather small (0.0016). It is therefore predicted that if this bond were twisted a very small hypsochromic shift of the *p* band would be observed. Experimentally it is observed that substitution of a single 12-methyl group brings about a bathochromic shift ($\Delta\nu = -920$ cm^{-1}) but that the 1-methyl derivative and the 1,12-dimethyl derivative absorb at very nearly the same frequency as the unsubstituted hydrocarbon. Cromartie and Murrell have therefore concluded that the hypsochromic shift due to the steric distortion cancels out the normal electronic bathochromic effect of the substituents.

So far the discussion has been restricted to the *p* band. Since the other bands cannot be well described in terms of one-electron approximation, it is difficult to correlate the directions of the steric shifts of the other bands with the sign of the partial π-bond order of the twisted bond in a particular orbital.

[8] D. M. Donaldson and J. M. Robertson, *Proc. Roy. Soc.* (*London*) **A220,** 157 (1953); *J. Chem. Soc.* **1953,** 17.

However, comparison of the spectrum of chrysene with the spectra of its sterically hindered methyl derivatives seems to show that some other bands are also shifted to longer wavelengths by the steric hindrance (see Table 18.2).

18.3. Effects of the Distortion of the Whole Aromatic System

In the cases where the steric strain is so severe that the distortion is spread almost evenly over the whole aromatic system, that is, in the cases where the aromatic system is warped or buckled as a whole, such a distortion will produce a general reduction of the absolute magnitudes of the relevant resonance integrals, with resulting bathochromic shifts of all the absorption bands.[9] If the resonance parameters k for all the bonds are reduced by a common factor by the distortion, this is equivalent to the reduction of the effective value of the negative of the standard resonance integral β by the same factor. Therefore, as a result of such a distortion, the energy separations between the molecular orbitals will become smaller, and hence the energies of all the electronic excitations will become smaller.

An interesting example is found in the series of paracyclophanes (XI), in which two benzene rings are bridged by a pair of methylene chains at *para* positions.

$$\begin{array}{ccc} & (CH_2)_m & \\ p\text{-}C_6H_4 & & p\text{-}C_6H_4 \\ & (CH_2)_n & \end{array}$$

XI

The paracyclophanes XI in which m and n are varied stepwise one at a time from $m = n = 2$ to $m = 5$ and $n = 6$ were prepared and their ultraviolet absorption spectra were studied by Cram and his co-workers.[10] According to the result, while the spectra of the paracyclophanes XI with $m \geq 4$ and $n \geq 4$ are almost identical with the spectrum of an open-chain analog p-C_2H_5—C_6H_4—$(CH_2)_4$—C_6H_4—C_2H_5-p (XII), the spectra of the lower homologs, XI with $m \leq 3$ and $n \leq 4$, are appreciably different from the spectrum of XII. Thus, as the values of m and n become smaller than 4, the intense band around 220 mμ (the p band) undergoes a bathochromic and a hypochromic shift, and the weak band around 270 mμ (the α band) undergoes similar shifts accompanied with loss of the vibrational fine structure

[9] C. A. Coulson, *in* "Steric Effects in Conjugated Systems" (G. W. Gray, ed.), p. 8. Butterworths, London, 1958.

[10] D. J. Cram, N. L. Allinger, and H. Steinberg, *J. Am. Chem. Soc.* **76,** 6132 (1954).

and with appearance of shoulders both at longer and at shorter wavelengths than the peak (see Table 18.3).

The spectral abnormalities characterizing the smaller paracyclophanes are attributed to a combination of the following two causes: (1) π-electron interactions between the two benzene rings (transannular interactions), and (2) distortion of the benzene rings from their normal planar geometry.

There must be severe steric strain in the molecule of the simplest member of the paracyclophane series, di-*p*-xylylene (XI with $m = n = 2$): if the two benzene rings were planar and if the methylene carbon atoms had normal tetrahedral angles, the two rings would be only 2.5 Å apart. It has been

TABLE 18.3

DIFFERENCE ($\Delta\lambda_{max}$) BETWEEN λ_{max} FOR PARACYCLOPHANES (XI) AND λ_{max} FOR AN OPEN-CHAIN ANALOG XII

(λ_{max} for XII) XI	The α band (273 mμ) $\Delta\lambda_{max}$	The p band (223 mμ) $\Delta\lambda_{max}$
$m = n = 2$	+29 mμ	+21 mμ
$m = n = 3$	+21	+23
$m = n = 4$	+1	+1

found by Brown[11] and by Lonsdale et al.[12] that the benzene rings are markedly distorted from the normal planar geometry. Thus, according to the result of the accurate X-ray analysis by Lonsdale and co-workers, which improved the result of Brown's earlier work, the maximum separation between the two benzene rings is 3.09 Å at the unsubstituted parts of the rings, reducing to 2.75 Å at the *para*-carbon atoms. This is achieved by increasing the valence angle at the methylene carbon atoms from the tetrahedral angle to 113° and by folding the benzene rings (through 14°) away from each other along the line of the *para*-carbon atoms. The benzene rings are thus in a "boat" form, the *para*-carbon atoms in each ring being 0.17 Å out of the plane of the remaining four carbon atoms. The value of 3.09 Å of the interplanar distance between the two parallel planes defined by the unsubstituted part of each benzene ring is much smaller than the values of approximately 3.40 Å found by X-ray analysis for the interplanar distances in the crystals of a number of aromatic compounds in which the aromatic

[11] C. J. Brown, *J. Chem. Soc.* **1953,** 3265, 3278.

[12] K. Lonsdale, H. J. Milledge, and K. V. K. Rao, *Proc. Roy. Soc.* (*London*) **A255,** 82 (1960).

nuclei occupy parallel planes. It seems to be certain that there should be more or less strong transannular interaction in the di-*p*-xylylene molecule.

With a view to separate and identify the effects of the transannular interaction and of the distortion of the benzene rings in the lower members of the paracyclophane series, Cram and co-workers investigated the spectra of *cis*- and *trans*-1,2-diphenylcyclopentane and of compounds of types XIII and XIV. Compounds of type XIII have a structure resulting from replacement of one of the *p*-phenylene groups of a paracyclophane by a 1,4-cyclohexylene group; compounds of type XIV have a structure in which the *para*-carbon atoms of a benzene ring are joined by a long polymethylene chain.

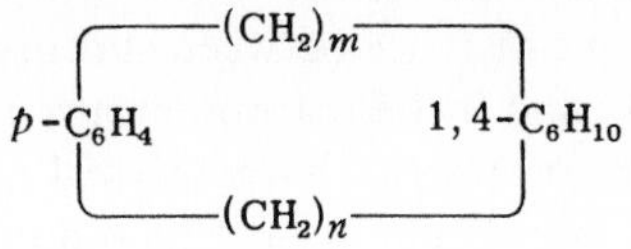

$(CH_2)_m$

p-C_6H_4 1,4-C_6H_{10}

$(CH_2)_n$

XIII

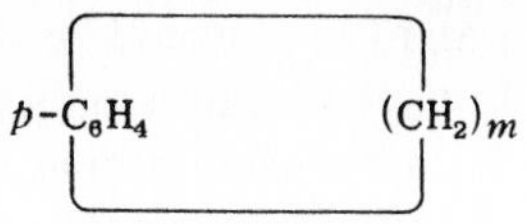

p-C_6H_4 $(CH_2)_m$

XIV

In *cis*-1,2-diphenylcyclopentane, the two benzene rings are considered to nearly face each other, but probably not completely because of the nonplanarity of the cyclopentane ring. In this compound there should be, of course, no distortion of the benzene rings. Therefore, this compound may be considered as a model which should show only the effect of the transannular interaction, being devoid of the effect of the distortion of the benzene rings. A very slight trend is observed for the absorption bands to move toward longer wavelengths and lower intensities in passing from *trans*-1,2-diphenylcyclopentane to the *cis* isomer. This trend is qualitatively similar to that observed for the paracyclophane series as m and n become smaller, but the magnitudes of the shifts are much smaller.

In compounds of types XIII and XIV there cannot be the transannular interaction, but in the smaller members of these series the benzene ring must be distorted as in the smaller members of the paracyclophane series. A trend similar to that observed with the paracyclophane series is found with these series. Thus the compound XIII with $m = 3$ and $n = 4$ and the compounds XIV with $m = 9$ and with $m = 10$ show abnormal spectra as compared with the spectrum of the open-chain reference compound XII, while higher homologs of these compounds show spectra similar to the spectrum of XII. However, the extent of the abnormality is considerably smaller than that of the spectra of the small paracyclophanes with $m \leq 3$ and $n \leq 4$. Thus, the extent of the bathochromic shifts of bands and the extent of the blurring of the α band in the small cycles of types XIII and XIV are smaller than in the lower members of the paracyclophane series. In addition, the shoulders or broad maxima appearing at both sides of the α band in the spectra of the lower members of the paracyclophane series do not appear in the spectra of the small cycles of types XIII and XIV.

From these spectral comparisons it is concluded that both the distortion of the benzene rings and the transannular interaction operate in the smaller paracyclophanes to give them their abnormal spectra.

Another important class of compounds which have a warped aromatic ring system is a series of angular *cata*-condensed aromatic hydrocarbons I (benzo-[*c*]phenanthrene), II (pentahelicene), and III (hexahelicene). If these compounds assumed regular planar geometries, the distance between carbon atoms 1 and 12 in I would be about 2.4 Å, the distance between carbon atoms 1 and 14 in II would be only about 1.4 Å, and in III, carbon atoms 1 and 2 would fall on carbon atoms 16 and 15, respectively. X-ray crystal analysis shows that compounds I and II have nonplanar helical geometries in which all the benzene rings are severely buckled, and that both the distances between atoms 1 and 12 in I and between atoms 1 and 14 in II are about 3.0 Å.[13] Hexahelicene, III, is much more overcrowded than these compounds, and is believed to have a nonplanar helical geometry, although no detailed X-ray crystal analysis appears to have been done on this compound. Such a helical geometry should be dissymmetrical and hence should lead to optical activity. Actually, hexahelicene was resolved into optical isomers, and the optical isomers were found to have a very high optical stability and a very high optical rotation ($[\alpha]_D$ $\pm 3640°$).[14]

Since in these compounds the distortion is distributed almost evenly over the whole molecule, the approximation in which the magnitudes of the effective resonance integrals for all the bonds are assumed to be evenly reduced seems to be applicable. Thus it is inferred that absorption bands of these helical compounds are at longer wavelengths than if these compounds were planar.

The effect of the nonplanarity of the geometry of these compounds on the electronic absorption spectra has been confirmed in two ways. First, it has been shown by Clar and Stewart[15] that for a very large number of unstrained planar aromatic hydrocarbons the ratio of the wavelength of the α band to that of the β band, $\lambda_\alpha/\lambda_\beta$, is very close to 1.35 (see Subsection 10.5.1). With strained aromatic hydrocarbons this ratio decreases. Thus, with compound I this ratio falls to 1.324, and with compound II it drops to only 1.27

[13] F. H. Herbstein and G. M. J. Schmidt, *J. Chem. Soc.* **1954,** 3302.

[14] M. S. Newman, W. B. Lutz, and D. Lednicer, *J. Am. Chem. Soc.* **77,** 3420, 4940 (1955).

[15] E. Clar and D. G. Stewart, *J. Am. Chem. Soc.* **74,** 6235 (1952).

and is accompanied with marked loss of intensity and vibrational fine structure of the bands.

The second line of evidence is found in the fact that in compound I there is a progressive bathochromic shift of a band with methylation at the 1- and 12-positions (see Table 18.4).[16] A similar order of shift occurs with other bands, particularly one around 314.5 mμ. These bathochromic shifts are greater than the shifts usually found as a result of hyperconjugation, and must

TABLE 18.4

THE SPECTRA OF BENZO[*c*]PHENANTHRENE AND ITS STERICALLY HINDERED METHYLATED DERIVATIVES[a]

Benzophenanthrene (I)	λ_{max}, mμ	$\Delta\nu$, cm^{-1}
Unsubstituted	281.5	—
1-Methyl-	286.5	−617
1,12-Dimethyl-	292.0	−1213

[a] Reference 16.

surely be the result of the increasing strain in the molecule. It has been mentioned in Section 18.1 that the spectrum of the 1,12-dimethyl derivative of compound I is broader than that of the parent compound. No accurate X-ray analysis of the geometry of this dimethyl derivative appears to have been done, but it has been shown that the molecular frame of this compound is severely distorted.[17] By the way, 1-methylbenzo[*c*]phenanthrene-4-acetic acid (the 1-methyl-4-carboxymethyl derivative of I)[18] and 1,12-dimethylbenzo[*c*]phenanthrene-5-acetic acid (the 1,12-dimethyl-5-carboxymethyl derivative of I)[19] have been resolved into optically active isomers.

[16] G. M. Badger and I. S. Walker, *J. Chem. Soc.* **1954,** 3238.
[17] F. L. Hirshfeld and G. M. J. Schmidt, *Acta Cryst.* **9,** 233 (1956).
[18] M. S. Newman and W. B. Wheatley, *J. Am. Chem. Soc.* **70,** 1913 (1948).
[19] M. S. Newman and R. M. Wise, *J. Am. Chem. Soc.* **78,** 450 (1956).

Chapter 19

Simple Composite-Molecule Method and a Classification of π–π^* Transitions

19.1. Introduction

It has already been mentioned that a twist of an essential single bond of an even alternant hydrocarbon brings about invariably a hypsochromic shift of the first intense π–π^* S–S band (the A band). In conjugated systems containing one or more heteroatoms and in nonalternant systems, however, a twist of an essential single bond does not necessarily cause a hypsochromic shift of the A band, but in some cases causes substantially no shift, and in some cases causes a bathochromic shift of the band.

In order to understand the different effects of twisting essential single bonds on absorption bands, it is convenient to separate the conjugated system (R–S) into the two mesomeric fragments (or subsystems) R and S on either sides of the essential single bond under consideration, and to describe the electronic wavefunctions of the conjugated system in terms of the wavefunctions for the separated fragments. Such methods are called methods of composite molecule or methods of building up molecules from fragments (see Subsection 7.3.2).

Methods of composite molecule can be divided into two main types. In the method of the first type, the π MO's of the composite conjugated system R–S are regarded as arising from mixing of the π MO's of the two fragments R and S under the influence of the effective one-electron Hamiltonian for the composite system. The method of this type, called the LCMO (linear-combinations-of-molecular-orbitals) method, has been developed mainly by Dewar.[1] The method of this type may be called the simple composite-molecule method.

[1] M. J. S. Dewar, *Proc. Cambridge Phil. Soc.* **45,** 639 (1949); *J. Chem. Soc.* **1950,** 2329; *J. Am. Chem. Soc.* **74,** 3341, 3345, 3350, 3353, 3355, 3357 (1952).

In the method of the second type, the π-electronic states of the composite system are regarded as arising from mixing of the ground electron configuration, the so-called locally excited (LE) electron configurations, which correspond to excited states of each fragment, and the so-called charge-transfer (CT) electron configurations, which result from transfer of one or more electrons from one fragment to another. The mixing may be considered to occur under the influence of the effective one-electron Hamiltonian for the composite system, or alternatively, under the influence of the total π-electronic Hamiltonian, which includes the electronic interaction terms. The method of this type including the electronic interaction, which may be called the advanced composite-molecule method, has been developed mainly by Longuet-Higgins and Murrell.[2]

In this chapter, the method of composite molecule in the one-electron approximation is described, and on the basis of this method π–π^* electronic transitions and steric effects on them are classified. The method of composite molecule using the total π-electronic Hamiltonian will be described in the next chapter.

19.2. Simple Composite-Molecule Method

19.2.1. The π MO's of Composite Conjugated Systems and Their Energies

Suppose that a conjugated system R–S is composed of two mesomeric subsystems R and S, which are linked to each other in R–S by a formally single bond between atom r belonging to R and atom s belonging to S. For example, for the conjugated diene system the two ethylenic bonds are taken as R and S; for a substituted benzene the benzene ring is taken as R and the substituent as S. Subsequently, the following notation is used: π MO's of R, ϕ_m or $\phi_{m(\mathrm{R})}$; π MO's of S, $\phi_{n'}$ or $\phi_{n(\mathrm{S})}$; π MO's of R–S, ψ_k or $\psi_{k(\mathrm{RS})}$; the energy of ϕ_m, $\eta_m = \alpha - v_m\beta$; the energy of $\phi_{n'}$, $\eta_{n'} = \alpha - v_{n'}\beta$; the energy of ψ_k, $\varepsilon_k = \alpha - w_k\beta$; the π AO of atom i, χ_i; the coefficient of χ_r in ϕ_m, $b_{m,r}$; the coefficient of χ_s in $\phi_{n'}$, $b_{n',s}$; the coefficient of χ_i in ψ_k, $c_{k,i}$; the Coulomb integral of χ_i, $\alpha_i = \alpha + h_i\beta$; the resonance integral between χ_i and χ_j, $\beta_{ij} = k_{ij}\beta$; the partial π-bond order of the i—j bond in ψ_k, $p_{ij}^{(k)} = c_{k,i}c_{k,j}$. Needless to say, α and β represent the standard Coulomb and resonance integrals, respectively.

[2] H. C. Longuet-Higgins and J. N. Murrell, *Proc. Phys. Soc.* (*London*) **A68,** 601 (1955); J. N. Murrell, *ibid.* **A68,** 969 (1955).

Unless the plane of R and that of S are perpendicular to each other, the resonance integral between χ_r and χ_s, β_{rs}, must be nonzero, and hence there must be interactions between ϕ_m's and $\phi_{n'}$'s. The effective one-electron Hamiltonians for R, S, and R–S, which are denoted by $\mathbf{h}^{\mathrm{R}}$, $\mathbf{h}^{\mathrm{S}}$, and $\mathbf{h}^{\mathrm{RS}}$, respectively, can be written as expressions (7.104), and the matrix elements of $\mathbf{h}^{\mathrm{RS}}$ between the π MO's of the fragments R and S, ϕ's, can be approximately given by expressions (7.106), (7.107), and (7.108). When $b_{m,r}b_{n',s}$ is denoted simply by $B_{mn'}$, expression (7.108) can be rewritten as follows:

$$h^{\mathrm{RS}}_{m,n'} = B_{mn'}\beta_{rs} = B_{mn'}k_{rs}\beta. \tag{19.1}$$

In the simple LCMO method, which is a modification of the simple LCAO MO method in its original form, with neglect of overlap integrals and of resonance integrals between nonadjacent AO's and without any of the subsequent refinements such as antisymmetrization and inclusion of configuration interaction, the π MO's of R–S, ψ's, are expressed as linear combinations of ϕ's, and are obtained as such by solving the proper secular equation. As mentioned in Subsection 7.3.2, when some requirements are fulfiled perturbation theory can be applied to this method.

We consider first the interaction between one MO of R, say ϕ_u, and one MO of S, say $\phi_{v'}$, assuming that $\eta_u \geq \eta_{v'}$ and hence that $v_u \geq v_{v'}$. The secular equation in the zero-overlap approximation is evidently as follows:

$$\begin{vmatrix} w - v_u & B_{uv'}k_{rs} \\ B_{uv'}k_{rs} & w - v_{v'} \end{vmatrix} = 0. \tag{19.2}$$

By solving this equation, we obtain

$$w = \tfrac{1}{2}\{v_u + v_{v'} \pm [(v_u - v_{v'})^2 + 4(B_{uv'}k_{rs})^2]^{1/2}\}. \tag{19.3}$$

The two resulting MO's of R–S and their energies can therefore be expressed as follows:

$$w(\text{upper}) = v_u + t_{uv'}, \tag{19.4}$$

$$w(\text{lower}) = v_{v'} + t_{v'u} = v_{v'} - t_{uv'}; \tag{19.5}$$

$$\psi(\text{upper}) = N(\phi_u - A^0_{uv'}k_{rs}\phi_{v'}), \tag{19.6}$$

$$\psi(\text{lower}) = N(\phi_{v'} - A^0_{v'u}k_{rs}\phi_u) = N(\phi_{v'} + A^0_{uv'}k_{rs}\phi_u). \tag{19.7}$$

In these expressions, N is the normalization factor, and $t_{uv'}$ and $A^0_{uv'}$ are positive quantities defined as follows:

$$t_{uv'} \equiv \tfrac{1}{2}\{-\Delta v_{uv'} + [(\Delta v_{uv'})^2 + 4(B_{uv'}k_{rs})^2]^{1/2}\} \equiv -t_{v'u}; \qquad (19.8)$$

$$A^0_{uv'} \equiv B_{uv'}/(\Delta v_{uv'} + t_{uv'}) = -A^0_{v'u}, \qquad (19.9)$$

where $\Delta v_{uv'}$ represents $(v_u - v_{v'})$.

In the case where $v_u = v_{v'}$, $t_{uv'}$ is equal to $B_{uv'}k_{rs}$, and the MO's and their energies can be expressed as follows:

$$w(\text{upper}) = v_u + B_{uv'}k_{rs} \equiv v_{(u-v')}, \qquad (19.10)$$

$$w(\text{lower}) = v_u - B_{uv'}k_{rs} \equiv v_{(u+v')}; \qquad (19.11)$$

$$\psi(\text{upper}) = 2^{-1/2}(\phi_u - \phi_{v'}) \equiv \phi_{(u-v')}, \qquad (19.12)$$

$$\psi(\text{lower}) = 2^{-1/2}(\phi_u + \phi_{v'}) \equiv \phi_{(u+v')}. \qquad (19.13)$$

In the case where $v_u > v_{v'}$, $t_{uv'}$ can be expressed as follows:

$$t_{uv'} = \tfrac{1}{2}[-\Delta v_{uv'} + \Delta v_{uv'}(1 + 4A^2_{uv'}k^2_{rs})^{1/2}], \qquad (19.14)$$

where $A_{uv'}$ represents $B_{uv'}/\Delta v_{uv'}$. When $|A_{uv'}k_{rs}| \ll 1$, the radical may be expanded so that the following expression is obtained:

$$t_{uv'} = \tfrac{1}{2}[-\Delta v_{uv'} + \Delta v_{uv'}(1 + 2A^2_{uv'}k^2_{rs} - 2A^4_{uv'}k^4_{rs} + \cdots)]$$
$$\doteqdot \Delta v_{uv'}A^2_{uv'}k^2_{rs} = B^2_{uv'}k^2_{rs}/\Delta v_{uv'}. \qquad (19.15)$$

In this case, $t_{uv'} \ll \Delta v_{uv'}$, and hence $A^0_{uv'} \doteqdot A_{uv'}$. Therefore, the MO's of R–S and their energies can be approximately expressed as follows:

$$w(\text{upper}) \doteqdot v_u + B^2_{uv'}k^2_{rs}/\Delta v_{uv'} \equiv v_u^{\text{P}}, \qquad (19.16)$$

$$w(\text{lower}) \doteqdot v_{v'} + B^2_{v'u}k^2_{rs}/\Delta v_{v'u} = v_{v'} - B^2_{uv'}k^2_{rs}/\Delta v_{uv'} \equiv v^{\text{P}}_{v'}; \qquad (19.17)$$

$$\psi(\text{upper}) \doteqdot N(\phi_u - A_{uv'}k_{rs}\phi_{v'}) \equiv \phi_u^{\text{P}}, \qquad (19.18)$$

$$\psi(\text{lower}) \doteqdot N(\phi_{v'} - A_{v'u}k_{rs}\phi_u) = N(\phi_{v'} + A_{uv'}k_{rs}\phi_u) \equiv \phi^{\text{P}}_{v'}. \qquad (19.19)$$

The upper MO may be said to be the MO resulting from perturbation of ϕ_u by interaction with $\phi_{v'}$, and similarly the lower MO may be said to be the MO resulting from perturbation of $\phi_{v'}$ by interaction with ϕ_u. According to perturbation theory, ϕ_u and $\phi_{v'}$ are the zeroth-order approximations to ψ(upper) and ψ(lower), respectively, and v_u and $v_{v'}$ are the first-order approximations to w(upper) and w(lower), respectively; ϕ_u^{P} and $\phi^{\text{P}}_{v'}$ are the first-order approximations to ψ(upper) and ψ(lower), respectively, and v_u^{P} and $v^{\text{P}}_{v'}$ are the second-order approximations to w(upper) and w(lower), respectively. The upper and the lower MO are displaced upward and downward, respectively, as a result of perturbation. Thus it may be said that two interacting energy levels always repel each other. It is evident that the repelling effect is proportional to the square of the overlap of the two interacting orbitals and inversely proportional to the energy difference of the two orbitals.

When more than two orbitals interact, and if the interactions are relatively weak, that is, if the relation $|A_{mn'}k_{rs}| \ll 1$ holds for each pair of orbitals of R and S, the effects of such interactions can be considered to be approximately additive. Subsequently, the following abbreviations, most of which have been used already, will be used:

$$B_{mn'} \equiv b_{m,r}b_{n',s} = B_{n'm}\,; \tag{19.20}$$

$$\Delta v_{mn'} \equiv v_m - v_{n'} = -\Delta v_{n'm}\,; \tag{19.21}$$

$$t_{mn'} \equiv \tfrac{1}{2}\{-\Delta v_{mn'} + [(\Delta v_{mn'})^2 + 4(B_{mn'}k_{rs})^2]^{1/2}\} \equiv -t_{n'm}\,; \tag{19.22}$$

$$A^0_{mn'} \equiv B_{mn'}/(\Delta v_{mn'} + t_{mn'}) = -A^0_{n'm}\,; \tag{19.23}$$

$$A_{mn'} \equiv B_{mn'}/\Delta v_{mn'} = -A_{n'm}\,; \tag{19.24}$$

$$G_{mn'} \equiv B^2_{mn'}/\Delta v_{mn'} = -G_{n'm}\,. \tag{19.25}$$

In the case where the relation $|A_{un'}k_{rs}| \ll 1$ holds for all the combinations of ϕ_u with $\phi_{n'}$'s except $\phi_{v'}$ and where the relation $|A_{v'm}k_{rs}| \ll 1$ holds for all the combinations of $\phi_{v'}$ with ϕ_m's except ϕ_u, the first-order perturbed MO corresponding to ϕ_u and its second-order perturbed energy can be expressed as follows:

$$\phi_u^{\mathrm{P}} = N\left[\phi_u - k_{rs}\sum_{n'\neq v'} A_{un'}\phi_{n'} - k_{rs}A^0_{uv'}\left(\phi_{v'} - k_{rs}\sum_{m\neq u} A_{v'm}\phi_m\right)\right]; \tag{19.26}$$

$$v_u^{\mathrm{P}} = v_u + t_{uv'} + [1 + (k_{rs}A^0_{uv'})^2]^{-1}k^2_{rs}\left[\sum_{n'\neq v'} G_{un'} + (k_{rs}A^0_{uv'})^2 \sum_{m\neq u} G_{v'm}\right]. \tag{19.27}$$

When u and v', and m and n', are interchanged, expressions for $\phi^{\mathrm{P}}_{v'}$ and $v^{\mathrm{P}}_{v'}$ are obtained.

When $v_u = v_{v'}$, evidently the following relations exist: $\Delta v_{uv'} = 0$, $t_{uv'} = k_{rs}B_{uv'}$, and $k_{rs}A^0_{uv'} = 1$. In this case, the zeroth-order MO's arising from the interaction between ϕ_u and $\phi_{v'}$ and their energies are given by expressions (19.10)–(19.13). Perturbation of these MO's by interactions with other orbitals of R and S gives rise to the following first-order perturbed MO's and second-order perturbed energies:

$$\psi(\text{upper}) \doteqdot \phi^{\text{p}}_{(u-v')} = N\left[\phi_u - \phi_{v'} - k_{rs}\left(\sum_{n' \neq v'} A_{un'}\phi_{n'} - \sum_{m \neq u} A_{v'm}\phi_m\right)\right], \tag{19.28}$$

$$w(\text{upper}) \doteqdot v^{\text{p}}_{(u-v')} = v_u + k_{rs}B_{uv'} + k_{rs}^2 \frac{1}{2}\left(\sum_{n' \neq v'} G_{un'} + \sum_{m \neq u} G_{v'm}\right); \tag{19.29}$$

$$\psi(\text{lower}) \doteqdot \phi^{\text{p}}_{(u+v')} = N\left[\phi_u + \phi_{v'} - k_{rs}\left(\sum_{n' \neq v'} A_{un'}\phi_{n'} + \sum_{m \neq n} A_{v'm}\phi_m\right)\right], \tag{19.30}$$

$$w(\text{lower}) \doteqdot v^{\text{p}}_{(u+v')} = v_u - k_{rs}B_{uv'} + k_{rs}^2 \frac{1}{2}\left(\sum_{n' \neq v'} G_{un'} + \sum_{m \neq u} G_{v'm}\right). \tag{19.31}$$

In the case where the two fragments R and S are identical mesomeric systems as in 1,3-butadiene and in biphenyl, it is convenient to take the corresponding orbitals of R and S in the symmetry combinations as follows:

$$2^{-1/2}(\phi_m - \phi_{m'}) \equiv \phi_{(m-m')} \equiv \phi_{m(-)}; \tag{19.32}$$

$$2^{-1/2}(\phi_m + \phi_{m'}) \equiv \phi_{(m+m')} \equiv \phi_{m(+)}\,. \tag{19.33}$$

The first-order perturbed energies of these zeroth-order "minus" and "plus" MO's are, of course, given by

$$v_{m(-)} = v_m + B_{mm'}k_{rs}\,; \tag{19.34}$$

$$v_{m(+)} = v_m - B_{mm'}k_{rs}\,. \tag{19.35}$$

Since $B_{mn'} = B_{m'n}$, $A_{mn'} = A_{m'n}$, and $G_{mn'} = G_{m'n}$, a minus MO can interact only with other minus MO's, and a plus MO can interact only with other plus MO's. Thus the first-order perturbed MO's arising from $\phi_{u(-)}$ and $\phi_{u(+)}$ and their second-order perturbed energies are given by

$$\phi^{\text{p}}_{u(-)} = N\left(\phi_{u(-)} + k_{rs}\sum_{m \neq u} A_{um'}\phi_{m(-)}\right), \tag{19.36}$$

$$\phi^{\text{p}}_{u(+)} = N\left(\phi_{u(+)} - k_{rs}\sum_{m \neq u} A_{um'}\phi_{m(+)}\right); \tag{19.37}$$

$$v^{\text{p}}_{u(-)} = v_u + k_{rs}B_{uu'} + k_{rs}^2 \sum_{m \neq u} G_{um'}\,, \tag{19.38}$$

$$v^{\text{p}}_{u(+)} = v_u - k_{rs}B_{uu'} + k_{rs}^2 \sum_{m \neq u} G_{um'}\,. \tag{19.39}$$

When $v_u > v_{v'}$, and when $|A_{uv'}k_{rs}| \ll 1$, the first-order perturbed MO's corresponding to ϕ_u and $\phi_{v'}$ and their second-order perturbed energies are evidently given by

$$\phi_u{}^{\mathrm{p}} = N\left(\phi_u - k_{rs} \sum_{n'} A_{un'}\phi_{n'}\right), \tag{19.40}$$

$$\phi_{v'}^{\mathrm{p}} = N\left(\phi_{v'} - k_{rs} \sum_{m} A_{v'm}\phi_m\right); \tag{19.41}$$

$$v_u{}^{\mathrm{p}} = v_u + k_{rs}^2 \sum_{n'} G_{un'} \equiv v_u + k_{rs}^2 g_u, \tag{19.42}$$

$$v_{v'}^{\mathrm{p}} = v_{v'} + k_{rs}^2 \sum_{m} G_{v'm} \equiv v_{v'} + k_{rs}^2 g_{v'}. \tag{19.43}$$

19.2.2. The Partial π-Bond Orders of the Essential Single Bond and the Extra-Resonance Energy

On the basis of the treatment developed in the preceding subsection, the partial π-bond orders of the essential single bond r—s connecting two fragments R and S in R–S can be expressed in analytical form. For example, in the case where $v_u = v_{v'}$, it is evident from expression (19.28) that the coefficients of χ_r and χ_s in $\phi_{(u-v')}^{\mathrm{p}}$ can be expressed as follows:

$$c_{(u-v'),r}^{\mathrm{p}} = N\left(b_{u,r} + k_{rs} \sum_{m \neq u} A_{v'm}b_{m,r}\right); \tag{19.44}$$

$$c_{(u-v'),s}^{\mathrm{p}} = -N\left(b_{v',s} + k_{rs} \sum_{n' \neq v'} A_{un'}b_{n',s}\right). \tag{19.45}$$

Accordingly, since N^2 is approximately equal to $\frac{1}{2}$ in this case, the partial π-bond order of the r—s bond in $\phi_{(u-v')}^{\mathrm{p}}$ can be approximately expressed as follows, the small terms which are of second order in k_{rs} being omitted:

$$\begin{aligned} p_{(u-v'),rs}^{\mathrm{p}} &= c_{(u-v'),r}^{\mathrm{p}} c_{(u-v'),s}^{\mathrm{p}} \\ &\doteqdot -\frac{1}{2}\left[B_{uv'} + k_{rs}\left(\sum_{n' \neq v'} G_{un'} + \sum_{m \neq u} G_{v'm}\right)\right]. \end{aligned} \tag{19.46}$$

Similarly, the partial π-bond order of the r—s bond in $\phi_{(u+v')}^{\mathrm{p}}$ can be approximately expressed as follows [see expression (19.30)]:

$$\begin{aligned} p_{(u+v'),rs}^{\mathrm{p}} &= c_{(u+v'),r}^{\mathrm{p}} c_{(u+v'),s}^{\mathrm{p}} \\ &\doteqdot \frac{1}{2}\left[B_{uv'} - k_{rs}\left(\sum_{n' \neq v'} G_{un'} + \sum_{m \neq u} G_{v'm}\right)\right]. \end{aligned} \tag{19.47}$$

When the expressions for the second-order perturbed energies of $\phi^{\mathrm{p}}_{(u-v')}$ and $\phi^{\mathrm{p}}_{(u+v')}$ [expressions (19.29) and (19.31)] are differentiated partially with respect to k_{rs}, the following relations are obtained:

$$\partial v^{\mathrm{p}}_{(u-v')}/\partial k_{rs} = -2p^{\mathrm{p}}_{(u-v'),rs}\,; \tag{19.48}$$

$$\partial v^{\mathrm{p}}_{(u+v')}/\partial k_{rs} = -2p^{\mathrm{p}}_{(u+v'),rs}\,. \tag{19.49}$$

These relations are identical with relation (7.63).

In the case where the relation $|A_{mn'}\,k_{rs}| \ll 1$ holds for all the pairs of ϕ_m and $\phi_{n'}$, the coefficients of χ_r and χ_s in $\phi_u{}^{\mathrm{p}}$ can be expressed as follows [see expression (19.40)]:

$$c^{\mathrm{p}}_{u,r} = Nb_{u,r}\,; \tag{19.50}$$

$$c^{\mathrm{p}}_{u,s} = -Nk_{rs}\sum_{n'} A_{un'}b_{n',s}\,. \tag{19.51}$$

Since N is approximately equal to 1, the partial π-bond order of the r—s bond in $\phi_u{}^{\mathrm{p}}$ can be approximately expressed as follows:

$$p^{\mathrm{p}}_{u,rs} = c^{\mathrm{p}}_{u,r}c^{\mathrm{p}}_{u,s} = -N^2k_{rs}\sum_{n'} G_{un'} \doteqdot -k_{rs}g_u\,. \tag{19.52}$$

When expression (19.42) is differentiated partially with respect to k_{rs}, the following relation is obtained:

$$\partial v_u{}^{\mathrm{p}}/\partial k_{rs} = 2k_{rs}g_u = -2p^{\mathrm{p}}_{u,rs}\,. \tag{19.53}$$

This relation is also identical with relation (7.63).

The resonance energy (RE) of R–S relative to (R + S), that is, the extra-resonance energy of R–S, is defined as the difference in π-electron energy of the ground electron configuration between (R + S) and R–S. The extra-resonance energy can be expressed as follows:

$$\begin{aligned} RE\ (\text{in } -\beta) &= \sum_m n_m v_m + \sum_{n'} n_{n'}v_{n'} - \sum_k n_k w_k \\ &\doteqdot \sum_m n_m(v_m - v_m{}^{\mathrm{p}}) + \sum_{n'} n_{n'}(v_{n'} - v^{\mathrm{p}}_{n'}), \end{aligned} \tag{19.54}$$

where n_m, for example, represents the number of electrons occupying ϕ_m in the ground electron configuration. In the ground electron configuration MO's of R, S, and R–S are usually either doubly occupied or empty. Since $B_{mn'} = B_{n'm}$ and $G_{mn'} = -G_{n'm}$, any raising of an occupied orbital of one fragment by interaction with an occupied orbital of the other fragment is always accompanied with lowering of the latter by the same amount. That is, the effects of the interactions between occupied orbitals of fragments on the total π-electron energy always cancel one another, and they make no net contribution to the extra-resonance energy: only the interactions between the unoccupied orbitals of fragments and the occupied orbitals of the other fragments, which lower the occupied orbitals and raise the unoccupied orbitals, make net contributions to it. Thus, when the relation $|A_{mn'}k_{rs}| \ll 1$ holds for each pair of an unoccupied orbital of one fragment and an occupied orbital of another fragment, the extra-resonance energy can be approximately expressed as follows:

$$RE\ (\text{in}\ -\beta) = -2k_{rs}^2 \left\{ \sum_{m}^{\text{occ}} \sum_{n'}^{\text{unocc}} - \sum_{m}^{\text{unocc}} \sum_{n'}^{\text{occ}} \right\} G_{mn'}, \qquad (19.55)$$

where $\sum^{\text{occ}}$ implies summation over occupied orbitals, $\sum^{\text{unocc}}$ summation over unoccupied orbitals.

19.3. Classification of π–π* Transitions in Composite Systems

As mentioned in Section 6.3, electronic transitions in molecular complexes can be classified into three main types: local, intermolecular charge-transfer, and charge-resonance. Analogously, π–π* transitions in composite molecules can be classified into three main types: local, intramolecular charge-transfer, and charge-resonance.

Suppose that ϕ_m and ϕ_u are, respectively, occupied and unoccupied π MO's of fragment R and that $\phi_{n'}$ and $\phi_{v'}$ are, respectively, occupied and unoccupied π MO's of fragment S. The one-electron transition from $\phi_m{}^{\text{P}}$ to $\phi_u{}^{\text{P}}$ in R–S can be considered to have a character closely similar to that of the one-electron transition from ϕ_m to ϕ_u in fragment R, and the one-electron transition from $\phi_{n'}^{\text{P}}$ to $\phi_{v'}^{\text{P}}$ in R–S can be considered to have a character closely similar to that of the one-electron transition from $\phi_{n'}$ to $\phi_{v'}$ in fragment S. Such transitions in composite molecules are called perturbed local transitions.

On the one-electron transition from $\phi_m{}^{\text{P}}$ to $\phi_{v'}^{\text{P}}$ there must occur a partial electron transfer from R to S. Similarly, the one-electron transition from $\phi_{n'}^{\text{P}}$ to $\phi_u{}^{\text{P}}$ must be accompanied with a partial electron transfer from S to R. Such transitions are called intramolecular electron-transfer or charge-transfer transitions.

When $v_m = v_{n'}$ and $v_u = v_{v'}$, transitions of another type can occur. On the one-electron transition from $\phi_{(m-n')}^{\text{P}}$ to $\phi_{(u+v')}^{\text{P}}$, for example, there must occur substantially no electron-transfer from one fragment to another. Such transitions are called charge-resonance transitions. In the case where the two fragments are identical, all the π–π* transitions are of this type.

19.4. Classification of Composite Conjugated Systems and Steric Effects on π–π* Transitions of Various Types†

19.4.1. Classification of Composite Conjugated Systems

Suppose that an even-membered conjugated system R–S is composed of two even-membered subsystems R and S.[3] The so-called two-directional notation is used here to describe the π MO's of R, S, and R–S. Thus, ϕ_{+1}, $\phi_{+1'}$, and ψ_{+1} refer to the highest bonding π MO's of R, S, and R–S, respectively, and ϕ_{-1}, $\phi_{-1'}$, and ψ_{-1} refer to the lowest antibonding π MO's of R, S, and R–S, respectively. With respect to the heights of the energies of $\phi_{-1'}$ and $\phi_{+1'}$ relative to the energies of ϕ_{-1} and ϕ_{+1}, the following three typical cases are distinguished:

$$\text{(A)} \quad v_{-1} = v_{-1'}; \qquad v_{+1} = v_{+1'}$$

$$\text{(B)} \quad v_{-1} < v_{-1'}; \qquad v_{+1} > v_{+1'}$$

$$\text{(C)} \quad v_{-1} > v_{-1'}; \qquad v_{+1} > v_{+1'}$$

These cases are exemplified by the following conjugated systems: (A) 1,3-butadiene and biphenyl; (B) styrene; (C) phenyl–carbonyl and vinyl–carbonyl.

19.4.2. Cases A and B

Because of the pairing properties of π MO's of even alternant hydrocarbon systems, any linear-conjugated even alternant hydrocarbon, in which both the fragments are also even alternant hydrocarbon systems, belongs to either case A or case B. In case A, ψ_{-1} and ψ_{+1} are regarded as $\phi^{\mathrm{p}}_{-1(+)}$ and $\phi^{\mathrm{p}}_{+1(-)}$, respectively, and the electronic transition from ψ_{+1} to ψ_{-1} is a charge-resonance transition. In case B, ψ_{-1} and ψ_{+1} are regarded as ϕ^{p}_{-1} and ϕ^{p}_{+1}, respectively, and the electronic transition from ψ_{+1} to ψ_{-1} is a perturbed local transition. In either case, the interaction of ϕ_{-1} with $\phi_{-1'}$ and that of ϕ_{+1} with $\phi_{+1'}$ make predominant contributions to the formation of ψ_{-1} and ψ_{+1}, respectively (see Fig. 19.1). As the value of k_{rs} becomes larger, evidently the energies of ψ_{-1} and ψ_{+1} become lower and higher, respectively, and consequently the energy difference between these orbitals becomes smaller. This means that a deviation from coplanarity of fragments

† This section is, so to speak, a revised version of Reference 3.

[3] H. Suzuki, *Bull. Chem. Soc. Japan* **35**, 1853 (1962).

R and S by a twist of the formally single r—s bond in any even alternant hydrocarbon system R–S, which is accompanied with a decrease in the value of k_{rs}, will result in a hypsochromic shift of the absorption band due to the one-electron transition from ψ_{+1} to ψ_{-1}. In particular, in case A the transition energy increases, to a first approximation, linearly with the decreasing value of k_{rs} [see expressions (19.38) and (19.39)]. In either case, ϕ_{-1} and $\phi_{-1'}$ are combined with the same sign in ψ_{-1}, and ϕ_{+1} and $\phi_{+1'}$ are combined

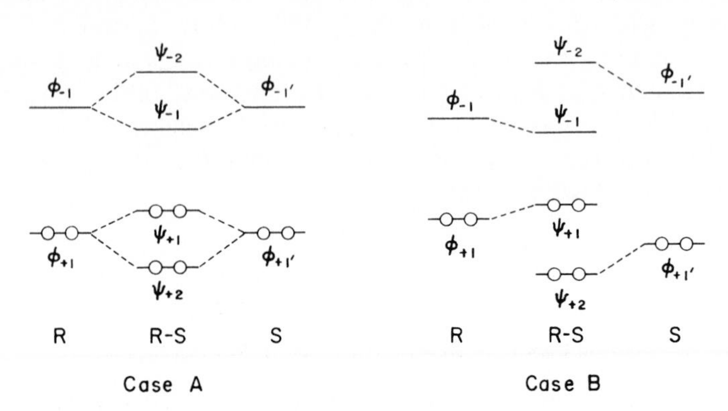

FIG. 19.1. Orbital energy level diagrams of composite π systems: case A and case B. The correlations between the energy levels of the fragments and those of the composite system are denoted with the broken lines.

with opposed signs in ψ_{+1}. Therefore, the partial π-bond orders of the r—s bond in ψ_{-1} and ψ_{+1} are positive and negative, respectively.

Biphenyl is taken as an example of composite conjugated systems belonging to case A: the two benzene rings are taken as fragments R and S, and the coannular bond (the 1—1′ bond) is taken as the r—s bond. By the use of perturbation theory, ψ_{-1} and ψ_{+1} of biphenyl and their energies can be expressed in terms of π MO's of benzene and their energies as follows [see expressions (19.36)–(19.39)]:

$$\psi_{-1} \doteqdot \phi^{\mathrm{P}}_{-1(+)} = N_{-1}(\phi_{-1(+)} - k_{rs} \sum_{m \neq -1} A_{-1,m'}\phi_{m(+)}), \tag{19.56}$$

$$\psi_{+1} \doteqdot \phi^{\mathrm{P}}_{+1(-)} = N_{+1}(\phi_{+1(-)} + k_{rs} \sum_{m \neq +1} A_{+1,m'}\phi_{m(-)}); \tag{19.57}$$

$$w_{-1} = -w_{+1} \doteqdot v^{\mathrm{P}}_{-1(+)} = v_{-1} - k_{rs}B_{-1,-1'} + k_{rs}^2 \sum_{m \neq -1} G_{-1,m'}. \tag{19.58}$$

The values of A's, B's, and G's can easily be calculated from the π MO's of benzene and their energies (see Table 3.5). The normalization factors, N_{-1} and N_{+1}, are given by

$$N_{-1} = N_{+1} = \left(1 + k_{rs}^2 \sum_{m \neq -1} A_{-1,m'}^2\right)^{-1/2} \doteqdot (1 + 0.090k_{rs}^2)^{-1/2}. \quad (19.59)$$

The energy of the one-electron transition from ψ_{+1} to ψ_{-1} is given by

$$\Delta E_{11} \text{ (in } -\beta) = w_{-1} - w_{+1} \doteqdot 2(1 - \tfrac{1}{3}k_{rs} + \tfrac{1}{54}k_{rs}^2). \quad (19.60)$$

The electric moment of this transition can be expressed as follows [see expressions (4.50) or (7.46)]:

$$\mathbf{M}_{11} = 2^{1/2}\int \psi_{-1}\mathbf{r}\psi_{+1}\, d\tau$$

$$\doteqdot 2^{1/2}N_{-1}N_{+1}\Bigg[\mathbf{I}_{-1(+),+1(-)} + k_{rs}\Bigg(\sum_{m \neq +1} A_{+1,m'}\mathbf{I}_{-1(+),m(-)}$$

$$- \sum_{m \neq -1} A_{-1,m'}\mathbf{I}_{m(+),+1(-)}\Bigg) - k_{rs}^2 \sum_{m \neq -1}\sum_{n \neq +1} A_{-1,m'}A_{+1,n'}\mathbf{I}_{m(+),n(-)}\Bigg], \quad (19.61)$$

where, for example,

$$\mathbf{I}_{-1(+),+1(-)} \equiv \int \phi_{-1(+)}\mathbf{r}\phi_{+1(-)}\, d\tau = \frac{1}{2}\left(\int \phi_{-1}\mathbf{r}\phi_{+1}\, d\tau - \int \phi_{-1'}\mathbf{r}\phi_{+1'}\, d\tau\right). \quad (19.62)$$

Components of each of the $\mathbf{I}$'s are zero except in the direction of the long axis of the molecule (the y axis), and hence $\mathbf{M}_{11}$ is in the direction of the y axis. When the length of a C—C bond in the benzene rings and that of the coannular bond are assumed to be 1.40 and 1.50 Å, respectively, the magnitude of $\mathbf{M}_{11}$ can be expressed as follows:

$$M_{11(y)} \text{ (in Å)} \doteqdot 2^{1/2}(1 + 0.090k^2)^{-1}(0.700 + 0.717k - 0.156k + 0.021k^2), \quad (19.63)$$

where k represents k_{rs}. The first of the terms in the last parentheses, 0.700, represents the magnitude of $\mathbf{I}_{-1(+),+1(-)}$. The second term, $0.717k$, corresponds to the term with $\mathbf{I}_{-1(+),-1(-)}$ and $\mathbf{I}_{+1(+),+1(-)}$ in expression (19.61), which originates from the contribution of $\phi_{-1(-)}$ in ψ_{+1} and that of $\phi_{+1(+)}$ in ψ_{-1}. The third term corresponds to the term with $\mathbf{I}_{-1(+),-3(-)}$ and $\mathbf{I}_{+3(+),+1(-)}$. It is noteworthy that the interactions between ϕ_{-1} and $\phi_{+1'}$ and between

$\phi_{-1'}$ and ϕ_{+1} give rise to the second term and thereby are mainly responsible for the increase in the magnitude of the transition moment associated with increase in the value of k, while they exert only a subsidiary effect of the second order in k on the transition energy. These interactions also play a predominant role in stabilizing the ground state of the molecule. It is expected from expression (19.63) that the dipole strength of the transition, $D_{11} = M_{11}^2$, will vary, to a very rough approximation, linearly with k and hence with cos θ, where θ represent the twist angle of the r—s bond.

Styrene belongs to case B. The resonance parameter k for the ethylenic bond is greater than that for a bond in the benzene ring, since the former bond is shorter than the latter. Accordingly, in this system the benzene ring is taken as fragment R and the ethylenic bond as fragment S. When the value of k_{rs} is decreased to zero, the electronic transition from ψ_{+1} to ψ_{-1} in this system is reduced to the electronic transition from ϕ_{+1} to ϕ_{-1} in the benzene ring. In case B expressions of the energy and moment of the $\psi_{+1} \rightarrow \psi_{-1}$ transition are much more complicated than in case A. However, since the energy difference between $\phi_{-1'}$ and ϕ_{-1} as well as that between ϕ_{+1} and $\phi_{+1'}$ is generally small, the situation is considered to be similar to that in case A, and conclusions similar to those described for the case of biphenyl will be drawn.

In general, any linear-conjugated even alternant hydrocarbon system which can be considered as being composed of different even alternant subsystems belongs to case B. The effect of the elongation of a linear conjugated system such as polyene and p-polyphenyl by adding one more mesomeric unit on the first π–π^* transition can be explained by regarding the elongated system as belonging to case B. The original conjugated system is regarded as R, and the added mesomeric unit as S. ϕ_{-1} is lowered as a result of interaction with $\phi_{-1'}$ and gives ψ_{-1} ; ϕ_{+1} is raised as a result of interaction with $\phi_{+1'}$ and gives ψ_{+1}. Consequently, the energy of the first π–π^* transition is decreased by the elongation of the conjugated system. Furthermore, since as the conjugated system R becomes longer coefficients $b_{-1,r}$ and $b_{+1,r}$ decrease and energy differences $v_{-1'} - v_{-1}$ and $v_{+1} - v_{+1'}$ increase (see Tables 10.5 and 12.9), the effect of the perturbation by the addition of one mesomeric unit will become smaller as the conjugated system becomes longer.

19.4.3. Case C

In case C ψ_{-1} and ψ_{+1} can be regarded as $\phi^{\mathrm{p}}_{-1'}$ and ϕ^{p}_{+1}, respectively. Therefore, in this case the first π–π^* transition is an intramolecular charge-transfer transition.

If the relation $|A_{mn'}k_{rs}| \ll 1$ holds for each pair of orbitals of R and S, first-order perturbation theory gives the following expressions for ψ_{-1} and ψ_{+1}:

$$\psi_{-1} \doteqdot \phi^{\mathrm{p}}_{-1'} = N_{-1}\left(\phi_{-1'} - k_{rs} \sum_{m} A_{-1',m}\phi_m\right); \tag{19.64}$$

$$\psi_{+1} \doteqdot \phi^{\mathrm{p}}_{+1} = N_{+1}\left(\phi_{+1} - k_{rs} \sum_{n'} A_{+1,n'}\phi_{n'}\right). \tag{19.65}$$

The second-order approximations to the energies of these orbitals are

$$w_{-1} \doteqdot v^{\mathrm{p}}_{-1'} = v_{-1'} + k^2_{rs} \sum_{m} G_{-1',m} = v_{-1'} + k^2_{rs}g_{-1'}; \tag{19.66}$$

$$w_{+1} \doteqdot v^{\mathrm{p}}_{+1} = v_{+1} + k^2_{rs} \sum_{n'} G_{+1,n'} = v_{+1} + k^2_{rs}g_{+1}. \tag{19.67}$$

Accordingly, the energy separation between ψ_{-1} and ψ_{+1} can be expressed as

$$\Delta E_{11} \text{ (in } -\beta) = \Delta w_{11} \doteqdot \Delta v_{-1',+1} + \Delta g_{-1',+1}k^2_{rs}, \tag{19.68}$$

where $\Delta v_{-1',+1}$ and $\Delta g_{-1',+1}$ represent $v_{-1'} - v_{+1}$ and $g_{-1'} - g_{+1}$, respectively. The partial π-bond orders of the r—s bond in ψ_{-1} and ψ_{+1} are approximately given by

$$p^{(-1)}_{rs} \doteqdot -k_{rs}N^2_{-1}g_{-1'}; \qquad p^{(+1)}_{rs} \doteqdot -k_{rs}N^2_{+1}g_{+1}. \tag{19.69}$$

Accordingly, the change in the π-bond order of the r—s bond on the one-electron transition from ψ_{+1} to ψ_{-1} is

$$\Delta P_{rs} = p^{(-1)}_{rs} - p^{(+1)}_{rs} \doteqdot -k_{rs}(N^2_{-1}g_{-1'} - N^2_{+1}g_{+1}). \tag{19.70}$$

The value of $\Delta g_{-1',+1}$ may be negative, zero, or positive, depending on the system. Now, for simplification, only four orbitals ϕ_{-1}, ϕ_{+1}, $\phi_{-1'}$, and $\phi_{+1'}$ are taken into account. Then, since $G_{mn'} = -G_{n'm}$, $g_{-1'}$, g_{+1}, and $\Delta g_{-1',+1}$ can be expressed as follows:

$$g_{-1'} = -G_{-1,-1'} + G_{-1',+1}; \tag{19.71}$$

$$g_{+1} = -G_{-1',+1} + G_{+1,+1'}; \tag{19.72}$$

$$\Delta g_{-1',+1} = 2G_{-1',+1} - (G_{-1,-1'} + G_{+1,+1'}). \tag{19.73}$$

In these expressions all the G's are positive quantities [see expression (19.25)]. The repulsive interaction between $\phi_{-1'}$ and ϕ_{+1} has an effect of increasing ΔE_{11} through $2G_{-1',+1}k^2_{rs}$, and the interaction between ϕ_{-1} and $\phi_{-1'}$ and that between ϕ_{+1} and $\phi_{+1'}$ have an effect of decreasing it through $-(G_{-1,-1'} + G_{+1,+1'})k^2_{rs}$ [see expression (19.68)]. It is to be remembered here that the strength of the repulsive interaction of two orbitals, say ϕ_m and $\phi_{n'}$, is represented by $G_{mn'}k^2_{rs}$ and that it is proportional to the square of the overlap, and is inversely proportional to the energy separation, of the two orbitals. The typical cases illustrated in Fig. 19.2 may be distinguished. In case C-1, $G_{-1,-1'} > G_{-1',+1}$, $G_{-1',+1} < G_{+1,+1'}$, and hence $g_{-1'} < 0$, $g_{+1} > 0$. In this case, the energy of $\phi^{\text{p}}_{-1'}$ becomes lower and the energy of ϕ^{p}_{+1} becomes higher as the perturbation increases. This means that increasing noncoplanarity of R and S by twist of the *r*—*s* bond will result in a hypsochromic shift of the band due to the one-electron transition from ϕ^{p}_{+1} to $\phi^{\text{p}}_{-1'}$. If it is assumed, for simplification, that the strength of the repulsive interaction between

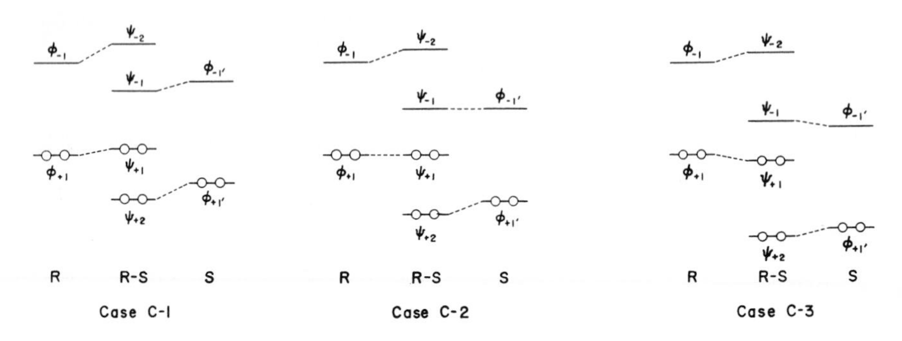

FIG. 19.2. Orbital energy level diagrams of composite π systems: cases C-1, C-2, and C-3.

two energy levels depends only on the energy separation between them, as the energy levels of S become lower, the magnitude of $G_{-1',+1}$ will increase, and the magnitudes of $G_{-1,-1'}$ and $G_{+1,+1'}$ will decrease. Case C-2 is the special case where $G_{-1,-1'} = G_{-1',+1} = G_{+1,+1'}$. In this case, $g_{-1'} = g_{+1} = 0$, and hence $\Delta g_{-1',+1} = 0$. In this case, and more generally, when $\Delta g_{-1',+1} = 0$, the energy separation between $\phi^{\text{p}}_{-1'}$ and ϕ^{p}_{+1} is equal to $\Delta v_{-1',+1}$, independently of the degree of the perturbation, and hence any twist of the *r*—*s* bond will cause no shift of the band due to the one-electron transition from ϕ^{p}_{+1} to $\phi^{\text{p}}_{-1'}$. In case C-3, the repulsive interaction between $\phi_{-1'}$ and ϕ_{+1} plays a predominant role in determining the dependence of the energy separation between $\phi^{\text{p}}_{-1'}$ and ϕ^{p}_{+1} on the degree of the perturbation. In this case a twist of the *r*—*s* bond will cause a bathochromic shift of the band due to the electronic transition from ϕ^{p}_{+1} to $\phi^{\text{p}}_{-1'}$.

Thus, the effect of twist of an essential single bond on the position of the first intramolecular charge-transfer band can be hypsochromic or bathochromic, depending on the manner in which orbitals of fragments linked by the bond interact, while the effect of twist of an essential single bond in an even alternant hydrocarbon on the position of the first π–π* band is always hypsochromic. Furthermore, it will be understood that the position of the intramolecular charge-transfer band is relatively insensitive to twist of the essential single bond, while the position of the first π–π^* band of an even alternant hydrocarbon is in general highly sensitive to the steric factor.

By the use of expressions (19.64) and (19.65) for ψ_{-1} and ψ_{+1}, the electric moment of the first π–π^* transition in case C can be approximately expressed as follows:

$$\mathbf{M}_{11} = 2^{1/2}\int \psi_{-1}\mathbf{r}\psi_{+1}\,d\tau \doteqdot 2^{1/2}\int \phi^{\mathrm{p}}_{-1'}\mathbf{r}\phi^{\mathrm{p}}_{+1}\,d\tau$$
$$= 2^{1/2}k_{rs}N_{-1}N_{+1}\left(\sum_{m} A_{-1',m}\mathbf{I}_{m,+1} + \sum_{n'} A_{+1,n'}\mathbf{I}_{-1',n'}\right), \qquad (19.74)$$

where

$$\mathbf{I}_{m,n} = \int \phi_m\mathbf{r}\phi_n\,d\tau. \qquad (19.75)$$

Transitions between orbitals of different fragments are forbidden in the zero-overlap approximation. The probability of the intramolecular charge-transfer transition from ϕ^{p}_{+1} to $\phi^{\mathrm{p}}_{-1'}$ just arises from the contribution of orbitals of R in $\phi^{\mathrm{p}}_{-1'}$ and that of orbitals of S in ϕ^{p}_{+1}. Evidently, the transition moment of this transition decreases to zero when the value of k_{rs} decreases to zero. Thus, it is expected that the intensity of an intramolecular charge-transfer transition will decrease rapidly when the essential single bond linking two fragments is twisted.

If the assumption is made that N_{-1}, N_{+1}, and **I**'s in expression (19.74) do not vary in magnitude to any appreciable extent with change in the value of k_{rs}, the dipole strength of the intramolecular charge-transfer transition, M_{11}^2, can be considered to be approximately proportional to k_{rs}^2, and hence to $\cos^2\theta_{rs}$, in which θ_{rs} represents the interplanar angle between R and S. Accordingly, since the position of an intramolecular charge-transfer band is in general comparatively insensitive to change in the value of k_{rs}, the oscillator strength of the transition can also be considered to be approximately proportional to k_{rs}^2 and to $\cos^2\theta_{rs}$. Furthermore, if the molar extinction coefficient at the absorption maximum, $\epsilon_{\max}$, can be taken to be proportional to the oscillator strength of the band, the following approximate relation is obtained:

$$\epsilon_{\max} \propto \cos^2\theta. \qquad (19.76)$$

This relation was assumed first by Braude[4] to explain the decrease in the intensity of the first intense band (at about 240 mμ) of acetophenone as well

[4] E. A. Braude, *in* "Determination of Organic Structures by Physical Methods" (E. A. Braude and F. C. Nachod, eds.), Vol. 1, p. 131. Academic Press, New York, 1955; E. A. Braude and F. Sondheimer, *J. Chem. Soc.* **1955,** 3754.

as benzaldehyde associated with introduction of *ortho*-substituents, although his reasoning for this relation is based on quite a wrong ground. As will be given a detailed discussion in Chapter 21, the 240-mμ band of acetophenone as well as that of benzaldehyde can be considered to be due to an intramolecular charge-transfer transition accompanied with a partial electron-transfer from the benzene ring to the carbonyl group. It should be noted that many assumptions have been needed to reach relation (19.76) by the present approach. This relation should be admitted as a very rough approximation to the relation between the intensity of an intramolecular charge-transfer band and the angle of twist of an essential single bond.

In case C the contribution of unoccupied orbitals of S in occupied orbitals of R–S is usually greater than that of unoccupied orbitals of R. That is, the contribution of $\phi_{-n'}$'s in ϕ^{p}_{+m}'s is usually greater than that of ϕ_{-m}'s in $\phi^{\mathrm{p}}_{+n'}$'s. Therefore, in the ground state of R–S there is usually a partial electron-transfer from R to S. Thus, if R is a hydrocarbon system, S is an electron-withdrawing or electron-accepting substituent. The lower the energies of low-lying unoccupied π orbitals of the substituent, the stronger will be the electron-accepting property of the substituent. Carbonyl and cyano groups are typical electron-accepting groups, and composite systems in which R is a hydrocarbon system and S is such a group are typical systems belonging to case C. Nitro and carboxyl groups are also strong electron-acceptors. Since these groups have three π centers, they have a π orbital corresponding to the nonbonding orbital of an odd alternant hydrocarbon, $\phi_{0'}$, in addition to $\phi_{-1'}$ and $\phi_{+1'}$. The lower two orbitals, $\phi_{+1'}$ and $\phi_{0'}$, are each doubly occupied. However, as will be described more fully in Chapter 22, the pseudo-nonbonding π orbital, $\phi_{0'}$, has no (or substantially no) contribution of the AO of the central atom (the nitrogen atom in the nitro group or the carbon atom in the carboxyl group), that is, $b_{0',s} = 0$, and hence, when such a group is substituted into a hydrocarbon, this orbital cannot interact with any π orbitals of the hydrocarbon. Therefore, composite conjugated systems R–S in which R is a hydrocarbon system and S is a nitro or a carboxyl group can be treated as belonging to case C, quite similarly to systems in which S is a carbonyl or a cyano group.

19.4.4. Case D

When R is an even alternant hydrocarbon system such as a vinyl group or a benzene ring and S is a substituent bearing lone-pair electrons of π character such as an amino, a hydroxyl, or a halogen group the energy level diagram for the composite molecule R–S may be written as in Fig. 19.3.

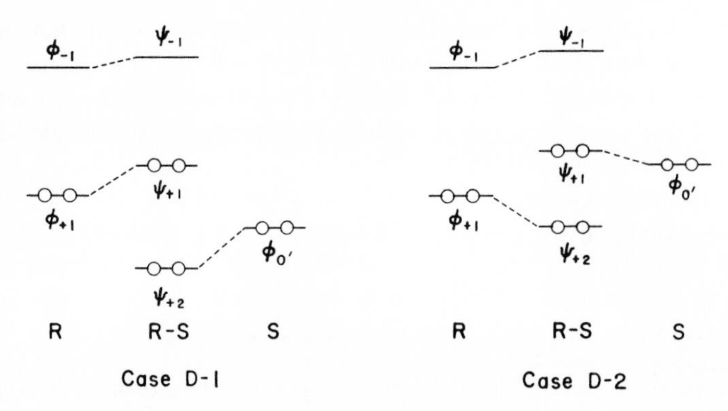

FIG. 19.3. Orbital energy level diagrams of composite π systems: cases D-1 and D-2.

In this figure, $\phi_{0'}$ represents the lone-pair orbital of the substituent. In case D-1, $v_{+1} > v_{0'}$; in case D-2, $v_{-1} > v_{0'} > v_{+1}$. As will be evident from this figure, when the conformation of R–S deviates from the coplanarity of R and S, in case D-1 the perturbed local transition $\phi^{\mathrm{p}}_{+1} \rightarrow \phi^{\mathrm{p}}_{-1}$ will undergo a hypsochromic shift and the intramolecular charge-transfer transition $\phi^{\mathrm{p}}_{0'} \rightarrow \phi^{\mathrm{p}}_{-1}$, which is accompanied with a partial electron-transfer from S to R, will undergo a bathochromic shift. On the other hand, in case D-2, a twist of the *r*—*s* bond will exert a bathochromic effect on the perturbed local transition and a hypsochromic effect on the intramolecular charge-transfer transition. In both the cases, the intensity of the intramolecular charge-transfer transition will rapidly decrease when the *r*—*s* bond is twisted.

Since unoccupied orbitals of R, ϕ_{-m}'s, contribute to the occupied orbital $\phi^{\mathrm{p}}_{0'}$, in the ground state of R–S there must be a partial electron-transfer from S to R. The substituent S is thus an electron-repelling or electron-donating mesomeric substituent. If it is assumed that the overlap of $\phi_{0'}$ with an orbital of R is constant independently of the kind of the substituent, the degree of the mixing of $\phi_{0'}$ with an orbital of R is determined by the energy separation between the two orbitals. The lower the energy of the lone-pair orbital of the substituent S, the smaller will be the degree of the mixing of the orbital and the unoccupied orbitals of the hydrocarbon system R, and hence the weaker will be the electron-donating power of the substituent.

19.4.5. INTERACTION OF TWO ODD ALTERNANT HYDROCARBON SUBSYSTEMS

The discussion has so far been made on cases where a conjugated system is divided into two fragments at an essential single bond. Let us consider here the case where an even alternant hydrocarbon is divided into two fragments R and S at an essential double bond. In this case, both the two fragments are odd alternant hydrocarbon systems, and hence each of the fragments has a nonbonding π orbital, which contains a single electron. The nonbonding orbital of R is denoted by ϕ_0, and that of S by $\phi_{0'}$. Then, the zeroth-order approximations to the lowest unoccupied and highest occupied orbitals of the composite molecule R–S are given by

$$\psi_{-1} \doteqdot 2^{-1/2}(\phi_0 - \phi_{0'}) \equiv \phi_{0(-)} ; \tag{19.77}$$

$$\psi_{+1} \doteqdot 2^{-1/2}(\phi_0 + \phi_{0'}) \equiv \phi_{0(+)} . \tag{19.78}$$

The first-order approximations to the energies of these orbitals are given by

$$w_{-1} \doteqdot v_{0(-)} = +b_{0,r}b_{0',s}k_{rs} ; \tag{19.79}$$

$$w_{+1} \doteqdot v_{0(+)} = -b_{0,r}b_{0',s}k_{rs} . \tag{19.80}$$

Accordingly, the energy of the one-electron transition from ψ_{+1} to ψ_{-1} is approximately given by $2b_{0,r}b_{0',s}k_{rs}$ (in $-\beta$). A twist of the essential double bond should therefore exert a bathochromic effect on that transition.

19.5. Treatment of the Effect of a Methyl Substituent on Absorption Bands by the Perturbation Method

We will consider in this section the case where R is a π system and S is a methyl group. The electron-donating inductive effect of the methyl group is usually treated by assigning a small negative value (−0.3–−0.5) to the Coulomb parameter h for the carbon atom to which the methyl group is attached (see Sections 9.2 and 9.4). As mentioned in Subsection 7.3.1, when the Coulomb integral for the carbon atom at position r is changed from α to $\alpha + h_r\beta$, the energy of a π orbital of the parent system, say ϕ_m, is changed from v_m approximately to $v_m - b_{m,r}^2 h_r$, in which $b_{m,r}$ is the coefficient of the AO at the r position, χ_r, in MO ϕ_m. Since the value of the Coulomb parameter for the methyl substituent is negative, this means that the energy of a π MO is raised by the methyl substitution in proportion to the square of the coefficient of the AO at the substitution position in the MO. The energy of the one-electron transition from ψ_{+1} to ψ_{-1} in the methyl-substituted system is therefore given approximately by

$$\Delta w_{11} \doteqdot v_{-1}^{\mathrm{p}} - v_{+1}^{\mathrm{p}} = v_{-1} - v_{+1} - (b_{-1,r}^2 - b_{+1,r}^2)h_r . \tag{19.81}$$

Since the absolute values of the coefficients of an AO in ϕ_{-1} and ϕ_{+1} in nonalternant hydrocarbons and conjugated systems containing one or more heteroatoms are in general different, the energy of the transition from ϕ_{+1} to ϕ_{-1} in such systems is in general sensitive to the inductive effect of the methyl substituent. The methyl substitution will increase or decrease the energy of the transition according as $b_{-1,r}^2$ is larger or smaller than $b_{+1,r}^2$. On the other hand, in even alternant hydrocarbons, $b_{-1,r}^2 = b_{+1,r}^2$, and hence the inductive effect of the methyl substituent exerts, to a first approximation, no net influence on the energy of the first π–π^* transition. Thus, whereas in nonalternant hydrocarbons and in heteroatom-containing π systems both the inductive and conjugative (hyperconjugative) effects of the methyl group are important, in even alternant hydrocarbons only the conjugative effect is important.

The conjugative effect of the methyl group has been treated in the simple LCAO MO method by using either the pseudo-heteroatom model or the modified vinyl model.[5] In the first model, the methyl group is regarded as a pseudo-heteroatom which contributes a pair of electrons to the π system to which the methyl group is attached.[6] The lone-pair orbital of π symmetry of this pseudo-heteroatom is denoted by $\phi_{0'}$, and as usual its energy is represented by $\eta_{0'} = \alpha - v_{0'}\beta$. Various values covering the range -1.4–-3.3 have been used as the value of $v_{0'}$, and values of about 0.7 to 0.8 have been used as the value of k_{rs}. Values of $v_{0'} = -2$ and $k_{rs} = 0.7$ are said to be most suitable for the simple LCAO MO method in the zero-overlap approximation.[5] When this model is used, the methyl-substituted π system R–S is considered to correspond to case D-1. Since $b_{0',s}^2 = 1$, the second-order approximations to the energies of ϕ_{-1}^{p} and ϕ_{+1}^{p} are given by

$$w_{-1} \doteqdot v_{-1}^{\mathrm{p}} = v_{-1} + k_{rs}^2 b_{-1,r}^2/(v_{-1} - v_{0'}); \tag{19.82}$$

$$w_{+1} \doteqdot v_{+1}^{\mathrm{p}} = v_{+1} + k_{rs}^2 b_{+1,r}^2/(v_{+1} - v_{0'}). \tag{19.83}$$

If R is an even alternant hydrocarbon system, since $b_{-1,r}^2 = b_{+1,r}^2$ and $v_{-1} = -v_{+1}$, the energy of the first π–π^* transition in R–S can be expressed as follows:

$$\Delta w_{11} \doteqdot v_{-1}^{\mathrm{p}} - v_{+1}^{\mathrm{p}} = v_{-1} - v_{+1} + 2k_{rs}^2 b_{+1,r}^2 v_{+1}/(v_{0'}^2 - v_{+1}^2). \tag{19.84}$$

Since v_{+1} is negative and $(v_{0'}^2 - v_{+1}^2)$ is positive, the last term, that is, the perturbation term, is a negative quantity. This means that the methyl substitution will cause a bathochromic shift of the first π–π^* transition.

[5] A. Streitwieser, Jr., "Molecular Orbital Theory for Organic Chemists," Sect. 5.7. Wiley, New York and London, 1961.

[6] F. A. Matsen, *J. Am. Chem. Soc.* **72,** 5243 (1950).

In the second model, the methyl group is regarded as a modified vinyl group, —Y=Z, in which Y corresponds to the carbon atom of the methyl group and Z is the pseudo-atom corresponding to the three hydrogen atoms, H_3.[7] The pseudo-π-orbitals of this modified vinyl group, $\phi_{-1'}$ and $\phi_{+1'}$, may be considered to be the orbitals composed of the $2p\pi$ AO of the carbon atom and the group orbital of π symmetry as a proper linear combination of three hydrogen $1s$ AO's, and alternatively, may be considered to be linear combinations of π symmetry of three antibonding C—H σ orbitals and of three bonding C—H σ orbitals. In their original calculations of hyperconjugation, in which overlap integrals were included, Mulliken, Rieke, and Brown[7] used values of 0.8 and 5 for k_{CY} (that is, k_{rs}) and k_{YZ}, respectively, and assumed that all the Coulomb parameters, h_C, h_Y, and h_Z, were zero. This method was adopted in some subsequent work, although most recent calculations have used different parameter values. For example, Peters[8] used values of 0.51 and 2.5 for k_{CY} and k_{YZ}, respectively, and, assuming all the Coulomb parameters to be zero, and neglecting overlap integrals, he accounted satisfactorily for the bathochromic shift of the p bands of aromatic hydrocarbons caused by introduction of a methyl substituent.

When the modified vinyl model is adopted, composite molecules R–S in which R is a π system and S is a methyl substituent are considered to correspond to case B. The second-order approximations to the energies of ψ_{-1} and ψ_{+1} of R–S are given by

$$w_{-1} \doteqdot v^{\mathrm{p}}_{-1} = v_{-1} - k^2_{rs}b^2_{-1,r}[b^2_{-1',s}/(v_{-1'} - v_{-1}) - b^2_{+1',s}/(v_{-1} - v_{+1'})]; \quad (19.85)$$

$$w_{+1} \doteqdot v^{\mathrm{p}}_{+1} = v_{+1} + k^2_{rs}b^2_{+1,r}[b^2_{+1',s}/(v_{+1} - v_{+1'}) - b^2_{-1',s}/(v_{-1'} - v_{+1})]. \quad (19.86)$$

If Coulomb parameters h_Y and h_Z are taken to be zero, $v_{-1'} = -v_{+1'} = k_{YZ}$, and $b^2_{-1',s} = b^2_{+1',s} = \frac{1}{2}$. Therefore, if R is an even alternant hydrocarbon system, the energy of the one-electron transition from ψ_{+1} to ψ_{-1} can be expressed as follows:

$$\Delta w_{11} \doteqdot v^{\mathrm{p}}_{-1} - v^{\mathrm{p}}_{+1} = v_{-1} - v_{+1} + 2k^2_{rs}b^2_{+1,r}v_{+1}/(v^2_{+1'} - v^2_{+1}). \quad (19.87)$$

Compare this expression with expression (19.84). Thus, whether we adopt the pseudo-heteroatom model or the modified vinyl model, we expect that the magnitude of the bathochromic shift of the p band of an aromatic compound due to the conjugative effect of the methyl substituent will be proportional to the square of the coefficient of the AO at the substitution position in the highest bonding π MO. This expectation has been approximately borne out.

[7] R. S. Mulliken, C. A. Rieke, and W. G. Brown, *J. Am. Chem. Soc.* **63,** 41 (1941); R. S. Mulliken and C. A. Rieke, *ibid.* **63,** 1770 (1941).

[8] D. Peters, *J. Chem. Soc.* **1957,** 646.

19.6. Mixing of Electron Configurations Formed of Orbitals of Fragments in the One-Electron Approximation

The electronic states of a composite molecule, R–S, can be considered as arising from mixing of electron configurations formed of orbitals localized on fragments, R and S. The present section deals with the mixing under the influence of an approximate total electronic Hamiltonian as a sum of one-electron Hamiltonians.

As mentioned already, electron configurations formed of orbitals localized on fragments fall into three classes: the ground, LE, and CT configurations. The effective one-electron Hamiltonian for R–S, $\mathbf{h}^{\mathrm{RS}}$, is given by expression (7.104). The approximate total π-electronic Hamiltonian used here, $\mathbf{H}^{\mathrm{RS}}$, is a sum of such one-electron Hamiltonians. Since $\mathbf{H}^{\mathrm{RS}}$ contains only one-electron terms, it has only the effect of mixing together electron configurations which differ by the position of just one electron. Furthermore, matrix elements of $\mathbf{h}^{\mathrm{RS}}$ between orbitals localized on the same fragment are approximately zero [see expressions (7.106) and (7.107)]. Therefore, in the present approximation the only effect of conjugation between the two fragments is to mix CT configurations with the ground configuration and with LE configurations. CT configurations cannot mix with one another, and LE configurations cannot mix either with one another or with the ground configuration.

As usual, the ground configuration is denoted by V_0, and the singly-excited singlet configuration which arises from excitation of one electron from an occupied orbital ϕ_k to an unoccupied orbital ϕ_r is denoted by V_1^{kr}. When ϕ_a and ϕ_b are orbitals of different fragments, the following matrix elements of $\mathbf{H}^{\mathrm{RS}}$ can be nonzero:

$$\langle V_0| \mathbf{H}^{\mathrm{RS}} |V_1^{ab}\rangle = 2^{1/2}\langle\phi_a| \mathbf{h}^{\mathrm{RS}} |\phi_b\rangle = 2^{1/2}B_{ab}k_{rs}\beta; \tag{19.88}$$

$$\langle V_1^{ca}| \mathbf{H}^{\mathrm{RS}} |V_1^{cb}\rangle = \langle\phi_a| \mathbf{h}^{\mathrm{RS}} |\phi_b\rangle = B_{ab}k_{rs}\beta; \tag{19.89}$$

$$\langle V_1^{ac}| \mathbf{H}^{\mathrm{RS}} |V_1^{bc}\rangle = -\langle\phi_a| \mathbf{h}^{\mathrm{RS}} |\phi_b\rangle = -B_{ab}k_{rs}\beta. \tag{19.90}$$

In the present approximation, neglecting electronic interactions, the energy difference between V_1^{ab} and V_0, that between V_1^{cb} and V_1^{ca}, and that between V_1^{ac} and V_1^{bc} are, of course, equal to the energy difference between ϕ_b and ϕ_a. The interaction of electron configurations can be treated by the use of perturbation theory. Thus, for example, the orbital energy diagrams for cases A and B in Fig. 19.1 and those for cases C-1 and C-3 in Fig. 19.2 can be replaced by the term energy diagrams shown in Fig. 19.4. In these diagrams energy levels for configurations or states of different classes are distinguished by the use of lines of different types. *LE*(R) and *LE*(S) represent V_1^{11} and $V_1^{1'1'}$, respectively, and *CT*(R → S) and *CT*(S → R) represent $V_1^{11'}$ and $V_1^{1'1}$, respectively.

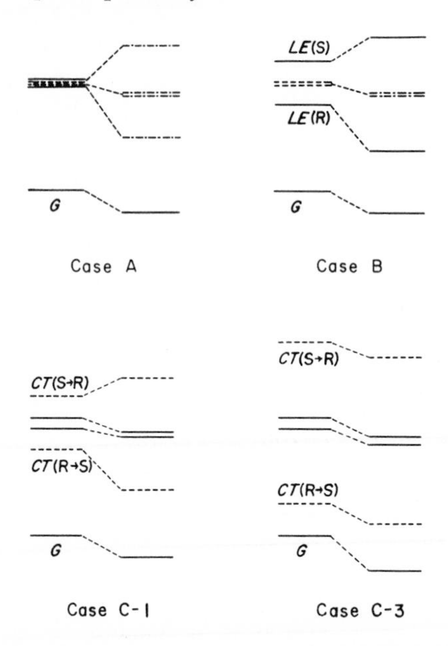

FIG. 19.4. Term energy level diagrams of composite π systems: cases A, B, C-1, and C-3. The energy levels before and after the interaction between the two fragments R and S of the composite system R–S are shown, and the correlations are denoted with the broken lines. Energy levels for electron configurations and states of different classes are distinguished by the use of lines of different types: ———, the ground configuration, the ground state, LE configurations, and perturbed LE states; – – – –, CT configurations and states; - - - - -, charge-resonance states.

In case A, the four electron configurations have the same energy, and hence there should be the first-order interactions between these configurations. Therefore, the lowest excited electronic state of the composite system will be lowered rapidly by increasing conjugation between the two fragments. It is convenient to take the two degenerate LE configurations and the two degenerate CT configurations in the symmetry combinations as follows:

$$2^{-1/2}(V_1^{11} \pm V_1^{1'1'}) \equiv {}^1LE^{11(\pm)} \equiv {}^1Ex^{11(\pm)}; \tag{19.91}$$

$$2^{-1/2}(V_1^{11'} \pm V_1^{1'1}) \equiv {}^1CT^{11'(\pm)} \equiv {}^1CR^{11'(\pm)}. \tag{19.92}$$

In the present approximation, these combinations have, of course, the same energy as the original configurations. However, as will be mentioned in the next chapter, when the interelectronic interactions are included the former combinations split into different energy levels. These combinations are called exciton configurations. The latter combinations are called charge-resonance configurations. When both the fragments are even alternant hydrocarbon systems, of the interaction matrix elements between these four combinations only the one between the "minus" exciton configuration, ${}^1Ex^{11(-)}$, and the "minus" charge-resonance configuration, ${}^1CR^{11'(-)}$, can be nonzero, and it is given by $2B_{+1,+1'}k_{rs}\beta$. The lowest excited electronic state formed from these configurations is evidently

$$2^{-1/2}({}^1Ex^{11(-)} + {}^1CR^{11'(-)}) = 2^{-1}(V_1^{11} - V_1^{1'1'} + V_1^{11'} - V_1^{1'1}),$$

independently of the value of k_{rs}. The ground electron configuration, V_0, can interact only with the "plus" charge-resonance configuration, ${}^1CR^{11'(+)}$. Evidently, the contribution of the "plus" charge-resonance configuration to the ground electronic state is responsible both for the stabilization of the ground state by the conjugation and for the intensification of the first π–π^* transition by the increasing conjugation.

In case B, the LE(R) configuration, V_1^{11}, is the lowest excited configuration, lying below the CT configurations. Therefore, the lowest excited state of the composite system has a predominant contribution of this LE configuration, and its energy is lowered increasingly by the increasing conjugation owing to the repulsive interaction with the higher-lying CT configurations, since there is no interaction between the LE configuration and the ground configuration. When both the fragments are even alternant hydrocarbon systems, the two CT configurations are degenerate and hence can form charge-resonance configurations. The LE(R) configuration can interact only with the "minus" charge-resonance configuration: the corresponding interaction matrix element is $2^{1/2}B_{+1,+1'}k_{rs}\beta$.

In case C, the CT(R → S) configuration is the lowest excited configuration, lying between the ground configuration and an LE configuration. The CT configuration and the ground configuration repel each other and so do the CT configuration and the LE configuration. Therefore, the energy of the lowest excited state of the composite system, which must has a predominant contribution of the CT configuration, relative to the ground state will be little changed by conjugation. If the CT configuration lies closer to the LE configurations as in case C-1, the energy of the CT state relative to the ground state will be lowered by the increasing conjugation. On the other hand, if the CT configuration lies closer to the ground configuration as in case C-3, the energy of the CT state relative to the ground state will be raised by the increasing conjugation.

Chapter 20

Advanced Composite-Molecule Method

20.1. Principle

The method of composite molecule in the one-electron approximation (the simple composite-molecule method), developed in the preceding chapter, is not a method basically different from the conventional simple LCAO MO method, but is only a modification or abbreviation of it. This method should have essentially the same limitations in interpreting electronic spectra as described for the simple MO method in Chapter 10. Thus, although this method has an advantage of being able to express in analytical form the effect of a mesomeric substituent and the effect of the decrease in conjugation across an essential single bond due to twist of that bond on absorption bands which can be considered as being approximately due to a one-electron transition from an occupied orbital to an unoccupied orbital, it cannot give an overall correct picture of electronic spectra. For example, we might be able to predict the effect of a mesomeric substituent on the p bands of aromatic hydrocarbons by using this method, but this is not likely to be successful for the α and β bands. As has been emphasized, a satisfactory interpretation of the whole features of electronic spectra can be given only by taking account of the interelectronic interactions.

In the method of composite molecule including the electronic interactions (the advanced composite-molecule method), which will be dealt with in the present chapter, the π-electronic states of a conjugated system are described as arising from mixing of electron configurations expressed in terms of π orbitals localized on its fragments under the influence of the correct total π-electronic Hamiltonian ($\mathbf{H}$) for the system. This method can be considered as a modification of the semiempirical LCAO ASMO CI method, which was described in Chapter 11.

As before, a conjugated system is represented by R–S, and its fragments by R and S. The correct total π-electronic Hamiltonian for the composite system R–S is denoted by $\mathbf{H}^{RS}$, and those for the separated fragments R and S are denoted by $\mathbf{H}^{R}$ and $\mathbf{H}^{S}$, respectively. An LE configuration is represented by R_mS_0 or $R_0S_{n'}$, in which R_m represents an excited state m of the separated fragment R, which is an eigenfunction of $\mathbf{H}^{R}$, obtained as a result of configuration interaction in R, and analogously $S_{n'}$ represents an excited state n' of the separated fragment S. R_0 and S_0 represent the ground states of R and S, respectively. The unperturbed ground electron configuration of the composite system is represented by R_0S_0. Alternatively, R_mS_0 is represented by ${}^1LE_m$ (or ${}^1LE_{m(\mathrm{R})}$) or ${}^3LE_m$ (or ${}^3LE_{m(\mathrm{R})}$) according as the configuration is singlet or triplet. Analogously, $R_0S_{n'}$ is represented by ${}^1LE_{n'}$ (or ${}^1LE_{n'(\mathrm{S})}$) or ${}^3LE_{n'}$ (or ${}^3LE_{n'(\mathrm{S})}$). The unperturbed ground configuration R_0S_0 is also represented by V_0 or G. The singlet CT(R → S) configuration which arises from the one-electron excitation from an occupied orbital ϕ_k of R to an unoccupied orbital $\phi_{r'}$ of S is represented by $V_1^{kr'}$ or ${}^1CT^{kr'}$, and similarly the singlet CT(S → R) configuration which arises from the one-electron excitation from an occupied orbital $\phi_{k'}$ of S to an unoccupied orbital ϕ_r of R is represented by $V_1^{k'r}$ or ${}^1CT^{k'r}$. The corresponding triplet CT configurations are represented by $T_1^{kr'}$ and $T_1^{k'r}$ or by ${}^3CT^{kr'}$ and ${}^3CT^{k'r}$.

The energies of LE configurations, R_mS_0's and $R_0S_{n'}$'s, and the dipole strengths of the transitions from the ground configuration to these LE configurations can be obtained from the spectra of the separated fragments R and S. Therefore, the method of composite molecule including the electronic interactions has an advantage that it can readily correlate the spectrum of a composite molecule to those of molecules corresponding to its fragments.

The general expressions of matrix elements of the total π-electronic Hamiltonian between electron configurations in terms of one- and two-electron integrals over MO's were given in Chapter 11 [see expressions (11.23)–(11.25)]. These general expressions can also be used in the present case, although in the present case the MO's are not the π MO's (ψ's) of the whole system but are the π MO's (ϕ's) localized on the fragments.

If ϕ_k and $\phi_{r'}$ are orbitals localized on different fragments R and S, respectively, the product of the coefficients of an AO χ_p in ϕ_k and $\phi_{r'}$, $b_{k,p}b_{r',p}$, is always zero for each AO, and hence in the zero-differential-overlap approximation two-electron integrals of type $[kr' \mid ll]$ or $[kr' \mid ls]$ are all zero. In this case, when core resonance integrals between nonneighboring AO's are neglected as usual, evidently $F_{[kr']}$ can be expressed as follows:

$$F_{[kr']} = h^{\mathrm{C}}_{[kr']} = b_{k,r}b_{r',s}\beta^{\mathrm{C}}_{rs} = B_{kr'}k_{rs}\beta^{\mathrm{C}}, \tag{20.1}$$

where β_{rs}^{C} is the core resonance integral for the r—s bond, which connects together the two fragments R and S, and β^{C} is the standard core resonance integral.

If ϕ_k and ϕ_r are orbitals localized on the same fragment R, $F_{[kr]}$ can be expressed as follows:

$$F_{[kr]} = {}^{R}h_{[kr]}^{C} + {}^{S}v_{[kr]}^{C} + \sum_{f}^{occ}\{2[kr\,|\,ff] - [kf\,|\,rf]\} + \sum_{f'}^{occ}\{2[kr\,|\,f'f'] - [kf'\,|\,rf']\}, \tag{20.2}$$

where ${}^{R}h_{[kr]}^{C}$ is the matrix element between ϕ_k and ϕ_r of the core Hamiltonian for the subsystem R, and ${}^{S}v_{[kr]}^{C}$ is that of the core potential operator for the subsystem S. The quantity ${}^{S}v_{[kr]}^{C}$ represents the energy of the attractive interaction between the π-electron density distribution $\phi_k\phi_r$ and the core potential of S, and the quantity $\sum_{f'}^{occ} 2[kr\,|f'f']$ represents the energy of the repulsive interaction between that π-electron density distribution and the total π-electron density distribution in S. These two quantities can be considered to cancel each other approximately in expression (20.2). The terms of $[kf'\,|\,rf']$ of course vanish. Therefore, expression (20.2) can be approximately rewritten as follows:

$$F_{[kr]} \doteqdot {}^{R}h_{[kr]}^{C} + \sum_{f}^{occ}\{2[kr\,|\,ff] - [kf\,|\,rf]\} = {}^{R}F_{[kr]}. \tag{20.3}$$

The ${}^{R}F_{[kr]}$ is $F_{[kr]}$ in the separated fragment R. Quite analogously, if $\phi_{k'}$ and $\phi_{r'}$ are orbitals localized on the fragment S, $F_{[k'r']}$ can be approximately expressed as follows:

$$F_{[k'r']} \doteqdot {}^{S}h_{[k'r']}^{C} + \sum_{f'}^{occ}\{2[k'r'\,|f'f'] - [k'f'\,|\,r'f']\} = {}^{S}F_{[k'r']}. \tag{20.4}$$

Let us consider first the interactions between LE configurations. The interactions between LE configurations arising from electronic excitations in the same fragment, that is, the interactions between LE(R) configurations or between LE(S) configurations, will be independent of whether the two fragments are juxtaposed or not. Therefore, it is convenient to begin by allowing electron configurations of each separated fragment to mix with one another under the influence of the Hamiltonian for the fragment and to use the resulting LE states of each fragment as R_m's or $S_{n'}$'s, so that the LE configurations arising from electronic excitations in the same fragment, R_mS_0's or $R_0S_{n'}$'s, will not interact further with one another and with the ground configuration R_0S_0 when the two fragments are brought to interact with each other.

If, as mentioned above, the LE(R) configuration $R_m S_0$ is an eigenfunction of $\mathbf{H}^{\mathrm{R}}$ and is expressed as a linear combination of V_1^{kr}'s or of T_1^{kr}'s, and if the LE(S) configuration $R_0 S_{n'}$ is an eigenfunction of $\mathbf{H}^{\mathrm{S}}$ and is expressed as a linear combination of $V_1^{l's'}$'s or of $T_1^{l's'}$'s, the matrix element of $\mathbf{H}^{\mathrm{RS}}$ between these two LE configurations can be expanded in terms of integrals of type $\langle V_1^{kr} | \mathbf{H}^{\mathrm{RS}} | V_1^{l's'} \rangle$ or $\langle T_1^{kr} | \mathbf{H}^{\mathrm{RS}} | T_1^{l's'} \rangle$. Evidently, the following expressions are derived from expressions (11.24) and (11.25):

$$\langle V_1^{kr} | \mathbf{H}^{\mathrm{RS}} | V_1^{l's'} \rangle = 2\, [kr \mid l's']; \tag{20.5}$$

$$\langle T_1^{kr} | \mathbf{H}^{\mathrm{RS}} | T_1^{l's'} \rangle = 0. \tag{20.6}$$

It is to be noted that two singlet LE configurations arising from electronic excitations in different fragments, that is, a ^{1}LE(R) and a ^{1}LE(S) configuration, can interact with each other only by virtue of the interelectronic interaction terms in the Hamiltonian, and that two triplet LE configurations arising from electronic excitations in different fragments, that is, a ^{3}LE(R) and a ^{3}LE(S) configuration, cannot interact with each other.

The transition moment of the local transition in R from V_0 to V_1^{kr}, $\mathbf{M}_{kr}$, and that of the local transition in S from V_0 to $V_1^{l's'}$, $\mathbf{M}_{l's'}$, are given by

$$\mathbf{M}_{kr} = 2^{1/2} \int \phi_k{}^*(i) \phi_r(i) \mathbf{r}_i \, d\tau_i \, ; \tag{20.7}$$

$$\mathbf{M}_{l's'} = 2^{1/2} \int \phi_{l'}^*(j) \phi_{s'}(j) \mathbf{r}_j \, d\tau_j \, . \tag{20.8}$$

The quantities $2^{1/2}\phi_k{}^*(i)\phi_r(i)$ and $2^{1/2}\phi_{l'}^*(j)\phi_{s'}(j)$ are real one-electron densities called the transition densities of the transitions $V_0 \rightarrow V_1^{kr}$ and $V_0 \rightarrow V_1^{l's'}$, respectively. The matrix element of $\mathbf{H}^{\mathrm{RS}}$ between V_1^{kr} and $V_1^{l's'}$, given by expression (20.5), is just the mutual electrostatic interaction energy of the two transition densities, one centered on R, the other on S. If this energy of the interaction of the two transition densities is expanded in a multipole series, the first term will be a dipole–dipole term, which represents the energy of the interaction of the two electric dipole moments $\mathbf{M}_{kr}$ and $\mathbf{M}_{l's'}$. The energy of the electrostatic interaction of two electric dipoles ***a*** of magnitude μ_a and ***b*** of magnitude μ_b with the relative orientations as illustrated in Fig. 20.1 is given by

$$\Gamma_{ab} = -(\mu_a \mu_b / r^3)(2 \cos \theta_a \cos \theta_b - \sin \theta_a \sin \theta_b \cos \varphi), \tag{20.9}$$

where r is the distance between the centers of the two dipoles. In this way, the energy of the electrostatic interaction of the two electric dipole moments $\mathbf{M}_{kr}$ and $\mathbf{M}_{l's'}$ depends on the magnitudes and relative orientations of the two moments and varies with the inverse cube of the distance between the centers of the two fragments R and S. If this interaction energy is nonzero, it can be taken as a good approximation to the interaction matrix element between V_1^{kr} and $V_1^{l's'}$. More generally, when both the two local transitions $R_0 \rightarrow R_m$ and $S_0 \rightarrow S_{n'}$ are allowed, having nonzero transition moments, even if there is no conjugative interaction between R and S, the two LE

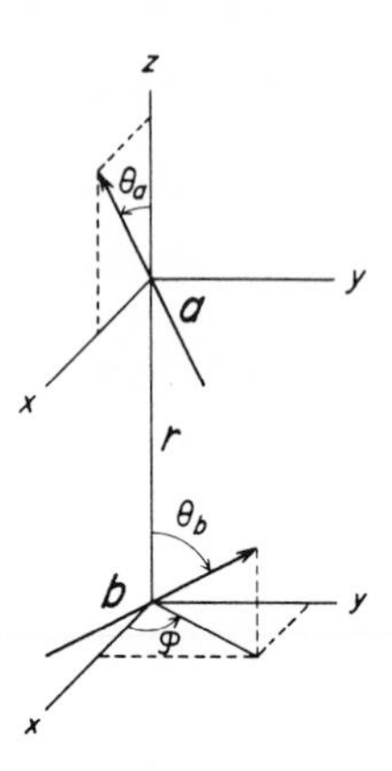

FIG. 20.1. Interaction of two electric dipoles.

configurations R_mS_0 and $R_0S_{n'}$ will interact with each other with an interaction matrix element of the magnitude approximately equal to the energy of the interaction of the two transition moments.

A particularly important case is when R and S are identical chromophores. In this case, two LE configurations R_mS_0 and $R_0S_{m'}$, which have the same energy, will undergo a first-order interaction and give the two states

$$2^{-1/2}(R_mS_0 \pm R_0S_{m'}) \equiv LE_m^{(\pm)} \equiv Ex_m^{(\pm)}, \tag{20.10}$$

separated by twice the interaction matrix element $\Gamma_{mm'}$. We have here a situation in which it is not possible to say that the excitation energy is localized on one fragment or the other. This is referred to as exciton delocalization. The exciton delocalization has an effect of mixing electronic states of different molecules as well as those of different groups of a molecule. Thus, the theory which has been described above, namely, the exciton theory, is important not only in understanding the spectra of molecules which have nonconjugated chromophoric groups (see Chapter 24) but also in understanding the spectra of molecular crystals (see Subsection 6.4.1) and of polymers in which different chromophores lie closely to one another.

Now let us consider the interactions between LE and CT configurations. Evidently, from expressions (11.24) and (11.25) the following expressions can be derived:

$$\langle V_1^{kr}|\, \mathbf{H}^{\mathrm{RS}}\, |V_1^{ls'}\rangle = \langle T_1^{kr}|\, \mathbf{H}^{\mathrm{RS}}\, |T_1^{ls'}\rangle = \delta_{kl}F_{[rs']} = \delta_{kl}B_{rs'}k_{rs}\beta^{\mathrm{C}}; \quad (20.11)$$

$$\langle V_1^{kr}|\, \mathbf{H}^{\mathrm{RS}}\, |V_1^{l's}\rangle = \langle T_1^{kr}|\, \mathbf{H}^{\mathrm{RS}}\, |T_1^{l's}\rangle = -\delta_{rs}F_{[kl']} = -\delta_{rs}B_{kl'}k_{rs}\beta^{\mathrm{C}}. \quad (20.12)$$

When LE configurations are expressed as linear combinations of V_1^{kr}'s or of T_1^{kr}'s, the interaction matrix element between an LE configuration and a CT configuration can be expanded in terms of integrals of these types, and hence can be easily evaluated by the use of these expressions.

The interaction matrix element between two CT configurations in which electrons have been transferred in the same direction, say two CT(R → S) configurations $V_1^{kr'}$ and $V_1^{ls'}$ or $T_1^{kr'}$ and $T_1^{ls'}$, is given by the following equation:

$$\begin{aligned} \langle V_1^{kr'}|\, \mathbf{H}^{\mathrm{RS}}\, |V_1^{ls'}\rangle - \delta_{kl}\,\delta_{r's'}E(V_0) &= \langle T_1^{kr'}|\, \mathbf{H}^{\mathrm{RS}}\, |T_1^{ls'}\rangle - \delta_{kl}\,\delta_{r's'}E(V_0) \\ &= \delta_{kl}F_{[r's']} - \delta_{r's'}F_{[kl]} - [kl \mid r's'] \\ &\doteqdot \delta_{kl}\,{}^{\mathrm{S}}F_{[r's']} - \delta_{r's'}\,{}^{\mathrm{R}}F_{[kl]} - [kl \mid r's']. \end{aligned} \quad (20.13)$$

When $k \neq l$ and $r' \neq s'$, this equation reduces to the following:

$$\langle V_1^{kr'}|\, \mathbf{H}^{\mathrm{RS}}\, |V_1^{ls'}\rangle = \langle T_1^{kr'}|\, \mathbf{H}^{\mathrm{RS}}\, |T_1^{ls'}\rangle = -[kl \mid r's']. \quad (20.14)$$

The interaction matrix element between two CT configurations in which electrons have been transferred in opposite directions, that is, a CT(R → S) and a CT(S → R) configuration, is evidently zero:

$$\langle V_1^{kr'}|\, \mathbf{H}^{\mathrm{RS}}\, |V_1^{l's}\rangle = \langle T_1^{kr'}|\, \mathbf{H}^{\mathrm{RS}}\, |T_1^{l's}\rangle = 0. \quad (20.15)$$

In the case where R and S are identical chromophores, $V_1^{kr'}$ and $V_1^{k'r}$ have the same energy, and $T_1^{kr'}$ and $T_1^{k'r}$ have also the same energy. As mentioned already (see Section 19.6), it is convenient to take such degenerate CT configurations in the symmetry combinations as follows:

$$2^{-1/2}(V_1^{kr'} \pm V_1^{k'r}) \equiv {}^1CT^{kr'(\pm)} \equiv {}^1CR^{kr'(\pm)}; \quad (20.16)$$

$$2^{-1/2}(T_1^{kr'} \pm T_1^{k'r}) \equiv {}^3CT^{kr'(\pm)} \equiv {}^3CR^{kr'(\pm)}. \quad (20.17)$$

Such combinations are called charge-resonance (abbreviated as CR) configurations. Since the interaction matrix elements between $V_1^{kr'}$ and $V_1^{k'r}$ and between $T_1^{kr'}$ and $T_1^{k'r}$ are zero, CR configurations have the same energy as their component CT configurations.

From expression (11.23) the following expression is derived for the interaction matrix element between a CT configuration and the ground configuration:

$$\langle V_1^{kr'} | \mathbf{H}^{RS} | V_0 \rangle = 2^{1/2} F_{[kr']} = 2^{1/2} B_{kr'} k_{rs} \beta^C. \qquad (20.18)$$

Thus for all the off-diagonal elements of an interaction matrix formed of the ground and singly-excited electron configurations have been formulated. In summarizing, it can be said as follows: between singlet LE configurations arising from electron excitations in different fragments [see expression (20.5)] and between CT configurations arising from electron transfers in the same direction [see expression (20.14)] there can be only interactions due to the electronic repulsive interaction terms in the total π-electronic Hamiltonian $\mathbf{H}^{RS}$; between LE and CT configurations [see expressions (20.11) and (20.12)] and between the ground and CT configurations [see expression (20.18)] there can be only conjugative interactions, which arise from the one-electron terms $\mathbf{h}^C$ in $\mathbf{H}^{RS}$; there can be no interactions between triplet LE configurations arising from electron excitations in different fragments [see expression (20.6)] and between CT configurations arising from electron transfers in opposite directions [see expression (20.15)]. It should be noted that the electronic repulsive interactions are in most part independent of the degree of coplanarity of fragments R and S in the molecule, whereas the conjugative interactions are very sensitively dependent on the degree of the coplanarity.

The diagonal matrix elements for LE configurations $R_m S_0$'s and $R_0 S_{n'}$'s, that is, the energies of LE configurations relative to the ground configuration $R_0 S_0$, can be considered as being equal to the energies of R_m's relative to R_0 and of $S_{n'}$'s relative to S_0. These energies can be evaluated from the calculations on the separated fragments R and S, or alternatively, when as mentioned already eigenfunctions of $\mathbf{H}^R$ and of $\mathbf{H}^S$ have been taken as R_m's and $S_{n'}$'s, respectively, these energies can be estimated from the spectra of the separated fragments R and S.

The diagonal matrix elements for CT configurations, that is, the energies of CT configurations relative to the ground configuration, are given by the following expression, which is obtained from expression (20.13) by setting $k = l$ and $r' = s'$:

$$E(V_1^{kr'}) - E(V_0) = E(T_1^{kr'}) - E(V_0) \doteqdot {}^S F_{[r'r']} - {}^R F_{[kk]} - [kk \mid r'r']. \qquad (20.19)$$

The ${}^S F_{[r'r']}$ is the zeroth-order SCF orbital energy of the acceptor orbital $\phi_{r'}$ of fragment S, and the ${}^R F_{[kk]}$ is that of the donor orbital ϕ_k of fragment R. The terms in this expression have a simple physical significance. The quantity $-{}^{\mathrm{R}}F_{[kk]}$ is the energy required to remove an electron from the doubly occupied orbital ϕ_k of the separated fragment R in the ground state, and, if ϕ_k is the highest occupied orbital of R, this is just the first ionization energy of the separated fragment R, I_{R}. The quantity $-{}^{\mathrm{S}}F_{[r'r']}$ is the energy gained by adding an electron to the vacant orbital $\phi_{r'}$ of the separated fragment S in the ground state, and, if $\phi_{r'}$ is the lowest vacant orbital of S, this is just the electron affinity of the separated fragment S, A_{S}. The last term, $[kk \mid r'r']$, can be interpreted as representing the energy of the Coulomb attraction between the transferred electron in $\phi_{r'}$ and the positive hole in ϕ_k it has left behind, $C(\mathrm{R}^+, \mathrm{S}^-)$. This term can be evaluated in the usual manner by expanding into the Coulomb repulsion integrals over AO's (see Section 11.3). Thus, the energies of CT configurations relative to the ground configuration can approximately be given by the following expression:

$$E(V_1^{kr'}) - E(V_0) = E(T_1^{kr'}) - E(V_0) \doteqdot I_k - A_{r'} - [kk \mid r'r']$$
$$\doteqdot I_{\mathrm{R}} - A_{\mathrm{S}} - C(\mathrm{R}^+, \mathrm{S}^-). \quad (20.20)$$

It is to be noted that the singlet CT configuration $V_1^{kr'}$ and the corresponding triplet CT configuration $T_1^{kr'}$ have the same energy.

In this way all the elements of the interaction matrix can be evaluated. By diagonalizing the interaction matrix we can obtain the electronic states of the composite system as linear combinations of the ground, LE, and CT configurations. It is evident that this method is quite analogous to the method developed by Mulliken to explain the spectra of the so-called charge-transfer molecular complexes (see Subsection 6.3.1).

20.2. Application

To illustrate the analysis of the spectra of "composite" molecules by the use of the method described in the preceding section, the analysis of the spectrum of 1,3-butadiene made by Longuet-Higgins and Murrell[1] is outlined here.

In this analysis the π-electronic states of the butadiene molecule are described in terms of the π MO's localized on the two ethylenic bonds, the 1—2 and the 1′—2′ bond (for the numbering of the carbon π centers, see Fig. 8.1). The 1—2 bond is denoted by R, and the 1′—2′ bond by S. The

[1] H. C. Longuet-Higgins and J. N. Murrell, *Proc. Phys. Soc.* (*London*) **A68**, 601 (1955); J. N. Murrell, *J. Chem. Phys.* **37**, 1162 (1962); J. N. Murrell, "The Theory of the Electronic Spectra of Organic Molecules," Sect. 7.2. Methuen, London, 1963.

π MO's of these fragments are taken as

$$\text{in R, } \phi_{\mp 1} = 2^{-1/2}(\chi_1 \mp \chi_2); \tag{20.21}$$

$$\text{in S, } \phi_{\mp 1'} = 2^{-1/2}(\chi_{1'} \mp \chi_{2'}). \tag{20.22}$$

In the ground electron configuration V_0 each of the two bonding π MO's, ϕ_{+1} and $\phi_{+1'}$, is, of course, occupied by two electrons. The singlet electron configuration arising from one-electron transition from an occupied orbital ϕ_{+k} to an unoccupied orbital ϕ_{-r} is denoted by V_1^{kr}, and the corresponding triplet electron configuration by T_1^{kr}. We have the following eight singly-excited configurations: two singlet LE configurations, V_1^{11} (or $^1LE^{11}$) and $V_1^{1'1'}$ (or $^1LE^{1'1'}$), two singlet CT configurations, $V_1^{11'}$ (or $^1CT^{11'}$) and $V_1^{1'1}$ (or $^1CT^{1'1}$), two triplet LE configurations, T_1^{11} (or $^3LE^{11}$) and $T_1^{1'1'}$ (or $^3LE^{1'1'}$), and two triplet CT configurations, $T_1^{11'}$ (or $^3CT^{11'}$) and $T_1^{1'1}$ (or $^3CT^{1'1}$). The π-electronic states of the molecule are constructed as linear combinations of these configurations.

The transition from V_0 to V_1^{11} has a nonzero transition moment in the direction of the 1—2 bond, and the transition from V_0 to $V_1^{1'1'}$ has a nonzero transition moment of the same magnitude in the direction of the 1′—2′ bond. The magnitude of these transition moments is taken as the same as the magnitude of the transition moment of the first π–π^* S–S transition (N–V transition) in ethylene, M_V. The two singlet LE configurations, V_1^{11} and $V_1^{1'1'}$, have the same energy, $E(^1LE)$, which is considered as being equal to the energy of the N–V transition in ethylene, E_V. These two degenerate singlet LE configurations interact with each other by virtue of exciton delocalization to give rise to two exciton configurations as follows:

$$2^{-1/2}(V_1^{11} + V_1^{1'1'}) \equiv {}^1Ex^{(+)}; \tag{20.23}$$

$$2^{-1/2}(V_1^{11} - V_1^{1'1'}) \equiv {}^1Ex^{(-)}. \tag{20.24}$$

If the interaction matrix element between the two singlet LE configurations is denoted by Γ, evidently the energy of $^1Ex^{(+)}$ is given by $E_V + \Gamma$, and the energy of $^1Ex^{(-)}$ by $E_V - \Gamma$. The value of E_V has been determined as 7.7 ev from the position of the first π–π^* S–S band (N–V band) of ethylene. The value of Γ has been calculated by the use of the approximation of dipole–dipole interaction to be +0.5 ev for the planar *s-trans* form (see Fig. 8.2) and +0.8 ev for the planar *s-cis* form (see Fig. 8.3).

The two triplet LE configurations, T_1^{11} and $T_1^{1'1'}$, have the same energy, $E(^3LE)$, which can be equated with the energy of the first π–π^* S–T transition (N–T transition) in ethylene, E_T. The value of E_T has been determined as 4.6 ev from the spectrum of ethylene. These two triplet LE configurations do not interact with each other. However, we can for convenience take these configurations in the symmetry combinations as follows:

$$2^{-1/2}(T_1^{11} \pm T_1^{1'1'}) \equiv {}^3Ex^{(\pm)}. \tag{20.25}$$

These symmetry combinations, the triplet exciton configurations, have of course the same energy as the component triplet LE configurations.

All the four CT configurations have the same energy, $E(CT)$, which can be evaluated by the use of Eq. (20.20). The difference between the ionization energy and electron affinity of ethylene, $I - A$, has been evaluated at 13.3 ev, and the Coulombic interaction energy $C(R^+, S^-)$ has been calculated by the use of the approximation of point-charge interactions to be 6.4 ev for the planar *s-trans* form and 6.7 ev for the planar *s-cis* form. It follows that the value of $E(CT)$ is 6.9 and 6.6 ev for the planar *s-trans* and the planar *s-cis* form, respectively. The two singlet CT configurations, as well as the two triplet CT configurations, do not interact with each other, since they are configurations in which electrons have been transferred in opposite directions. However, we can also for convenience take these CT configurations in the symmetry combinations as follows:

$$2^{-1/2}(V_1^{11'} \pm V_1^{1'1}) \equiv {}^1CR^{(\pm)}; \tag{20.26}$$

$$2^{-1/2}(T_1^{11'} \pm T_1^{1'1}) \equiv {}^3CR^{(\pm)}. \tag{20.27}$$

Of course, these symmetry combinations, the charge-resonance configurations, have the same energy $E(CT)$.

Now our problem reduces to the problem of determining the π-electronic states of butadiene as linear combinations of the ground configuration, the four exciton configurations, and the four charge-resonance configurations. All the matrix elements of $\mathbf{h}^{\mathrm{C}}$ and hence of $\mathbf{F}$ between π MO's localized on different fragments are equal to $\frac{1}{2}\beta_{11'}^{\mathrm{C}}$, in which $\beta_{11'}^{\mathrm{C}}$ is the core resonance integral across the central bond (the 1—1′ bond) of the molecule. Therefore, by the use of Eqs. (20.11), (20.12), and (20.18), the following expressions for the off-diagonal elements of the interaction matrix are obtained:

$$\langle {}^1CR^{(+)}|\, \mathbf{H}^{\mathrm{RS}}\, |V_0\rangle = 2^{-1/2}(\langle V_1^{11'}|\, \mathbf{H}^{\mathrm{RS}}\, |V_0\rangle + \langle V_1^{1'1}|\, \mathbf{H}^{\mathrm{RS}}\, |V_0\rangle) = \beta_{11'}^{\mathrm{C}}\,; \tag{20.28}$$

$$\begin{aligned}\langle {}^1CR^{(-)}|\, \mathbf{H}^{\mathrm{RS}}\, |{}^1Ex^{(-)}\rangle = 2^{-1}(&\langle V_1^{11'}|\, \mathbf{H}^{\mathrm{RS}}\, |V_1^{11}\rangle + \langle V_1^{1'1}|\, \mathbf{H}^{\mathrm{RS}}\, |V_1^{1'1'}\rangle \\ &- \langle V_1^{11'}|\, \mathbf{H}^{\mathrm{RS}}\, |V_1^{1'1'}\rangle - \langle V_1^{1'1}|\, \mathbf{H}^{\mathrm{RS}}\, |V_1^{11}\rangle) = \beta_{11'}^{\mathrm{C}}\,;\end{aligned} \tag{20.29}$$

$$\langle {}^3CR^{(-)}|\, \mathbf{H}^{\mathrm{RS}}\, |{}^3Ex^{(-)}\rangle = \beta_{11'}^{\mathrm{C}}\,. \tag{20.30}$$

All the other off-diagonal elements are zero. Therefore, we can write down separately the interaction matrices for singlet and triplet configurations as follows:

Singlet

$$\begin{array}{c} V_0 \\ ^1CR^{(+)} \\ ^1CR^{(-)} \\ ^1Ex^{(-)} \\ ^1Ex^{(+)} \end{array} \begin{bmatrix} 0 & \beta_{11'}^{C} & 0 & 0 & 0 \\ \beta_{11'}^{C} & E(CT) & 0 & 0 & 0 \\ 0 & 0 & E(CT) & \beta_{11'}^{C} & 0 \\ 0 & 0 & \beta_{11'}^{C} & E_V - \Gamma & 0 \\ 0 & 0 & 0 & 0 & E_V + \Gamma \end{bmatrix} \tag{20.31}$$

Triplet

$$\begin{array}{c} ^3CR^{(+)} \\ ^3CR^{(-)} \\ ^3Ex^{(-)} \\ ^3Ex^{(+)} \end{array} \begin{bmatrix} E(CT) & 0 & 0 & 0 \\ 0 & E(CT) & \beta_{11'}^{C} & 0 \\ 0 & \beta_{11'}^{C} & E_T & 0 \\ 0 & 0 & 0 & E_T \end{bmatrix} \tag{20.32}$$

These matrices are diagonalized by the use of a value of -1.68 ev for $\beta_{11'}^{C}$, a value which was used by Pariser and Parr (see Subsection 11.3.3), and of the above-mentioned values for E_V, E_T, $E(CT)$, and Γ. The energy levels before and after the configuration interaction are schematically shown in Fig. 20.2.

The wavefunctions of the lowest singlet and triplet excited states of the planar *s-trans* and *s-cis* forms and the energies of these states relative to the ground configuration are as follows:

		Wavefunction	Energy, ev
Singlet	*s-trans*	$0.74\ {}^1CR^{(-)} + 0.67\ {}^1Ex^{(-)}$	5.4
	s-cis	$0.74\ {}^1CR^{(-)} + 0.67\ {}^1Ex^{(-)}$	5.1
Triplet	*s-trans*	$0.88\ {}^3Ex^{(-)} + 0.48\ {}^3CR^{(-)}$	3.7
	s-cis	$0.87\ {}^3Ex^{(-)} + 0.50\ {}^3CR^{(-)}$	3.6

The energies of the first S–S and S–T transitions of 1,3-butadiene have been found to be 5.8 and 3.5 ev, respectively. According to Murrell, in comparing the calculated transition energies with the observed ones, the depression of the ground state by singly-excited states rather should be neglected, since in the calculation doubly-excited states have been neglected. The depression of the ground state by singly-excited states is considered to be to a large extent cancelled out by the depression of the singly-excited states by doubly-excited states.

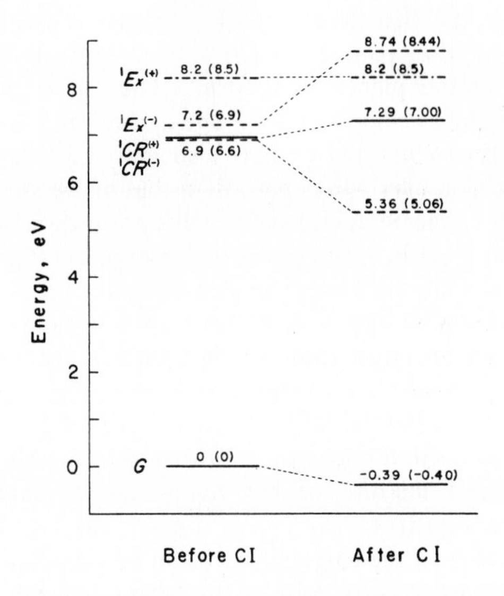

FIG. 20.2. The energy level diagram of the planar *s-trans* form of 1,3-butadiene as a composite π system. The energy levels of singlet states before and after the configuration interactions are shown, and the correlations are denoted with the dotted lines. The numerals written outside the parentheses are the energy values (in electron volts units) of configurations and states relative to the ground configuration; the numerals written inside the parentheses are the energy values for the planar *s-cis* form.

The transition from the ground state to the first singlet excited state is allowed both in the *s-trans* and *s-cis* forms. If the contribution of the plus charge-resonance configuration $^1CR^{(+)}$ to the ground state is neglected, the intensity of this transition just arises from the contribution of the minus exciton configuration $^1Ex^{(-)}$ to the first excited state, since transitions from the ground configuration to CT configurations are forbidden in the zero-overlap approximation. The transition moment of the transition from the ground configuration to the exciton configuration is calculated by compounding vectorially the transition moments of the N–V transitions of the two ethylenic bonds. Thus, if the magnitude of the transition moment of the ethylene N–V transition is denoted by M_V, in the planar *s-trans* form the moment of the transition from the ground configuration to the minus singlet exciton configuration is in the direction along the ethylenic bonds and has a magnitude of $2^{1/2}M_V$, and in the planar *s-cis* form it is in the direction along the central "single" bond (the 1—1′ bond) and has a magnitude of $2^{-1/2}M_V$. Therefore, the magnitude of the moment of the transition from the ground state to the lowest singlet excited state is approximately $0.67 \times 2^{1/2}M_V$ in the planar *s-trans* form, and it is approximately $0.67 \times 2^{-1/2}M_V$ in the planar *s-cis* form. Thus, it is predicted that the first S–S band of the *s-cis* diene will be one-quarter as intense as that of the *s-trans* diene. The calculation predicts also that the first S–S band of the *s-cis* diene will be at longer wavelength than that of the *s-trans* diene. These conclusions are in agreement with the conclusions drawn from the orthodox calculation by the P-method and also with observation (see Section 16.1).

We have found that the lowest singlet excited state is an almost equal mixture of the minus exciton and minus charge-resonance configurations. In the one-electron approximation, as described in the preceding chapter, the lowest energy electronic transition can be considered approximately as the one-electron transition from $2^{-1/2}(\phi_{+1} - \phi_{+1'})$ to $2^{-1/2}(\phi_{-1} + \phi_{-1'})$. It is evident that the excited configuration arising from this transition corresponds to an equal mixture of the minus exciton and minus charge-resonance configurations.

When the geometry of the conjugated system deviates from the coplanarity of the two ethylenic bonds by twist of the central "single" bond, the value of $-\beta^{C}_{11'}$ will decrease, and consequently the lowest singlet excited state will be raised with a resulting hypsochromic shift of the first S–S band. This conclusion is the same as the one obtained already.

Longuet-Higgins and Murrell[1] have made a similar analysis of the spectrum of biphenyl. In this case, of course, the biphenyl molecule is considered as a composite molecule composed of two benzene rings, B and B′. As mentioned in Subsection 12.2.5, eight singlet LE configurations corresponding to the α, p, β, and β' states of the two benzene rings and eight singlet CT configurations arising from one-electron transfer from the doubly degenerate highest bonding π MO's of each benzene ring to the doubly degenerate lowest antibonding π MO's of another benzene ring are taken into account. Now, as usual, the notation for the benzene π MO's shown in Table 3.5 is used, and for the electron configurations the following notation is used: the LE configuration corresponding to the α state of benzene ring B, for example, is denoted by ${}^1LE_{\alpha(\mathrm{B})}$, and that corresponding to the α state of another benzene ring by ${}^1LE_{\alpha(\mathrm{B'})}$; the symmetry combinations of ${}^1LE_{\alpha(\mathrm{B})}$ and ${}^1LE_{\alpha(\mathrm{B'})}$, that is, the α exciton configurations, are denoted by ${}^1Ex_{\alpha}^{(\pm)}$;

the CT configuration arising from one-electron transfer from ϕ_{+1} (localized on benzene ring B) to $\phi_{-1'}$ (localized on benzene ring B′) is denoted by ${}^1CT^{11'}$, and that arising from one-electron transfer from $\phi_{+1'}$ (localized on benzene ring B′) to ϕ_{-1} (localized on benzene ring B) by ${}^1CT^{1'1}$; the symmetry combinations of ${}^1CT^{11'}$ and ${}^1CT^{1'1}$, that is, the charge-resonance configurations formed of ${}^1CT^{11'}$ and ${}^1CT^{1'1}$, are denoted by ${}^1CR^{11'(\pm)}$.

The eight CT configurations can be considered to have the same value of $I - A$ of 10.88 ev, since I and A of benzene are 9.25 and -1.63 ev, respectively, but they must have different Coulomb contributions to their energy. By the use of the point-charge approximation, the Coulomb terms, $C(B^+, B'^-)$, for $CT^{22'}$, $CT^{11'}$, and $CT^{12'}$ (as well as $CT^{21'}$) have been calculated to be 3.35, 4.17, and 3.61 ev, respectively. Therefore, the energies of these CT configurations as well as the energies of the corresponding charge-resonance configurations are 7.53, 6.71, and 7.27 ev, respectively. The energies of the α, p, β, and β' states of benzene are, respectively, 4.8, 6.0, 6.7, and 6.7 ev.

The intense band appearing at about 250 mμ has been assigned to the transition to the state which arises from mixing of ${}^1Ex_p^{(-)}$ with ${}^1Ex_{\beta'}^{(-)}$, ${}^1CR^{11'(-)}$, and ${}^1CR^{22'(-)}$. The intensity of this band is considered to just arise from the contribution of the minus β' exciton configuration to the excited state. The transition from the ground configuration to this exciton configuration is allowed and polarized in the direction along the long axis of the molecule. The so-called hidden bands in the region 270–290 mμ have been ascribed to the transitions to the states which arise from slight perturbation of the α exciton configurations by the charge-resonance configurations formed of ${}^1CT^{12'}$ and ${}^1CT^{1'2}$ and of ${}^1CT^{21'}$ and ${}^1CT^{2'1}$.

According to the classification made in the preceding chapter, the examples described hitherto, in which the two fragments are identical chromophores, belong to case A. In this case, since CT configurations contribute to states always in the form of charge-resonance configurations, electronic transitions cannot be accompanied with any actual charge transfer from one fragment to another.

On the other hand, in case C and in case D, CT configurations which arise from electron transfers in opposite directions are no longer equivalent, and hence electronic transitions should be accompanied with actual charge transfer from one fragment to another. The intensity of a transition originates from the contribution of allowed local transitions to that transition, or in other words, from the contribution of LE configurations to the excited state. When the energy separation between an LE configuration and a CT configuration is large enough, according to perturbation theory, the coefficient of the LE configuration in the CT state function resulting from perturbation of the CT configuration by the LE configuration is approximately proportional to the interaction matrix element between the two configurations, and hence to the resonance integral across the bond (the *r—s* bond) between the fragments R and S, β_{rs}^{C}. This in turn is approximately proportional to $\cos \theta_{rs}$, in which θ_{rs} is the interplanar angle between R and S, that is, the angle of twist of the *r—s* bond. Therefore, as mentioned in Subsection 19.4.3, the intensity of a CT band will be roughly proportional to $\cos^2 \theta_{rs}$. It is also expected that, in such a case, the change in the energy of the CT state caused by the perturbation will be small, and hence that the position of the CT band will be comparatively insensitive to the twist of the *r—s* bond. Such a situation actually occurs in the spectra of substituted benzenes such as acetophenone, nitrobenzene, and aniline, as will be mentioned in detail in the following two chapters. The application of the advanced composite-molecule method to such compounds will be illustrated in Chapter 22 by taking nitrobenzene (case C) and aniline (case D) as examples.

Chapter 21

Carbonyl Compounds

21.1. The Carbonyl Group

Before proceeding to discuss the electronic absorption spectra of various carbonyl compounds and their relations to the geometry of the molecules, we consider first the electronic structure and light absorption properties of the carbonyl group.

A Cartesian coordinate system is set for the carbonyl group as illustrated in Fig. 4.1: the C—O internuclear axis is taken as the x axis, and the plane of the group as the xy plane. Formaldehyde, the simplest carbonyl compound, has twelve valence electrons: one from each hydrogen atom, four from the carbon atom, and six from the oxygen atom. In the ground state of the molecule, these valence electrons are distributed in such a manner that each of the following six orbitals is occupied by two electrons: the two C—H bonding σ orbitals, the C—O bonding σ orbital (σ), the C—O bonding π orbital (π or ϕ_{+1}), and the two nonbonding orbitals on the oxygen atom. One of the two nonbonding orbitals on the oxygen atom is probably the $2p_y$ orbital (n_p or ϕ_{n}), and the other is probably an sp hybrid orbital (n_{sp}), which is formed of the $2s$ and $2p_x$ orbitals. The highest occupied orbital is the nonbonding p orbital, and the second highest one is probably the bonding π orbital. By electron impact studies formaldehyde has been found to have three ionization potentials 10.8 ± 0.1, 11.8 ± 0.2, and 13.1 ± 0.2 ev.[1] These almost certainly correspond to the removal of one electron from the nonbonding p orbital, the bonding π orbital, and the C—O bonding σ orbital, respectively. The lowest unoccupied orbital of the carbonyl group is the antibonding π orbital (π^* or ϕ_{-1}), and the second lowest one is probably the C—O antibonding σ orbital (σ^*).

[1] T. M. Sugden and W. C. Price, *Trans. Faraday Soc.* **44**, 116 (1948).

Aliphatic simple aldehydes and ketones show a weak absorption band with fine structure which covers the range 230–350 mμ and has its maximum in the region 280–300 mμ ($\epsilon_{max} \simeq 10, f \simeq 10^{-4}$). For example, data on this band of formaldehyde in the vapor state are as follows: $\lambda_{onset} = 353$ mμ, $\lambda_{max} = 295$ mμ, $\epsilon_{max} = 10, f = 2.4 \times 10^{-4}$. This band is attributed to the one-electron singlet–singlet transition from the nonbonding p orbital to the antibonding π orbital, namely, the S–S n–π^* transition. This transition is forbidden by local symmetry, since the product of a p_y and a p_z orbital has no dipole moment (see Subsection 4.3.3). This transition is considered to be made allowed by coupling with a molecular vibration. By analyzing the rotational structure of this n–π^* band of formaldehyde, Dieke and Kistiakowsky[2] have shown that this band is polarized in the direction of the y axis. A theoretical study by Pople and Sidman[3] shows that if the originally forbidden n_p–π^* transition gains most of its intensity through coupling with the out-of-plane bending vibration of the CH_2 group this transition can have a nonzero transition moment in the direction of the y axis. As mentioned already (see Subsection 6.2.2), this n–π^* band shifts to shorter wavelength in going from vapor to solution, and the magnitude of the shift increases with the increasing polarity of the solvent.

Formaldehyde shows a very weak absorption band ($\epsilon_{max} \simeq 10^{-3}$) in the region of wavelengths longer than 395 mμ. This band is attributed to the S–T (singlet–triplet) n_p–π^* transition.[4] The separation between the 0–0 subband of the S–S n–π^* band and that of the S–T n–π^* band is thus only 3100 cm^{-1}. This separation is very small compared with the singlet–triplet separation of π–π^* bands. As mentioned in Section 10.2, this is considered to be due to the fact that the exchange repulsion integral $K_{n\pi^*}$ (that is, $[n_p, \pi^* \mid n_p, \pi^*]$) is very small [see Eq. (10.12) or (11.20)]. The hypsochromic shift by polar and especially hydroxylic solvents and the small singlet–triplet separation are characteristics of n–π^* transitions. While in the ground state the formaldehyde molecule has a planar equilibrium geometry, in the excited state resulting from the S–S and S–T n_p–π^* transitions, it has been inferred to have a nonplanar, pyramidal, equilibrium geometry (see Section 5.3).

Simple carbonyl compounds show two absorption bands in the vacuum ultraviolet region, one in the region 170–200 mμ and the other in the region 150–170 mμ. The longer wavelength band (acetaldehyde: $\lambda_{max} = 182$ mμ, $\log \epsilon_{max} = 4.01$, $f = 0.037$; acetone: $\lambda_{max} = 195$ mμ, $\log \epsilon_{max} = 3.96$,

[2] C. H. Dieke and G. B. Kistiakowsky, *Phys. Rev.* **45**, 4 (1934).
[3] J. A. Pople and J. W. Sidman, *J. Chem. Phys.* **27**, 1270 (1957).
[4] G. W. Robinson, *Can. J. Phys.* **34**, 699 (1956).

$f = 0.046$)[5] has been attributed to the allowed S–S n_p–σ^* transition.[6] The shorter wavelength band (formaldehyde: $\lambda_{max} = 156$ mμ; acetaldehyde: $\lambda_{max} = 167$ mμ, $\log \epsilon_{max} = 4.3$, $f = 0.13$) has been attributed to the S–S π–π^* transition. Rydberg bands appear in the region of wavelengths shorter than about 160 mμ, and the first ionization potential of formaldehyde, which, as mentioned already, probably corresponds to the removal of an electron from the nonbonding p orbital on the oxygen atom, has been determined to be 10.83 ev.[7]

Now attention is focussed on the π orbitals of the carbonyl group. The $2p_z$ atomic orbitals of the carbon and oxygen atoms of the carbonyl group are denoted by χ_1 and χ_2, respectively. Then, the π MO's of the carbonyl group can be expressed as follows:

$$\phi_{-1} = b_{-1,1}\chi_1 + b_{-1,2}\chi_2 ; \tag{21.1}$$

$$\phi_{+1} = b_{+1,1}\chi_1 + b_{+1,2}\chi_2 . \tag{21.2}$$

As usual, the Coulomb integrals of χ_1 and χ_2, that is, α_1 and α_2, are expressed as α and $\alpha + h_2\beta$, respectively, and the resonance integral between χ_1 and χ_2, β_{12}, is expressed as $k_{12}\beta$. Since oxygen is more electronegative than carbon, the Coulomb parameter h_2 should be positive. As mentioned in Section 9.2, a value of about 1.0–1.3 seems to be suitable for this parameter. Of course, the resonance parameter k_{12} is positive. The energy of the antibonding π MO ϕ_{-1}, η_{-1}, is expressed as $\alpha - v_{-1}\beta$, and that of the bonding π MO ϕ_{+1}, η_{+1}, as $\alpha - v_{+1}\beta$. In the zero-overlap approximation, as will be easily verified, the energies and AO coefficients of these π MO's can be expressed as follows:

$$v_{-1} = \tfrac{1}{2}[(h_2{}^2 + 4k_{12}^2)^{1/2} - h_2] > 0; \tag{21.3}$$

$$v_{+1} = -(v_{-1} + h_2) < 0; \tag{21.4}$$

$$b_{+1,1} = -b_{-1,2} = [v_{-1}/(v_{-1} - v_{+1})]^{1/2}; \tag{21.5}$$

$$b_{+1,2} = b_{-1,1} = [(v_{-1} + h_2)/(v_{-1} - v_{+1})]^{1/2}. \tag{21.6}$$

It is evident that $b_{+1,2} > b_{+1,1} > 0$. This means that the bonding π MO is localized more on the oxygen atom and the antibonding π MO more on the carbon atom.

The length of the carbonyl bond in a number of compounds has an average value of 1.23 ± 0.01 Å. By the use of the value of 1.23 Å, the resonance

[5] J. S. Lake and A. J. Harrison, *J. Chem. Phys.* **30**, 361 (1959).

[6] H. L. McMurry, *J. Chem. Phys.* **9**, 231, 241 (1941); P. G. Wilkinson, *J. Mol. Spectroscopy* **2**, 387 (1958).

[7] A. D. Walsh, *Proc. Roy. Soc.* (*London*) **A185,** 176 (1946).

parameter for this bond is evaluated at 0.844, on the basis of the usual assumption of the proportionality of the resonance parameter k to the overlap integral S [see Eq. (9.11), and also Table 9.5]. When the values $h_2 = 1.3$ and $k_{12} = 0.844$ are used, the following values are obtained:

$$v_{-1} = +0.415, v_{+1} = -1.715, b_{+1,1} = -b_{-1,2} = 0.441, b_{-1,1} = b_{+1,2} = 0.897.$$

21.2. Effects of Substituents on the Absorption Bands of the Carbonyl Group

When a methyl group is attached to the carbon atom of the carbonyl group, the methyl group will raise the antibonding π MO of the carbonyl group both through the inductive and the hyperconjugative effect (see Section 19.5), while it will exert no effect on the nonbonding p orbital. Accordingly,

TABLE 21.1.

THE n–π^* BANDS OF SIMPLE CARBONYL COMPOUNDS

	Formaldehyde		Acetaldehyde		Acetone	
Solvent	λ_{max}, mμ	log ϵ_{max}	λ_{max}, mμ	log ϵ_{max}	λ_{max}, mμ	log ϵ_{max}
(Vapor)	295	1.0	290	1.0	280	1.05
Heptane	—	—	290	1.22	279	1.11
Water	288	1.13	277	0.90	265	1.25

it is expected that the methyl substitution will increase the energy of the n–π^* transition, with a consequent hypsochromic shift of the n–π^* band. This prediction is borne out by comparison of data on the n–π^* bands of formaldehyde, acetaldehyde, and acetone, which are shown in Table 21.1.

The n–π^* band of cyclohexanone ($\lambda_{max} = 285$ mμ, log $\epsilon_{max} = 1.2$, in ethanol) is at longer wavelength than that of acetone. The n–π^* band of cyclopentanone ($\lambda_{max} = 300$ mμ, log $\epsilon_{max} = 1.2$, in ethanol) is at further longer wavelength and shows a more well-developed vibrational fine structure. Both the bathochromic shift and the development of fine structure may be related to the steric strain in these ring systems (see Subsection 16.1.4).

Substitution of a strongly electron-donating group having a lone pair of electrons of π character, such as NR_2, OR, or halogen, into the carbonyl group, as in an amide, acid, ester, or acid halide, results in a large hypsochromic shift of the n–π^* band (see Table 21.2). This is presumably because the antibonding π MO of the carbonyl group, ϕ_{-1}, will be raised by interaction with the lone-pair orbital of the substituent, $\phi_{0'}$, while the nonbonding p orbital, ϕ_n, is not affected. That is, if the carbonyl group and the electron-donating group are taken as R and S, respectively, the situation belongs to case D-2 (see Subsection 19.4.4).

The first S–S π–π^* bands of simple (nonconjugated) aliphatic carboxylic acids, esters, and amides are at slightly longer wavelengths than the S–S π–π^* bands of simple carbonyl compounds. For example, while the π–π^*

TABLE 21.2

EFFECT OF ELECTRON-DONATING GROUPS ON THE n–π^* BAND OF THE CARBONYL GROUP

Compound	Solvent	λ_{max}, mμ	ϵ_{max}
Acetone	Ethanol	272.5	18.6
Acetone	Hexane	279	14.8
Acetic acid	None	204	45.5
Acetic acid	Ethanol	204	41
Ethyl acetate	Water	204	60
Acetic anhydride	None	217	56
Dimethyl carbonate	None	<180	>50
Acetamide	Water	214	—
Acetyl chloride	None	234.5	50
Acetyl chloride	Hexane	235	53

band of acetone in the gaseous state is at 156 mμ, the π–π^* bands of acetic acid, ethyl acetate, and formamide in the gaseous state are at 160.5 mμ ($\epsilon_{max} = 4200$), 164 mμ ($\epsilon_{max} = 3600$), and 171 mμ, respectively.[8] The π–π^* bands of simple carboxylic acids, esters, and amides are considered to be the CT bands due to the one-electron transition from the occupied orbital $\phi^{p}_{0'}$, that is, the orbital that arises from perturbation of the nonbonding orbital of π character of —OH, —OR, or —NH_2, to the unoccupied orbital ϕ^{p}_{-1}, that is, the orbital that arises from perturbation of the antibonding π orbital of the carbonyl group (see the energy level diagram for case D-2 in Fig. 19.3). Nagakura[8] calculated the energies and wavefunctions of π-electronic states of acetic acid, formamide, and some related compounds by the method of composite molecule described in Chapter 20, considering these states as arising from interactions of the ground configuration V_0, the CT configuration $^1CT^{0',-1}$, and the LE configuration $^1LE^{+1,-1}$. Some of the results are shown in Table 21.3.

When the carbonyl group is conjugated with an ethylenic bond or with an

[8] S. Nagakura, *J. Chem. Phys.* **23**, 1441 (1955); *Mol. Phys.* **3**, 105 (1960); *Pure Appl. Chem.* **7**, 79 (1963).

aromatic group, as will be mentioned in detail in a later section, the n–π^* band of the carbonyl group is shifted to longer wavelength, presumably because the antibonding π orbital of the carbonyl group will become lower as a result of interaction mainly with the lowest antibonding π orbital of the unsaturated hydrocarbon group.

By the way, the S–S n–π^* band of a thiocarbonyl compound is in general at much longer wavelength than that of the corresponding carbonyl compound, and is responsible for the color of the compound. For example, the

TABLE 21.3

THE CT BANDS OF ACETIC ACID AND FORMAMIDE

Molecule	Transition energy, ev obs.	theoret.	Contribution of the CT configuration ${}^1CT^{0',-1}$ (%) ground state	upper state
Acetic acid	7.72	7.73	13.2	44.7
Formamide	7.19	7.08	20.8	60.2

band of cyclohexanethione, which is a pink liquid, has the following characteristics in cyclohexane solution: $\lambda_{\text{max}} = 504$ mμ, $\epsilon_{\text{max}} \simeq 10$.[9] The effect of substitution on the thiocarbonyl group is similar to that on the carbonyl group. Thus, conjugation of the thiocarbonyl group with mesomeric substituents bearing lone-pair electrons shifts the n–π^* band to shorter wavelength (for example, Me(EtO)C=S; $\lambda_{\text{max}} = 377$ mμ, $\epsilon_{\text{max}} = 19$, in cyclohexane), and conjugation with aromatic groups shifts the band to longer wavelength (for example, $Ph_2C{=}S$: $\lambda_{\text{max}} = 605$ mμ, $\epsilon_{\text{max}} = 66$, in diethyl ether).[9]

21.3. Conjugated Dicarbonyl Compounds (α,β-Dicarbonyls)

In compounds containing two carbonyl groups adjacent to each other, the π MO's of the conjugated dicarbonyl system may be considered as being formed by interaction of the π MO's of the two carbonyl groups, as illustrated in the schematic energy level diagram shown in Fig. 21.1. When the two carbonyl groups are coplanar, the conjugative interaction should operate fully, and hence the separation between the two antibonding π MO's, ψ_{-2} (that is, $\phi^{\text{p}}_{-1(-)}$) and ψ_{-1} (that is, $\phi^{\text{p}}_{-1(+)}$), and that between the two bonding π MO's, ψ_{+1} (that is, $\phi^{\text{p}}_{+1(-)}$) and ψ_{+2} (that is, $\phi^{\text{p}}_{+1(+)}$), should become maximum.

[9] S. F. Mason, *Quart. Rev.* (*London*) **15**, 287 (1961).

The two nonbonding *p* orbitals on the two oxygen atoms, ϕ_n and $\phi_{n'}$, will interact weakly with each other in the planar *s-cis* conformation of the dicarbonyl system, and will give rise to the two MO's of different energies, $\phi_{n(-)}$ and $\phi_{n(+)}$, which are the antisymmetric and symmetric combinations of the two nonbonding orbitals, respectively. Of course, $\phi_{n(-)}$ must be higher than $\phi_{n(+)}$. The energy separation between these two "nonbonding" MO's

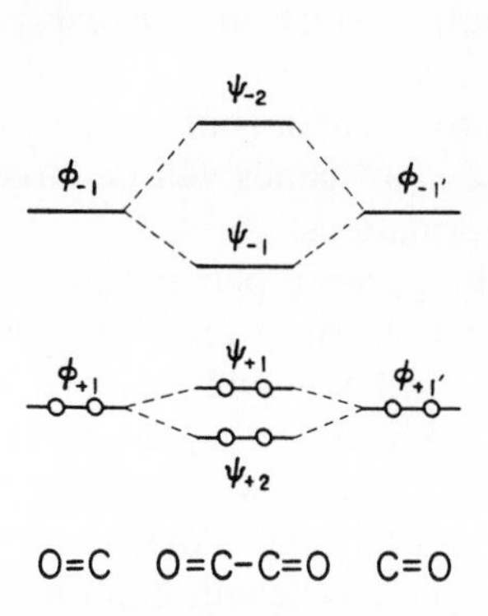

FIG. 21.1. Orbital energy levels of the conjugated dicarbonyl system.

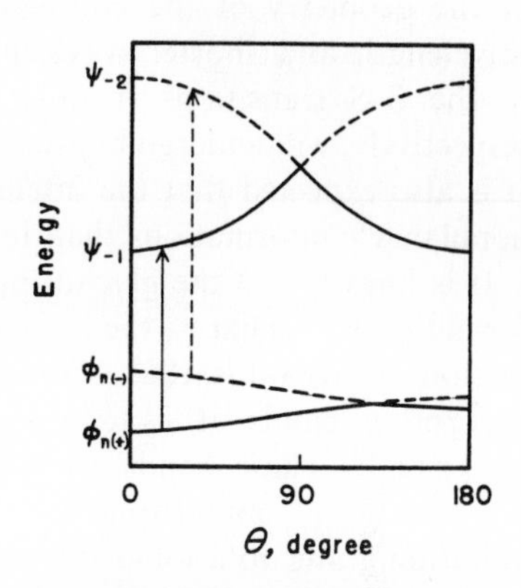

FIG. 21.2. Orbital energies of the conjugated dicarbonyl system as functions of the interplanar angle (θ) between the two carbonyl bonds.

is considered to be very small, presumably less than 1000 cm^{-1}. The energy separation will decrease rapidly with increasing deviation from the planar *s-cis* conformation. In the planar *s-trans* conformation the energy separation must be very much small and is considered to be less than 100 cm^{-1}.

Thus, the energies of the two antibonding π MO's (ψ_{-2} and ψ_{-1}) and two nonbonding MO's ($\phi_{n(-)}$ and $\phi_{n(+)}$) will vary with the interplanar angle θ between the two carbonyl groups in manners illustrated in Fig. 21.2, in which θ is taken as 0° for the planar *s-cis* conformation.

If the two nonbonding orbitals ϕ_n and $\phi_{n'}$ are pure 2*p* AO's, transitions from these nonbonding orbitals, and hence from their symmetry combinations, to the antibonding π MO's are all forbidden in planar conformations. In nonplanar conformations, however, weak interaction will occur between $\phi_{n(+)}$ and $\phi^{p}_{+1(-)}$ and between $\phi_{n(-)}$ and $\phi^{p}_{+1(+)}$. Consequently, the two nonbonding MO's $\phi_{n(+)}$ and $\phi_{n(-)}$ will obtain a small contribution of $\phi_{+1(-)}$ and that of $\phi_{+1(+)}$, respectively, and thereby the transitions from $\phi_{n(+)}$ to

$\phi^{p}_{-1(+)}$ (that is, ψ_{-1}) and from $\phi_{n(-)}$ to $\phi^{p}_{-1(-)}$ (that is, ψ_{-2}) will become weakly allowed. If each of the two nonbonding orbitals ϕ_n and $\phi_{n'}$ is not purely 2*p* but has a contribution of the 2*s* AO, these transitions will be allowed even in planar conformations, having nonzero transition moments in the direction perpendicular to the molecular plane.

From the above considerations it is expected that with increasing deviation of the geometry of the conjugated system from the planarity the longer wavelength and shorter wavelength n–π^* bands, which are probably due to the S–S transitions from $\phi_{n(+)}$ to $\phi^{p}_{-1(+)}$ and from $\phi_{n(-)}$ to $\phi^{p}_{-1(-)}$, respectively, will undergo hypsochromic and bathochromic shifts, respectively. It is also expected that the intensities of these n–π^* bands will be larger in nonplanar conformations than in planar conformations.

It is known that the glyoxal molecule, $(CHO)_2$, has a planar *s-trans* conformation.[10,11] That is, the interplanar angle θ in the most preferred conformation of glyoxal is 180°. Glyoxal, which is a yellow liquid, has two weak absorption bands at $\lambda_{max} = 450$ mμ, $\epsilon_{max} \simeq 5$ and at $\lambda_{max} = 267.5$ mμ, $\epsilon_{max} \simeq 6$.[12] These bands are almost certainly due to the transitions from $\phi_{n(+)}$ to $\phi^{p}_{-1(+)}$ and from $\phi_{n(-)}$ to $\phi^{p}_{-1(-)}$, respectively. Analyses of the vibrational and rotational structure of the longer wavelength band indicate that the transition is polarized perpendicularly to the molecular plane.[11,13] As mentioned above, this is what is expected if the intensity of this band originates from contribution of the 2*s* AO's to the nonbonding orbitals. According to Brand,[11] the 0–0 subband of this band is at 21,977 cm^{-1} (455 mμ), and the 0–0 subband of the corresponding S–T band is at 19,544 cm^{-1} (512 mμ).

The interplanar angle θ in the most preferred conformation of biacetyl, $(CH_3CO)_2$, in solution has been determined to be about 160°.[14] This compound in ethanol solution shows two n–π^* bands at $\lambda_{max} = 417$ mμ, $\epsilon_{max} = 9.6$ and at $\lambda_{max} = 286$ mμ, $\epsilon_{max} = 24.6$.[15] Thus, as is expected, the longer wavelength and shorter wavelength n–π^* bands of biacetyl are at shorter and longer wavelengths, respectively, than the corresponding bands of glyoxal. The 0–0 subband of the longer wavelength n–π^* band of biacetyl in the crystalline state at 4°K is at 22,873 cm^{-1} (437 mμ), and that of the

[10] J. E. Lu Valle and V. Schomaker, *J. Am. Chem. Soc.* **61,** 3520 (1939).

[11] J. C. D. Brand, *Trans. Faraday Soc.* **50,** 431 (1954).

[12] A. Luthy, *Z. physik. Chem.* **107,** 285 (1923); G. Mackinney and O. Temmer, *J. Am. Chem. Soc.* **70,** 3586 (1948).

[13] G. W. King, *J. Chem. Soc.* **1957,** 5054.

[14] P. H. Cureton, C. G. LeFèvre, and R. J. W. LeFèvre, *J. Chem. Soc.* **1961,** 4447.

[15] N. J. Leonard, H. A. Laitinen, and E. H. Mottus, *J. Am. Chem. Soc.* **75,** 3300 (1953).

corresponding S–T band is at 20,421 cm^{-1} (490 $m\mu$).[16] The longer wavelength n–π^* band of biacetyl shows only a small hypsochromic shift on increasing the polarity of the solvent, and the shorter wavelength n–π^* band shows almost no shift. There is, however, a noticeable change in the intensity of the bands, especially the longer wavelength band: the intensity of this band is diminished when biacetyl is dissolved in solvents that contain oxygen, nitrogen, or fluorine atoms.[17] This effect has been ascribed to intermolecular dispersion forces.

The spectrum of oxalyl chloride, $(COCl)_2$, markedly varies with temperature, probably indicating the prevalence of rotational isomerism.[18]

When the α,β-dicarbonyl system is part of an alicyclic system, the interplanar angle θ between the two carbonyl groups will depend on the ring size. For small rings the θ will be nearly 0°, and as the ring size increases, the θ will increase. Leonard and Mader[19] have examined the spectra of a series of alicyclic 1,2-diketones. Data on the n–π^* bands of the alicyclic 1,2-diketones and an open-chain analog and the values of θ for these compounds deduced from models are shown in Table 21.4. As is seen in this table, as the value of θ increases from 0°, the wavelength λ_{max} of the longer wavelength n–π^* band decreases to a minimum at $\theta = 90°$ and increases again at larger values of θ, in agreement with the expectation (see Fig. 21.2), while the shorter wavelength n–π^* band remains at nearly constant wavelength. The fact that the intensities of these bands are larger in more nonplanar conformations than in more planar conformations is also in agreement with the expectation.

A variety of evidence indicates that the benzil molecule, $(C_6H_5CO)_2$, has a skew structure in which the two planar benzoyl groups lie in planes approximately at right angles (about 97°) to each other.[14,20] Consistently with the skew structure, benzil and some of its derivatives have spectra which considerably resemble those of benzaldehyde and its correspondingly substituted derivatives. For example, benzil exhibits a strong band (probably a CT band) at 259 $m\mu$ (log ϵ_{max} = 4.31) and a weak band (probably an n–π^* band) at 370 $m\mu$ (log ϵ_{max} = 1.89), which are to be compared with the corresponding bands of benzaldehyde, which appear at 240 $m\mu$ (log ϵ_{max} = 4.12) and at 320 $m\mu$ (log ϵ_{max} = 1.7), respectively.[20] When benzil is substituted

[16] J. W. Sidman and D. S. McClure, *J. Am. Chem. Soc.* **77,** 6461 (1955).

[17] L. S. Forster, *J. Am. Chem. Soc.* **77,** 1417 (1955).

[18] J. W. Sidman, *J. Am. Chem. Soc.* **78,** 1527 (1956).

[19] N. J. Leonard and P. M. Mader, *J. Am. Chem. Soc.* **72,** 5388 (1950).

[20] N. J. Leonard, R. T. Rapala, H. L. Herzog, and E. R. Blout, *J. Am. Chem. Soc.* **71,** 2997 (1949).

TABLE 21.4

THE n–π^* BANDS AND INTERPLANAR ANGLES (θ) BETWEEN THE TWO CARBONYL GROUPS OF 1,2-DIKETONES

Compound	θ, deg.	The first n–π^* band		The second n–π^* band	
		λ_{max}, mμ	ϵ_{max}	λ_{max}, mμ	ϵ_{max}
Camphorquinone (I)	0–10	466	30.7	—	—
II, $n = 6$	0–60	380	11.1	297.5	28.9
II, $n = 7$	90–110	337	33.8	299	34.5
II, $n = 8$	100–140	343	21.4	295.5	43.2
Dipivaloyl (III)	90–180	365	21.0	285	52.5
II, $n = 18$	100–180	384	21.4	286.5	59

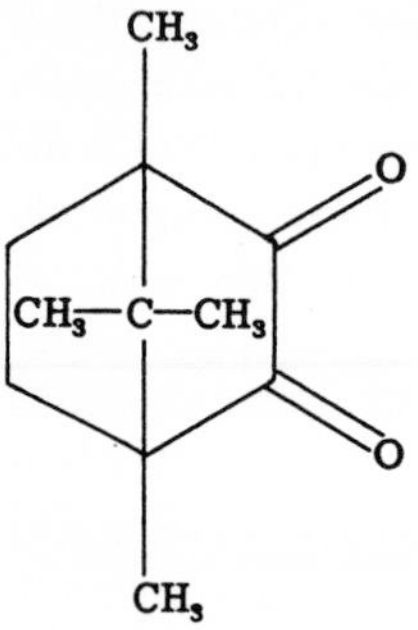

$(CH_3)_3C$—CO—CO—$C(CH_3)_3$

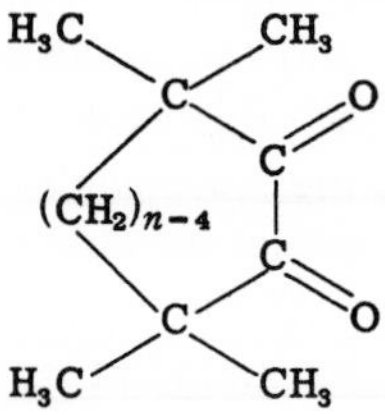

I II III

in the *ortho* positions, the n–π^* band at 370 mμ is successively shifted to longer wavelengths, to 400 mμ for mesityl phenyl diketone and finally to two peaks at 467 and 493 mμ in mesitil.[21] Presumably, with increasing hindrance at the *ortho* positions, each phenyl ring is twisted out of the plane of the carbonyl group to which it is attached, and the two carbonyl groups are forced into the coplanar *s-trans* conformation.

21.4. Vinyl–Carbonyl and Phenyl–Carbonyl Compounds

21.4.1. GENERAL

Compounds in which an ethylenic bond or a benzene ring is conjugated with a carbonyl group show spectra quite different from those of compounds containing the insulated chromophores. For example, acrolein, CH_2=CHCHO, shows an intense band at about 208 mμ (ϵ_{max} = 13,000) and a

[21] N. J. Leonard and E. R. Blout, *J. Am. Chem. Soc.* **72**, 484 (1950).

weak band at about 330 mμ (ϵ_{max} = 13). The former band is the first π–π* S–S band, which has a character of an intramolecular CT band, and the latter is the first n–π* S–S band. Benzaldehyde, C_6H_5CHO, as well as acetophenone, $C_6H_5COCH_3$, shows four bands in the near-ultraviolet region. The weakest band, which appears at about 320 mμ ($\epsilon_{max} \doteqdot 10$), is almost certainly the first n–π* S–S band. The weak band with fine structure, which appears in the 270–290 mμ region ($\epsilon_{max} \doteqdot 1500$), is probably a perturbed LE band corresponding to the benzene α band. The intense band at about

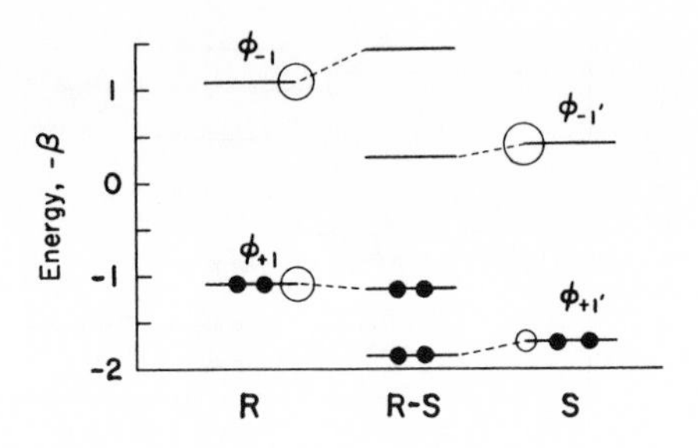

FIG. 21.3. The orbital energy level diagram of the vinyl–carbonyl system as a composite π system: $\overset{2}{C}{=}\overset{1}{C}{-}\overset{1'}{C}{=}\overset{2'}{O}$, with C=C as R and C=O as S.

240 mμ ($\epsilon_{max} \doteqdot 15{,}000$) is interpreted as a π–π* S–S band which has a character of an intramolecular CT band.[22] The intense band at about 200 mμ is probably a perturbed LE band corresponding to the benzene p band.

The vinyl–carbonyl system, C=C—C=O, can be considered as the composite system in which the fragments R and S are an ethylenic bond and a carbonyl bond, respectively, and the phenyl–carbonyl system, C_6—C=O, can be considered as the composite system in which the fragments R and S are a benzene ring and a carbonyl bond, respectively. The energy level diagrams for these systems are shown in Figs. 21.3 and 21.4. The carbon and oxygen atoms of the carbonyl group are numbered as 1′ and 2′, respectively, and the carbon atom (of R) to which the carbonyl group is attached is numbered as 1. The orbitals localized on the carbonyl group are indicated by an attached prime. In the energy level diagrams, the circle drawn on the end of each of the lines representing the orbital energy levels of the fragments indicates the relative magnitude of the coefficient of AO χ_1 or $\chi_{1'}$ in the corresponding localized orbital: it is drawn in such a way that its radius is proportional to the AO coefficient. The energy levels of the carbonyl group are

[22] J. Tanaka, S. Nagakura, and M. Kobayashi, *J. Chem. Phys.* **24,** 311 (1956).

drawn on the basis of the following parameter values: $h_{2'} = 1.3$, $k_{1'2'} = 0.844$. The nonbonding p orbital of the carbonyl group, $\phi_{n'}$, is considered to be unaffected by the conjugation.

The acrolein 208-mμ band and the benzaldehyde 240-mμ band can be thought of as being due to the S–S one-electron transition from ϕ^{p}_{+1} to $\phi^{p}_{-1'}$. This transition is evidently accompanied with a partial electron transfer from

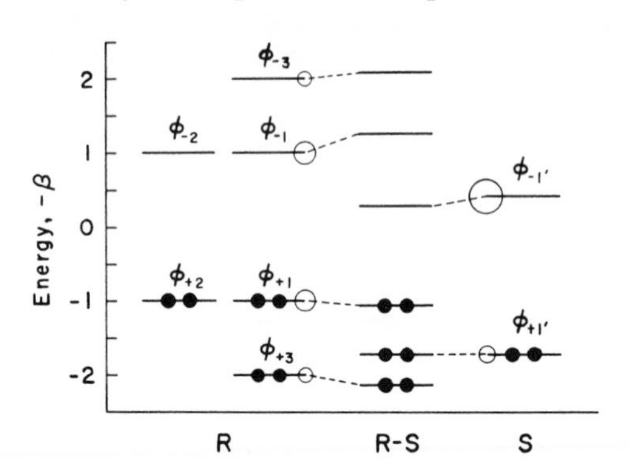

FIG. 21.4. The orbital energy level diagram of the phenyl–carbonyl system as a composite π system:

1 1′ 2′
—C═O

R S

R to S. Since the acceptor orbital $\phi^{p}_{-1'}$ is lower than $\phi_{-1'}$ and the donor orbital ϕ^{p}_{+1} is higher than $\phi_{+1'}$, these CT π–π* bands are at longer wavelength than the π–π* band of the isolated carbonyl group.

The acrolein 330-mμ band and the benzaldehyde 320-mμ band can be considered to be due to the S–S one-electron transition from $\phi_{n'}$ to $\phi^{p}_{-1'}$. Since $\phi^{p}_{-1'}$ is lower than $\phi_{-1'}$, these n–π* bands are at longer wavelength than the n–π* band of the isolated carbonyl groups.

When the carbonyl group is conjugated with a conjugated polyene system, both the n–π* and the π–π* bands shift to longer wavelength as the length of the conjugated system increases. However, the energy of the n–π* band does not decrease so rapidly as does the energy of the π–π* band, since for the π–π* band the energy is lowered both by the raising of the donor orbital ϕ^{p}_{+1} and by the lowering of the acceptor orbital $\phi^{p}_{-1'}$ when the conjugated system is extended, while for the n–π* band the energy is lowered only by the lowering of the acceptor orbital $\phi^{p}_{-1'}$. If the conjugated system becomes long enough the n–π* band becomes swamped under the more intense π–π* band.

The locations of the n–π^* and π–π^* bands of the conjugated enone or enal system (that is, the vinyl–carbonyl system C═C—C═O) are strongly affected by solvents: in passing from one solvent to a more polar solvent, the n–π^* band is displaced to shorter wavelength and the π–π^* band to longer wavelength (see Subsection 6.2.2.).

21.4.2. Effects of Substituents on the Absorption Bands of the Vinyl–Carbonyl System

A set of empirical rules for determining the location of the first π–π^* CT band of the conjugated enone or enal system has been proposed by Woodward[23] and by Fieser and Fieser.[24]

TABLE 21.5

Solvent Corrections for the CT Band of Conjugated Enals and Enones to Give the Location in Ethanol, a Standard Solvent

Solvent	Woodward[a]	Fieser and Fieser[b]
Methanol	−1 mμ	0 mμ
Chloroform	0	+1
Ether	+6	+7
Hexane	+7	+11
Water	—	−8
Dioxan	—	+5

[a] Reference 23. [b] Reference 24.

Since the solvent effect on the location of the CT band is considerably large, in order to correlate the spectral data in the literature it is necessary to know the solvent in which the spectra have been determined and to correct the observed wavelengths of the band to a standard solvent. The corrections proposed by Woodward and by Fieser and Fieser are shown in Table 21.5.

Woodward has deduced from examination of data for some fifty compounds the following generalizations with regard to the effect of alkyl substitution on the location (λ_{max}) of the CT band in ethanol. Starting with a value of 215 mμ for the λ_{max} of methyl vinyl ketone (IV, $X = CH_3$, $\alpha = \beta = \beta' = H$), when one, two, and three alkyl groups are introduced, the CT band is shifted

[23] R. B. Woodward, *J. Am. Chem. Soc.* **63,** 1123 (1941); **64,** 72, 76 (1942).

[24] L. F. Fieser and M. Fieser, "Steroids," Reinhold, New York, 1959.

$$\begin{array}{c} \beta \quad \alpha \quad X \\ \diagdown \quad | \quad | \\ C_2=C_1-C_{1'}=O_{2'} \\ \diagup \\ \beta' \end{array}$$

IV

to longer wavelength by about 10, 20, and 32 mμ, respectively. When the ethylenic bond is exocyclic to a six-membered alicyclic system, that is, when the compound has a 2,2-five-atom bridge or a 1,1′-four-atom bridge, an additional bathochromic shift of about 5 mμ occurs. When the ethylenic bond is exocyclic to two six-membered rings, that is, when the compound has a 2,2-five-atom bridge and a 1,1′-four-atom bridge, an additional bathochromic shift of about 10 mμ occurs. The CT band of a conjugated enal (IV, X = H) is located at shorter wavelength by about 5 mμ than that of the corresponding conjugated enones (IV, X = alkyl).

A more careful examination of data seems to suggest that the above generalizations should be partly modified as follows: the magnitudes of the bathochromic shift per an α-alkyl substitution and per a β-alkyl substitution are about 5 and 10 mμ, respectively.

On the basis of examination of the spectral data on steroids containing the conjugated enone system, Fieser and Fieser have modified Woodward's generalizations as follows: in the steroid system the magnitudes of the bathochromic shift of the CT band per an α-alkyl and a β-alkyl substitution are about 10 and 12 mμ, respectively.

When the ethylenic bond is exocyclic to a five-membered ring, the additional bathochromic shift is slightly larger (by about 1–2 mμ) than that for the case where the ethylenic bond is exocyclic to a six-membered ring.

The above generalizations can be applied even to 2-cyclohexenones, in which the carbonyl bond is exocyclic and the ethylenic bond is endocyclic to a six-membered alicyclic system, without invoking any special structural effect. Thus, the observed wavelength of the CT band of 2-cyclohexenone is in good agreement with the value of 225 mμ predicted for an open-chain β-monoalkyl substituted enone. However, when the ethylenic bond is in a five-membered ring and the carbonyl group is exocyclic to the ring, the CT band is at considerably shorter wavelength (by about 5–10 mμ) than that of the corresponding open-chain or six-membered ring ketone (for example: 2-cyclopentenone, λ_{max} = 218 mμ, ϵ_{max} = 9500; 2-cyclohexenone, λ_{max} = 224.5 mμ, ϵ_{max} = 10,300).[25] In contrast, a bathochromic shift of the CT

[25] W. M. Schubert and W. A. Sweeney, *J. Am. Chem. Soc.* **77**, 2297 (1955).

band on passing from a six- to a five-membered ring is observed in conjugated enones and enals in which the carbonyl group is outside and the ethylenic bond in the ring (see Table 21.6).[25] While the values of λ_{max} for 1-acetyl- and 1-formylcyclohexene are in fairly good agreement with the values predicted for α,β-dialkyl substituted enones and enals, respectively, the values for the five-membered analogs are considerably larger.

The first n–π^* bands of conjugated enones and enals appear mostly in the 300–330 mμ region in ethanol solution, and they exhibit vibrational fine

TABLE 21.6

THE CT BANDS OF 1-ACETYL- AND 1-FORMYLCYCLOALKENES[a]

Compound	λ_{max}, mμ	ϵ_{max}
1-Acetylcyclopentene	239	13,000
1-Acetylcyclohexene	233	12,500
1-Acetylcycloheptene	236	10,000
1-Formylcyclopentene	237	12,000
1-Formylcyclohexene	229	12,000

[a] The data were taken from Reference 25.

structure. The most rigid molecules (for example, some polycyclic compounds) show the best-defined fine structure. The location of the main peak of the n–π^* band appears to be displaced to shorter wavelength by alkyl substitution. For example, in the case of 2-cyclohexenones, one α (that is, 2-) or β (that is, 3-) alkyl substituent causes a hypsochromic shift of about 6 mμ and two alkyl substituents cause a hypsochromic shift of about 20 mμ.[26] This probably means that the alkyl substitution on the ethylenic bond raises the energy of the π^* orbital ($\phi^{p}_{-1'}$), since the nonbonding p orbital ($\phi_{n'}$) is considered to be practically unaffected.

Lastly, the effect of introduction of a hydroxyl substituent into the C=C—C=O system is discussed. When X in the appended formula IV is changed from the hydrogen atom to the hydroxyl group, with a change from an $\alpha\beta$-unsaturated aldehyde to an $\alpha\beta$-unsaturated carboxylic acid, a small hypsochromic shift of the first π–π^* band is observed. For example, while the π–π^* band of crotonaldehyde (IV with X = H, $\beta = CH_3$, $\alpha = \beta' = H$) is at 217 mμ (ϵ_{max} = 15,650) in ethanol solution, the π–π^* band of crotonic acid is at 204 mμ (ϵ_{max} = 11,700) in ethanol solution, and is at 208 mμ (ϵ_{max} = 12,500) in hexane solution. In general, the first π–π^* bands of

[26] R. C. Cookson and S. H. Dandegaonker, *J. Chem. Soc.* **1955,** 1651.

$\alpha\beta$-unsaturated carboxylic acids are in the 204–220 mμ region (ϵ_{max} = ca. 10,000–20,000). On the other hand, when α or β in the formula IV with X = H or alkyl is replaced by the hydroxyl group, the first π–π^* band is strongly displaced to longer wavelength, to about 270 ± 5 mμ.

These facts may be explained in terms of interaction of the lowest vacant and highest occupied π orbitals of the C═C—C═O system, ψ_{-1} and ψ_{+1}, with the nonbonding $2p\pi$ orbital of the hydroxyl group, $\phi_{0'}$. Since $\phi_{0'}$ is probably lower than ψ_{+1}, both ψ_{+1} and ψ_{-1} will be raised by interaction with $\phi_{0'}$. Of course, $\phi_{0'}$ is much closer in energy to ψ_{+1} than to ψ_{-1}. Therefore,

TABLE 21.7

THE LOWEST ANTIBONDING AND HIGHEST BONDING π ORBITALS OF THE C═C—C═O SYSTEM

Orbital (ψ_m)	Energy w_m	AO coefficient $c_{m,2}$	$c_{m,1}$	$c_{m,1'}$	$c_{m,2'}$
ψ_{-1}	+0.271	−0.592	+0.148	+0.698	−0.375
ψ_{+1}	−1.146	+0.587	+0.623	+0.093	−0.508

if the AO coefficients of ψ_{-1} and ψ_{+1} at the position to which the hydroxyl group is attached were comparable, ψ_{+1} would be much more raised than is ψ_{-1}, and hence the π–π^* band would be shifted to longer wavelength. As mentioned already, ψ_{-1} is considered to be the orbital arising from perturbation of the antibonding π orbital of the carbonyl group, and ψ_{+1} is considered to be the orbital arising from perturbation of the bonding π orbital of the ethylenic bond. Therefore, ψ_{-1} is considered to have a predominant contribution of the $2p\pi$ AO of the carbon atom of the carbonyl group, $\chi_{1'}$, and ψ_{+1} is considered to have predominant contributions of the $2p\pi$ AO's of the carbon atoms of the ethylenic bond, χ_1 and χ_2. Table 21.7 shows the energies and AO coefficients of ψ_{-1} and ψ_{+1} of the C═C—C═O system, calculated by the simple LCAO MO method on the basis of the following parameter values: $h_{2'} = 1.3$, $k_{1'2'} = 0.8434$, $k_{11'} = 0.858$, and $k_{12} = 1.080$. When the hydroxyl group is attached to the 1′ position, since the square of the AO coefficient of ψ_{-1} at the 1′ position, $c^2_{-1,1'}$, is much larger than that of ψ_{+1}, $c^2_{+1,1'}$, ψ_{-1} will be raised by a larger amount by interaction with $\phi_{0'}$ than is ψ_{+1}, and hence the π–π^* band will be shifted to shorter wavelength. On the other hand, $c^2_{-1,2}$ is comparable to $c^2_{+1,2}$, and $c^2_{-1,1}$ is much smaller than $c^2_{+1,1}$. Therefore, when the hydroxyl group is attached to the 2 or the 1 position, ψ_{+1} will be much more raised by interaction with $\phi_{0'}$ than is ψ_{-1}, and hence the π–π^* band will be shifted to longer wavelength.

21.4.3. Effects of Substituents on the Absorption Bands of the Phenyl–Carbonyl System

The spectra of benzaldehyde and acetophenone are shown in Fig. 21.5, along with the spectrum of benzophenone. The data on the spectra of benzaldehyde, acetophenone, and their mono-methylated derivatives are listed in Table 21.8.

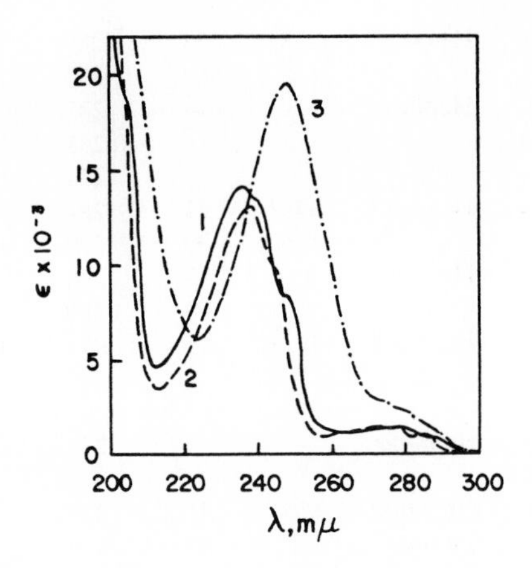

Fig. 21.5. The absorption spectra of benzaldehyde (1), acetophenone (2), and benzophenone (3) in heptane.

The CT band of acetophenone is at about the same wavelength as that of benzaldehyde, but the intensity of the former is appreciably smaller than that of the latter. Both in benzaldehyde and in acetophenone, introduction of a methyl substituent to the benzene ring causes a bathochromic shift of the CT band. However, the magnitude of the bathochromic shift caused by the *o*-methyl substituent is considerably smaller than that caused by the *p*-methyl substituent. In addition, the intensity of the CT band of the *o*-methyl compound is appreciably smaller than that of the *p*-methyl compound: the intensity of the CT band of *o*-methylacetophenone is smaller than that of acetophenone, while that of *p*-methylacetophenone is greater than that of acetophenone. The hypsochromic and hypochromic effect of the *o*-methyl substituent as compared with the *p*-methyl substituent is conceivably attributed at least partly to the steric effect.

The steric effect in the spectra of vinyl–carbonyl, phenyl–carbonyl, and some other conjugated carbonyl compounds will be discussed in the next section.

TABLE 21.8

THE SPECTRA OF BENZALDEHYDE, ACETOPHENONE, AND THEIR MONO-METHYLATED DERIVATIVES[a]

Compound	Solvent	The n–π^* band λ_{max}, mμ	ϵ_{max}	The α band λ_{max}, mμ	ϵ_{max}	The CT band λ_{max}, mμ	ϵ_{max}
Benzaldehyde							
Unsubstituted	Hexane[b]	338	53	289	1200	248*	12,500
		328	57	280	1400	242	14,000
	Heptane[c]	—	—	289	890	248	8600
				281	1410	241*	13,400
				274	1520	237	14,100
p-Methyl-	Hexane[b]	340*	23	284*	1000	257	12,500
		326	30	279	1200	251	15,000
m-Methyl-	Hexane[b]	—	—	290	800	251	12,000
				280	1000	245	13,500
o-Methyl-	Hexane[b]	322	45	291	1700	251	13,000
						243	12,500
Acetophenone							
Unsubstituted	Heptane[c]	—	—	278	1480	246*	9690
						239	13,130
	Heptane[d]	320	41	279	890	238	12,600
	Ethanol[b]	315	55	279	1200	243	13,200
p-Methyl-	Ethanol[d]	—	—	280	1590	252	13,500
o-Methyl-	Ethanol[b]	331	50	281	1200	245	8300
						242	8700

[a] Asterisk indicates a shoulder. [b] Reference 27. [c] Reference 36. [d] Reference 22.

21.5. The Steric Effect in the Spectra of Conjugated Carbonyl Compounds

21.5.1. PRELIMINARY CONSIDERATIONS

As mentioned above, while introduction of a *p*-methyl substituent into benzaldehyde or acetophenone brings about bathochromic shift and small increase in intensity of the CT band, introduction of an *o*-methyl substituent causes much smaller bathochromic shift and definite decrease in intensity of the CT band. The difference between the effects of *p*- and *o*-methyl substituents is attributed at least partly to the steric effect of the *o*-substituent, which will hinder the benzene ring and the carbonyl group from assuming a coplanar arrangement. If it is assumed that the electronic effect of the *o*-methyl substituent is approximately equivalent to that of the *p*-methyl substituent, the net steric effect of the *o*-methyl substituent on the location of the CT band can be said to be slightly hypsochromic. The hypochromic effect of the *o*-methyl substituent is greater in acetophenone than in benzaldehyde. Of course, the steric hindrance to the coplanarity of the benzene ring and the carbonyl group must be much severer with the acetyl group than with the formyl group.

In general, the effect of twisting the essential single bond (the 1—1′ bond) in the vinyl–carbonyl as well as the phenyl–carbonyl system on the CT band seems to be slightly hypsochromic and highly hypochromic. The apparent small bathochromic shift of the CT band observed sometimes with sterically hindered compounds as compared with sterically unhindered compounds is almost certainly due to the electronic bathochromic effect of the substituents outweighing the steric hypsochromic effect. As mentioned in Subsection 19.4.3, such a comparative insensibleness of the location and such a marked sensibleness of the intensity to steric factors are characteristics of CT bands.

The energy level diagrams for the vinyl–carbonyl system and for the phenyl–carbonyl system were already shown in Figs. 21.3 and 21.4. Evidently, according to the classification made in Chapter 19, these systems belong to case C-1. The forms and energies of the lowest vacant and highest occupied π orbitals of these systems are given approximately by expressions (19.64)–(19.67), in which k_{rs} represents $k_{11'}$.

When the fragment R is an even alternant hydrocarbon system as in this case, $v_{+m} = -v_{-m}$, and $b_{+m,1} = b_{-m,1}$. Since $v_{-1'}$ is positive, $-G_{-1',-m}$ is therefore greater than $G_{-1',+m}$. This means that the contribution of the antibonding orbitals of R to $\phi^{\mathrm{p}}_{-1'}$ is greater than that of the bonding orbitals of R. It follows that $g_{-1'}$ in expression (19.66) is negative and hence that the partial π-bond order $p_{11'}^{(-1)}$ is positive [see expression (19.69)].

As is easily verified, when $h_{2'} = -v_{+1}$ the following relations exist: $-G_{+1,-1'} = G_{+1,+1'}$, $g_{+1} = 0$ [see expression (19.67)], and $p_{11'}^{(+1)} = 0$ [see expression (19.69)]. In this case the energy of ϕ^{p}_{+1} is independent of $k_{11'}$. As the value of $h_{2'}$ becomes smaller, the value of $-G_{+1,-1'}$ decreases owing to increase in the energy separation $v_{-1'} - v_{+1}$ and to decrease in $b_{-1',1'}$, and the value of $G_{+1,+1'}$ increases owing to decrease in the energy separation $v_{+1} - v_{+1'}$ and to increase in $b_{+1',1'}$. Thus, if $h_{2'} < -v_{+1}$, the following relations exist: $-G_{+1,-1'} < G_{+1,+1'}$, $g_{+1} > 0$, and $p_{11'}^{(+1)} < 0$. In this case the energy of ϕ^{p}_{+1} becomes higher as the value of $k_{11'}$ increases. If $h_{2'} > -v_{+1}$, the following relations exist: $-G_{+1,-1'} > G_{+1,+1'}$, $g_{+1} < 0$, and $p_{11'}^{(+1)} > 0$. In this case, the energy of ϕ^{p}_{+1} becomes lower as the value of $k_{11'}$ increases.

Putting the above discussions together, we can say as follows. If $h_{2'}$ is smaller than, or nearly equal to, $-v_{+1}$, then $\Delta g_{-1',+1}$ [$= g_{-1'} - g_{+1}$: see expression (19.68)] is negative. In this case, the π-bond order of the 1—1′ bond increases on the one-electron transition from ϕ^{p}_{+1} to $\phi^{p}_{-1'}$, and decrease in $k_{11'}$ associated with twist of the 1—1′ bond causes hypsochromic shift of the corresponding CT band. Of course, the smaller the value of $h_{2'}$ is, the larger is the absolute value of $\Delta g_{-1',+1}$, and hence the more effective will be the steric hypsochromic shift. The extreme case where $h_{2'}$ is zero reduces to case A or case B. If $h_{2'}$ has a moderately larger value than $-v_{+1}$, $\Delta g_{-1',+1}$ may be nearly zero. In this case, any twist of the 1—1′ band will result in substantially no shift of the CT band. In the case where $h_{2'}$ is much larger than $-v_{+1}$, $\Delta g_{-1',+1}$ is positive. In this case, the π-bond order of the 1—1′ bond decreases on the one-electron transition from ϕ^{p}_{+1} to $\phi^{p}_{-1'}$, and twist of the 1—1′ bond will cause bathochromic shift of the CT band.

As mentioned already, a value of 1.2–1.3 seems to be pertinent to the Coulomb parameter h for the oxygen $2p\pi$ AO. When $h_{2'}$ is 1.2–1.3, $\Delta g_{-1',+1}$ has a small negative value. This means that twist of the 1—1′ bond in the vinyl–carbonyl system as well as the phenyl–carbonyl system will cause small hypsochromic shift of the CT band.

Since nitrogen is less electronegative than oxygen, the value pertinent to the Coulomb parameter for the nitrogen $2p\pi$ AO must be smaller than that for the oxygen $2p\pi$ AO. A value of about 0.6 seems to be pertinent to the Coulomb parameter for the nitrogen $2p\pi$ AO. Therefore, if the fragment S is —C=N— instead of —C=O, the steric hypsochromic shift of the CT band will be more effective. The fact that the steric hypsochromic effect in semicarbazones of *ortho*-substituted acetophenones is appreciably more effective than in the acetophenones[27] seems to suggest the validity of this surmise.

If the oxygen atom of the carbonyl group in the vinyl–carbonyl system or in the phenyl–carbonyl system could be replaced by a much more electronegative atom, a steric bathochromic shift of the CT band would be expected in the resulting composite system. A similar effect will be expected when a strongly electron-donating mesomeric substituent, such as the amino group, is introduced into the fragment R, since the orbitals ϕ_{+1} and ϕ_{-1} will be raised by interaction with the nonbonding orbital of the substituent.

In Figs. 21.6 and 21.7 are shown the relations of the energy (Δw_{11}) and dipole strength (M^2_{11}) of the one-electron $\psi_{+1} \rightarrow \psi_{-1}$ transition to the angle of twist ($\theta_{11'}$) of the 1—1′ bond in the vinyl–carbonyl system and in the phenyl–carbonyl system, respectively.[28] [In the calculation, a value of 1.3 was used for $h_{2'}$, and the relations among the twist angle (θ), length (R), and resonance

[27] E. A. Braude and F. Sondheimer, *J. Chem. Soc.* **1955,** 3754.
[28] H. Suzuki, *Bull. Chem. Soc. Japan* **35,** 1853 (1962).

parameter (k) of the 1—1′ bond were assumed to be given by expressions (9.38) and (9.39).] In these figures, the corresponding relations in the conjugated diene system and in the styrene system are also shown for comparison. It is seen that the intensity of the CT band will be highly sensitive,

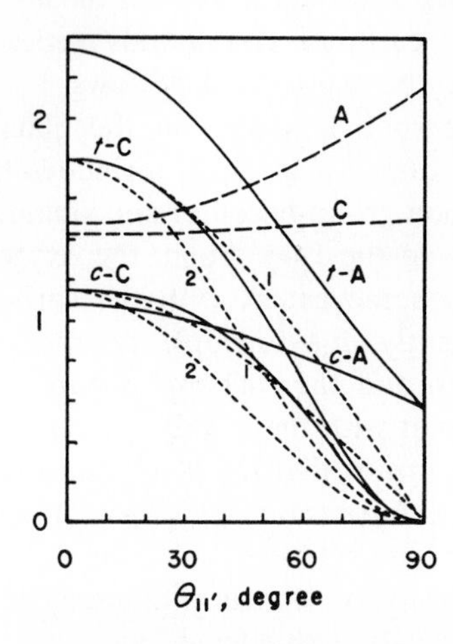

FIG. 21.6. The energy (Δw_{11}) and dipole strength (M_{11}^2) of the one-electron transition from ψ_{+1} to ψ_{-1} as functions of the angle of twist of the 1—1′ bond ($\theta_{11'}$) in the vinyl–carbonyl system and in the conjugated diene system. ----, Δw_{11} (in $-\beta$); solid lines, M_{11}^2 (in Å²). A, the conjugated diene system; C, the vinyl–carbonyl system. *t*, *s-trans;* *c*, *s-cis;* ....., 1, $M_{11}^2(0°) \cdot \cos\ \theta_{11'}$; 2, $M_{11}^2(0°) \cdot \cos^2 \theta_{11'}$.

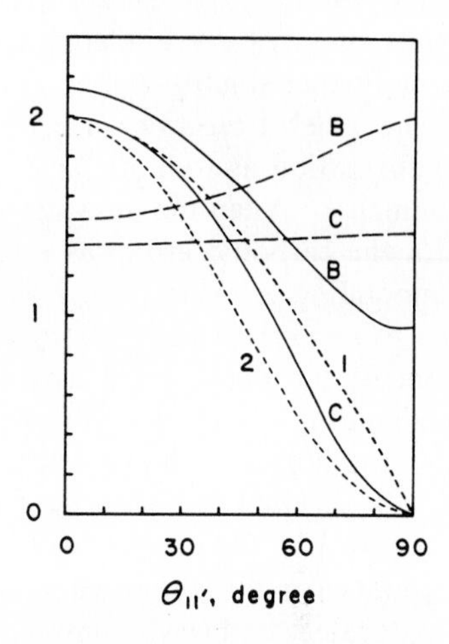

FIG. 21.7. The energy (Δw_{11}) and dipole strength (M_{11}^2) of the one-electron transition from ψ_{+1} to ψ_{-1} as functions of the twist angle of the 1—1′ bond ($\theta_{11'}$) in the phenyl–carbonyl system and in the styrene system. ----, Δw_{11} (in $-\beta$); solid lines, M_{11}^2 (in Å²). B, the styrene system; C, the phenyl–carbonyl system; ....., 1, M_{11}^2 (0°) $\cdot \cos\ \theta_{11'}$; 2, $M_{11}^2(0°) \cdot \cos^2 \theta_{11'}$.

and the position of the band will be by far less sensitive, to change in the angle of twist of the 1—1′ bond.

So far the discussion has been restricted to the steric effect on the CT band. Now we turn to the n–π^* band. The steric effect on the n–π^* band appears to be slightly hypsochromic and slightly hyperchromic. As mentioned above, the energy of $\phi^{p}_{-1'}$ is almost certainly lower than that of $\phi_{-1'}$. This means that the energy of $\phi^{p}_{-1'}$ will become higher as the angle of twist of the 1—1′ bond increases. The steric hypsochromic shift of the n–π^* band is probably mainly due to the raising of the acceptor orbital $\phi^{\text{p}}_{-1'}$ by twist of the 1—1′ bond. The intensity of the n–π^* bands of sterically unhindered $\alpha\beta$-unsaturated carbonyl compounds (such as vinyl–carbonyl and phenyl–carbonyl compounds) appears to be slightly higher than the intensity of the n–π^* band of the isolated carbonyl group ($\epsilon_{\max} = 10$–20), and the intensity of the n–π^* bands of sterically hindered $\alpha\beta$-unsaturated carbonyl compounds appears to be further slightly higher. These facts may be explained as follows.

The n–π^* transition in $\alpha\beta$-unsaturated carbonyl systems in the planar conformation as well as that in the isolated carbonyl group is forbidden by symmetry. When the unsaturated hydrocarbon group becomes noncoplanar with the carbonyl group as a result of twist of the 1—1′ bond, the oxygen nonbonding p orbital ($\phi_{\text{n}'}$) will overlap to some extent with the carbon $2p\pi$ AO's at the 1 and 2 positions. Consequently, in a noncoplanar arrangement of the unsaturated hydrocarbon group and the carbonyl group, the nonbonding orbital $\phi_{\text{n}'}$ will mix to some extent with the π MO's, especially with the highest bonding π MO (ψ_{+1}). This means that the n–π^* transition ($\phi_{\text{n}'} \rightarrow \psi_{-1}$) will mix with the allowed π–π^* CT transition ($\psi_{+1} \rightarrow \psi_{-1}$), and consequently the former will borrow the intensity from the latter. This accounts for the enhanced n–π^* band intensity of sterically hindered $\alpha\beta$-unsaturated carbonyl compounds. In sterically unhindered compounds, torsional vibrations about the 1—1′ bond will allow to a smaller extent the mixing of the n–π^* and π–π^* transitions.

Some of $\beta\gamma$-unsaturated carbonyl compounds and $\gamma\delta$-unsaturated carbonyl compounds in which the ethylenic bond and the carbonyl bond are not coplanar show intensified n–π^* bands. This intensification can also be accounted for in terms of the mixing of the n–π^* and π–π^* transitions.

21.5.2. Conjugated Enones and Enals

In Table 21.9 data on the first CT bands of some conjugated enones are shown. It is evident that in the series of 1-acylcyclohexenes (entry numbers 3–6) the intensity of the band decreases rapidly as the steric hindrance to the coplanarity of the ethylenic bond and the carbonyl bond increases.

The intensity of the CT band of the C═C—C═O system depends not only on the angle of twist of the 1—1′ bond but also on whether the conformation about the 1—1′ bond is *s-trans* or *s-cis*. In general, as Fig. 21.6 shows, the intensity of the CT band is expected to be higher in an *s-trans* conformation than in the *s-cis* conformation with the same twist angle.

In the absence of steric effects, it is generally agreed that conjugated enones possess the *s-trans* conformation. Thus, 1-acetylcyclohexene (3) is believed to exist predominantly in the *s-trans* conformation. However, the introduction of a methyl substituent into the 2 position should cause a smaller steric interference in the *s-cis* than in the *s-trans* conformation. Turner and Voitle[29] have proposed that 1-acetyl-2-methylcyclohexene (4) exists in a nearly planar *s-cis* conformation and have ascribed the small intensity of the CT band of this compound as compared with that of 1-acetylcyclohexene to the difference

TABLE 21.9

THE CT BANDS OF CONJUGATED ENONES IN ETHANOL

No.	Compound	λ_{max}, mμ	ϵ_{max}
1	Mesityl oxide $(CH_3)_2C{=}CH{-}CO{-}CH_3$[a]	231	12,000
2	3,4-Dimethyl-3-penten-2-one $(CH_3)_2C{=}C(CH_3){-}CO{-}CH_3$[a]	245.5	6000
3	1-Acetylcyclohexene[b]	232	12,500
4	1-Acetyl-2-methylcyclohexene[b]	245	6500
5	1-Trimethylacetyl-2-methylcyclohexene[b]	239	1300
6	1-Acetyl-2,6,6-trimethylcyclohexene[b]	243	1400
7	2-Cyclohexenone[c]	224.5	10,300
8	2,3,5,5-Tetramethyl-2-cyclohexenone[a]	247	9600
9	2-Isopropylidene-5-methylcyclohexanone[a]	252	6500

[a] Reference 31. [b] Reference 30. [c] Reference 25.

in the conformation. On the other hand, Braude and Timmons[30] have proposed that this 2-methyl compound exists in a nonplanar *s-trans* conformation, and have ascribed the reduction in the intensity of the CT band to the nonplanarity of the conjugated system. The question of whether this compound has a nearly planar *s-cis* or a nonplanar *s-trans* conformation is now rather firmly resolved in favor of the *s-cis* conformation. Thus, from a consideration of the infrared absorption spectra, Waight and Erskine[31] have concluded that this compound exists predominantly in the *s-cis* conformation and that not more than 15% of the molecules is present in the *s-trans* conformation at room temperature.

Results of dipole moment measurements suggest that the C=C—C=O system in mesityl oxide (that is, 4-methyl-3-penten-2-one, entry number 1)

[29] R. B. Turner and D. M. Voitle, *J. Am. Chem. Soc.* **73,** 1403 (1951).

[30] E. A. Braude and C. J. Timmons, *J. Chem. Soc.* **1955,** 3766.

[31] E. S. Waight and R. L. Erskine, *in* "Steric Effects in Conjugated Systems" (G. W. Gray, ed.), p. 73. Butterworths, London, 1958.

and 3,4-dimethyl-3-penten-2-one (2) is predominantly in the *s-cis* conformation.[32] Of course, the C═C—C═O system in pulegone (2-dimethylmethylene-5-methylcyclohexanone, entry number 9) is in the fixed *s-cis* conformation. The intensities of the CT bands of *s-cis* enones 2, 4, and 9 are very similar, having alike ϵ_{max} values of about 6500, and are much lower than the intensity of the CT band of the *s-cis* mesityl oxide (1). Probably the *s-cis* C═C—C═O system is not planar even in mesityl oxide, and it deviates more largely from the planarity in the former three compounds.

The C═C—C═O system in 2-cyclohexenones (entry numbers 7 and 8) is of course in the fixed *s-trans* conformation. It is evident that the conjugated

TABLE 21.10

THE CT BANDS OF 1-FORMYLCYCLOHEXENES IN ETHANOL[a]

1-Formylcyclohexene	λ_{max}, mμ	ϵ_{max}
Unsubstituted	229	12,100
2-Methyl-	242	11,200
2,6,6-Trimethyl-	249	11,600

[a] The data were taken from Reference 33.

enone system is more largely nonplanar in 2,3,5,5-tetramethyl-2-cyclohexenone (8) than in 2-cyclohexenone (7).

For 2,6,6-trimethyl-1-acetylcyclohexene (6) the *s-cis* conformation is as strongly hindered as the *s-trans* conformation, and the interplanar angle between the ethylenic bond and the carbonyl bond in this compound is thought to be larger at least than 70°. It is noteworthy that even in this highly hindered compound the CT band is still detectable though with very low intensity. As is expected, the n–π^* band (λ_{max} = 305 mμ) of this compound has a markedly enhanced intensity (ϵ_{max} = 90).

Data on the first CT bands of some 1-formylcyclohexenes are shown in Table 21.10. It will be seen that the steric effect of the 2- and 6-methyl substituents is much smaller in the formyl compounds than in the acetyl analogs.

21.5.3. CONJUGATED DIENONES AND TRIENONES

Conjugated dienones show the first CT band at longer wavelength by about 30 mμ or more than that of the corresponding conjugated enones.[33] Data on

[32] J. B. Bentley, R. B. Everard, R. J. B. Marsden, and L. E. Sutton, *J. Chem. Soc.* **1949,** 2957; G. H. Estok and J. S. Dehn, *J. Am. Chem. Soc.* **77,** 4769 (1955).

[33] E. A. Braude, E. R. H. Jones, H. P. Koch, R. W. Richardson, F. Sondheimer, and J. B. Toogood, *J. Chem. Soc.* **1949,** 1890.

the CT bands of some cyclohexene derivatives containing the side chain —CH═CHCOCH$_3$ at the 1 position are shown in Table 21.11. In this conjugated system, introduction of a methyl group into the 2 position of the cyclohexene ring shifts the CT band to longer wavelength, but does not

TABLE 21.11

THE CT BANDS OF SOME CONJUGATED DIENONES IN ETHANOL[a]

1-2′-Acetylvinylcyclohexene	λ_{max}, mμ	ϵ_{max}
Unsubstituted	281	20,800
2-Methyl-	296	20,300
6,6-Dimethyl-	281	13,000
2,6,6-Trimethyl- (β-ionone)	296	10,700
2,5,6,6-Tetramethyl- (β-irone)	295	11,200

[a] The data were taken from Reference 33.

significantly change the intensity. On the other hand, introduction of two methyl groups into the 6 position reduces the intensity by about half.

Extending the conjugated system with an additional ethylenic bond results in a further 30 mμ or more bathochromic shift of the CT band.[33] Data on the

TABLE 21.12

THE CT BANDS OF SOME CONJUGATED TRIENONES IN ETHANOL[a]

1-2′-Acetylvinyl-1,3-cyclohexadiene	λ_{max}, mμ	ϵ_{max}
6-Methyl-	337	17,700
2,6,6-Trimethyl-	338	9100

[a] The data were taken from Reference 33.

CT bands of some 1,3-cyclohexadiene derivatives containing the side chain —CH═CH—COCH$_3$ are shown in Table 21.12. In this conjugated trienone system, the 6-*gem*-dimethyl substituent exerts also a marked steric hypochromic effect.

21.5.4. PHENYL–CARBONYL COMPOUNDS

Benzaldehyde as well as acetophenone shows the first π–π^* CT band at about 240 mμ. Introduction of a methyl substituent into the *para* position of these compounds increases the intensity of the CT band. On the other hand, introduction of a methyl substituent into the *ortho* position decreases the intensity of the CT band. The hypochromic effect of the *ortho* substituent is evidently due to the steric interference of the substituent with the formyl or acetyl group, which will rotate the group out of the plane of the benzene ring. Data on the CT bands of benzaldehyde, acetophenone, and their methylated derivatives are shown in Table 21.13. The CT band of each of these compounds except the compounds bearing two *o*-methyl substituents has two

TABLE 21.13

THE CT BANDS OF BENZALDEHYDE, ACETOPHENONE, AND THEIR METHYLATED DERIVATIVES IN SATURATED HYDROCARBON SOLVENTS[a]

Compound	Solvent	λ_{max}, mμ	ϵ_{max}
Benzaldehyde[b]			
Unsubstituted	Hexane	242	14,000
4-Methyl-	Hexane	251	15,000
2-Methyl-	Hexane	243	12,500
2,6-Dimethyl-	Hexane	251	12,500
2,4,6-Trimethyl-	Hexane	264	14,500
Acetophenone[c]			
Unsubstituted[d]	Heptane	239	13,130
Unsubstituted	Isopentane	237	12,050
4-Methyl-	Isopentane	246	18,100
3-Methyl-	Isopentane	241	12,400
2-Methyl-	Isopentane	237	10,700
2,4-Dimethyl-	Isopentane	246	12,350
2,5-Dimethyl-	Isopentane	240	9750
2,6-Dimethyl-	Isopentane	235*	2010
2,4,6-Trimethyl-	Isopentane	238	2960

[a] Asterisk denotes a shoulder.

[b] The data for benzaldehyde and its derivatives were taken from Reference 35.

[c] The data for acetophenone and its derivatives except the data for acetophenone in heptane solution were taken from Reference 31.

[d] Reference 36.

peaks. In most cases, the shorter wavelength peak is more intense than the longer wavelength one (see Table 21.8 and Fig. 21.5). The data listed in Table 21.13 are for the shorter wavelength peak. It will be seen that the steric interference is severer in the *ortho*-substituted acetophenones than in the correspondingly substituted benzaldehydes. Waight and Erskine[31] have found that the locations and intensities of the first CT bands of the acetophenones are substantially unchanged when the temperature is lowered to $-196°$C.

As is seen in Table 21.13, the CT band is still detectable even in 2,6-dimethyl- and 2,4,6-trimethylacetophenone, which are strongly hindered compounds. Introduction of two methyl groups into the 3 and 5 positions of these compounds appears to enhance the steric inhibition of the conjugative interaction between the benzene ring and the carbonyl group. Thus, 2,3,5,6-tetramethyl- and 2,3,4,5,6-pentamethylacetophenone exhibit substantially no CT band, and their spectra rather resemble the spectra of the correspondingly methyl-substituted benzenes.[34] This marked enhancement of the steric interference by *meta*-substitution is attributed to the so-called buttressing effect: the indirect steric effect of the *meta*-substituents which "buttress" the *ortho*-substituents and in this way enhance the steric interference between the *ortho*-substituents and the substituent at the 1 position.

Braude and his co-workers[35] have assumed that there is no significant steric hindrance to the planarity of the conformation in acetophenone as well as in benzaldehyde. However, acetophenone is probably slightly hindered.[36] As mentioned in Section 13.1, the α-methyl group in α-methylstyrene exerts a definite steric effect on the spectrum, and the most probable conformation of this compound is inferred to be nonplanar. While the geometry of the relevant part of benzaldehyde is considered to be similar to that of styrene, the geometry of acetophenone is considered to be similar to that of α-methylstyrene. Accordingly, it is presumed that there should be appreciable steric interaction between the methyl group and the phenyl group in acetophenone as in α-methylstyrene. The value of the exaltation of molar refraction, EM_D, for acetophenone (+0.52) is considerably smaller than the value for benzaldehyde (+0.80).[37] This fact seems to indicate that acetophenone is less conjugated than benzaldehyde, and therefore seems to support the view that the most probable conformation of acetophenone is probably nonplanar, while that of benzaldehyde is probably planar.

Data on the first CT bands of benzaldehyde, acetophenone, and their *p*-biphenylyl and 2-fluorenyl analogs are compared in Table 21.14. The CT band of each acetyl compound is at shorter wavelength and has a lower intensity than that of the corresponding formyl compound. This fact seems to indicate the presence of the steric interference between the methyl group and the aromatic ring in each acetyl compound.

[34] W. F. Forbes and W. A. Mueller, *J. Am. Chem. Soc.* **79,** 6495 (1957).

[35] E. A. Braude, F. Sondheimer, and W. F. Forbes, *Nature* **173,** 117 (1954); E. A. Braude and F. Sondheimer, *J. Chem. Soc.* **1955,** 3754.

[36] H. Suzuki, *Bull. Chem. Soc. Japan* **33,** 613 (1960).

[37] K. v. Auwers and F. Eisenlohr, *Ber.* **43,** 806 (1910).

TABLE 21.14

THE CT BANDS OF SOME ARYL–CARBONYL COMPOUNDS, R–S, IN HEPTANE[a,b]

	S = Formyl		S = Acetyl	
R	λ_{max}, mμ	ϵ_{max}	λ_{max}, mμ	ϵ_{max}
Phenyl	248.1	8600	245.5*	9690
	240.5*	13,400	238.5	13,130
	236.9	14,100		
p-Biphenylyl	283.5	24,150	276.2	20,650
2-Fluorenyl	317.2	34,600	312.8	31,200
	312.2*	28,200	305.1	21,850
	304.4	26,300	299.7	22,400
	297.7	24,200	291.8	22,450

[a] The data were taken from Reference 36.
[b] Asterisk denotes a shoulder.

Table 21.15 shows data on the first CT bands of benzocyclanones (V).[38] The values of ϵ_{max} and f decrease with the increasing size of the alicyclic ring. Thus, in the series of unsubstituted benzocyclanones, the values of ϵ_{max} and f are high for the compounds with $n = 5$ and 6, indicating that the

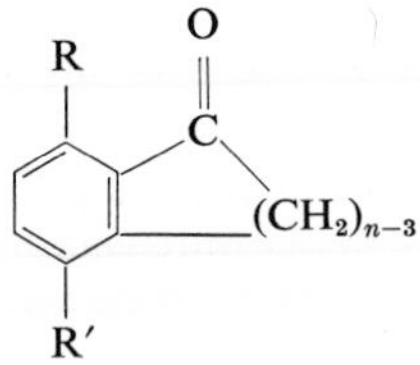

V

TABLE 21.15

THE CT BANDS OF BENZOCYCLANONES (V)[a]

n	R	R′	λ_{max}, mμ	ϵ_{max}	f
5	H	H	239	12,720	0.208
6	H	H	243	11,450	0.196
7	H	H	240	9000	0.153
8	H	H	243	6500	0.115
5	CH_3	H	238	12,800	0.208
5	CH_3	CH_3	244.5	12,300	0.179
6	CH_3	CH_3	246.5	10,390	0.169
7	CH_3	CH_3	244.5	4950	0.088

[a] The data were taken from Reference 38.

[38] G. D. Hedden and W. G. Brown, *J. Am. Chem. Soc.* **75**, 3744 (1953).

carbonyl group is constrained in a conformation coplanar or nearly coplanar with the benzene ring. When $n = 7$ and 8, the values decrease markedly, indicating that the carbonyl group is twisted out of the plane of the benzene ring. A similar phenomenon, but with a more abrupt decrease in the intensity at $n = 7$, appears in the series of the methyl-substituted benzocyclanones.

Lastly in this subsection, the spectrum of benzophenone (see Fig. 21.5) is briefly referred to. The two phenyl groups in this compound cannot be simultaneously coplanar with the carbonyl bond. This situation is very similar to that in the case of 1,1-diphenylethylene (see Section 13.3). However, while 1,1-diphenylethylene shows the first intense π–π^* S–S band (the so-called A band) at almost the same position as that of styrene, benzophenone shows the first CT band ($\lambda_{max} = 248.5$ mμ, $\epsilon_{max} = 19{,}550$, in *n*-heptane) at appreciably longer wavelength than that of benzaldehyde. This probably indicates that the angle through which each phenyl group in benzophenone is rotated out of the plane of the carbonyl bond is not so large as the corresponding twist angle in 1,1-diphenylethylene. The n–π^* band of benzophenone ($\lambda_{max} = 330$ mμ, $\epsilon_{max} = 160$) is much more intense than the n–π^* bands of benzaldehyde and acetophenone (ϵ_{max} = about 50). This intensification of the n–π^* band is almost certainly due to the mixing of the n–π^* transition with the π–π^* CT transition, which will occur as a result of the noncoplanarity of the carbonyl group and the benzene rings (see Subsection 21.5.1).

21.5.5. Acetylazulenes

The azulenyl–carbonyl system is known as a typical example of conjugated systems in which important absorption bands undergo bathochromic shifts when the conjugated system becomes nonplanar owing to twist of the essential single bond. The steric effects in the spectra of a series of acetylazulenes have been extensively studied by Heilbronner and Gerdil.[39]

1-Acetylazulene, 1,3-diacetylazulene (for the skeleton and numbering of carbon atoms of azulene, see Fig. 8.5), and their alkyl-substituted derivatives show four band systems in the visible and near-ultraviolet region, which have all more or less resolved vibrational fine structure. The longest wavelength band (A band), appearing at about 550 mμ with log ϵ_{max} of about 2.7, the second band (B band), appearing in the 370–400 mμ region with log ϵ_{max} of about 3.9, and the third band (C band), appearing in the 270–310 mμ region with log ϵ_{max} of about 4.5, are considered to be bands essentially localized in

[39] E. Heilbronner and R. Gerdil, *Helv. Chim. Acta* **39**, 1996 (1956).

the azulene ring corresponding to the p, α, and β (and β') bands of azulene, respectively (see Subsection 10.3.4). The shortest wavelength band (D band), appearing in the 200–250 mμ region with log ϵ_{max} of about 4.3, is probably an intramolecular CT band accompanied with a partial electron transfer from the azulene ring to the carbonyl group.

TABLE 21.16

THE A BANDS OF ACETYLAZULENES IN CYCLOHEXANE[a]

No.	Azulene	λ_{max}, mμ	ϵ_{max}	$f \times 10^3$
1	1-Acetyl-	546	440	8.5
2	1-Acetyl-3-methyl-	577	436	9.0
3	1-Acetyl-2-methyl-	548	322	7.6
4	1-Acetyl-3,8-dimethyl-5-isopropyl-	584	481	10.1
5	1,3-Diacetyl-	516	574	10.9
6	1,3-Diacetyl-2-methyl-	526	433	9.1

[a] The data were taken from Reference 39.

Data on the A bands of acetylazulenes are shown in Table 21.16, and data on the p bands of azulene and alkylazulenes are shown in Table 21.17. In these tables, λ_{max} and ϵ_{max} refer to the values for the most intense peak of the vibrational fine structure. It is seen that the p band of azulenes is shifted to shorter wavelength and is increased in the intensity by introduction of the acetyl group.

TABLE 21.17

THE p BANDS OF AZULENE AND ALKYLAZULENES IN CYCLOHEXANE OR PETROLEUM ETHER[a]

No.	Azulene	λ_{max}, mμ	ϵ_{max}	$f \times 10^3$
1′	Unsubstituted	580	329	4.5
2′	1-Methyl- (3-Methyl-)	608	294	5.1
3′	2-Methyl-	566	260	4.8
4′	1,4-Dimethyl-7-isopropyl- (3,8-Dimethyl-5-isopropyl-)	605	425	8.0

[a] The data were taken from Reference 39.

Introduction of a methyl substituent into the 2 or the 8 position of 1-acetylazulene should bring about a steric hindrance to the coplanarity of the acetyl group with the azulene ring. From consideration of scale models, the interplanar angle θ between the acetyl group and the azulene ring in compounds

3 and 4 has been estimated at about 30–40°. In order to evaluate the steric effect on the A band, it is necessary to make allowance for the electronic effects of the alkyl substituents. It should be recalled here that alkyl substitution in azulene can produce either a bathochromic or a hypsochromic shift of the *p* band, depending on the position of substitution (see Subsection 10.6.2). This allowance is made by comparing each acetylated alkylazulene with the parent alkylazulene, not with azulene itself. Such comparisons are made in Table 21.18. In this table, $\Delta\lambda_{max}$, $\Delta\epsilon_{max}$, and Δf are the differences between

TABLE 21.18

COMPARISON OF THE A BANDS OF ACETYLAZULENES AND THE *p* BANDS OF AZULENES

Ketone	Hydrocarbon	$\Delta\lambda_{max}$[a], mμ	$\Delta\epsilon_{max}$[b]	Δf[c] $\times 10^3$
1	1′	−34	+111	+4.0
2	2′	−31	+142	+3.9
3	3′	−18	+62	+2.8
4	4′	−21	+56	+2.1
5	1′	−64	+245	+6.4
6	3′	−40	+173	+4.3

[a] $\Delta\lambda_{max} = \lambda_{max}$ (the A band of the ketone) $- \lambda_{max}$ (the *p* band of the hydrocarbon).
[b] $\Delta\epsilon_{max} = \epsilon_{max}$ (the A band of the ketone) $- \epsilon_{max}$ (the *p* band of the hydrocarbon).
[c] $\Delta f = f$ (the A band of the ketone) $- f$ (the *p* band of the hydrocarbon).

the values of λ_{max}, ϵ_{max}, and f of the A band of a ketone and those of the *p* band of the corresponding azulene, respectively. While the sterically unhindered acetyl group in compounds 1, 2, and 5 exerts a comparatively large hypsochromic and hyperchromic effect on the *p* band, the sterically hindered acetyl group in compounds 3, 4, and 6 exerts a much smaller hypsochromic and hyperchromic effect. Thus, it can be said that the introduction of steric hindrance has resulted in a net bathochromic and hypochromic effect. It is of interest that the values of $\Delta\lambda_{max}$, $\Delta\epsilon_{max}$, and Δf for the sterically unhindered diacetyl compound 5 and those for the sterically hindered diacetyl compound 6 are respectively about twice as large as those for the corresponding mono-acetyl compounds, for example, compounds 1 and 3, indicating that the effects of the acetyl groups are nearly additive.

1-Formyl-3,8-dimethyl-5-isopropylazulene, which has a formyl group in place of the acetyl group in compound 4, has the λ_{max} value of 572 mμ and the $\Delta\lambda_{max}$ value of −33 mμ. This $\Delta\lambda_{max}$ value is in good agreement with the values for the unhindered acetyl compounds, indicating that this formyl compound is also sterically unhindered, in contrast with the sterically hindered acetyl analog, 4.

The steric hindrance to the coplanarity of the acetyl group and the azulene ring appears also to exert a bathochromic and hypochromic effect on the D band (see Table 21.19).

Summarizing the foregoings, we can say that the steric hindrance to the coplanarity of the carbonyl group with the azulene ring exerts "abnormal" bathochromic effects both on the A band, which is the LE band corresponding to the *p* band of azulene, and on the D band, which is the first CT band, while

TABLE 21.19

THE D BANDS OF ACETYLAZULENES IN CYCLOHEXANE[a]

Compound	λ_{max}, mμ	ϵ_{max}	log ϵ_{max}
1	236	19,500	4.29
2	239	20,900	4.32
3	240	17,800	4.25
4	275	13,500	4.13
5	241	31,600	4.50
6	243	20,900	4.32

[a] The data were taken from Reference 39.

its effects on the intensities of these bands are hypochromic as would be normally expected. The "abnormal" steric bathochromic effects in the azulenyl–carbonyl system can be interpreted as below by the use of the method of composite molecule, that is, in terms of interaction of π MO's localized on the azulene ring and on the carbonyl bond.[28]

The π-electronic system in 1-acetylazulene and related compounds is represented as R–S, R and S being taken to refer to the azulene ring and the carbonyl group, respectively. As usual, π MO's in R are represented as ϕ_m, and those in S, as $\phi_{n'}$. As mentioned in Subsection 10.3.4, the *p* band of azulene is considered to be almost completely due to the one-electron transition from the highest occupied MO, ϕ_{+1}, to the lowest vacant MO, ϕ_{-1}. Accordingly, the A band of 1-acetylazulenes is considered to be approximately due to the one-electron transition from ϕ^{p}_{+1} to ϕ^{p}_{-1}. Needless to say, ϕ^{p}_{+1} and ϕ^{p}_{-1} are the orbitals arising from perturbation of ϕ_{+1} and ϕ_{-1}, respectively, by interaction with the π MO's of the carbonyl group. The D band is considered to be due to the one-electron transition from ϕ^{p}_{+1} to $\phi^{p}_{-1'}$. Of course, $\phi^{p}_{-1'}$ is the orbital arising from perturbation of the antibonding π MO in the carbonyl group by interaction with the π MO's in the azulene ring.

The energies (v_m) of the Hückel π MO's of azulene, and the coefficients ($b_{m,1}$) of the carbon $2p\pi$ AO at the 1 position in these MO's, calculated by Pariser,[40] are shown in Table 21.20. It is to be noted that $b_{+1,1}$ is much larger than $b_{-1,1}$. On the other hand, $b_{-1',1'}$ is much larger than $b_{+1',1'}$: as mentioned in Section 21.1, when the values $h_{2'} = 1.3$ and $k_{1'2'} = 0.844$ are used, the energies and AO coefficients of the carbonyl π MO's are evaluated as follows: $v_{-1'} = +0.415$, $v_{+1'} = -1.715$; $b_{-1',1'} = 0.897$, $b_{+1',1'} = 0.441$. Therefore, when R and S undergo the conjugative interaction, the interaction

TABLE 21.20

THE ENERGIES AND AO COEFFICIENTS AT THE 1 POSITION OF THE HÜCKEL MO'S (ϕ_m) OF AZULENE

m	v_m	$b_{m,1}$
−5	+2.095	0.259
−4	+1.869	0.250
−3	+1.579	0.436
−2	+0.738	0.299
−1	+0.400	0.063
+1	−0.477	0.543
+2	−0.887	0.259
+3	−1.356	0.221
+4	−1.652	0.268
+5	−2.310	0.323

between ϕ_{+1} and $\phi_{-1'}$ is probably predominantly important. Thus, since $-v_{+1}$ is smaller than $h_{2'}$, the perturbed orbital ϕ^{P}_{+1} will become lower with the increasing perturbation, the depressing effect of $\phi_{-1'}$ outweighing the raising effect of $\phi_{+1'}$ (see Subsection 21.5.1). The perturbed orbital ϕ^{P}_{-1} will also become lower with the increasing perturbation. However, since $b_{+1,1}$ is much larger than $b_{-1,1}$, the repulsive interaction of ϕ_{+1} with $\phi_{-1'}$ will be stronger than the repulsive interaction of ϕ_{-1} with $\phi_{-1'}$, notwithstanding that $\phi_{-1'}$ is closer in energy to ϕ_{-1} than to ϕ_{+1}. It follows that the depression of ϕ^{P}_{+1} is probably greater than that of ϕ^{P}_{-1}. This means that the energy separation between ϕ^{P}_{-1} and ϕ^{P}_{+1} will increase as the conjugative interaction between R and S increases. The antibonding π orbital of the carbonyl group, $\phi_{-1'}$, is subject to the raising effect of ϕ_{-1} and lower orbitals of R, and to the depressing effect of ϕ_{-2} and higher orbitals of R. As a result, the energy of perturbed orbital $\phi^{\text{P}}_{-1'}$ will be almost insensitive to changes in the extent of

[40] R. Pariser, *J. Chem. Phys.* **25,** 1112 (1956).

the conjugative interaction. This means that the energy separation between $\phi^{\mathrm{p}}_{-1'}$ and ϕ^{p}_{+1} will also increase as the conjugative interaction increases. It will thus be evident that both the perturbed local transition from ϕ^{p}_{+1} to ϕ^{p}_{-1} and the intramolecular CT transition from ϕ^{p}_{+1} to $\phi^{\mathrm{p}}_{-1'}$ will undergo bathochromic shifts when the conjugative interaction between the azulene ring and the carbonyl group decreases as a result of twist of the essential single bond connecting together the two fragments.

Chapter 22

Nitrobenzene, Benzoic Acid, Aniline, and Related Compounds

22.1. Nitrobenzene and Related Compounds

22.1.1. The Nitro Group

The nitro group has, as is well known, a planar structure. The nitrogen $2p\pi$ AO (χ_N) and two oxygen $2p\pi$ AO's (χ_{O1} and χ_{O2}) combine into three π MO's, ϕ_{-1}, ϕ_0, and ϕ_{+1}. The nitrogen atom contributes two π electrons, and each of the oxygen atoms contributes one π electron: hence, there are four π electrons in the nitro group.

The nitro group has a symmetry plane which is perpendicular to the plane of the group and bisects the angle between the two N—O bonds. This means that the two oxygen atoms, O_1 and O_2, are equivalent to each other. It is convenient to form the following group orbitals from the two oxygen $2p\pi$ AO's:

$$\chi_{O(s)} = 2^{-1/2}(\chi_{O1} + \chi_{O2}); \tag{22.1}$$

$$\chi_{O(a)} = 2^{-1/2}(\chi_{O1} - \chi_{O2}). \tag{22.2}$$

The symmetrical group orbital $\chi_{O(s)}$ interacts with the nitrogen $2p\pi$ AO χ_N, and the antisymmetrical group orbital $\chi_{O(a)}$ does not. Therefore, in the zero-overlap approximation, the three π MO's of the nitro group can be expressed as follows:

$$\phi_{-1} = b_{-1,1}\chi_N - b_{+1,1}\chi_{O(s)}; \tag{22.3}$$

$$\phi_0 = \chi_{O(a)}; \tag{22.4}$$

$$\phi_{+1} = b_{+1,1}\chi_N + b_{-1,1}\chi_{O(s)}. \tag{22.5}$$

Tanaka[1] has evaluated the coefficients as follows: $b_{-1,1} = 0.7009$, $b_{+1,1} =$

[1] J. Tanaka, *Nippon Kagaku Zasshi* **79**, 1373 (1958).

0.7133. Of course, ϕ_{-1}, ϕ_0, and ϕ_{+1} are the antibonding, nonbonding, and bonding π orbitals, respectively. In the ground state, each of the two lower orbitals, ϕ_{+1} and ϕ_0, is occupied by two electrons, and the antibonding orbital, ϕ_{-1}, is, of course, vacant.

Simple aliphatic nitro compounds, such as nitromethane, show a moderately intense band at about 200 mμ ($\epsilon_{max} \simeq 10{,}000$) and a very weak band at about 275 mμ ($\epsilon_{max} \simeq 20$). The former is probably due to the lowest energy π–π^* S–S transition, that is, the one-electron S–S transition from ϕ_0 to ϕ_{-1}, and the latter is due to the n–π^* S–S transition, that is, the one-electron S–S transition from the oxygen nonbonding AO's to ϕ_{-1}. The n–π^* transition is similar to the carbonyl n–π^* transition and in fact occurs at approximately the same wavelength with a similar intensity. It shows a prominent hypsochromic shift in polar solvents, similarly to the carbonyl n–π^* transition.

22.1.2. Nitrobenzene

The molecule of nitrobenzene has a planar structure, and it can be represented as a composite molecule R–S, in which R and S represent the benzene ring and the nitro group, respectively. As usual, the MO's localized on R are denoted by ϕ_m, and the MO's localized on S are distinguished from them by attaching a prime. Since the nonbonding π orbital of the nitro group, $\phi_{0'}$, has no contribution of the nitrogen $2p\pi$ AO, χ_N (or $\chi_{1'}$), it does not interact with the π MO's of the benzene ring. Therefore, the orbital energy level diagram for this conjugated system can be considered to be similar to that for the phenyl–carbonyl system, shown in Fig. 21.4, except for rather small differences in the following two points: the bonding π MO of the nitro group is probably much lower than that of the carbonyl group; the AO coefficients $b_{-1',1'}$ and $b_{+1',1'}$ in the nitro group are comparable to each other, while in the carbonyl group, $b_{-1',1'}$ is much larger than $b_{+1',1'}$. From consideration of the energy level diagram, it is inferred that the energies of perturbed orbitals ϕ^{p}_{+1} and $\phi^{\mathrm{p}}_{-1'}$ in the phenyl–nitro system will be slightly lowered by increasing the conjugative interaction between the benzene ring and the nitro group.

The spectrum of nitrobenzene in saturated hydrocarbon solution (see Fig. 22.1) shows a weak band at about 330 mμ ($\epsilon_{max} = 140$), a shoulder at about 280 mμ ($\epsilon = 640$), a moderately intense band at about 250 mμ ($\epsilon_{max} = 8700$), and an intense band at about 200 mμ ($\epsilon_{max} = 15{,}500$).[2] All these bands undergo more or less large hypsochromic shifts on going from the solution spectrum to the vapor spectrum. In the vapor spectrum,

[2] S. Nagakura, M. Kojima, and Y. Maruyama, *J. Mol. Spectroscopy* **13**, 174 (1964).

the 280-mμ band, which is almost hidden under the tail of the 250-mμ band in the solution spectrum, shows a well-developed vibrational fine structure.

Labhart[3] measured the spectrum of nitrobenzene in hexane solution in a strong electric field for the purpose of determining the direction of electric polarization of bands. According to his result, the three bands at about 330, 280, and 250 mμ are polarized, respectively, parallel, perpendicular, and

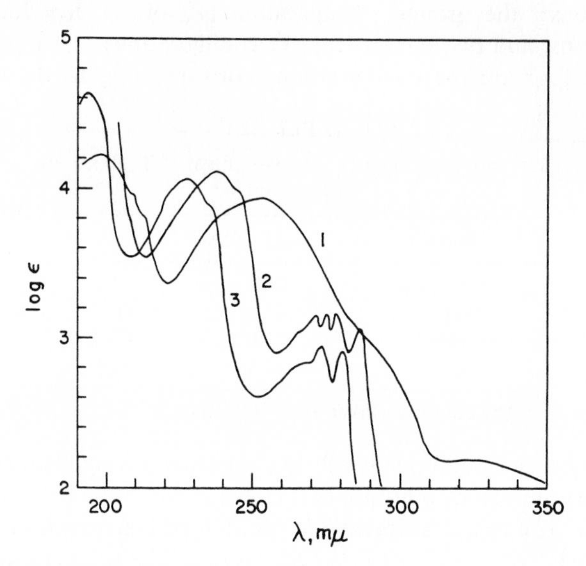

FIG. 22.1. The absorption spectra of nitrobenzene (1), acetophenone (2), and methyl benzoate (3) in heptane.

parallel to the twofold symmetry axis of the molecule, that is, the C—N bond axis. The 330-mμ band is almost certainly the n–π^* band which is due to the one-electron transition from the oxygen n orbitals to $\phi^{p}_{-1'}$. The 280-mμ band is probably the LE band corresponding to the α band of benzene. The 250-mμ band is interpreted as an intramolecular CT band,[2] which is probably due to the one-electron transition from ϕ^{p}_{+1} to $\phi^{p}_{-1'}$. Most of the relatively high intensity of the n–π^* band as compared with the n–π^* band of nitromethane is considered to come from the mixing of the originally forbidden n–π^* transition with the allowed intramolecular CT transition, which will be made possible by torsional vibrations about the C—N bond.

The 250-mμ band is very sensitive to the solvent effect.[2] Thus, as will be seen in Table 22.1, this band undergoes a large bathochromic shift when the

[3] H. Labhart, *Tetrahedron* **19**, Suppl. 2, 223 (1963).

polarity of the solvent increases. This means that the upper state of the transition responsible for this band is an extremely polar state, supporting the assignment of this band as a CT band.

Nagakura and his co-workers[2,4] calculated the energies and wavefunctions of π-electronic states of nitrobenzene by the method of composite molecule including electronic interactions, taking account of eight singlet electron configurations: the ground configuration (V_0 or G), five locally excited configurations, and two charge-transfer configurations. Of the five LE configurations, four are the configurations corresponding to the α, p, β, and

TABLE 22.1

THE SOLVENT EFFECT ON THE FIRST CT BAND OF NITROBENZENE[a]

Solvent	λ_{max}, mμ	ϵ_{max}
(Vapor phase)	240	8500
Heptane	251.7	8660
Methanol	259.5	8270
Water	268.2	7710

[a] The data were taken from Reference 2.

β' states of benzene, and one is the configuration arising from the π–π^* S–S transition within the nitro group, that is, the one-electron S–S transition from $\phi_{0'}$ to $\phi_{-1'}$. The energies (relative to the ground configuration) of these LE configurations, LE_α, LE_p, LE_β, $LE_{\beta'}$, and $LE(NO_2)$, were determined as 4.89 ev (265 mμ), 6.17 ev (200 mμ), 6.98 ev (177.5 mμ), 6.98 ev, and 6.26 ev (198 mμ), respectively, from the observed positions of the corresponding absorption bands of benzene and nitromethane. The two CT configurations originate from the one-electron S–S transition from each of the doubly degenerate highest occupied orbitals of the benzene ring, ϕ_{+1} and ϕ_{+2}, to the vacant orbital of the nitro group, $\phi_{-1'}$. The CT configuration arising from the one-electron transition from ϕ_{+1} to $\phi_{-1'}$, $CT^{11'}$, is symmetric with respect to the symmetry plane that is along the C—N bond axis and vertical to the molecular plane, and hence it can be designated as CT_s. Another CT configuration, $CT^{21'}$, which arises from the one-electron transition from ϕ_{+2} to $\phi_{-1'}$, is antisymmetric with respect to the symmetry plane, and can be designed as CT_a. The energies of these CT configurations were evaluated as 4.93 and 5.06 ev, respectively, by the use of the equation

$$E(CT) = I_B - A_{NO_2} - C, \tag{22.6}$$

[4] S. Nagakura, *Pure Appl. Chem.* **7**, 79 (1963).

where I_{B} and A_{NO_2} are the ionization potential of benzene (9.24 ev) and the electron affinity of the nitro group (0.40 ev), respectively, and C is the energy of the Coulombic interaction between the negative charge distributed in the nitro group and the positive charge distributed in the benzene ring in the CT configuration [see Eq. (20.20)]. The C can be evaluated from the wavefunctions of the MO's concerned and the molecular geometry by the Pariser–Parr method [see Section 11.3].

The off-diagonal matrix elements of the total π-electronic Hamiltonian (**H**) necessary for the calculation of configuration interaction can be evaluated by the use of the general expressions given in Section 20.1, in terms of $\beta_{\text{CN}}^{\text{C}}$, the core resonance integral between the adjacent carbon and nitrogen $2p\pi$ AO's. As the wavefunctions of excited states of benzene those given in Table 3.7 are used, and as the wavefunctions of the π MO's of benzene those given in Table 3.5 are used. As the wavefunctions of the π MO's of the nitro group those given by expressions (22.3)–(22.5) with $b_{-1',1'} = 0.7009$ and $b_{+1',1'} = 0.7133$ are used. Then, the interaction matrix element between LE_p and CT_{s}, for example, can be expressed as follows:

$$\begin{aligned}\langle LE_p|\,\mathbf{H}\,|CT_{\text{s}}\rangle &= \langle 2^{-1/2}(V_1^{11} + V_1^{22})|\,\mathbf{H}\,|V_1^{11'}\rangle = 2^{-1/2}\langle V_1^{11}|\,\mathbf{H}\,|V_1^{11'}\rangle \\ &= 2^{-1/2}b_{-1,1}b_{-1',1'}\beta_{\text{CN}}^{\text{C}} = -6^{-1/2} \times 0.7009\beta_{\text{CN}}^{\text{C}}. \end{aligned} \tag{22.7}$$

Analogously, we obtain

$$\langle G|\,\mathbf{H}\,|CT_{\text{s}}\rangle = 2^{1/2}b_{+1,1}b_{-1',1'}\beta_{\text{CN}}^{\text{C}} = (\tfrac{2}{3})^{1/2} \times 0.7009\beta_{\text{CN}}^{\text{C}}\,; \tag{22.8}$$

$$\langle LE_{\beta'}|\,\mathbf{H}\,|CT_{\text{s}}\rangle = 2^{-1/2}b_{-1,1}b_{-1',1'}\beta_{\text{CN}}^{\text{C}} = -6^{-1/2} \times 0.7009\beta_{\text{CN}}^{\text{C}}\,; \tag{22.9}$$

$$\langle LE_{\alpha}|\,\mathbf{H}\,|CT_{\text{a}}\rangle = -2^{-1/2}b_{-1,1}b_{-1',1'}\beta_{\text{CN}}^{\text{C}} = 6^{-1/2} \times 0.7009\beta_{\text{CN}}^{\text{C}}\,; \tag{22.10}$$

$$\langle LE_{\beta}|\,\mathbf{H}\,|CT_{\text{a}}\rangle = 2^{-1/2}b_{-1,1}b_{-1',1'}\beta_{\text{CN}}^{\text{C}} = -6^{-1/2} \times 0.7009\beta_{\text{CN}}^{\text{C}}. \tag{22.11}$$

The other off-diagonal elements are zero. Thus, the symmetrical CT configuration, CT_{s}, interacts only with the p and β' LE configurations and with the ground configuration, and the antisymmetrical CT configuration, CT_{a}, interacts only with the α and β LE configurations. The LE configuration arising from the electronic excitation within the nitro group, $LE(NO_2)$, does not interact with any other configurations, since $\phi_{0'}$ has no contribution of the nitrogen $2p\pi$ AO. A value of -2.40 ev was used for $\beta_{\text{CN}}^{\text{C}}$ in the sterically unhindered nitrobenzene molecule.

The term energy level diagrams before and after the configuration interaction are shown in Fig. 22.2. The wavefunction of the ground state can be expressed as follows:

$$^1\Phi_0 = [G] = 0.9657G + 0.2572CT_{\text{s}} - 0.0271LE_p - 0.0241LE_{\beta'}. \tag{22.12}$$

Thus, the ground state has the contributions of G and CT_S, amounting to 93.3 and 6.6%, respectively, and very small contributions of LE_p and $LE_{\beta'}$. The first excited state, ${}^1\Phi_1$, has almost equal contributions of LE_α and CT_a, amounting to 48.5% for each configuration, and the transition to this state is almost certainly responsible for the 280-mμ band. The second excited

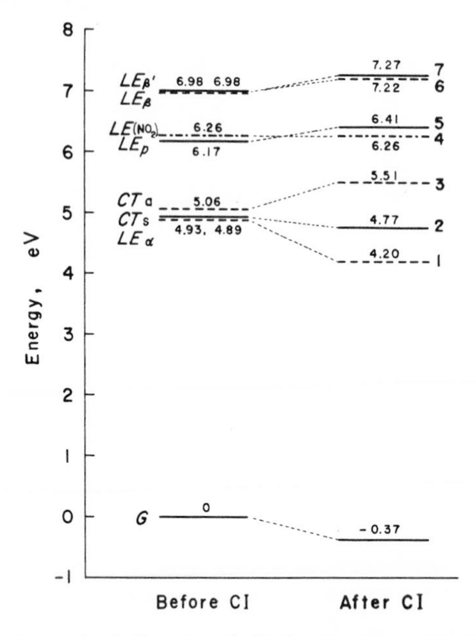

FIG. 22.2. Term energy level diagram of nitrobenzene as a composite π system. The energy levels of singlet states before and after the configuration interactions are shown, and the correlations are denoted with the dotted lines.

state, ${}^1\Phi_2$, has primarily a character of the symmetrical CT configuration, CT_S, its contribution amounting to 70.3%. The transition to this CT state is almost certainly responsible for the 250-mμ band. The calculated and observed values of the transition energies and oscillator strengths of absorption bands are compared in Table 22.2.

22.1.3. GEOMETRY AND SPECTRA OF NITROBENZENE AND RELATED COMPOUNDS

When the nitro group is twisted out of the plane of the benzene ring, the absolute value of the core resonance integral across the C—N bond, β_{CN}^C, will decrease, approximately in proportion to cos θ, in which θ is the interplanar angle between the benzene ring and the nitro group. From the energy level diagram in Fig. 22.2, it is predicted that twist of the C—N bond will cause hypsochromic shift of the 280-mμ band, substantially no shift of the 250-mμ band, and very small bathochromic shift of the 200-mμ band. Furthermore, since the 250-mμ band is a CT band, the intensity of this band

TABLE 22.2

COMPARISON OF THE THEORETICAL AND EXPERIMENTAL VALUES OF THE TRANSITION ENERGIES AND OSCILLATOR STRENGTHS OF ABSORPTION BANDS OF NITROBENZENE

Absorption band (experimental)[a]		Transition energy, ev		Oscillator strength		The upper state of the transition; the electron configuration that makes the largest contribution to the upper state (%)
λ_{max}, mμ	ϵ_{max}	obs.	theoret.	obs.	theoret.	
(340)	(140)	(3.65)	—	(0.003)	—	(n–π^* transition)
(282.5)	(640)	(4.39)	4.57	(0.01)	0.02 (x)[b]	$^1\Phi_1$; CT_a(48.5), LE_α(48.5)
288	500	4.38				
240	8500	5.11	5.14	0.17	0.48 (y)	$^1\Phi_2$; CT_s(70.3)
			5.87		0.08 (x)	$^1\Phi_3$; LE_α(50.6), CT_a(40.6)
193	17,300	6.42	6.63	0.38	0.60 (x)	$^1\Phi_4$; $LE(NO_2)$(100)
(with a shoulder)			6.77		0.31 (y)	$^1\Phi_5$; LE_p(77.4)
164	27,800	7.56	7.59	0.87	1.03 (x)	$^1\Phi_6$; LE_β(88.2)
			7.63		0.40 (y)	$^1\Phi_7$; $LE_{\beta'}$(80.2)

[a] The experimental values are those for the vapor spectrum except for the values in parentheses, which are the values for the spectrum of the hexane solution measured by Labhart (Ref. 3).

[b] The letters in parentheses following the theoretical values of oscillator strength indicate the direction of the transition moment. The molecular plane has been taken as the *xy* plane, and the twofold symmetry axis has been taken as the *y* axis.

is expected to decrease rapidly with the increasing angle of twist: when the twist angle is 90°, there must be no mixing of the CT configuration with LE configurations, and hence the CT band will lose its intensity completely. These predictions appear to be borne out experimentally.

According to the results of measurement of spectra of isooctane solutions by Wepster and co-workers,[5,6,7] in the spectra of nitrobenzene and 2-methylnitrobenzene the first CT band has a maximum at about 250 mμ, and in the

[5] B. M. Wepster, *Rec. trav. chim.* **76,** 335 (1957).

[6] J. Burgers, M. A. Hoefnagel, P. E. Verkade, H. Visser, and B. M. Wepster, *Rec. trav. chim.* **77,** 491 (1958).

[7] B. M. Wepster, *in* "Progress in Stereochemistry 2" (W. Klyne and P. B. D. de la Mare, eds.), p. 99. Butterworths, London, 1958.

spectra of 2-ethyl- and 2-isopropylnitrobenzene it has a maximum at slightly shorter wavelength than 250 mμ. In the spectra of 2-*tert*-butylnitrobenzene and of 2,6-dialkylnitrobenzenes, there is no maximum, and the first CT band reveals itself as an inflection or shoulder at slightly shorter wavelength than 250 mμ. The values of the molar extinction coefficient at the maximum or inflection for these compounds are shown in Table 22.3. It is seen that the

TABLE 22.3

THE INTENSITIES OF THE CT BANDS OF NITROBENZENE AND ITS ALKYLATED DERIVATIVES IN ISOOCTANE[a,b]

Nitrobenzene	ϵ_{max}	θ, deg.
Unsubstituted	8900	(0)
2-Me	6070	34
2-Et	5300	40
2-*i*-Pr	4150	47
2-*tert*-Bu	1540*	65
2,3-di-Me	4210	47
2,6-di-Me	1500*	66
2,4,6-tri-Me	2170*	60
2,3,5,6-tetra-Me	990*	71
2,6-di-*tert*-Bu	640*	74

[a] The data were taken from References 5, 6, and 7.
[b] Asterisk denotes no maximum; ϵ at 250 mμ is given.

intensity of the CT band decreases rapidly as the steric hindrance to the coplanarity of the nitro group and the benzene ring increases. It is noteworthy that the steric effect in 2,3-dimethylnitrobenzene is greater than that in 2-methylnitrobenzene and that the steric effect in 2,3,5,6-tetramethylnitrobenzene is greater than that in 2,6-dimethylnitrobenzene. The enhancement of the steric effect of the *ortho* substituents by the *meta* substituents is attributed to the so-called buttressing effect. In this connection, it is of interest that the *m*-dinitrobenzene molecule is nonplanar, each nitro group inclining at about 11° to the plane of the benzene ring,[8] whereas the nitrobenzene molecule is planar.[9] In the *m*-dinitrobenzene molecule there should occur a buttressing effect of the nitro groups at positions 1 and 3 on the hydrogen atom at position 2.

[8] J. Trotter, *Acta Cryst.* **14,** 244 (1961).
[9] J. Trotter, *Acta Cryst.* **12,** 884 (1959).

The values of the interplanar angle θ listed in the last column of Table 22.3 are the values calculated by Wepster by means of the relation

$$\epsilon/\epsilon_0 = \cos^2 \theta, \tag{22.13}$$

which was assumed originally by Braude (see Subsection 19.4.3). In this relation, ϵ is the molar extinction coefficient at the maximum of the CT band of the compound under consideration, and ϵ_0 is that of the planar reference compound. As mentioned already, this relation should be regarded only as a rough approximation. However, this assumed relation appears to be to some extent justified by the fact that the calculated value of θ for 2,6-dimethylnitrobenzene, 66°, is in good agreement with the value determined by X-ray crystal analysis for nitromesitylene (2,4,6-trimethylnitrobenzene), 66.4°,[10] and by the fact that the θ value calculated by means of this relation is approximately the same for the corresponding substituted nitrobenzene and *p*-nitroaniline, as will be shown later.

It is expected that the intensity of the n–π^* band will increase with the increasing interplanar angle between the nitro group and the benzene ring, owing to the increasing mixing of allowed π–π^* transitions with the n–π^* transition. Actually, the n–π^* bands of some sterically hindered nitrobenzene derivatives are known to be considerably stronger than the n–π^* band of nitrobenzene ($\epsilon_{max} = 140$). For example, the ϵ_{max} of the n–π^* band of 2,6-dimethylnitrobenzene is 250 ($\lambda_{max} = 330$ mμ), and that of nitromesitylene is about 400 ($\lambda_{max} = 335$ mμ).

X-ray crystal analysis shows that 1,5-dinitronaphthalene has a nonplanar structure, in which each nitro group is twisted by 49° out of the plane of the naphthalene ring.[11] This compound shows a comparatively low-intensity band at 323 mμ ($\epsilon_{max} \simeq 6000$), which is polarized in the direction parallel to the C—N bonds.[12] This band has been interpreted as the superposition of the CT band and the LE band corresponding to the α band of naphthalene. The spectrum of 9-nitroanthracene rather resembles the spectrum of anthracene, no CT band being detectable. This compound has been found by X-ray crystal analysis to have a nonplanar structure, in which the interplanar angle between the nitro group and the anthracene ring is as large as 85°.[13]

The C—N bond length of nitrobenzene has been found to be 1.486 Å,[14] and that of nitromesitylene has been found to be 1.476 Å.[10] It is of interest that the C—N bond of nitrobenzene is longer than that of nitromesitylene, notwithstanding that the π-bond order of the

[10] J. Trotter, *Acta Cryst.* **12,** 605 (1959).

[11] J. Trotter, *Acta Cryst.* **13,** 95 (1960).

[12] M. Kojima, J. Tanaka, and S. Nagakura, *Theoret. chim. Acta* (*Berl.*) **3,** 432 (1965).

[13] J. Trotter, *Acta Cryst.* **12,** 237 (1959).

[14] J. Trotter, *Tetrahedron* **8,** 13 (1960).

former must be higher than that of the latter. In view of this fact, some doubt was thrown on the existence of the resonance interaction between the phenyl and nitro groups in the ground state of nitrobenzene. However, Nagakura and his co-workers[2] have argued for the existence of the resonance interaction, on the basis of the existence of the CT band in the spectrum of nitrobenzene and of the calculation result which shows that the ground state of nitrobenzene has the contribution of the symmetric CT configuration amounting to 6.6%. Since the calculated stabilization energy due to the conjugation is only 8.4 kcal/mole, the resonance interaction may be said to be rather small. This may be in conformity with the fact that, as was pointed out by Mulliken,[15] nitrobenzene belongs to the so-called sacrificial conjugation system. The calculation shows that in the symmetrical CT configuration both the adjacent carbon and nitrogen atoms are positively charged to some extent. Nagakura and his co-workers have suggested that the bond lengthening due to the electrostatic repulsion between the carbon and nitrogen atoms possibly compensates the shrinkage due to the double-bond character.

22.2. Benzoic Acid and Related Compounds

22.2.1. The Carboxyl Group

The carboxyl group is planar, and has three π orbitals, ϕ_{-1}, ϕ_0, and ϕ_{+1}, which can be considered to be formed from one carbon and two oxygen $2p\pi$ AO's. The carbon and oxygen atoms of the carbonyl group contribute each one π electron, and the oxygen atom of the hydroxyl group contributes two π electrons. In the ground state, each of the two lower π orbitals, ϕ_0 and ϕ_{+1}, is occupied by two electrons, and the highest π orbital, ϕ_{-1}, is vacant. Thus the carboxyl group can be considered to be iso-π-electronic with the nitro group, although the two oxygen atoms of the carboxyl group are not equivalent, while the two oxygen atoms of the nitro group are equivalent.

The three π orbitals of the carboxyl group can also be considered to arise from the interactions of the π orbitals of the carbonyl group with the lone-pair orbital of $p\pi$ character of the oxygen atom of the hydroxyl group (see Section 21.2). The antibonding and bonding π orbitals of the carbonyl group are raised and lowered, respectively, by interaction with the lone-pair orbital to give rise to the highest and lowest π orbitals of the carboxyl group, ϕ_{-1} and ϕ_{+1}, respectively, and the lone-pair orbital is perturbed by the interaction to give rise to the intermediate π orbital of the carboxyl group, ϕ_0. As mentioned in Section 21.2, the n–π^* band of the carboxyl group (acetic acid, $\lambda_{max} = 204$ mμ), which is considered to be due to the one-electron transition from the nonbonding orbital (ϕ_n) of the oxygen atom

[15] R. S. Mulliken, *Tetrahedron* **5**, 253 (1959); **6**, 68 (1959).

of the carbonyl group to ϕ_{-1}, is at much shorter wavelength than the n–π^* band of the carbonyl group (acetone, $\lambda_{max} = 279$ mμ), and the first π–π^* band of the carboxyl group (acetic acid, $\lambda_{max} = 160.5$ mμ), which is considered to be the CT band due to the one-electron transition from ϕ_0 to ϕ_{-1}, is at slightly longer wavelength than the π–π^* band of the carbonyl group (acetone, $\lambda_{max} = 155$ mμ).

The contribution of the carbon $2p\pi$ AO, χ_1, to ϕ_0 is probably very small, since the two π orbitals of the carbonyl group contribute with opposite signs to ϕ_0. If the two oxygen atoms of the carboxyl group were assumed to be equivalent, the orbital ϕ_0 would be expressed simply as the antisymmetrical combination of the two oxygen $2p\pi$ AO's, just as the corresponding orbital of the nitro group [see expressions (22.2) and (22.4)].

22.2.2. Benzoic Acid

The spectrum of benzoic acid as well as that of methyl benzoate (see Fig. 22.1) shows three bands at about 280, 230, and 200 mμ. The 280-mμ band ($\log \epsilon_{max} = 2.95$, $f = 0.013$) has vibrational fine structure, and has been found to have the transition moment in the direction that is in the molecular plane and is perpendicular to the phenyl–carboxyl bond (the 1—1′ bond) axis.[16] This band is certainly the LE band corresponding to the benzene α band. The 230-mμ band ($\log \epsilon_{max} = 4.10$, $f = 0.19$) has been found to be polarized in the direction that is in the molecular plane and makes an angle of 6.5° with the 1—1′ bond axis.[16] This band is interpreted as an intramolecular CT band, which is probably accompanied with a partial electron transfer from the benzene ring to the carboxyl group. In the simple MO approximation, this band may be interpreted as being due to the one-electron transition from ϕ^{p}_{+1} to $\phi^{p}_{-1'}$, where ϕ^{p}_{+1} is the orbital arising from the perturbation of the highest bonding π orbital ϕ_{+1} of benzene (R), and $\phi^{p}_{-1'}$ is the orbital arising from the perturbation of the antibonding π orbital of the carboxyl group (S), $\phi_{-1'}$. The $\phi^{p}_{-1'}$ in the phenyl–carboxyl system is probably higher in energy than the corresponding orbital in the phenyl–carbonyl system since, as mentioned in the preceding subsection, the antibonding π orbital of the carboxyl group is probably higher than that of the carbonyl group. This probably accounts for the fact that the CT band of benzoic acid is at appreciably shorter wavelength than the CT band of benzaldehyde. The 200-mμ band ($\log \epsilon_{max} = 4.65$, $f = 0.93$) has been found to be polarized approximately in the direction of the 1—1′ bond axis. This band is probably the LE band corresponding to the p band of benzene.

[16] J. Tanaka, *Nippon Kagaku Zasshi* **79,** 1114, 1379 (1958).

Hosoya, Tanaka, and Nagakura[17] studied carefully the effects of chain-like hydrogen bonding of benzoic acid with various solvents and of ring dimer formation on the CT band near 230 mμ. According to them, the maximum of the CT band of benzoic acid, which is at 229 mμ in the spectrum of the *n*-heptane solution of a concentration of about 10^{-5} *M*, continuously shifts toward longer wavelength with the increasing concentration and finally reaches 233 mμ in the spectrum of a concentration higher than 10^{-2} *M*. At the same time, the intensity at this maximum, ϵ_{max}, increases with the increasing concentration of benzoic acid, from about 11,000 in the solution of a concentration of about 1.04×10^{-5} *M* to about 15,500 in the solution of a concentration of about 1.04×10^{-2} *M*.

It is well known that there exists a monomer–dimer equilibrium represented by expression (22.14) for benzoic acid in inert solvents such as *n*-heptane.

$$2C_6H_5\text{—}C{\overset{\displaystyle O}{\underset{\displaystyle OH}{\lessgtr}}} \rightleftharpoons C_6H_5\text{—}C{\overset{\displaystyle O\cdots HO}{\underset{\displaystyle OH\cdots O}{\lessgtr}}}C\text{—}C_6H_5 \tag{22.14}$$

The equilibrium constant *K* is defined as [dimer]/[monomer]^2, in which [dimer] and [monomer] represent the concentrations of the dimer and monomer, respectively. The "dimer fraction" *X*, which is defined as the ratio of the number of "molecules" in the dimeric form to the total number of the "molecules" present, should, of course, increase with increase in the total concentration of the benzoic acid molecules. The bathochromic and hyperchromic shift of the CT band caused by the increase in the concentration of benzoic acid is undoubtedly due to the increase in the dimer fraction. The CT band of the free benzoic acid is considered to have the maximum at 229 mμ ($f = 0.19$), and that of benzoic acid in the dimeric form is considered to have the maximum at 233 mμ ($\epsilon_{max} = 16{,}200, f = 0.27$). From the change of the CT band with the total concentration, Hosoya et al. succeeded in evaluating the dimer fractions at various concentrations and the equilibrium constant. According to the result, the dimer fractions at 30°C at the concentrations of 1.04×10^{-5} and 1.04×10^{-2} *M* are 0.136 and 0.920, respectively, and the equilibrium constant at 30°C is 7.95×10^3.

Hosoya et al. explained the bathochromic and hyperchromic shift of the CT band of benzoic acid caused by the dimer formation in terms of the exciton-type dipole–dipole interaction between the component molecules in the dimer (see Section 20.1). Owing to the interaction, the CT states of the two

[17] H. Hosoya, J. Tanaka, and S. Nagakura, *J. Mol. Spectroscopy* **8,** 257 (1962).

component molecules split into the symmetrical and antisymmetrical combinations, which are higher and lower in energy, respectively, than the CT state of the monomer. The transition to the upper symmetrical exciton state is forbidden, and the transition to the lower antisymmetrical exciton state is allowed. Therefore, this splitting accounts at least qualitatively for the bathochromic shift of the CT band by the dimer formation. The largest contribution to the intensity increment arises from the mixing of the CT transition in a component molecule with the nearest shorter wavelength transition ($\lambda_{max} = 195$ mμ, $f = 0.93$) in the other component molecule, which results from the exciton-type interaction between the two transition moments. It can be said that the CT band is intensified mainly by borrowing the intensity from the nearest shorter wavelength strong band in the neighboring molecule.

In solutions in alcohols, benzoic acid forms chain-like hydrogen bonding with the solvent molecules, which can act as proton donors as well as proton acceptors. In solutions in ethers such as dioxan and tetrahydrofuran, benzoic acid forms also chain-like hydrogen bonding with the solvent molecule, which acts as a proton acceptor. Such chain-like hydrogen bonding appears not to affect greatly the position of the CT band: the CT band of benzoic acid in various alcohols and ethers appears at 226–228 mμ.

Methyl benzoate is of course incapable of forming a hydrogen-bridged ring dimer, but is capable of forming hydrogen bonding with alcohols, acting as a proton acceptor. The CT band of methyl benzoate appears at almost the same wavelength ($\lambda_{max} = 228$ mμ) and with a nearly constant intensity ($\epsilon_{max} = 11{,}700$–$13{,}000$) in various solvents including *n*-heptane, ethers, and alcohols.

22.2.3. Geometry and Spectra of Substituted Benzoic Acids

From the energy-level diagram for the phenyl–carboxyl system as a composite system, it is expected that the first CT band will shift slightly to shorter wavelength and decrease rapidly in intensity when the carboxyl group is twisted out of the plane of the benzene ring.

The ultraviolet absorption spectra of mono- and 2,6-disubstituted benzoic acids having substituents such as methyl, chloro, bromo, methoxyl, and hydroxyl were measured with 95% ethanol solutions by Moser and Kohlenberg.[18] As in the spectrum of benzoic acid itself, three bands are generally observed. While the position of the 280-mμ band is only slightly

[18] C. M. Moser and A. I. Kohlenberg, *J. Chem. Soc.* **1951,** 804.

dependent upon, and that of the 200-mμ band is almost independent of the nature and position of the substituent, the position of the CT band, which appears at about 230 mμ in the spectrum of benzoic acid itself, is very sensitive to them. The data on the CT band are shown in Table 22.4.

The *para* substitution shifts the CT band to longer wavelength. The magnitudes of the shift with various substituents are in the order OH > OCH_3 > Br > CH_3 > Cl.

In *ortho*-substituted compounds the CT band is at shorter wavelength and of lower intensity than that of the corresponding *para*-substituted compound.

TABLE 22.4

THE EFFECTS OF SUBSTITUENTS ON THE CT BAND OF BENZOIC ACID

(Solvent: 95% ethanol)[a]

Substituent	Benzoic acid: λ_{max} = 228 mμ (log ϵ_{max} = 4.00) Position of substitution			
	2-	3-	4-	2,6-di-
CH_3	228 (3.71)	232 (3.95)	236 (4.14)	**
Cl	229 (3.71)*	230 (3.92)	234 (4.18)	**
Br	224 (3.80)*	225 (3.92)*	240 (4.10)	**
CH_3O	230 (3.79)	230 (3.83)	249 (4.14)	**
HO	236 (3.86)	236 (3.78)	251 (4.09)	250 (3.80)

[a] A single asterisk denotes no distinct maximum; values at an approximate point of inflection are given. A double asterisk denotes no maximum.

On the basis of a simple resonance consideration alone, it would be expected that the position of the CT band of an *ortho*-substituted compound should be nearly the same as that of the CT band of the *para* isomer. The hypsochromic and hypochromic shift of the CT band observed in the spectra of the *ortho*-substituted acids when compared with the spectra of the *para* isomers is conceivably, partly at least, due to the effect of the steric hindrance to the coplanarity of the carboxyl group with the benzene ring. The CT band disappears in almost all the 2,6-disubstituted acids, in which the steric hindrance is probably very serious.

The spectra of the *meta*-substituted acids are considerably similar to the spectra of their *ortho* isomers: the CT band in the *meta* isomers is at nearly the same wavelength as in their *ortho* isomers, although its intensity is usually higher in the *meta* isomers than in their *ortho* isomers. In general, if benzene is substituted by one electron-donating group (for example, NH_2, OH) and one electron-accepting group (for example, NO_2, COOH), the spectra of the *ortho*- and *meta*-substituted compounds are apparently similar to each other but different from that of the *para* compound.[19]

The noncoplanarity of the carboxyl group and the benzene ring in *ortho*-substituted benzoic acids has been confirmed by Ferguson and Sim[20] by X-ray crystal analyses of some *ortho*-substituted benzoic acids. According to their results, the interplanar angles (θ) between the carboxyl group and the benzene ring in 2-chloro- and 2-bromobenzoic acid and 2-chloro-5-nitrobenzoic acid are 13.7, 18.3, and 23.0°, respectively. The increase in the inclination angle θ of 9.3° in 2-chloro-5-nitrobenzoic acid over the angle in 2-chlorobenzoic acid has been attributed to the buttressing effect of the nitro group at the 5-position, which as a direct consequence is also inclined with an angle of 7° to the benzene ring plane. In all the three compounds, in addition to the rotation of the carboxyl group out of the benzene ring plane, a small out-of-plane bending of the exocyclic carbon–carbon bond (of about 2°) and significant in-plane splaying-out of the exocyclic carbon–carbon and carbon–halogen bonds were observed. Thus, the angle between the two exocyclic bonds is increased beyond the value of 60° appropriate to a regular planar model to 67.4° in 2-chlorobenzoic acid, to 68.3° in 2-bromobenzoic acid, and to 67.2° in 2-chloro-5-nitrobenzoic acid.

The spectrum of *o,o′*-diphenic acid (*o,o′*-dicarboxybiphenyl) is considerably similar to that of benzoic acid.[21] Evidently, the carboxyl groups in the *ortho* positions interfere to such an extent that the coplanar arrangement of the two benzene rings is impossible and π-electronic interaction across the coannular bond is completely hindered.

22.3. Aniline and Related Compounds

22.3.1. The Amino Group

The ammonia molecule is known to have a pyramidal structure with the valence angle of about 107°. In this molecule the nitrogen valences probably have a nearly sp^3 hybridization. The first ionization potential of ammonia, which corresponds to the removal of one of the lone-pair electrons,

[19] J. N. Murrell, "The Theory of the Electronic Spectra of Organic Molecules," Sect. 10.1. Methuen, London, 1963.

[20] G. Ferguson and G. A. Sim, *Proc. Chem. Soc.* **1961,** 162; *Acta Cryst.* **14,** 1262 (1961); *ibid.* **15,** 346 (1962); *J. Chem. Soc.* **1962,** 1767. See also G. Ferguson and J. M. Robertson, *in* "Advances in Physical Organic Chemistry," (V. Gold, ed.), Vol. 1, p. 203. Academic Press, New York, 1963.

[21] E. C. Dunlop, B. Williamson, W. H. Rodebush, and A. M. Buswell, *J. Am. Chem. Soc.* **63,** 1167 (1941).

has been determined by photo-ionization to be 10.15 ev.[22] In aniline the amino group will probably approach a planar structure with sp^2 hybridization and become nearly coplanar with the benzene ring plane, so that the nitrogen lone-pair orbital approaching the pure $2p\pi$ orbital does maximally interact with the π orbitals of the benzene ring. In *N,N*-dimethylaniline, the nitrogen valences are expected to be almost completely planar at about 120° to one another and to have the lone-pair electrons in a nearly pure $2p\pi$ orbital. Actually, for some *N,N*-dimethylamino-substituted aromatic compounds, the dimethylamino group has been found to be planar.

Substitution at the *ortho* positions of *N,N*-dimethylaniline will cause the dimethylamino group to twist out of the benzene ring plane, and the nitrogen valences will become increasingly more pyramidal as the dimethylamino group is twisted. The rotation of the symmetry axis of the lone-pair orbital out of the position parallel to the symmetry axes of the carbon $2p\pi$ orbitals of the benzene ring and the accompanying decrease in p character of the lone-pair orbital will reduce the overlap, and hence, the interaction between the lone-pair orbital and the π orbitals of the benzene ring.

22.3.2. Aniline

The spectrum of aniline in saturated hydrocarbon solvents shows three bands in the near ultraviolet region. The first or longest wavelength band is a comparatively weak band having vibrational fine structure in the 270–295 mμ region, and has the most intense peak at 287 mμ ($\log \epsilon_{max} = 3.2$). The second band is a moderately intense, structureless band appearing in the 220–250 mμ region, and has its maximum at 235 mμ ($\log \epsilon_{max} = 3.9$). The third band is an intense band appearing in the 185–210 mμ region, and its maximum is at 196 mμ ($\log \epsilon_{max} = 4.3$).

The conjugated system of aniline can be regarded as a composite system R–S, in which R and S represent the benzene ring and the nitrogen atom bearing lone-pair electrons, respectively. In this system, the nitrogen atom acts as an electron donor, and the benzene ring acts as an electron acceptor. According to the classification made in Section 19.4, this system belongs probably to case D-2, and the orbital energy level diagram for this system may be given qualitatively by the one for case D-2 shown in Fig. 19.3. Of course, $\phi_{0'}$ represents the lone-pair orbital on the nitrogen atom. The first band is probably the perturbed LE band corresponding to the benzene α band. The second band, that is, the moderately intense 235-mμ band, is

[22] K. Watanabe, *J. Chem. Phys.* **26,** 542 (1957).

probably the intramolecular CT band due to the transition that may be approximated as the one-electron transition from $\phi^{\mathrm{p}}_{0'}$ to ϕ^{p}_{-1}. This transition should of course be accompanied with a partial electron transfer from the amino group to the benzene ring.

Murrell[23,24] calculated the energies of π-electronic states of aniline by the method of composite molecule, by taking account of the configuration interaction among the ground electron configuration (denoted by G), the four singlet LE electron configurations corresponding to the α, p, β, and β' states of benzene, and the two singlet CT electron configurations obtained by taking an electron out of the lone-pair orbital of the nitrogen atom, $\phi_{0'}$, and placing it in either of the degenerate vacant π orbitals of the benzene ring, ϕ_{-1} and ϕ_{-2}. As usual, the singlet LE configurations corresponding to the α, p, β, and β' states of benzene are denoted by LE_α, LE_p, LE_β, and $LE_{\beta'}$, respectively (for the wavefunctions of the excited states of benzene, see Table 3.7; for the π orbitals of benzene, see Table 3.5). Of course, LE_p and $LE_{\beta'}$, as well as G, are symmetrical with respect to the symmetry plane (the yz plane) that is perpendicular to the molecular plane (the xy plane) and along the C—N bond axis (the y axis), and LE_α and LE_β are antisymmetrical with respect to that symmetry plane. The singlet CT configuration arising from the one-electron transfer from $\phi_{0'}$ to ϕ_{-1}, $V_1^{0'1}$, is symmetrical with respect to the yz plane, and hence it is designated as CT_{S}. On the other hand, the singlet CT configuration arising from the one-electron transfer from $\phi_{0'}$ to ϕ_{-2}, $V_1^{0'2}$, is antisymmetrical with respect to the yz plane, and hence it is designated as CT_{a}.

The energies of the LE configurations (relative to the ground configuration) are assumed to be equal to the energies of the corresponding excited states of benzene. Murrell used values of 4.88, 6.14, 6.74, and 6.74 ev for the energies of LE_α, LE_p, LE_β, and $LE_{\beta'}$, respectively. The energies of the CT configurations are given as usual by

$$E(CT) = I_{\mathrm{D}} - A_{\mathrm{A}} - C, \qquad (22.15)$$

where I_{D} is the ionization potential of the electron-donating group (in this case, that of the nitrogen atom in the amino group, I_{N}), A_{A} is the electron affinity of the electron-accepting group (in this case, that of benzene, A_{B}), and C is the energy of the electrostatic interaction between the positive charge on the electron-donating group and the negative charge on the electron-accepting group in the CT configuration. The C for CT_{S}, C_{S}, is

[23] J. N. Murrell, *Proc. Phys. Soc.* (*London*) **68A,** 969 (1955).
[24] J. N. Murrell, *Quart. Rev.* (*London*) **15,** 191 (1961).

$$C_s = [-1, -1 \mid 0', 0'] = b^2_{-1,1}(C_1C_1 \mid NN) + 2b^2_{-1,2}(C_2C_2 \mid NN) + 2b^2_{-1,3}(C_3C_3 \mid NN) + b^2_{-1,4}(C_4C_4 \mid NN)$$
$$= \tfrac{1}{6}\{2(C_1C_1 \mid NN) + (C_2C_2 \mid NN) + (C_3C_3 \mid NN) + 2(C_4C_4 \mid NN)\}, \quad (22.16)$$

where $(C_1C_1 \mid NN)$, for example, represents the two-center Coulomb repulsion integral between the $2p\pi$ AO on the carbon atom at the 1 position, χ_1, and the lone-pair orbital on the nitrogen atom, $\chi_{1'}$, that is, $\phi_{0'}$. Analogously, the C for CT_a, C_a, is given by

$$C_a = [-2, -2 \mid 0', 0'] = \tfrac{1}{2}\{(C_2C_2 \mid NN) + (C_3C_3 \mid NN)\}. \quad (22.17)$$

Murrell calculated the values of C_s and C_a to be 6.03 and 4.84 ev, respectively, by assuming that the length of the C—N bond is 1.46 Å. Formerly[23] he used the value of 10.52 ev for I_D, which is the value found by Morrison and Nicholson[25] by electron impact for the ionization potential of ammonia, $I(NH_3)$, and the value of -0.54 ev for A_A, which is the theoretical value calculated by Pople and Hush[26] for the electron affinity of benzene, A_B. However, more recently[24] he adopted the value of 10.15 ev as I_D, which is the value found by Watanabe[22] by photo-ionization for $I(NH_3)$, and the value of -1.63 ev as A_A, which is the theoretical value calculated by Hedges and Matsen[27] for A_B. Accordingly, the value of $(I_D - A_A)$ is calculated to be 11.06 or 11.78 ev, the value of $E(CT_s)$ to be 5.03 or 5.75 ev, and the value of $E(CT_a)$ to be 6.22 or 6.94 ev. There will be a small error here, because the ionization potential of the amino group in aniline must be different from that of ammonia, since, as mentioned in the preceding subsection, probably the amino group in aniline is planar and its lone-pair orbital has a nearly pure p character, while the ammonia molecule has a pyramidal structure and its lone-pair orbital is nearly an sp^3 hybrid orbital. At any rate, it is to be noted that CT_s is almost certainly the second lowest excited configuration, lying above LE_α and below LE_p.

The off-diagonal matrix elements of the total π-electronic Hamiltonian are given by [see expressions (20.12) and (20.18)]

$$\langle LE_p | \mathbf{H} | CT_s \rangle = \langle LE_{\beta'} | \mathbf{H} | CT_s \rangle = \langle LE_\alpha | \mathbf{H} | CT_a \rangle = \langle LE_\beta | \mathbf{H} | CT_a \rangle$$
$$= -2^{-1/2}\langle \phi_{+1} | \mathbf{h}^C | \phi_{0'} \rangle = -2^{-1/2} b_{+1,1}\beta_1{}^C = -6^{-1/2}\beta_1{}^C; \quad (22.18)$$

$$\langle G \ \mathbf{H} | CT_s \rangle = 2^{1/2}\langle \phi_{-1} | \mathbf{h}^C | \phi_{0'} \rangle = 2^{1/2} b_{-1,1}\beta_1{}^C = -(\tfrac{2}{3})^{1/2}\beta_1{}^C. \quad (22.19)$$

[25] J. D. Morrison and A. J. C. Nicholson, *J. Chem. Phys.* **20**, 1021 (1952).
[26] J. A. Pople and N. S. Hush, *Trans. Faraday Soc.* **51**, 600 (1955).
[27] R. M. Hedges and F. A. Matsen, *J. Chem. Phys.* **28**, 950 (1958).

All the other off-diagonal elements are zero. In these expressions, β_1^{C} represents the core resonance integral between χ_1 and $\chi_{1'}$, that is, the core resonance integral across the C—N bond. Murrell used a value of -1.60 ev for β_1^{C}.

The lowest excited configuration LE_α is slightly lowered as a result of interaction with CT_{a}, and gives the lowest excited state, ${}^1\Phi_1$ or $[LE]_\alpha$. The first band ($\lambda_{\max} = 287$ mμ) is attributed to the transition from the ground state, $[G]$, to this excited state. The intensity of this band comes mainly from the contribution of LE_β in this excited state. The second lowest excited configuration CT_{s} is lowered by interaction with LE_p and $LE_{\beta'}$, and gives the second lowest excited state, ${}^1\Phi_2$ or $[CT]_{\mathrm{s}}$. The second band ($\lambda_{\max} = 235$ mμ) is attributed to the transition to this excited state. The intensity of this band originates from the conbtribution of $LE_{\beta'}$ in this state.

Kimura, Tsubomura, and Nagakura[28] measured the absorption spectra of aniline and some *N*-alkyl-substituted anilines in the vapor phase in the wavelength region longer than 150 mμ, and calculated the energies and oscillator strengths of electronic transitions in these compounds by improving the above calculation by Murrell.

Kimura et al. used values of 4.89, 6.17, 6.98, and 6.98 ev for the energies of LE_α, LE_p, LE_β, and $LE_{\beta'}$, respectively. These are the values taken from the electronic absorption spectrum data of benzene measured by them. In order to calculate C_{s} and C_{a} by the use of Eqs. (22.16) and (22.17), the two-center integrals (CC | NN) were calculated by the use of a quadratic equation with the atomic distance R_{CN} (in angstrom units)

$$(\mathrm{CC} \mid \mathrm{NN}) = 0.1922R_{\mathrm{CN}}^2 - 2.502R_{\mathrm{CN}} + 10.83 \quad \text{(in electron volts)}, \tag{22.20}$$

which was derived in the way suggested by Pariser and Parr[29] (see Subsection 11.3.2) by the use of the data for the valence state ionization potentials and electron affinities for carbon and nitrogen taken from the table recently given by Pilcher and Skinner.[30] No detailed structural data on the anilines, even on aniline itself, have yet been reported. Kimura et al. assumed that the C—C bonds had the same length as in benzene (1.397 Å) and that the length of the C—N bond was equal to that in 2,5-dichloroaniline (1.407 Å[31]), determined by X-ray crystal analysis. From this molecular geometry, C_{s} and C_{a} were calculated to be 5.44 and 5.00 ev, respectively.

In the calculation of the off-diagonal interaction matrix elements between configurations, Kimura et al. took account of β_2^{C}, which represents the

[28] K. Kimura, H. Tsubomura, and S. Nagakura, *Bull. Chem. Soc. Japan* **37**, 1336 (1964).
[29] R. Pariser and R. G. Parr, *J. Chem. Phys.* **21**, 767 (1953).
[30] G. Pilcher and H. A. Skinner, *J. Inorg. Nucl. Chem.* **24**, 937 (1962).
[31] T. Sakurai, M. Sundaralingam, and G. A. Jefferey, *Acta Cryst.* **16**, 354 (1963).

core resonance integral between the $2p\pi$ AO on an *ortho*-position carbon atom (χ_2 or χ_6) and the lone-pair orbital on the nitrogen atom ($\chi_{1'}$, that is, $\phi_{0'}$), in addition to $\beta_1{}^{\mathrm{C}}$. This means that expressions (22.18) and (22.19) are replaced by the following expressions:

$$\begin{aligned}
&\langle LE_p | \mathbf{H} | CT_{\mathrm{s}} \rangle \\
&\quad = \langle LE_{\beta'} | \mathbf{H} | CT_{\mathrm{s}} \rangle = \langle LE_{\alpha} | \mathbf{H} | CT_{\mathrm{a}} \rangle = \langle LE_{\beta} | \mathbf{H} | CT_{\mathrm{a}} \rangle \\
&\quad = -2^{-1/2} \langle \phi_{+1} | \mathbf{h}^{\mathrm{C}} | \phi_{0'} \rangle = -2^{-1/2} (b_{+1,1}\beta_1{}^{\mathrm{C}} + b_{+1,2}\beta_2{}^{\mathrm{C}} + b_{+1,6}\beta_2{}^{\mathrm{C}}) \\
&\quad = -6^{-1/2} (\beta_1{}^{\mathrm{C}} + \beta_2{}^{\mathrm{C}});
\end{aligned} \tag{22.21}$$

$$\begin{aligned}
&\langle G | \mathbf{H} | CT_{\mathrm{s}} \rangle \\
&\quad = 2^{1/2} \langle \phi_{-1} | \mathbf{h}^{\mathrm{C}} | \phi_{0'} \rangle = 2^{1/2} (b_{-1,1}\beta_1{}^{\mathrm{C}} + b_{-1,2}\beta_2{}^{\mathrm{C}} + b_{-1,6}\beta_2{}^{\mathrm{C}}) \\
&\quad = -(\tfrac{2}{3})^{1/2} (\beta_1{}^{\mathrm{C}} - \beta_2{}^{\mathrm{C}}).
\end{aligned} \tag{22.22}$$

Of course, all the other off-diagonal matrix elements are zero. The ratio of $\beta_2{}^{\mathrm{C}}$ to $\beta_1{}^{\mathrm{C}}$ was determined to be 0.116, by assuming that it is equal to the ratio of the corresponding overlap integrals, S_2/S_1. Then, $\beta_1{}^{\mathrm{C}}$ was assumed to be a parameter common to all the anilines under consideration, and $(I_{\mathrm{D}} - A_{\mathrm{A}})$ in expression (22.15), that is, $(I_{\mathrm{N}} - A_{\mathrm{B}})$, was regarded as a parameter whose value varies from compound to compound. These parameters were evaluated in such a way that the calculated transition energies fit as well as possible to the experimental values for the anilines. The best values for $\beta_1{}^{\mathrm{C}}$ and $\beta_2{}^{\mathrm{C}}$ were found to be -1.94 and -0.25 ev, respectively, and the best values of $(I_{\mathrm{D}} - A_{\mathrm{A}})$ for the individual anilines were determined as follows (in ev units): 11.34 for aniline, 11.04 for *N*-methylaniline, 10.94 for *N*-ethylaniline, 10.84 for *N,N*-dimethylaniline, and 10.64 for *N,N*-diethylaniline. Accordingly, the energies of CT_{s} and CT_{a} of aniline are calculated to be 5.90 and 6.34 ev, respectively. It is of interest that the determined value of $(I_{\mathrm{N}} - A_{\mathrm{B}})$ decreases with the increasing alkyl substitution. This tendency is parallel to the general tendency in the effect of alkyl substitution on the ionization potential of ammonia.

The energy levels of aniline before and after the configuration interaction, calculated by Kimura et al., are shown in Fig. 22.3, in which the energy levels of the configurations and states symmetrical with respect to the symmetry plane vertical to the molecular plane are drawn with solid lines, the energy levels of the configurations and states antisymmetrical with respect to the symmetry plane are drawn with broken lines, and the correlations between configurations and states are denoted with dotted lines. The calculated energies (ΔE) and oscillator strengths (f) of the electronic transitions in aniline from the ground state to the various excited states are compared with the experimental values in Table 22.5. In this table, the states symmetrical

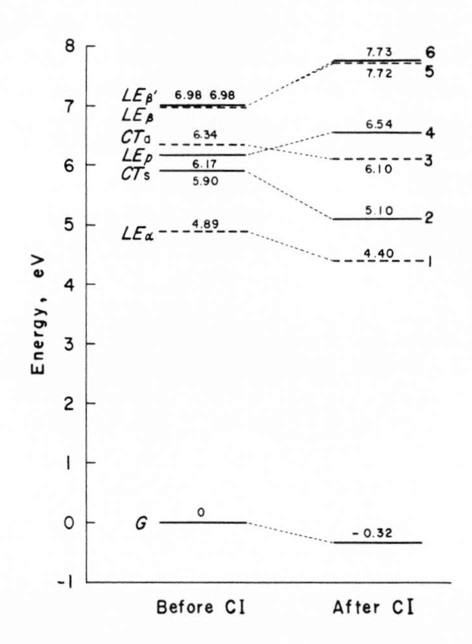

FIG. 22.3. Term energy level diagram of aniline as a composite π system. The energy levels of singlet states before and after the configuration interactions are shown, and the correlations are denoted by the dotted lines.

TABLE 22.5

COMPARISON OF THE THEORETICAL AND EXPERIMENTAL VALUES OF THE TRANSITION ENERGIES (ΔE) AND OSCILLATOR STRENGTHS (f) OF ABSORPTION BANDS OF ANILINE

Observed (gas) ΔE, ev	Observed (gas) f	Calculated ΔE, ev	Calculated f	The upper state of the transition; the electron configurations that make large contribution to the upper state (%)	The symmetry of the upper state
4.40	0.028	4.72	0.021	$^1\Phi_1$; LE_α(74.6)	A
5.39	0.144	5.42	0.174	$^1\Phi_2$; CT_S(50.0), LE_p(34.9)	S
6.40	0.510	6.42	0.416	$^1\Phi_3$; LE_β(39.9), CTa(38.7)	A
6.88	0.570	6.86	0.505	$^1\Phi_4$; LE_p(53.1), $LE_{\beta'}$(37.5)	S
7.87	(0.68)	8.04	0.745	$^1\Phi_5$; LE_β(57.3), CTa(38.7)	A
		8.05	0.416	$^1\Phi_6$; $LE_{\beta'}$(51.1), CT_S(35.9)	S

with respect to the symmetry plane vertical to the molecular plane are indicated with symbol S, and those antisymmetrical with respect to the symmetry plane are indicated with symbol A. Of course, the ground state belongs to symmetry species S. The transitions from the ground state to the states belonging to symmetry species S are polarized in the direction along the symmetry axis of the molecule, that is, along the C—N bond axis (the y axis). The transitions to the states belonging to symmetry species A are polarized in the direction along the x axis, which is in the molecular plane and perpendicular to the y axis. The wavefunctions of the ground and lowest two excited states are as follows:

$$^1\Phi_0 = [G] = 0.974G - 0.225CT_{\mathrm{s}} + 0.031LE_p + 0.028LE_{\beta'}; \tag{22.23}$$

$$^1\Phi_1 = [LE]_\alpha = 0.864LE_\alpha - 0.475CT_{\mathrm{a}} + 0.165LE_\beta; \tag{22.24}$$

$$^1\Phi_2 = [CT]_{\mathrm{s}} = 0.707CT_{\mathrm{s}} - 0.591LE_p - 0.337LE_{\beta'} + 0.192G. \tag{22.25}$$

The intensity of the transition to the lowest excited state $[LE]_\alpha$ originates mainly from the contribution of LE_β (amounting to 2.72%) in this state. The intensity of the transition to the second lowest excited state $[CT]_{\mathrm{s}}$ originates mainly from the contribution of $LE_{\beta'}$ (amounting to 11.36%) in this state. Similarly, the intensity of the transition to the third lowest excited state $[CT]_{\mathrm{a}}$ originates mainly from the contribution of LE_β (amounting to 39.94%) in this state, and the intensity of the transition to the fourth lowest excited state $[LE]_p$ originates mainly from the contribution of $LE_{\beta'}$ (amounting to 37.45%) in this state.

When the amino group is twisted out of the benzene ring plane and the symmetry axis of the long-pair orbital on the nitrogen atom is rotated out of the position parallel to the symmetry axes of the carbon $2p\pi$ AO's of the benzene ring, the absolute value of β_1^{C} will decrease and hence the extent of the mixing of configurations will decrease. When the twist angle of the C—N bond (θ) increases, the following behaviors of various bands of the aniline system are expected from the aforementioned results of calculation, especially from the term energy level diagram shown in Fig. 22.3: the first, second, and third bands will shift more or less markedly to shorter wavelength, and the fourth band will shift slightly to longer wavelength; the intensities of these bands will decrease more or less rapidly. These expectations are generally borne out experimentally, as will be seen in the succeeding subsections. The second band, which is a CT band in nature, is conceivably most sensitive to the steric factor. In the following subsections the steric effect on this band will be mainly discussed.

22.3.3. Nuclear-Alkylated Anilines

The data on the first CT bands of aniline and some nuclear-alkylated anilines collected by Wepster[32] are shown in Table 22.6. It is seen that the CT band of aniline is affected only slightly by introduction of methyl, or even *tert*-butyl, groups at the *ortho* positions. Considering the possible electronic bathochromic and hyperchromic effect of alkyl substituents, it may be said that the steric effect of the *ortho* substituents on this band is hypochromic and

TABLE 22.6

The First CT Bands of Anilines in Isooctane

Aniline	λ_{max}, mμ	ϵ_{max}	f
Unsubstituted	234	9130	0.175
2-Me	234	8800	0.164
2-*tert*-Bu	237	7850	0.158
2,4,6-tri-Me	237	8600	0.150
2,6-di-*tert*-Bu	240	6900	0.133
2,4,6-tri-*tert*-Bu	242	8750	0.171

perhaps hypsochromic. The fact that the steric effect in the anilines is much smaller than that in *N*,*N*-dimethylanilines, which will be discussed in the succeeding subsection, is conceivably due to the fact that the amino group is much less bulky than the dimethylamino group.

22.3.4. *N*-Methylanilines and *N*,*N*-Dimethylanilines

As will be expected, the steric effect of *ortho*-substitution in *N*-methylaniline is much clearer than in aniline. For example, the intensity of the CT band of 2,4,6-tri-*tert*-butyl-*N*-methylaniline (λ_{max} = 247 mμ, ϵ_{max} = 4900 in isooctane) is much lower than that of *N*-methylaniline itself (λ_{max} = 243 mμ, ϵ_{max} = 13,200).[32]

In contrast to the spectrum of aniline, that of *N*,*N*-dimethylaniline is markedly affected by *ortho*-substitution. The data on the spectra of *N*,*N*-dimethylaniline and some methylated derivatives, measured by Klevens and Platt,[33] are shown in Table 22.7, in which the data on the spectrum of benzene are also shown for the purpose of comparison. It will be seen that the predictions made in Subsection 22.3.2 as to the steric effects on various bands of the aniline system are well borne out. Especially, it is to be noted that the

[32] B. M. Wepster, *in* "Steric Effects in Conjugated Systems" (G. W. Gray, ed.), p. 82. Butterworths, London, 1958.

[33] H. B. Klevens and J. R. Platt, *J. Am. Chem. Soc.* **71**, 1714 (1949).

second band (that is, the first CT band) is markedly decreased in intensity by *ortho*-substitution. In the spectrum of 2,6-dimethyl-*N*,*N*-dimethylaniline the first CT band almost completely disappears. The weak band appearing at about 260 mμ in the spectrum of this compound is probably the perturbed benzene α band (the first band). The first CT band, which has been markedly reduced in intensity, may be masked under this band.

TABLE 22.7

THE SPECTRA OF *N*,*N*-DIMETHYLANILINES IN *n*-HEPTANE

N,*N*-Dimethyl-aniline	The 1st band λ_{max}, mμ	$\epsilon_{max} \times 10^{-2}$	f	The 2nd band λ_{max}, mμ	$\epsilon_{max} \times 10^{-2}$	f
Unsubstituted	296	23	0.04	250	138	0.28
4-Methyl-	304	21	0.03	254	151	0.32
2-Methyl-	288	12	0.02	248	63	0.12
2,6-Dimethyl-	259	22	0.05	—	—	—
(Benzene)	254	2	0.001	—	—	—

N,*N*-Dimethyl-aniline	The 3rd band λ_{max}, mμ	$\epsilon_{max} \times 10^{-2}$	f	The 4th band λ_{max}, mμ	$\epsilon_{max} \times 10^{-2}$	f
Unsubstituted	200	222	0.54	176	366	0.79
4-Methyl-	203	228	0.46	183	208	0.40
2-Methyl-	207	114	0.23	185	308	0.69
2,6-Dimethyl-	209	86	0.15	194	361	0.66
(Benzene)	202	69	0.10	184	460	0.69

Klevens and Platt[33] found that the intensities of the second, third, and fourth bands of *N*,*N*-dimethylanilines decrease gradually as the size of the *ortho* substituent increases, and they related the changes in these intensities to the angles of twist of the dimethylamino group out of the benzene ring plane calculated from the van der Waals radii of *ortho* substituents. Table 22.8 shows the data on the second bands (that is, the first CT bands) of *N*,*N*-dimethylaniline and some *ortho*-substituted derivatives and the minimum angles of twist of the dimethylamino group in these compounds calculated for planar and pyramidal models of the dimethylamino group. When the calculated minimum twist angle (θ) is plotted against the oscillator strength (f) of the second, the third, or the fourth band, the resulting points lie, although very roughly, on a $\cos^2 \theta$ curve, which at $\theta = 0°$ passes through the f value for the unsubstituted *N*,*N*-dimethylaniline and at $\theta = 90°$ through the f value for the o,o'-dimethyl derivative. This means that the f value of the

TABLE 22.8
THE FIRST CT BANDS AND TWIST ANGLES OF THE DIMETHYLAMINO GROUP OF *ORTHO*-SUBSTITUTED *N,N*-DIMETHYLANILINES[a]

N,N-Dimethyl-aniline	The first CT band λ_{max}, mμ	ϵ_{max}	f	Computed minimum angle: Planar model, deg.	Computed minimum angle: Pyramidal model, deg.	Effective spectroscopic angle, deg.
Unsubstituted	250.0	13,750	0.28	0	0	0
2-F	250.0	11,600	0.21	29	22	33
2-Cl	255.8	7600	0.14	58	44	44
2-Br	254.5	5900	0.12	65	49	54
2-Me	247.5	6300	0.12	75	56	60
2,6-di-Me	259.1[b]	2200	0.05	75	90	90 (assumed)

[a] The data were taken from Reference 33.
[b] This maximum is probably not associated with the CT band, but belongs to the perturbed benzene α band.

second band (the first CT band) is approximately proportional to the cosine-square of the angle of twist, since the f value of this CT band should be zero at $\theta = 90°$:

$$f \propto \cos^2 \theta. \tag{22.26}$$

The curve for the fourth band is rather flat, and the curves for the second and third bands are considerably steep. It is possible to use the f values and the cosine-square curves to compute values called "effective spectroscopic twist angles" for the individual compounds. Such values derived from averaging the second and third band data are shown in the last column of Table 22.8.

Wepster and co-workers[6] evaluated the twist angle θ in some *ortho*-alkylated *N,N*-dimethylanilines from the ϵ_{max} values of the second band, by assuming that $\epsilon_{max} \propto \cos^2 \theta$. The results are summarized in Table 22.9.

TABLE 22.9
THE FIRST CT BANDS OF *N,N*-DIMETHYLANILINES IN 2,2,4-TRIMETHYLPENTANE AND THE CALCULATED TWIST ANGLES OF THE DIMETHYLAMINO GROUP

N,N-Dimethylaniline	λ_{max}, mμ	ϵ_{max}	ϵ/ϵ_0	θ, deg.
Unsubstituted	251	15,500 (ϵ_0)	(1)	(0)
2-Me	248	6360	0.41	50
2-Et	249	4950	0.32	56
2-*i*-Pr	248	4300	0.28	58
2-*tert*-Bu	—	630[a]	0.04	78

[a] No maximum; ϵ at 250 mμ is given.

22.3.5. BENZOCYCLAMINES

Benzocyclamines (I) can be considered to be analogs of *o*-methyl-*N*,*N*-dimethylaniline. The data on the spectra of benzocyclamines I with $n = 1, 2$, and 3, measure by Remington,[34] are shown in Table 22.10, in which the data for *N*,*N*-dimethylaniline are included for the purpose of comparison.

CH_3 N $(CH_2)_n$ C H_2

I

The spectra of *N*-methylindoline (I, $n = 1$) and *N*-methyltetrahydroquinoline (I, $n = 2$) are quite similar to each other, and they are rather similar to the spectrum of *N*,*N*-dimethylaniline. Undoubtedly, in these benzocyclamines the nitrogen valences are constrained in such a configuration

TABLE 22.10

THE SPECTRA OF BENZOCYCLAMINES (I) IN 2,2,4-TRIMETHYLPENTANE

Compound	The first band		The second band	
	λ_{max}, mμ	ϵ_{max}	λ_{max}, mμ	ϵ_{max}
N,*N*-Dimethylaniline	297	2400	250	13,700
N-Methylindoline (I, $n = 1$)	303	2750	253	10,200
N-Methyltetrahydroquinoline (I, $n = 2$)	303	2700	257	10,200
N-Methyl-*homo*-tetrahydroquinoline (I, $n = 3$)	284	2000	252	8500

that the lone-pair orbital does effectively interact with the π orbitals of the benzene ring. The spectrum of *N*-methyl-*homo*-tetrahydroquinoline (I, $n = 3$) is considerably different from the spectra of the two lower homologs, both the first and the second band being at shorter wavelengths and of lower intensities. This difference arises almost certainly from the larger deviation from planarity of the structure of the seven-membered alicyclic system,

[34] W. R. Remington, *J. Am. Chem. Soc.* **67**, 1838 (1945).

which inhibits to a larger extent the conjugative interaction between the nitrogen lone-pair orbital and the π orbitals of the benzene ring.

In this connection, it is of interest that the spectrum of benzoquinuclidine (II) shows no indication of the conjugative interaction between the nitrogen lone-pair orbital and the benzene π orbitals, closely resembling the spectrum of benzene.[35]

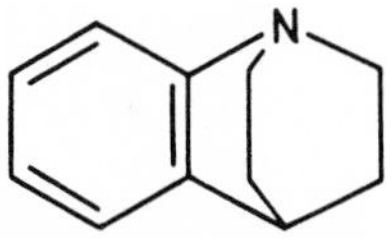

II

22.3.6. ACETANILIDES

The spectrum of acetanilide in ethanol shows a moderately intense band at 242 mμ (ϵ_{max} = 14,400),[32] which is probably an intramolecular CT band. This band disappears almost completely in the spectrum of 2,4,6-trimethylacetanilide, which is closely similar to the spectrum of mesitylene.[32] This means probably that the conjugative interaction between the amide nitrogen atom and the benzene ring has been almost completely eliminated by the steric effect of the two methyl substituents at the *ortho* positions.

22.4. Generalization of the Substituent Effects in the Spectra of Monosubstituted Benzenes—Weak and Strong Substituent Effects

Nagakura and Kimura[36] applied the method mentioned in Subsection 22.3.2 to various monosubstituted benzenes C_6H_5—X in which X is an electron-donating group bearing lone-pair electrons of $p\pi$ character, such as hydroxyl, alkoxyl, mercapto, and halogen. Table 22.11 shows the best values of $(I_X - A_B)$ found for various compounds and the values of the electrostatic interaction energies for the symmetrical and antisymmetrical CT configurations, C_s and C_a, calculated from Eqs. (22.16) and (22.17) by the use of the appropriate values found in the literature for the C—X bond lengths. The values of $(I_X - A_B)$ are almost parallel to the values of the ionization potentials of CH_3X, $I_X(CH_3X)$, recorded in the literature. When the values of $I_X(CH_3X)$ are assumed to be equal to the values of I_X and are subtracted from the values of $(I_X - A_B)$, values ranging from −0.68 to

[35] B. M. Wepster, *Rec. trav. chim.* **71,** 1159 (1952).
[36] S. Nagakura and K. Kimura, *Nippon Kagaku Zasshi* **86,** 1 (1965).

-1.36 ev except for the cases where $X = NH_2$ and $N(CH_3)_2$* are obtained as approximate values of A_B (the electron affinity of benzene). The mean value of about -1.0 ev is probably a good approximation to the value of A_B, which was calculated theoretically, as mentioned already, by Pople and Hush,[26] by Hedges and Matsen,[27] and most recently by Scott[37] to be

TABLE 22.11

THE VALUES OF $(I_X - A_B)$, C_s, C_a, $E(CT_s)$, AND $E(CT_a)$ CALCULATED FOR C_6H_5—X (Unit: ev)

X	$I_X - A_B$	C_s	C_a	$E(CT_s)$	$E(CT_a)$
OH	12.16	5.26	4.71	6.90	7.45
OCH_3	11.86	5.26	4.71	6.60	7.15
OC_2H_5	11.76	5.26	4.71	6.50	7.05
SH	10.49	4.59	4.24	5.90	6.25
F	13.32	5.72	4.83	7.60	8.49
Cl	12.17	4.77	4.32	7.40	7.85
Br	11.54	4.54	4.18	7.00	7.36
I	10.23	4.23	3.93	6.00	6.30
NH_2	11.25	5.44	5.00	5.81	6.25
NH_2[a]	11.34	5.44	5.00	5.90	6.34
$N(CH_3)_2$[a]	10.84	5.44	5.00	5.40	5.84

[a] Reference 28.

-0.54, -1.63, and -1.1 ev, respectively. Table 22.11 also shows the values of $E(CT_s)$ and $E(CT_a)$ calculated by

$$E(CT_s) = (I_X - A_B) - C_s \, ; \tag{22.27}$$

$$E(CT_a) = (I_X - A_B) - C_a \, . \tag{22.28}$$

It is to be noted that CT_s is always lower than CT_a.

It should be remembered now that CT_s can interact only with G, LE_p, and $LE_{\beta'}$ and that the energy of LE_p, $E(LE_p)$, is 6.17 ev. When $E(CT_s)$ is

* There is a considerable scatter in the values of the ionization potentials of CH_3NH_2 and $N(CH_3)_3$. In addition, as mentioned already, $I_X(C_6H_5X)$ for $X = NH_2$ and $N(CH_3)_2$ must be different from $I_X(CH_3X)$ for $X = NH_2$ and $N(CH_3)_2$, since the hybridization of the nitrogen valences in $C_6H_5NH_2$ and $C_6H_5N(CH_3)_2$ must be different from that in CH_3NH_2 and $N(CH_3)_3$.

[37] D. R. Scott, private communication to the authors of the paper cited in Reference 36, which was quoted in that paper.

smaller than $E(LE_p)$ as in the case where the substituent is $N(CH_3)_2$, NH_2, SH, or I, the intramolecular CT band due to the transition to the perturbed CT state corresponding to CT_S, $[CT]_S$, will appear at longer wavelength than the perturbed LE band corresponding to the benzene *p* band. That is, in such a case, the spectrum in the accessible near ultraviolet region is so different from the spectrum of benzene that it cannot be understood only in terms of perturbation of the electronic states of benzene. In such a case, according to Murrell,[38] the substituent is said to be exerting a strong mesomeric effect on the spectrum. In such a case, the perturbed *p* and β' bands will appear at shorter wavelengths than the corresponding benzene bands.

On the other hand, when $E(CT_S)$ is greater than $E(LE_p)$ as in the case where the substituent is OR, OH, F, or Cl, the intramolecular CT band will not appear in the accessible ultraviolet region, and the perturbed *p* band will appear at longer wavelength than the benzene *p* band. In such a case, the substituent is said to be exerting a weak mesomeric effect on the spectrum.

The antisymmetrical CT configuration, CT_a, can interact only with the antisymmetrical LE configurations, LE_α and LE_β. In all the cases, CT_a is much higher than LE_α, whose energy is 4.89 ev. Therefore, the benzene α band will be slightly shifted to longer wavelength and will be slightly intensified by the substituent, since LE_α undergoes repulsive interaction with CT_a and indirectly with LE_β. Second-order perturbation theory may be applied to the interaction between CT_a and LE_α, since the energy separation between these configurations is usually much greater than the interaction matrix element between these configurations. We may say that even the strongest of the simple electron-donating substituents exerts a weak mesomeric effect on the benzene α band.

A similar generalization can be made of the effects of electron-accepting substituents, such as nitro, carbonyl, and carboxyl. In monosubstituted benzenes in which the substituent is such an electron-accepting group, the CT configuration that arises from the one-electron transition from the highest occupied symmetrical π orbital of the benzene ring (ϕ_{+1}) to the lowest vacant π orbital of the substituent ($\phi_{-1'}$), $CT^{11'}$, lies below LE_p (see the energy-level diagram for nitrobenzene shown in Fig. 22.2). Therefore, in such systems, the intramolecular CT band due to the transition to the perturbed CT state corresponding to this CT configuration appears at longer wavelength than the perturbed *p* band, which appears at shorter wavelength than the original benzene *p* band (see Subsection 22.1.2 and also Subsections

[38] J. N. Murrell, "The Theory of the Electronic Spectra of Organic Molecules," Chap. 10. Methuen, London, 1963.

21.4.1 and 22.2.2). The substituent can be said to exert a strong effect on the spectrum. The CT configuration arising from the one-electron transition from the highest occupied antisymmetrical π orbital of the benzene ring (ϕ_{+2}) to $\phi_{-1'}$, $CT^{21'}$, is usually much higher than LE_α. The substituent can be said to exert a weak effect on the benzene α band.

In styrene, the lowest singlet excited configuration is LE_α, and the second lowest one is $CT^{11'}$, that is, the CT configuration arising from the one-electron transition from the highest occupied symmetrical π orbital of the benzene ring (ϕ_{+1}) to the vacant π orbital of the ethylenic bond ($\phi_{-1'}$). After the configuration interaction, the lowest excited state is $[LE]_\alpha$, and the second lowest state is $[CT]^{11'}$, in which the contribution of $CT^{11'}$ amounts to 58.22% and the contributions of LE_p, $LE_{\beta'}$, and LE(C=C) amount to 20.07, 9.55, and 11.02%, respectively.[36] Therefore, in this system the intramolecular CT band due to the transition to $[CT]^{11'}$ should appear as the second band at shorter wavelength than the perturbed α band and longer wavelength than the perturbed p band, which should appear at shorter wavelength than the original benzene p band. The intense styrene band appearing in the 220–260 mμ region (λ_{max} = 248 mμ in solution in saturated hydrocarbon solvents; λ_{max} = 238 mμ in the vapor phase) (see Chapter 13) is almost certainly this CT band. Thus, the vinyl group in the styrene system can also be said to be exerting a strong effect on the spectrum.

In general, if the spectrum of a composite molecule R–S is so different from the spectra of compounds corresponding to fragments R and S that it cannot be understood only in terms of perturbation of electronic states of the fragments, showing a new band, which is an intramolecular CT band in nature, in the accessible near ultraviolet region, the interaction between the fragments in the composite molecule may be said to be strong.

In a monosubstituted benzene C_6H_5—X in which X is a strong substituent, either electron-donating or electron-accepting, if the repulsive interaction of CT_S (or $CT^{11'}$) with LE_p and $LE_{\beta'}$ is stronger than that with G as is usually the case, the energy of $[CT]_S$ (or $[CT]^{11'}$) relative to the ground state will be smaller than the energy of CT_S (or $CT^{11'}$) relative to the ground configuration, and hence, as mentioned already in Section 19.6, twist of the substituent out of the benzene ring plane will exert a hypsochromic effect on the CT band. On the other hand, if the repulsive interaction of CT_S (or $CT^{11'}$) with LE_p and $LE_{\beta'}$ is weaker than that with G because CT_S (or $CT^{11'}$) is very low and lies closer to G than to LE_p, the energy of $[CT]_S$ (or $[CT]^{11'}$) relative to the ground state will be greater than the energy of CT_S (or $CT^{11'}$) relative to the ground configuration, and hence twist of the substituent will exert a bathochromic effect on the CT band.

22.5. Nitroanilines and Related Compounds

22.5.1. 4-Nitroaniline and Its Nuclear-Alkylated Derivatives

The spectrum of 4-nitroaniline shows two intense bands in the accessible ultraviolet region: one band appears at $\lambda_{max} = 375$ mμ (log $\epsilon_{max} = 4.18$) and the other at $\lambda_{max} = 227$ mμ (log $\epsilon_{max} = 3.86$) in ethanol solution.[5] The first band at 375 mμ is probably an intramolecular CT band due to a transition accompanied with a partial electron transfer from the electron-donating amino group to the electron-accepting nitro group through the π system of the benzene ring. The steric effect on this band was studied by Arnold and co-workers[39,40] and by Wepster and co-workers.[5,6,7,32,41] Data on this band of 4-nitroaniline and some nuclear-alkylated derivatives are collected in Table 22.12. Examination of these data leads to the following generalizations.

The values of ϵ_{max} and f decrease in a way which is comprehensively related to the size and number of the substituents in the *ortho* positions to the nitro group. As in the aniline series described in Subsection 22.3.3, methyl groups in the *ortho* positions to the amino group do not appear to interfere seriously with the coplanarity of the amino group with the benzene ring, as will be seen from comparison of the data on the 2,5-dimethyl compound (compound 3) with those on the 3-methyl compound (compound 4). Therefore, the decrease in the intensity is considered to be caused mainly by the rotation of the nitro group out of the benzene ring plane. The values of the angle of twist θ calculated by assuming that $\epsilon_{max} \propto \cos^2 \theta$ are also shown in Table 22.12. When the nitro group is twisted out of the benzene ring plane, the conjugative electron-withdrawing effect of the nitro group will decrease and hence the basic strength of the amino group will increase. This expectation was well borne out experimentally.[32] The values of pK_b in water at 25°C for some of compounds listed in Table 22.12 are as follows: (1) 12.89, (4) 12.50, (7) 12.04, (9) 11.41, (12) 11.64, (13) 11.07. Thus, it is seen that the basic strength increases as the value of θ increases. In 3,5-di-*tert*-butyl-4-nitroaniline (compound 13) the value of θ is probably approximately 90°. The difference between the pK_b value of 11.07 for this compound and the pK_b value of 9.03 for 3,5-di-*tert*-butylaniline is attributed to the pure inductive electron-withdrawing effect of the nitro group.

[39] R. T. Arnold and J. Richter, *J. Am. Chem. Soc.* **70,** 3505 (1948).

[40] R. T. Arnold and P. N. Craig, *J. Am. Chem. Soc.* **72,** 2728 (1950); R. T. Arnold, V. J. Webers, and R. M. Dodson, *ibid.* **74,** 368 (1952).

[41] R. van Helden, P. E. Verkade, and B. M. Wepster, *Rec. trav. chim.* **73,** 39 (1954).

TABLE 22.12

THE FIRST BANDS OF 4-NITROANILINES IN ETHANOL AND THE CALCULATED VALUES OF THE TWIST ANGLE OF THE NITRO GROUP (θ)

No.	4-Nitroaniline (NH_2-1)	λ_{max}, mμ	f	ϵ_{max}	f/f_0[a]	ϵ/ϵ_0[a]	θ, deg.
1	Unsubstituted[b]	376	0.376	15,500	(1)	(1)	(0)
2	2,3-Trimethylene-[c]	376	—	13,900	—	0.90	19
3	2,5-Dimethyl-[d]	378	0.347	13,600	0.92	0.88	21
4	3-Methyl-[b]	374	—	13,200	—	0.85	23
5	2,3-5,6-Bistrimethylene-[e]	372	—	12,800	—	0.83	25
6	2,3-Tetramethylene-[d]	383	0.306	11,200	0.81	0.72	32
7	2,3-Dimethyl-[d]	382	0.286	9750	0.76	0.63	38
8	2,3-Trimethylene-5,6-tetramethylene-[e]	387	—	7300	—	0.47	47
9	3,5-Dimethyl-[b]	385	0.126	4840	0.34	0.31	56
10	2,3-5,6-Bistetramethylene-[e]	397	—	2240	—	0.14	68
11	2,3-Dimethyl-5,6-tetramethylene-[e]	393	—	1970	—	0.13	69
12	2,3,5,6-Tetramethyl-[b]	396	0.039	1560	0.10	0.10	72
13	3,5-Di-*tert*-butyl-[f]	401	0.015	540	0.04	0.03	79

[a] f_0 and ϵ_0 refer to the values for the unsubstituted 4-nitroaniline.
[b] Reference 5.
[c] Reference 39.
[d] Reference 41.
[e] Reference 40.
[f] Reference 6.

A single methyl group in the *ortho* position to the nitro group does not greatly hinder the coplanarity of the nitro group with the benzene ring, as will be seen from comparison of the data on compound 4 with those on compound 1. On the other hand, di-*ortho*-substitution greatly enhances the steric hindrance (see data on compound 9). Similar phenomena were observed in the nitrobenzene series (see Subsection 22.1.3), in the benzoic acid series (see Subsection 22.2.3), in the *N,N*-dimethylaniline series (see Subsection 22.3.4), and in the acetophenone series (see Subsection 21.5.4). In molecular models of the rigid Stuart type, if the functional group is a symmetrical one, such as nitro, amino, and dimethylamino, the presence of two identical substituents in the *ortho* positions to the functional group would not require the twist angle of the functional group to be greater than when only one of them causes noncoplanarity of the functional group with the benzene ring. However, while the delocalization energy curve is the same in both cases, the steric repulsion energies should be twice as large with di-*ortho*-substitution as with mono-*ortho*-substitution. Moreover, the effective size of each *ortho* substituent will be larger with di-substitution, since bending away of the functional group is no longer of help. Therefore, the angle of twist of the functional group should be expected to be greater with di-*ortho*-substitution than with mono-*ortho*-substitution.

The intensity in the 2,3-dimethyl compound (compound 7) is much smaller than that in the electronically equivalent 2,5-dimethyl analog (3). The greater steric effect of the two adjacent methyl groups is a buttressing effect. The much lower intensity in the 2,3,5,6-tetramethyl compound (12) as compared with the intensity in the 3,5-dimethyl compound (9) is also probably, partly at least, due to the buttressing effect.

The intensity decreases in the order the 2,3-trimethylene compound (2) > the 2,3-tetramethylene compound (6) > the 2,3-dimethyl compound (7), and in the order the 2,3-5,6-bistrimethylene compound (5) > the 2,3-trimethylene-5,6-tetramethylene compound (8) > the 2,3-5,6-bistetramethylene compound (10) > the 2,3-dimethyl-5,6-tetramethylene compound (11) > the 2,3,5,6-tetramethyl compound (12). This means that the steric effect of substituents decreases in the order the *ortho*-dimethyl grouping > the *ortho*-tetramethylene bridge > the *ortho*-trimethylene bridge. In the bridged compounds, the interfering alkyl groups are tied back and hence produce smaller steric effects.

While the intensity of the CT band is highly sensitive to the steric factors, the wavelength λ_{max} of the band is comparatively insensitive. Even when the possible electronic bathochromic effect of the alkyl substituents is taken into consideration, the steric effect on the λ_{max} seems to be rather bathochromic. This is not unexpected from the following consideration. When the 4-nitroaniline system is divided into two fragments, the aminophenyl group as R and the nitro group as S, the CT band may be considered to be due to the one-electron transition from the orbital arising from perturbation of the highest occupied π orbital of R, ϕ^{p}_{+1}, to the orbital arising from perturbation of the vacant π orbital of S, $\phi^{p}_{-1'}$. The highest occupied π orbital of R, ϕ_{+1}, is considered to arise mainly from mixing of the lone-pair orbital of the amino group ($\phi_0(NH_2)$) with the highest bonding symmetrical π orbital of the benzene ring ($\phi_{+1}(B)$). Since ϕ_{+1} in the aminophenyl group must be higher in energy than ϕ_{+1} in the benzene ring, the repulsive interaction between $\phi_{-1'}$ and ϕ_{+1} in the 4-nitroaniline system may be more effective than the corresponding interaction in the nitrobenzene system. Furthermore, since the vacant π orbitals in the aminophenyl group are probably higher than the corresponding π orbitals in the benzene ring, the lowering of $\phi_{-1'}$ by interaction with the vacant π orbitals of R may be less effective in the 4-nitroaniline system than in the nitrobenzene system. Therefore, since the CT band of nitrobenzene undergoes a very small hypsochromic or substantially no shift on twisting the nitro group (see Subsection 22.1.3), it is understandable that the CT band of 4-nitroaniline undergoes a small bathochromic shift on twisting the nitro group.

22.5.2. 4-Nitro-*N*,*N*-Dimethylaniline and Its Alkylated Derivatives

Data on the spectra of 4-nitro-*N*,*N*-dimethylaniline and its derivatives methylated at the *ortho* positions to the nitro group, measured by Remington,[34] are shown in Table 22.13, in which data on the spectrum of *N*,*N*-dimethylaniline are also shown for the purpose of comparison. As in

TABLE 22.13

The Spectra of 4-Nitro-*N*,*N*-Dimethylaniline and Its Derivatives Methylated at the *Ortho* Positions to the Nitro Group in Ethanol

4-Nitro-*N*,*N*-dimethylaniline ($N(CH_3)_2$-1)	λ_{max}, mμ	$\epsilon_{max} \times 10^{-2}$	λ_{max}, mμ	$\epsilon_{max} \times 10^{-2}$	λ_{max}, mμ	$\epsilon_{max} \times 10^{-2}$
Unsubstituted	388	183	—	—	231	86
3-Methyl-	386	165	—	—	235	82
3,5-Dimethyl-	394	63	305	32	250	100
(*N*,*N*-Dimethylaniline)	—	—	297	24	250	137

TABLE 22.14

THE FIRST BANDS OF 4-NITRO-*N*,*N*-DIMETHYLANILINE AND SOME DERIVATIVES ALKYLATED AT THE *ORTHO* POSITIONS TO THE DIMETHYLAMINO GROUP IN 96% ETHANOL

4-Nitro-*N*,*N*-dimethylaniline ($N(CH_3)_2$-1)	λ_{max}, mμ	ϵ_{max}
Unsubstituted	ca. 390	20,000
2-Methyl-	ca. 375	10,200
2,6-Dimethyl-	ca. 380	5450
2-*tert*-Butyl-	—	570[a]

[a] No maximum; ϵ at 380 mμ.

the case of 4-nitroaniline, the intensity of the first (CT) band of the nitrodimethylaniline is slightly decreased by one methyl substituent at the *ortho* position to the nitro group and is markedly decreased by two methyl substituents at the *ortho* positions to the nitro group. It is to be noted that as the conjugative interaction of the nitro group with the benzene ring is decreased by the sterically-hindering methyl substituents the second band appearing at 231 mμ in the spectrum of the unsubstituted nitrodimethylaniline comes more closely to resemble the second band (that is, the first CT band) of *N*,*N*-dimethylaniline. The spectrum of the 3,5-dimethyl derivative resembles as a whole the spectrum of *N*,*N*-dimethylaniline.

When the *ortho* positions to the dimethylamino group in 4-nitro-*N*,*N*-dimethylaniline are substituted with alkyl groups, the dimethylamino group will be twisted out of the benzene ring plane. The twist of the dimethylamino group exerts a marked hypochromic effect and a moderate hypsochromic effect on the first band (see Table 22.14).[42] In the spectrum of the 2-*tert*-butyl derivative, which is highly sterically hindered, the first band disappears almost completely, and the spectrum is closely similar to the spectrum of the correspondingly alkylated nitrobenzene.

22.5.3. POSITIONAL ISOMERS OF NITROANILINE AND NITRO-*N*,*N*-DIMETHYLANILINE

Data on the first (CT) band of *o*- and *p*-nitroaniline and of *o*- and *p* nitro-*N*,*N*-dimethylaniline, obtained by Remington,[34] are shown in Table 22.15. It is seen that in each case the first band of the *ortho* compound is much less intense and at considerably longer wavelength than that of the

[42] See References 5, 6, and 7. See also J. N. Murrell, "The Theory of the Electronic Spectra of Organic Molecules," Chap. 11. Methuen, London, 1963.

TABLE 22.15

THE FIRST BANDS OF NITROANILINES AND NITRO-*N*,*N*-DIMETHYLANILINES IN ETHANOL

Compound	λ_{max}, mμ	ϵ_{max}
o-Nitroaniline	400	6310
p-Nitroaniline	370	16,200
o-Nitro-*N*,*N*-dimethylaniline	415	2460
p-Nitro-*N*,*N*-dimethylaniline	388	18,290

para isomer. The decrease in the intensity of the band in going from the *para* to the *ortho* isomer is far more pronounced in the case of nitrodimethylaniline than in the case of nitroaniline, and is almost certainly, partly at least, due to the steric hindrance to the coplanarity of the functional groups with the benzene ring in the *ortho* isomer and due to the shorter distance between the two end groups of the conjugated system in the *ortho* isomer than in the *para* isomer.

The difference in λ_{max} between the *ortho* and *para* isomers has been attributed to the difference in the contribution of the energy of the electrostatic interaction (C) between the two end groups, the amino (or the dimethylamino) group as an electron donor and the nitro group as an electron acceptor, to the energy of the CT excited state.[24] Since the two end groups are closer together in the *ortho* isomer than in the *para* isomer, the C should be larger for the *ortho* isomer than for the *para* isomer, and hence the energy of the CT excited state relative to the ground state in the *ortho* isomer will be lower than that in the *para* isomer.

In this connection, it is of interest that the spectrum of *m*-nitro-*N*,*N*-dimethylaniline is rather similar to that of the *ortho* isomer, but is different from that of the *para* isomer.[43] In general, as mentioned already (see Subsection 22.2.3), if benzene is substituted with one electron-donating group and one electron-accepting group, the spectra of the *ortho*- and *meta*-substituted compounds are similar to each other but different from the spectrum of the *para* isomer, the first (CT) bands of the *ortho* and *meta* compounds being at longer wavelengths and of lower intensities than that of the *para* isomer. This fact is diametrically opposed to the expectation from the resonance theory that the *meta* compound would be the odd one out, but this fact can be easily explained by the molecular orbital theory incorporating LE and CT states.[43]

[43] R. Grinter, E. Heilbronner, M. Godfrey, and J. N. Murrell, *Tetrahedron Letters* **1961,** 771; see also R. Grinter and E. Heilbronner, *Helv. Chim. Acta* **45,** 2496 (1962) and Reference 19.

Chapter 23

Azobenzenes and Related Compounds

23.1. The Azo Group

The azo group, —N=N—, is structurally very similar to the vinylene group, —CH=CH—, and these two groups are iso-π-electronic. In place of the two C—H bonds of the vinylene group, the azo group has two lone-pair orbitals, which, like the carbon valence orbitals of the vinylene group forming the C—H bonds, are probably approximately sp^2 hybrid orbitals.

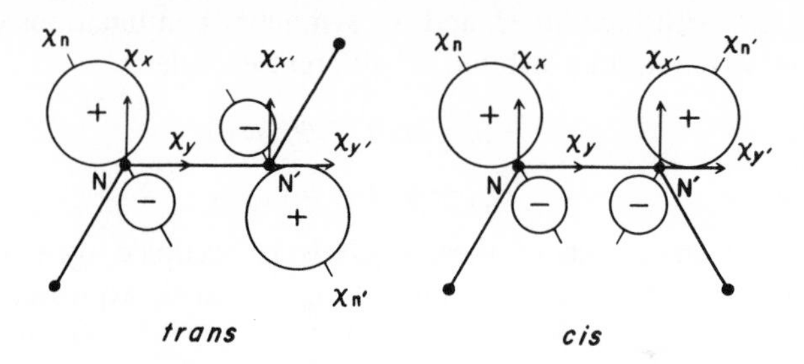

FIG. 23.1. The *trans*- and *cis*-azo groups.

As illustrated in Fig. 23.1, the plane of the azo group is taken as the xy plane, and the direction along the N=N bond axis is taken as the direction of the y axis. The $2s$, $2p_x$, $2p_y$, and $2p_z$ orbitals of one of the two nitrogen atoms (N) are denoted by χ_s, χ_x, χ_y, and χ_z, respectively, and those of the other nitrogen atom (N′) by $\chi_{s'}$, $\chi_{x'}$, $\chi_{y'}$, and $\chi_{z'}$, respectively. Then, the antibonding and bonding π orbitals of the azo group can be expressed as follows:

$$\phi_{-1} = 2^{-1/2}(\chi_z - \chi_{z'}) \equiv \phi_{A\pi(-)}\,; \tag{23.1}$$

$$\phi_{+1} = 2^{-1/2}(\chi_z + \chi_{z'}) \equiv \phi_{A\pi(+)}\,. \tag{23.2}$$

The one-electron S–S transition from $\phi_{A\pi(+)}$ to $\phi_{A\pi(-)}$, the π–π^* S–S transition in the azo group, is allowed and is polarized along the y axis. The N=N bond length ($R_{N=N}$) in most azo compounds is about 1.23 Å. From this bond length the length of the transition moment of this transition is calculated as follows:

$$M_{\pi,\pi^*} = 2^{1/2}\int \phi_{A\pi(+)} y \phi_{A\pi(-)}\, d\tau = 2^{-1/2}\left(\int \chi_z^2\, y\, d\tau - \int \chi_{z'}^2\, y\, d\tau\right)$$

$$= 2^{-1/2}(y_N - y_{N'}) = 2^{-1/2} R_{N=N} \doteqdot 0.870 \text{ Å}. \tag{23.3}$$

The lone-pair orbital of atom N is expressed as follows, when it is assumed to be a pure sp^2 orbital:

$$\chi_n = 3^{-1/2}\chi_s + 2^{-1/2}\chi_x - 6^{-1/2}\chi_y . \tag{23.4}$$

Similarly, the lone-pair orbital of atom N′ in the *trans*-azo group and that in the *cis*-azo group are expressed as follows:

$$\textit{trans} : \chi_{n'} = 3^{-1/2}\chi_{s'} - 2^{-1/2}\chi_{x'} + 6^{-1/2}\chi_{y'} ; \tag{23.5}$$

$$\textit{cis} : \chi_{n'} = 3^{-1/2}\chi_{s'} + 2^{-1/2}\chi_{x'} + 6^{-1/2}\chi_{y'} . \tag{23.6}$$

The two lone-pair orbitals on atoms N and N′, χ_n and $\chi_{n'}$, will overlap to some extent with each other, and the symmetry combinations of them will give rise to two molecular orbitals of different energies:

$$2^{-1/2}(\chi_n - \chi_{n'}) \equiv \phi_{An(-)} ; \tag{23.7}$$

$$2^{-1/2}(\chi_n + \chi_{n'}) \equiv \phi_{An(+)} . \tag{23.8}$$

In the ground state each of these orbitals is occupied by two electrons. The overlap integral between χ_n and $\chi_{n'}$, $S_{n,n'}$, can be expressed as follows:

$$\textit{trans} : S_{n,n'} = 3^{-1}S_{2s-2s} + 6^{-1}S_{2p\sigma-2p\sigma} - 2^{-1}S_{2p\pi-2p\pi} - 2^{1/2} \times 3^{-1}S_{2s-2p\sigma} ; \tag{23.9}$$

$$\textit{cis} : S_{n,n'} = 3^{-1}S_{2s-2s} + 6^{-1}S_{2p\sigma-2p\sigma} + 2^{-1}S_{2p\pi-2p\pi} - 2^{1/2} \times 3^{-1}S_{2s-2p\sigma} . \tag{23.10}$$

From the N=N bond length of 1.23 Å and Mulliken's tables of overlap integrals (see Subsection 9.3.2), S_{2s-2s}, $S_{2p\sigma-2p\sigma}$, $S_{2p\pi-2p\pi}$, and $S_{2s-2p\sigma}$ for the azo group are evaluated as 0.369, 0.333, 0.215, and 0.387, respectively. Accordingly, the value of $S_{n,n'}$ is calculated to be −0.111 for the *trans* configuration and +0.103 for the *cis* configuration. The resonance integral between χ_n and $\chi_{n'}$ is denoted by $k_{n,n'}\beta$, in which β represents, as usual, the standard resonance integral, that is, the π–π resonance integral for a C—C bond in benzene. Then, by the use of the usual assumption of the proportionality of the resonance integral to the overlap integral, the value of $k_{\mathrm{n,n'}}$ is calculated to be -0.447 for the *trans* configuration and $+0.415$ for the *cis* configuration. Although too much reliance is not to be placed on these numerical values, it seems to be quite reasonable that the value of $k_{\mathrm{n,n'}}$ is negative for the *trans* configuration and positive for the *cis* configuration. Accordingly, for the *trans* configuration $\phi_{\mathrm{An}(+)}$ should be higher than $\phi_{\mathrm{An}(-)}$, and for the *cis* configuration the reverse should be true.

The one-electron transitions from $\phi_{\mathrm{An}(+)}$ and $\phi_{\mathrm{An}(-)}$ to $\phi_{\mathrm{A}\pi(-)}$ are subsequently referred to as the $n(+)$–π^* and $n(-)$–π^* transitions, respectively. Since the n orbitals are linear combinations of sp^2 hybrid orbitals and the antibonding π orbital is a linear combination of $2p\pi$ orbitals, the transition moments of these n–π^* transitions, $\mathbf{M}_{n(+),\pi^*}$ and $\mathbf{M}_{n(-),\pi^*}$, can be expanded into several terms with the transition moments involving atomic orbitals, of which only the transition moment involving the χ_s component of the n orbitals and the χ_z component of the π^* orbital and that involving the $\chi_{s'}$ component of the n orbitals and the $\chi_{z'}$ component of the π^* orbital are nonzero. The one-electron transition from χ_s to χ_z as well as that from $\chi_{s'}$ to $\chi_{z'}$ has a nonzero transition moment along the z axis, whose magnitude is calculated to be 0.554 Å by the use of the Slater atomic orbitals with the effective nuclear charge of 3.9 (which corresponds to the Slater μ-value of 1.95) for χ_s and χ_z:

$$M_{s,z} = 2^{1/2}\int \chi_s z \chi_z \, d\tau \doteqdot 2^{1/2} \times 0.392 \doteqdot 0.554\ (\text{Å}). \qquad (23.11)$$

$\mathbf{M}_{s,z}$ and $\mathbf{M}_{s',z'}$ contribute with the same sign to $\mathbf{M}_{n(-),\pi^*}$. Therefore, the $n(-)$–π^* transition is allowed, having a nonzero transition moment in the direction of the z axis, that is, in the direction perpendicular to the plane of the azo group, whose magnitude is calculated to be 0.320 Å:

$$M_{n(-),\pi^*} = 3^{-1/2} \times 2^{-1}(M_{s,z} + M_{s',z'}) \doteqdot 3^{-1/2} \times 0.554 \doteqdot 0.320\ (\text{Å}). \qquad (23.12)$$

On the other hand, $\mathbf{M}_{s,z}$ and $\mathbf{M}_{s',z'}$ contribute with opposite signs to $\mathbf{M}_{n(+),\pi^*}$ and hence cancel out each other. Therefore, the $n(+)$–π^* transition is forbidden. Thus, in the *trans*-azo group, the longer wavelength n–π^* transition should be forbidden, and the shorter wavelength n–π^* transition should be allowed; on the other hand, in the *cis*-azo group, the longer wavelength n–π^* transition should be allowed, and the shorter wavelength n–π^* transition should be forbidden.

The resonance integral between χ_z and $\chi_{z'}$ is denoted by $k_{\mathrm{N}\pi,\mathrm{N}\pi'}\beta$. The value of $k_{\mathrm{N}\pi,\mathrm{N}\pi'}$ is determined to be 0.862 by the use of the N=N bond length of 1.23 Å and the usual assumption that the resonance integral is proportional to the overlap integral. The Coulomb integrals for χ_z (as well as $\chi_{z'}$) and χ_n (as well as $\chi_{n'}$) are expressed as $\alpha_{N\pi} = \alpha + h_{N\pi}\beta$ and $\alpha_{Nn} = \alpha + h_{Nn}\beta$, respectively, in which α represents, as usual, the standard Coulomb integral, that is, the Coulomb integral for the carbon $2p\pi$ atomic orbital. The values of $h_{N\pi}$ and h_{Nn} are determined to be 0.60 and 1.09 (or 1.56),

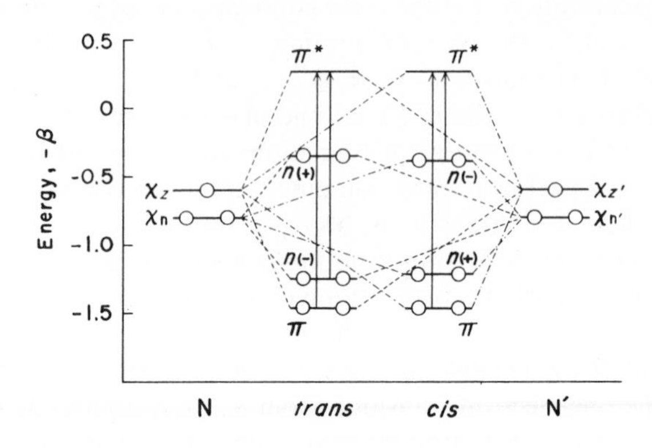

FIG. 23.2. Orbital energy levels of the *trans*- and *cis*-azo groups. The vertical arrows indicate allowed transitions.

respectively, by the use of the assumption that the Coulomb integral is proportional to the valence state ionization potential (see Subsection 9.2.2, especially Table 9.3). However, in order to produce a better agreement of calculated transition energies with observed ones in various azo compounds, a lower value, probably a value of about 0.8, seems to be more pertinent to h_{Nn}. By the use of these parameter values ($h_{N\pi} = 0.600$, $h_{Nn} = 0.800$, $k_{N\pi,N\pi'} = 0.862$, $k_{Nn,Nn'}(trans) = -0.447$, $k_{Nn,Nn'}(cis) = 0.415$) the orbital energy-level diagram for the azo group can be constructed as shown in Fig. 23.2.

23.2. Aliphatic Azo Compounds

The simplest aliphatic azo compound, azomethane $CH_3—N{=}N—CH_3$, has a weak absorption band at about 350 mμ (in the vapor phase, $\lambda_{max} = 347$ mμ; in ethanol solution, $\lambda_{max} = 340$ mμ, $\epsilon_{max} = 4.5$;[1] in methyl chloride solution at -40°C, $\lambda_{max} = 350$ mμ, $\epsilon_{max} = 25$).[2] This band is almost

[1] H. C. Ramsperger, *J. Am. Chem. Soc.* **50,** 123 (1928).
[2] O. Ruff and W. Willenberg, *Ber.* **73B,** 724 (1940).

certainly an n–π^* band. Hexafluoroazomethane, $CF_3—N{=}N—CF_3$, in methyl chloride solution at $-40°C$ has a similar band at $\lambda_{max} = 365$ mμ ($\epsilon_{max} = 120$).[2] Azoisopropane, $(CH_3)_2CH—N{=}N—CH(CH_3)_2$, has also a similar band at $\lambda_{max} = 357$ mμ ($\epsilon_{max} = 17$), which has an integrated intensity corresponding to a transition moment length of 0.03 Å.[3] In the spectrum of this compound in the vapor phase there are two other bands at shorter wavelengths, one at $\lambda_{max} = 196$ mμ with a transition moment length of 0.35 ± 0.05 Å and the other at 163 mμ with a transition moment length of 0.62 ± 0.07 Å.[3]

These simple azo compounds have undoubtedly a *trans* configuration. The very weak band at about 350 mμ is presumably due to the symmetry-forbidden $n(+)$–π^* transition, and the weak band at 196 mμ and the moderately intense band at 163 mμ are presumably due to the symmetry-allowed $n(-)$–π^* and π–π^* transitions, respectively.

As elucidated in the preceding section, it is expected that, while the longer-wavelength n–π^* band in *trans*-azo compounds should be forbidden, that in *cis*-azo compounds should be allowed. In accordance with this expectation, the longer wavelength n–π^* band of alicyclic azo compounds in which the azo group has a fixed *cis* configuration has a much higher intensity than that of open-chain azo compounds such as azomethane and azoisopropane. For example, the ϵ_{max} of the n–π^* band appearing at $\lambda_{max} = 329$ mμ in the spectrum of 3,5-diphenyl-1,2-dehydropyrazolidine (I) in ethanol solution has a value of 291,[4] which is to be compared with the value of 5–20 for the 350-mμ band of open-chain azo compounds.

N=N
C_6H_5—CH CH—C_6H_5
C
H_2

I

23.3. Azobenzenes

23.3.1. Spectra of *Trans*- and *Cis*-Azobenzene

When the azo group conjugates with one and then two benzene rings, the longer-wavelength n–π^* band progressively shifts to longer wavelength and

[3] M. B. Robin and W. T. Simpson, *J. Chem. Phys.* **36,** 580 (1962).
[4] C. G. Overberger and J. Anselme, *J. Am. Chem. Soc.* **84,** 869 (1962).

TABLE 23.1

SPECTRA OF SOME AZO COMPOUNDS IN ETHANOL

	The first n–π^* band			The first π–π^* band		
Compound	λ_{max}, mμ	ϵ_{max}	f	λ_{max}, mμ	ϵ_{max}	f
$CH_3N{=}NCH_3$ (*trans*)	340	4.5	—	—	—	—
$C_6H_5N{=}NCH_3$ (*trans*)	404	90	—	260	7800	—
$C_6H_5N{=}NC_6H_5$ (*trans*)	443	510	0.016	320	21,300	0.561
$C_6H_5N{=}NC_6H_5$ (*cis*)	433	1520	0.025	281	5260	0.172

increases in intensity. At the same time, π–π^* bands characteristic of the conjugated systems appear at fairly long wavelengths (see Table 23.1).

The spectra of *trans*- and *cis*-azobenzene in ethanol solution and the spectrum of *trans*-azobenzene in the solid state measured by the KCl-disk technique are shown in Fig. 23.3. The values of λ_{max} of the first n–π^* and first π–π^* bands of *trans*-azobenzene (measured by the author) and those of the first π–π^* band of *cis*-azobenzene (measured by Birnbaum et al.[5]) in various solvents are shown in Table 23.2.

Apart from the n–π^* band, which is present in azobenzenes but not in stilbenes, the spectra of *trans*- and *cis*-azobenzene are closely similar to the spectra of *trans*- and *cis*-stilbene, respectively, as might be expected from the

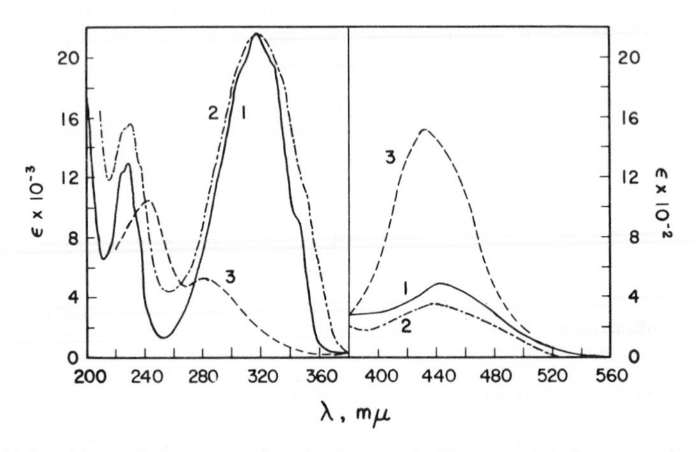

FIG. 23.3. Absorption spectra of azobenzenes. 1, the spectrum of *trans*-azobenzene in ethanol; 2, the KCl-disk spectrum of *trans*-azobenzene; 3, the spectrum of *cis*-azobenzene in ethanol.

[5] P. P. Birnbaum, J. H. Linford, and D. W. G. Style, *Trans. Faraday Soc.* **49**, 735 (1953).

similarity of nuclear framework of these compounds. The first π–π^* band (the A band) of the *cis* isomer is at considerably shorter wavelength and has a markedly lower intensity than that of the *trans* isomer. Furthermore, the A band of the *cis* isomer is structureless, while that of the *trans* isomer has a rather well-developed vibrational fine structure.

In contrast to the case of *trans*-stilbene (see Subsection 14.4.2), the KCl-disk spectrum of *trans*-azobenzene is not significantly different from the solution spectrum, except for small wavelength displacement of bands. Especially,

TABLE 23.2

THE WAVELENGTHS OF THE FIRST n–π^* AND FIRST π–π^* BANDS OF *TRANS*- AND *CIS*-AZOBENZENE IN VARIOUS SOLVENTS

	Trans-azobenzene		*Cis*-azobenzene
Solvent	The n–π^* band λ_{max}, mμ	The π–π^* band λ_{max}, mμ	The π–π^* band λ_{max}, mμ
Hexane	444	317	280
Ethanol	440	317	285
Chloroform	440	319	290
Benzene	443	321	—
(KCl-disk)	440	319	—

it is to be noted that no marked redistribution of intensity among the vibrational subbands of the A band appears to occur with *trans*-azobenzene on going from the solution spectrum to the KCl-disk spectrum, the third subband appearing as the most intense peak of the A band in both the spectra.

X-ray crystal analysis shows that the *trans*-azobenzene molecule in the crystalline state has a planar conformation, in which the length of the N═N bond is 1.23 Å and the length of each C—N bond is 1.41 Å.[6] Judging from the fact that the solution spectrum is very similar to the KCl-disk spectrum and that the wavelength shifts of bands on going from the solution to the KCl-disk spectrum are small, we may infer that the most probable conformation of *trans*-azobenzene in solution is also planar or nearly planar.

On the other hand, the *cis*-azobenzene molecule cannot be planar. Actually, X-ray crystal analysis shows that the *cis*-azobenzene molecule has a nonplanar conformation, in which the length of the N═N bond is 1.23 Å, the length of each C—N bond is 1.46 Å, and the interplanar angle between each benzene

[6] J. M. Robertson, *J. Chem. Soc.* **1939,** 232.

ring and the azo group is 56°.[7] Evidently, as in the case of stilbene, the difference in the geometry of *trans*- and *cis*-azobenzene should be responsible for the differences in the position, intensity, and shape of the A bands of these compounds.

The much higher intensity of the first n–π^* band of *cis*-azobenzene as compared with that of *trans*-azobenzene suggests that a large part of the intensity has been borrowed from allowed π–π^* transitions through mixing of the n orbital with π orbitals, which can occur in nonplanar conformations but cannot occur in planar conformations.

23.3.2. Simple MO Treatment of Azobenzenes

The π system of azobenzene, B–A–B′, is similar to that of stilbene, B–E–B′: models of *trans*, *cis*, and abstract linear forms of the skeleton of the system can be considered to be given by models for stilbene shown in Fig. 14.2, if the vinylene group (E) has been replaced by the azo group (A). The π orbitals of the two benzene rings, B and B′, are denoted by ϕ_m and $\phi_{m'}$, respectively, and the antibonding and bonding π orbitals of the ethylenic bond are denoted by $\phi_{E\pi(-)}$ and $\phi_{E\pi(+)}$, respectively. The forms of the benzene π orbitals shown in Table 3.5 are also adopted here. As mentioned in Section 23.1, the antibonding and bonding π orbitals of the azo group are denoted by $\phi_{A\pi(-)}$ and $\phi_{A\pi(+)}$, respectively. The π orbitals of the stilbene system can be considered as arising from interaction of the π orbitals localized on fragments B, E, and B′, that is, from interaction of ϕ_m's, $\phi_{E\pi}$'s, and $\phi_{m'}$'s. Analogously, when the interaction with the n orbitals of the azo group in nonplanar conformations is neglected, the π orbitals of the azobenzene system can be considered as arising from interaction of the π orbitals localized on fragments B, A, and B′, that is, from interaction of ϕ_m's, $\phi_{A\pi}$'s, and $\phi_{m'}$'s.

It is convenient to take corresponding π orbitals localized on B and B′ in the symmetry combinations:

$$2^{-1/2}(\phi_m - \phi_{m'}) \equiv \phi_{m(-)}\,; \tag{23.13}$$

$$2^{-1/2}(\phi_m + \phi_{m'}) \equiv \phi_{m(+)}\,. \tag{23.14}$$

These symmetry combinations are of course degenerate and have the same energy as the original localized orbitals, ϕ_m and $\phi_{m'}$, since ϕ_m and $\phi_{m'}$ do not interact with each other. In the abstract linear models, with respect to the rotation of 180° about the z axis ($C_2(z)$) $\phi_{m(-)}$'s as well as $\phi_{A\pi(-)}$ and $\phi_{E\pi(-)}$ are antisymmetric, and $\phi_{m(+)}$'s as well as $\phi_{A\pi(+)}$ and $\phi_{E\pi(+)}$ are symmetric.

[7] G. C. Hampson and J. M. Robertson, *J. Chem. Soc.* **1941**, 409.

Evidently, $\phi_{m(-)}$'s can interact only with $\phi_{A\pi(-)}$ and $\phi_{E\pi(-)}$, and $\phi_{m(+)}$'s can interact only with $\phi_{A\pi(+)}$ and $\phi_{E\pi(+)}$. In the azobenzene system, the interaction matrix element between $\phi_{m(-)}$ and $\phi_{A\pi(-)}$, as well as that between $\phi_{m(+)}$ and $\phi_{A\pi(+)}$, can be expressed as follows:

$$\begin{aligned}\langle\phi_{m(-)}|\,\mathbf{h}\,|\phi_{A\pi(-)}\rangle = \langle\phi_{m(+)}|\,\mathbf{h}\,|\phi_{A\pi(+)}\rangle &= 2^{-1}(b_{m,1}k_{N\pi,1} + b_{m',1'}k_{N\pi',1'})\beta \\ &= b_{m,1}k_{N\pi,1}\beta, \end{aligned} \tag{23.15}$$

where $b_{m,1}$ is the coefficient of the carbon $2p\pi$ AO at the 1 position (χ_1) in ϕ_m, and $k_{N\pi,1}$ is the resonance parameter between the nitrogen $2p\pi$ AO at

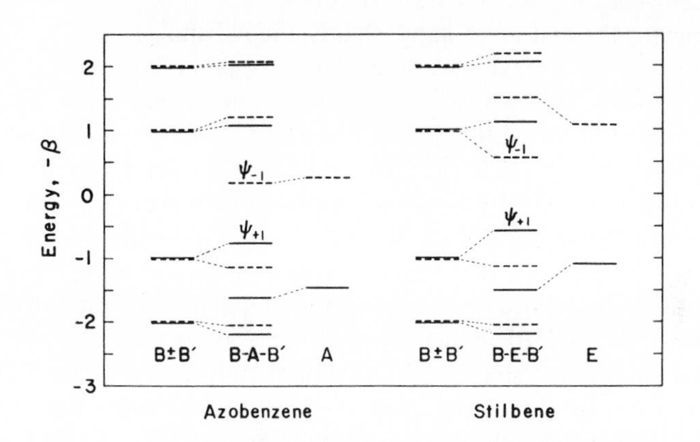

FIG. 23.4. Orbital energy level diagrams of azobenzene (B–A–B′) and stilbene (B–E–B′). The energy levels of the orbitals which are symmetrical with respect to $C_2(z)$ in the abstract linear models are shown with solid lines, and those of the orbitals which are antisymmetrical are shown with broken lines.

the α position ($\chi_{N\pi}$) and χ_1. Quite analogously, in the stilbene system, the interaction matrix element between $\phi_{m(-)}$ and $\phi_{E\pi(-)}$ as well as that between $\phi_{m(+)}$ and $\phi_{E\pi(+)}$ can be expressed as follows:

$$\langle\phi_{m(-)}\,|\mathbf{h}|\,\phi_{E\pi(-)}\rangle = \langle\phi_{m(+)}\,|\mathbf{h}|\,\phi_{E\pi(+)}\rangle = b_{m,1}k_{\alpha,1}\beta, \tag{23.16}$$

where $k_{\alpha,1}$ is the resonance parameter between the carbon $2p\pi$ AO's at the α and 1 positions.

The orbital energy level diagrams for azobenzene and stilbene as composite conjugated systems are shown in Fig. 23.4. In construction of these diagrams, the following parameter values were used: for the azo group, $h_{N\pi} = 0.600$, $k_{N\pi,N\pi'} = 0.862$; for the ethylenic bond, $k_{\alpha,\alpha'} = 1.080$. Solid lines represent the energy levels of the orbitals symmetrical with respect to $C_2(z)$, and broken lines represent the energy levels of the orbitals antisymmetrical with respect to $C_2(z)$. Benzene π orbitals ϕ_{-2}, $\phi_{-2'}$, ϕ_{+2}, and $\phi_{+2'}$ do not interact with the π orbitals of the azo group and with the π orbitals of the ethylenic bond, since these benzene π orbitals have no contribution of χ_1 and $\chi_{1'}$. The energy levels of these benzene π orbitals have been omitted from the energy level diagrams in Fig. 23.4.

The highest occupied π orbital (ψ_{+1}) of the azobenzene system as well as that of the stilbene system is $\phi^{\text{p}}_{+1(+)}$, that is, the orbital arising from perturbation of $\phi_{+1(+)}$. On the other hand, the lowest vacant π orbital (ψ_{-1}) of the azobenzene system is $\phi^{\text{p}}_{\text{A}\pi(-)}$, while that of the stilbene system is $\phi^{\text{p}}_{-1(-)}$. In each system the one-electron transition from ψ_{+1} to ψ_{-1} is considered to be responsible for the first π–π^* band, that is, the A band. It is to be noted that, while the $\psi_{+1} \rightarrow \psi_{-1}$ transition in the stilbene system is a perturbed LE transition, the $\psi_{+1} \rightarrow \psi_{-1}$ transition in the azobenzene system is a CT transition, which is accompanied with a partial electron transfer from the two benzene rings to the azo group. Evidently, the probability of the $\psi_{+1} \rightarrow \psi_{-1}$ transition in the azobenzene system will rapidly decrease as the two benzene rings are rotated out of the plane of the azo group, and it will vanish when the interplanar angle (θ) between each benzene ring and the azo group becomes 90°.

In the stilbene system, as the extent of the conjugative interaction among the fragments B, E, and B′ increases, the energy of ψ_{-1} (that is, $\phi^{\text{p}}_{-1(-)}$) will rapidly become lower, since $\phi_{-1(-)}$ is subject only to the lowering effect by $\phi_{\text{E}\pi(-)}$, and the energy of ψ_{+1} (that is, $\phi^{\text{p}}_{+1(+)}$) will rapidly become higher, since $\phi_{+1(+)}$ is subject only to the raising effect by $\phi_{\text{E}\pi(+)}$. That is, as the extent of the conjugative interaction increases, the energy of the $\psi_{+1} \rightarrow \psi_{-1}$ transition will rapidly decrease. It follows that twist of the benzene rings out of the ethylenic bond plane will exert a strong hypsochromic effect on the A band.

On the other hand, in the azobenzene system, as the extent of the conjugative interaction among the fragments B, A, and B′ increases, the energy of ψ_{-1} (that is, $\phi^{\text{p}}_{\text{A}\pi(-)}$) will become lower only slightly, since $\phi_{\text{A}\pi(-)}$ is subject to the depressing effect by $\phi_{-1(-)}$ and $\phi_{-3(-)}$ and to the slightly smaller raising effect by $\phi_{+1(-)}$ and $\phi_{+3(-)}$, while the energy of ψ_{+1} (that is $\phi^{\text{p}}_{+1(+)}$) will rapidly become higher, similarly to the energy of the corresponding orbital of the stilbene system. This means that the steric effect on the A band of the azobenzene system will also be hypsochromic but will be less effective than that on the A band of the stilbene system.

Thus it will be understood that the values of λ_{max} and ϵ_{max} of the A band of *cis*-azobenzene are appreciably smaller than those of the A band of *trans*-azobenzene and that the difference between the λ_{max} values of *trans*- and *cis*-azobenzene is comparatively small, in spite of the large interplanar angle in *cis*-azobenzene. It seems to be important that the variation in the energy of the $\psi_{+1} \rightarrow \psi_{-1}$ transition in the azobenzene system is mainly due to the variation in the energy of ψ_{+1} and that the energy of ψ_{-1} is comparatively insensitive to the variation in the extent of the conjugative interaction among the fragments.

The first n–π^* band of *trans*-azobenzene is probably due to the one-electron transition from $\phi_{\mathrm{An}(+)}$ to ψ_{-1}, and that of *cis*-azobenzene is probably due to the one-electron transition from $\phi_{\mathrm{An}(-)}$ to ψ_{-1}. The fact that the first n–π^* bands of *trans*- and *cis*-azobenzene are at nearly the same wavelength is considered to indicate that the energies of ψ_{-1} in these two isomers are close to each other, since the energy of $\phi_{\mathrm{An}(+)}$ in the *trans* isomer and that of $\phi_{\mathrm{An}(-)}$

TABLE 23.3

RESULT OF THE SIMPLE LCAO MO CALCULATION ON *TRANS*- AND *CIS*-AZOBENZENE

	w_{-1}	w_{+1}	Δw_{11}	M_{11}^2, Å	RE, $-\beta$
trans	+0.183	−0.757	0.940	3.528 (x, y)	0.570
cis	+0.235	−0.914	1.149	0.752 (y)	0.150

in the *cis* isomer are probably close to each other (see Section 23.1). Furthermore, from the above considerations it is expected that the energy of ψ_{-1} will not markedly vary and the energy of ψ_{+1} will appreciably vary when one or two phenyl groups of azobenzene are replaced by *para*-substituted phenyl groups or other aryl groups such as naphthyl groups. This expectation seems to have been borne out experimentally, as will be shown in Section 23.4.

The value of $k_{\mathrm{N}\pi,1}$ should depend on the length (R) and twist angle (θ) of the C—N bond. The value of this parameter for *trans*-azobenzene, $k_{\mathrm{N}\pi,1}$ (1.41 Å, 0°), is determined, from the usual assumption of the proportionality of the resonance integral to the overlap integral, to be 0.749. Analogously, the value of $k_{\mathrm{N}\pi,1}$ for the case where $R = 1.46$ Å and $\theta = 0°$ is determined to be 0.681, and hence the value of $k_{\mathrm{N}\pi,1}$ for *cis*-azobenzene, $k_{\mathrm{N}\pi,1}$ (1.46 Å, 56°), is determined to be 0.681 cos 56° $\doteqdot$ 0.381. As mentioned already, the values of the parameters for the azo group are determined as follows: $h_{\mathrm{N}\pi} = 0.600$, $k_{\mathrm{N}\pi,\mathrm{N}\pi'} = 0.862$. The simple LCAO MO calculation by the use of these parameter values leads to the result summarized in Table 23.3.

The transition moment of the $\psi_{+1} \rightarrow \psi_{-1}$ transition, $\mathbf{M}_{11}$, in *trans*-azobenzene has nonzero components in the directions of the x and y axes. The squares of the x and y components are as follows: $M_{11,x}^2 = 1.638$ Å^2, $M_{11,y}^2 = 1.890$ Å^2. This means that the transition moment is in the molecular plane (the xy plane) and in the direction making an angle of about 43° with the N=N bond axis (the y axis), that is, in the direction approximately along the line connecting the two *para* positions of the molecule. Robin and Simpson[3] found by measuring the spectrum of *trans*-azobenzene in the crystalline state with polarized light that the transition moment of the A band of this compound is in the molecular plane and in the direction making an angle of $46 \pm 3°$ with the N=N bond axis. It is rather surprising that the direction of the transition moment predicted from the simple MO theory is in such good agreement with that found experimentally.

The difference in the extra-resonance energy (RE) between *trans*- and *cis*-azobenzene is calculated to be $0.420\,(-\beta)$. This value is to be compared with the actual energy difference between these two isomers, which has been determined to be 12 kcal/mole[8] (the heat of *cis–trans* isomerization) or 9.9 kcal/mole[9] (the difference in the heat of combustion at 25°C).

23.3.3. Interaction between n Orbitals and π Orbitals in the Azobenzene System in Nonplanar Conformations

Up to this point the interaction between the n orbitals and π orbitals has been neglected. When $\theta = 0°$, this interaction does not occur. However, when $\theta \neq 0°$, carbon $2p\pi$ AO's χ_1 and $\chi_{1'}$ should interact with the appropriate p components of nitrogen n AO's χ_{n} and $\chi_{\mathrm{n'}}$, respectively, and consequently, π orbitals should interact with the n orbitals. Thus, strictly speaking, the symmetry distinction between π and n orbitals is lost in nonplanar conformations.

Now the direction of the N_α—C_1 bond axis is defined as the direction of the a axis, and the direction that is in the plane of the azo group (the xy plane) and is perpendicular to the a axis is defined as the direction of the b axis. The $2p$ AO on the nitrogen atom at the α position having the symmetry axis in the direction of the a axis is denoted by χ_a, and that having the symmetry axis in the direction of the b axis is denoted by χ_b. Then, the sp^2 lone-pair orbital on the nitrogen atom, χ_{n}, which is given by expression (23.4), can also be expressed as follows:

$$\chi_{\mathrm{n}} = 3^{-1/2}\chi_s + 2^{-1/2}\chi_b - 6^{-1/2}\chi_a\,. \qquad (23.17)$$

Since χ_1 cannot interact with χ_s and χ_a, the resonance parameter between χ_1 and χ_{n}, $k_{\mathrm{n},1}$, can be expressed as follows:

$$\begin{aligned} k_{\mathrm{n},1} = \langle\chi_{\mathrm{n}}|\,\mathbf{h}\,|\chi_1\rangle/\beta &= 2^{-1/2}\langle\chi_b|\,\mathbf{h}\,|\chi_1\rangle/\beta \\ &= 2^{-1/2}k_{\mathrm{N}\pi,1}(R, 0°)\sin\theta. \end{aligned} \qquad (23.18)$$

[8] G. S. Hartley, *J. Chem. Soc.* **1938,** 633.

[9] R. J. Corruccini and E. C. Gilbert, *J. Am. Chem. Soc.* **61,** 2925 (1939).

When the two phenyl groups are symmetrically twisted out of the plane of the azo group, just as in the symmetrically twisted models of the stilbene system illustrated in Fig. 14.3, the upper n MO of the azo group, $\phi_{\mathrm{An}(u)}$, which is $\phi_{\mathrm{An}(+)}$ in the *trans* form and is $\phi_{\mathrm{An}(-)}$ in the *cis* form, interacts only with the symmetrical π orbitals, and the lower n MO of the azo group, $\phi_{\mathrm{An}(l)}$, which is $\phi_{\mathrm{An}(-)}$ in the *trans* form and is $\phi_{\mathrm{An}(+)}$ in the *cis* form, interacts only with the antisymmetrical π orbitals. The interaction between $\phi_{\mathrm{An}(u)}$ and ψ_{+1} is important. The matrix element for this interaction is given by $c_{+1,1(+)}k_{\mathrm{n},1}\beta$, in which $c_{+1,1(+)}$ represents the coefficient of the group orbital $\chi_{1(+)} \equiv 2^{-1/2}(\chi_1 + \chi_{1'})$ in ψ_{+1}. For *cis*-azobenzene ($R = 1.46$ Å, $\theta = 56°$) this interaction matrix element is calculated to be $0.210\,\beta$. As a result of this interaction, $\phi_{\mathrm{An}(u)}$ is raised, and ψ_{+1} is lowered. The resulting perturbed orbitals can be expressed as follows:

$$\begin{aligned}\phi^{\mathrm{p}}_{\mathrm{An}(u)} &= (1 + \lambda^2)^{-1/2}(\phi_{\mathrm{An}(u)} - \lambda\psi_{+1}) \\ &\doteqdot (1 - \lambda^2)^{1/2}(\phi_{\mathrm{An}(u)} - \lambda\psi_{+1}); \end{aligned} \tag{23.19}$$

$$\begin{aligned}\psi^{\mathrm{p}}_{+1} &= (1 + \lambda^2)^{-1/2}(\psi_{+1} + \lambda\phi_{\mathrm{An}(u)}) \\ &\doteqdot (1 - \lambda^2)^{1/2}(\psi_{+1} + \lambda\phi_{\mathrm{An}(u)}). \end{aligned} \tag{23.20}$$

If values of 0.800 and 0.415 are used for h_{Nn} and $k_{\mathrm{n,n'}}$, respectively, the w value of $\phi_{\mathrm{An}(u)}$ in the *cis*-azo group is calculated to be -0.385. The w value of ψ_{+1} in *cis*-azobenzene has been calculated to be -0.914, as shown in Table 23.3. From these values the w values of $\phi^{\mathrm{p}}_{\mathrm{An}(u)}$ and ψ^{p}_{+1} are calculated to be -0.312 and -0.987, respectively, and the value of λ in expressions (23.19) and (23.20) is calculated to be 0.329. Accordingly, since $w_{-1} = +0.235$, the energies of the one-electron transitions from $\phi^{\mathrm{p}}_{\mathrm{An}(u)}$ and ψ^{p}_{+1} to ψ_{-1} are calculated to be 0.547 ($-\beta$) and 1.222 ($-\beta$), respectively. The transition moment of the one-electron transition from $\phi_{\mathrm{An}(u)}$ to ψ_{-1} is in the z direction, and the square of its magnitude is 0.090 Å^2. The transition moment of the one-electron transition from ψ_{+1} to ψ_{-1} is in the y direction, and the square of its magnitude is 0.752 Å^2. Accordingly, the dipole strength of the one-electron transition from $\phi^{\mathrm{p}}_{\mathrm{An}(u)}$ to ψ_{-1} is calculated to be 0.162 Å^2 ($\doteqdot 0.090(1 - \lambda^2) + 0.752\lambda^2$), and that of the one-electron transition from ψ^{p}_{+1} to ψ_{-1} is calculated to be 0.680 Å^2 ($\doteqdot 0.090\lambda^2 + 0.752(1 - \lambda^2)$). Thus it can be said that the first n–π^* band of *cis*-azobenzene has a comparatively high intensity not only because the unperturbed n–π^* transition in the *cis* configuration is symmetry-allowed but also because the transition borrows a part of the intensity from the first π–π^* transition.

The first n–π^* band of *trans*-azobenzene is much weaker than that of *cis*-azobenzene, but is considerably stronger than that of azomethane. Since the first n–π^* transition in the *trans* configuration is symmetry-forbidden, this fact suggests that the intensity of the n–π^* band of *trans*-azobenzene has been borrowed from the first π–π^* band (the A band). As mentioned already, *trans*-azobenzene has a planar equilibrium conformation in the crystalline state. However, torsional vibrations of the benzene rings about the C—N bonds will allow $\phi_{An(u)}$ to mix with ψ_{+1}, as expressed by Eqs. (23.19) and (23.20). The dipole strength of the one-electron transition from $\phi^{P}_{An(u)}$ to ψ_{-1} is approximately given by $\lambda^2 M_{11}^2$, in which M_{11}^2 represents the dipole strength of the unperturbed first π–π^* transition, that is, the one-electron transition from ψ_{+1} to ψ_{-1}. Thus the torsional vibrations allow the first n–π^* transition to appropriate a small fraction of the probability of the first π–π^* transition. When the parameter values $h_{Nn} = 0.800$ and $k_{n,n'}(trans) = -0.447$ are used, the w value of $\phi_{An(u)}$ is calculated to be -0.353. The w value of ψ_{-1} in *trans*-azobenzene ($R = 1.41$ Å, $\theta = 0°$) is $+0.183$, as shown in Table 23.3. By the use of these values, for various values of θ the energy (ΔE, in units of $-\beta$) and dipole strength (M^2, in units of Å^2) of the first n–π^* transition in *trans*-azobenzene are calculated to be as follows: for $\theta = 0°$, $\Delta E = 0.536$, $M^2 = 0$; for $\theta = 5°$, $\Delta E = 0.535$, $M^2 = 0.010$; for $\theta = 10°$, $\Delta E = 0.532$, $M^2 = 0.036$.

Having the first n–π^* band become allowed through mixing with the first π–π^* band implies that the two bands should have the same polarization. Actually, Robin and Simpson[3] found that the transition moment of the first n–π^* band of *trans*-azobenzene is in the direction that is in the molecular plane and makes an angle of $47 \pm 3°$ with the N=N bond axis. This direction agrees almost completely with the direction of the transition moment of the first π–π^* band.

23.4. Azobenzene Analogs and Derivatives

23.4.1. Naphthyl Analogs of Azobenzene

Data on the first n–π^* and first π–π^* bands of *trans*-azobenzene and its naphthyl analogs, measured by Schulze et al.,[10] are shown in Table 23.4. In this table, 1-Naph and 2-Naph represent the 1-naphthyl and 2-naphthyl groups, respectively. As is seen in this table, while the position of the first π–π^* band (the A band) considerably varies from compound to compound, the position of the first n–π^* band remains nearly constant at about 444–455

[10] J. Schulze, F. Gerson, J. N. Murrell, and E. Heilbronner, *Helv. Chim. Acta* **44,** 428 (1961).

$m\mu$. These facts are well accounted for on the basis of the considerations made in Subsection 23.3.2. The upper MO of the transitions responsible for these two bands, ψ_{-1}, is the perturbed orbital originating from the antibonding π orbital of the azo group, $\phi^{P}_{A\pi(-)}$, and its energy is considered to be almost independent of the kind of the aryl groups. Of course, the lower orbital of the first n–π^* transition, $\phi_{An(+)}$, is unaffected by the aryl groups. On the other hand, the lower orbital of the first π–π^* transition, ψ_{+1}, is the orbital

TABLE 23.4

SPECTRA OF *TRANS*-AZOBENZENE AND ITS NAPHTHYL ANALOGS IN CYCLOHEXANE

Compound	The first n–π^* band		The first π–π^* band	
	λ_{max}, $m\mu$	ϵ_{max}	λ_{max}, $m\mu$	ϵ_{max}
Ph—N=N—Ph	444	450	316	22,900
Ph—N=N—1-Naph	454	890	371	12,000
Ph—N=N—2-Naph	446	910	329	22,500
1-Naph—N=N—1-Naph	464	1470	397	16,500
1-Naph—N=N—2-Naph	455	1500	381	19,600
2-Naph—N=N—2-Naph	445	1490	334	28,100

arising from perturbation of the highest bonding π orbitals of the aryl groups, and hence its energy largely depends on the kind of the aryl groups. By the way, the fact that the first n–π^* bands of azonaphthalenes are much more intense than that of azobenzene suggests that the azonaphthalenes will undergo larger torsional vibrations of the aryl groups about the C—N bonds than does azobenzene.

23.4.2. SUBSTITUTED AZOBENZENES

Data on the first n–π^* and first π–π^* bands of *trans*- and *cis*-azobenzene and their *para*-monosubstituted derivatives, measured by Birnbaum et al.,[5] are shown in Table 23.5. The relations between the spectra of *trans*- and *cis*-azobenzene are almost exactly repeated between the spectra of the *trans* and *cis* isomers of each derivative. Thus, the first n–π^* band of a *cis* compound is more intense than that of its *trans* isomer, and the first π–π^* band of a *cis* compound is at considerably shorter wavelength and has a lower intensity than that of its *trans* isomer. Both in the *trans* series and in the *cis* series, the first π–π^* band shifts to longer wavelength as the *para* substituent is changed in the order H, F, Cl, Br, I, CH_3, C_2H_5O, but the position of the first n–π^* band remains almost unchanged. This indicates that introduction of the

TABLE 23.5

SPECTRA OF *PARA*-MONOSUBSTITUTED AZOBENZENES IN ETHANOL

	Trans				*Cis*			
	the n–π^* band		the π–π^* band		the n–π^* band		the π–π^* band	
p-Substituent	λ_{max}, mμ	ϵ_{max}	λ_{max}, mμ	ϵ_{max}	λ_{max}, mμ	ϵ_{max}	λ_{max}, mμ	ϵ_{max}
H	443.3	510	319.6	21,300	432.7	1518	280.6	5260
F	440.5	512	323.2	20,300	433.5	1285	287.0	4750
Cl	443.1	606	327.8	23,600	434.0	1598	289.0	5750
Br	444.8	635	328.8	27,000	437.1	1730	288.5	6200
I	445.4	710	331.3	28,000	434.8	1689	292.5	6500
CH_3	450.9	680	333.3	23,500	446.4	1665	295.8	5450
C_2H_5O	—	—	348.3	26,100	440.3	2432	316.7	7600

para substituent affects the energy of ψ_{+1} but does not substantially affect the energy of ψ_{-1}. This fact can also be well accounted for by considering the nature of these orbitals.

Data on the spectra of *trans*-azobenzene and some methylated derivatives are shown in Table 23.6. These data were estimated from spectral curves reproduced in the literature.[11] Therefore, too much reliance is not to be placed on these numerical values. These data seem, however, to show that methyl substituents at the *ortho* positions exert an appreciable bathochromic effect on the first n–π^* band and a small hypsochromic and hypochromic effect on the first π–π^* band. There is probably not so severe steric interference in *ortho*-substituted derivatives of azobenzene as in *ortho*-substituted derivatives of stilbene, since azobenzene has no hydrogen atoms at the α and α' positions,

TABLE 23.6

SPECTRA OF *TRANS*-AZOBENZENE AND ITS METHYLATED DERIVATIVES IN ETHANOL

	The first n–π^* band		The first π–π^* band	
Compound	λ_{max}, mμ	log ϵ_{max}	λ_{max}, mμ	log ϵ_{max}
Azobenzene	436	2.79	316	4.29
p,p'-Azotoluene	435	2.96	338	4.31
m,m'-Azotoluene	438	2.75	323	4.24
o,o'-Azotoluene	460	2.73	331	4.17
Azomesitylene	466	2.96	328	4.12

[11] P. Grammaticakis, *Bull. soc. chim. France* [5] **18,** 951 (1951).

TABLE 23.7

THE FIRST π–π^* BANDS OF *p,p'*-AZOPHENOL AND ITS METHYLATED DERIVATIVES IN ETHANOL

Compound	λ_{max}, mμ	log ϵ_{max}	ϵ_{max}
4,4'-Dihydroxyazobenzene	365	4.3	—
2,6-Dimethyl-4,4'-dihydroxyazobenzene	366	4.36	22,700
2,6,2',6'-Tetramethyl-4,4'-dihydroxyazobenzene	357	4.25	17,900

in contrast to stilbene. However, the foregoing effects of *ortho* methyl substituents seem to suggest the existence of some steric interference in the *ortho*-substituted azobenzenes.

A small hypsochromic and hypochromic effect of *ortho* methyl substituents on the first π–π^* band is also found in comparison of the spectra of 4,4'-dihydroxyazobenzene (*p,p'*-azophenol) and its *ortho*-methylated derivatives (see Table 23.7).

23.5. Azoxy Compounds

Spectral data on azomethane,[1] its mono-oxide (azoxymethane),[12] and its dioxide (nitrosomethane dimer)[13] are compared in Table 23.8. As is seen, the π–π^* band of azomethane is markedly shifted to longer wavelength when one and then two oxygen atoms are introduced. The formal positive charge

TABLE 23.8

SPECTRA OF AZOMETHANE AND ITS OXIDES

Compound	Solvent	The first n–π^* band		The first π–π^* band	
		λ_{max}, mμ	ϵ_{max}	λ_{max}, mμ	ϵ_{max}
CH_3—N=N—CH_3 (*trans*)	Ethanol	340	4.5	—	—
CH_3—N(→O)=N—CH_3 (*trans*)	Ethanol	274	43	217	7250
CH_3—N(→O)=N(→O)—CH_3 (*trans*)	Water	—	—	276	10,700
(*cis*)	Water	—	—	265	10,000

[12] B. W. Langley, B. Lythgoe, and L. S. Rayner, *J. Chem. Soc.* **1952,** 4191.

[13] H. T. J. Chilton, B. G. Gowenlock, and J. Trotman, *Chem. & Ind.* (*London*) **1955,** 538; B. G. Gowenlock and J. Trotman, *J. Chem. Soc.* **1956,** 1670.

brought about on the nitrogen atom by the formation of an N—O coordinate bond will exert almost equal lowering effects on the antibonding and the bonding π orbital of the azo group, that is, $\phi_{A\pi(-)}$ and $\phi_{A\pi(+)}$. However, if the lone-pair orbital of π character on the oxygen atom (ϕ_{On}) is lower in energy than $\phi_{A\pi(+)}$, it will exert a larger raising effect on $\phi_{A\pi(+)}$ than on $\phi_{A\pi(-)}$. In this case, the lowest-energy π–π^* transition in the whole π system is the transition from $\phi^p_{A\pi(+)}$ (the orbital arising from perturbation of $\phi_{A\pi(+)}$) to $\phi^p_{A\pi(-)}$ (the orbital arising from perturbation of $\phi_{A\pi(-)}$), and its energy will be smaller than the energy of the π–π^* transition in the azo group, that is, the transition from $\phi_{A\pi(+)}$ to $\phi_{A\pi(-)}$. If ϕ_{On} is higher in energy than $\phi_{A\pi(+)}$, lying between $\phi_{A\pi(-)}$ and $\phi_{A\pi(+)}$, then the lowest-energy π–π^* transition in the whole π system is the transition from ϕ^p_{On} (the orbital arising from perturbation of ϕ_{On}) to $\phi^p_{A\pi(-)}$. In this case too, the energy of the transition will be smaller than the energy of the π–π^* transition in the azo group.

The n–π^* band of azoxymethane is at considerably shorter wavelength and has a considerably higher intensity than that of azomethane. While the n–π^* band of azomethane is due to the symmetry-forbidden one-electron transition from the higher one of the two n MO's formed from the two nitrogen n AO's, $\phi_{An(+)}$, to $\phi_{A\pi(-)}$, the n–π^* band of azoxymethane is probably due to the weakly allowed one-electron transition from the nitrogen n AO to $\phi^p_{A\pi(-)}$. In nitrosomethane dimer, both nitrogen atoms have no n orbital, and hence there appears no n–π^* band.

The relations between the spectra of *trans*- and *cis*-azoxybenzene are somewhat similar to the relations between the spectra of *trans*- and *cis*-azobenzene. For example, the intensity of the first π–π^* band of *cis*-azoxybenzene (λ_{max} = 328 mμ, ϵ_{max} = 3530 in ethanol solution) is much lower than that of the first π–π^* band of the *trans* isomer (λ_{max} = 324 mμ, ϵ_{max} = 15,220 in ethanol solution). However, in the spectra of azoxybenzenes there appear no n–π^* bands. Just as the n–π^* band of azoxymethane is at considerably shorter wavelength than that of azomethane, the n–π^* bands of azoxybenzenes are probably at considerably shorter wavelength than the first n–π^* bands of azobenzenes, and hence they are probably masked under intense π–π^* bands.

23.6. Hydrazo Compounds

In saturated compounds having lone-pair electrons, the highest occupied orbital is the lone-pair orbital (χ_n), and the longest-wavelength band arises probably from the promotion of a nonbonding electron in the lone-pair orbital to an antibonding σ orbital, namely, an n–σ^* transition. For example, data on the spectra of ammonia and some saturated amines measured by Tannenbaum et al.[14] are shown in Table 23.9. The first bands of these compounds, which appear with comparatively low intensities in the 190–230 mμ region, are considered to be due to the first n–σ^* transition. As will be expected, these bands disappear in acidic solutions. The moderately intense bands appearing at shorter wavelength are probably due to σ–σ^* transitions.

TABLE 23.9

VAPOR SPECTRA OF AMMONIA AND AMINES

	The n–σ^* band		The σ–σ^* band	
Compound	λ_{max}, mμ	log ϵ_{max}	λ_{max}, mμ	log ϵ_{max}
NH_3	194.2	3.75	151.5	—
CH_3NH_2	215.0	2.77	173.7	3.34
$(CH_3)_2NH$	220.0	2.00	190.5	3.52
$(CH_3)_3N$	227.3	2.95	199.0	3.60

A compound containing two heteroatoms whose lone-pair orbitals can overlap with each other, such as hydrazine, absorbs at longer wavelength than the corresponding compound containing only one heteroatom. For example, the the first band of N,N'-dimethylhydrazine (λ_{max} = 245 mμ, ϵ_{max} = 1000 in the vapor state)[15] is at considerably longer wavelength than the first band of methylamine (see Table 23.9). As in the azo group, the overlap of the lone-pair AO's on the two nitrogen atoms in the hydrazo group will give rise to two MO's of different energies, one of which is higher and another which is lower in energy than the original lone-pair AO's. The first band of the dimethylhydrazine might be considered to be due to the one-electron transition from the higher one of the two MO's formed from the nitrogen lone-pair AO's to an antibonding σ orbital.

However, it is doubtful whether the hydrazo group in the ground state has such a planar or nearly planar structure that the two nitrogen lone-pair AO's can most effectively overlap with each other. It is well known that the hydrogen peroxide molecule has a nonplanar equilibrium conformation in which the two H—O—O planes make an angle close to 90° with each other.[16]

[14] E. Tannenbaum, E. M. Coffin, and A. J. Harrison, *J. Chem. Phys.* **21,** 311 (1953).
[15] W. L. Kay and H. A. Taylor, *J. Chem. Phys.* **10,** 497 (1942).
[16] S. C. Abrahams, R. L. Collin, and W. N. Lipscomb, *Acta Cryst.* **4,** 15 (1951).

Similarly, the hydrazine molecule is considered to have a nonplanar equilibrium conformation in which the two amino groups are attached to each other with the torsional angle of about 90°, although the potential energy curve for the torsional vibration about the N—N bond is considered to be rather shallow.[17] In general, according to the LCAO MO treatment with inclusion of overlap integrals, when two lone-pair orbitals combine to form two MO's, which are each occupied by two electrons, the stabilization by the low-energy MO is always less effective than the destabilization by the high-energy MO.

TABLE 23.10

COMPARISON OF THE SPECTRA OF ANILINE AND HYDRAZOBENZENE IN ETHANOL

Compound	The α band		The CT band	
	λ_{max}, mμ	log ϵ_{max}	λ_{max}, mμ	log ϵ_{max}
$C_6H_5NH_2$	284.5	3.24	234	4.06
$C_6H_5NHNHC_6H_5$	290	3.6	245	4.3

Accordingly, the total energy of a planar conformation, in which the lone-pair AO's interact fully with each other and hence the splitting of the two MO's formed from the lone-pair AO's is at its maximum, must be higher than that of a nonplanar conformation in which the *p* components of the lone-pair AO's do not overlap at all and hence the splitting of the two MO's formed from the lone-pair AO's is small.

In this connection, it is of interest that the spectrum of hydrazobenzene, C_6H_5—NH—NH—C_6H_5, is fairly similar to that of aniline (see Table 23.10). This spectral similarity between hydrazobenzene and aniline is reminiscent of the spectral similarity between benzil and benzaldehyde (see Section 21.3).

23.7. Disulfides

The relation between saturated aliphatic sulfides R—S—R and disulfides R—S—S—R resembles the relation between saturated aliphatic amines R—NH_2 and hydrazo compounds R—NH—NH—R. Dialkyl sulfides have

[17] T. Kasuya and T. Kojima, a paper presented at International Symposium on Molecular Structure and Spectroscopy, held in Tokyo, September, 1962.

a weak band at about 210 mμ ($\epsilon_{max} \simeq 1000 \sim 2000$).[18] This band is probably due to a one-electron transition from a sulfur nonbonding $3p$ orbital to a sulfur vacant orbital (probably the $4s$ orbital or perhaps one of the $3d$ orbitals) or to an antibonding σ orbital. Dialkyl disulfides show a fairly broad, weak band at considerably longer wavelength ($\lambda_{max} \simeq 250$ mμ, $\epsilon_{max} \simeq 500$).[19]

TABLE 23.11

THE FIRST BANDS OF ALIPHATIC DISULFIDES IN ETHANOL

Compound	λ_{max}, mμ	ϵ_{max}
Di-*n*-propyl disulfide	250	480
4:8-Thioctic acid (III, $n = 7$)	252	475
Tetramethylene disulfide (II, $n = 6$)	295	300
5:8-Thioctic acid (III, $n = 6$)	293	300
Trimethylene disulfide (II, $n = 5$)	334	160
6:8-Thioctic acid (III, $n = 5$)	334	160

Data on the first bands of some cyclic disulfides, polymethylene disulfides (II) and thioctic acids (III), measured by Barltrop et al.,[19] are shown in Table 23.11, in which data on the first band of an open-chain disulfide are also shown for the purpose of comparison.

$(CH_2)_{n-4}$
H_2C CH_2
S—S

II

$(CH_2)_{n-4}$
H_2C $CH-(CH_2)_{9-n}-COOH$
S—S

III

As is seen in this table, the first band of a compound having a dithio group in a seven-membered ring (III with $n = 7$) is at almost the same wavelength and has almost the same intensity as that of an open-chain disulfide, and it progressively shifts to longer wavelength and decreases in intensity as the ring size becomes smaller. This phenomenon can be explained as follows. In open-chain dialkyl disulfides, the repulsion between the nonbonding p electrons on the two adjacent sulfur atoms will probably force the molecule into a staggered equilibrium conformation in which the two C—S—S planes are inclined at an angle as large as 90°. In cyclic dithio ethers, however, as

[18] E. A. Fehnel and M. Carmack, *J. Am. Chem. Soc.* **71,** 84 (1949).

[19] J. A. Barltrop, P. M. Hayes, and M. Calvin, *J. Am. Chem. Soc.* **76,** 4348 (1954).

the ring size is reduced, steric requirements will make smaller the interplanar angle between the two C—S—S planes and hence will increase the overlap of the two sulfur nonbonding *p* orbitals, with a consequent increase in the energy of the higher one of the two MO's formed from the two nonbonding *p* orbitals. The situation is somewhat similar to that encountered with the first n–π^* bands of cyclic 1,2-diketones (see Section 21.3).

Chapter 24

Interactions between Nonneighboring Atomic Orbitals

24.1. Introduction

In the preceding chapters we saw that the simple LCAO MO theory neglecting the resonance integrals between nonneighboring atomic orbitals can interpret fairly well the spectra of conjugated compounds. In addition, we saw that the spectrum of a molecule that contains two chromophores which are not conjugated together is essentially the same as the combined spectra of the two molecules, each of which has only one of these chromophores in the absence of the other. There are, however, some compounds in which the interactions between nonneighboring atomic orbitals exert a small but important effect on the spectrum. Of course, for such interactions to occur, it is necessary for the molecule to assume a geometry which is favorable for overlap of the nonneighboring atomic orbitals. For example, as mentioned in Section 21.3, the two nonbonding p orbitals on the two oxygen atoms in 1,2-diketones can most effectively interact with each other when the diketone system has the planar *s-cis* conformation, and the interaction becomes less effective as the interplanar angle between the two carbonyl planes increases.

Even if the two chromophores in a molecule are so far apart that electronic overlapping and exchange between them are negligibly small, an electronic excitation in one chromophore is not necessarily independent of that in the other chromophore. In some cases the long-range interaction of exciton type between excited states of the two nonconjugated chromophores in a molecule is considered to exert an important effect on the spectrum of the molecule. As mentioned in Section 20.1, the exciton-type interaction is highly sensitive to the relative orientations of the electric moments of the two transitions, and hence to the relative orientations of the two chromophores in the molecule. Thus, the extent of the interaction is at its maximum when the two transition moments are parallel to each other or lie on a straight line, and it is zero when the two transition moments make a right angle with each other.

This last chapter will deal with the effects of such "nonneighbor interactions," that is, interactions between nonneighboring atomic orbitals and between nonconjugated chromophores on the spectra, with special references to the relations between the effects and the geometry of molecules.

24.2. Nonneighbor Interactions in Carbonyl Compounds

24.2.1. Interaction between the n and π Orbitals in Sterically Hindered $\alpha\beta$-Unsaturated Carbonyl Compounds

As mentioned in Subsection 21.5.1, the n–π^* bands of sterically hindered $\alpha\beta$-unsaturated carbonyl compounds are higher in intensity than those of sterically unhindered compounds, and the enhanced n–π^* band intensities of the former compounds are considered to arise from mixing of allowed π–π^* transitions with the originally forbidden n–π^* transition, which occurs by virtue of the nonplanar conformation of the conjugated system. In a planar conformation of the conjugated system, the oxygen nonbonding $2p$ orbital (χ_{n}) is orthogonal to the π orbitals. In a nonplanar conformation, however, χ_{n} interacts with the β- or o-carbon $2p\pi$ orbital and hence with the π orbitals. The interaction between χ_{n} and the highest occupied π orbital (ψ_π) is most important. As a result of the interaction, χ_{n} is perturbed as follows:

$$\chi_{\mathrm{n}}^{\mathrm{p}} = (1 + \lambda^2)^{-1/2}(\chi_{\mathrm{n}} - \lambda\psi_\pi) \doteqdot (1 - \lambda^2)^{1/2}(\chi_{\mathrm{n}} - \lambda\psi_\pi), \qquad (24.1)$$

where λ is a constant much smaller than unity. If the one-electron transition from the unperturbed n orbital to a vacant π orbital (ψ_{π^*}) is forbidden, the oscillator strength of the one-electron transition from the perturbed n orbital $\chi_{\mathrm{n}}^{\mathrm{p}}$ to the vacant π orbital ψ_{π^*}, $f(n^{\mathrm{p}}$–$\pi^*)$, is given approximately by

$$f(n^{\mathrm{p}}–\pi^*) = \lambda^2 f(\pi–\pi^*), \qquad (24.2)$$

where $f(\pi$–$\pi^*)$ represents the oscillator strength of the one-electron transition from ψ_π to ψ_{π^*}. In this way in nonplanar conformations of the conjugated system the originally forbidden n–π^* transition appropriates a small fraction of the intensity of the allowed π–π^* transition.

When the energy difference and interaction matrix element between χ_{n} and ψ_π are denoted by $\Delta E_{\mathrm{n},\pi}$ and $H_{\mathrm{n},\pi}$, respectively, the first-order perturbation theory gives the following expression to the constant λ:

$$\lambda = -H_{\mathrm{n},\pi}/\Delta E_{\mathrm{n},\pi}\,. \qquad (24.3)$$

Accordingly, expression (24.2) can be rewritten as follows:

$$f(n^{\mathrm{p}}\text{–}\pi^*) = (H_{\mathrm{n},\pi}/\Delta E_{\mathrm{n},\pi})^2 f(\pi\text{–}\pi^*). \tag{24.4}$$

Of course, $H_{\mathrm{n},\pi}$ is proportional to the overlap of the two orbitals, χ_{n} and ψ_π .

24.2.2. Interaction between a Carbonyl Group and an Ethylenic Bond Separated by One Saturated Carbon Atom

The spectra of compounds containing a carbonyl group and an unsaturated group that is not conjugated with the carbonyl group are often marked by the existence of absorption bands which do not belong to either of the two "isolated" chromophores. Thus some $\beta\gamma$-unsaturated ketones absorb with variable intensity in the 200–260 mμ region. For example, the spectrum of bicyclooctenone I*a* in cyclohexane solution shows a band at $\lambda_{\max} = 202$ mμ with $\epsilon_{\max} = 3000$, in addition to the intensified n–π^* band at $\lambda_{\max} = 290$ mμ with $\epsilon_{\max} = 110$.[1] The spectrum of bicycloheptenone I*b* in cyclohexane solution shows a similar moderate-intensity band at $\lambda_{\max} = 214$ mμ,[1] and the spectrum of bicycloheptenone I*c* in isooctane solution also shows a similar band at $\lambda_{\max} = 225$ mμ.[2,3]

1 2 2′ 1′ O

$(CH_2)_m$ $(CH_2)_n$

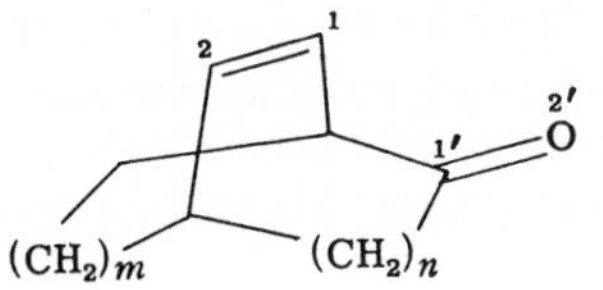

I*a* : $m = 1, \ n = 1$ I*b* : $m = 0, \ n = 1$ I*c* : $m = 1, \ n = 0$

In the 200–260 mμ region high-intensity bands of $\alpha\beta$-unsaturated ketones (conjugated enones) occur. As mentioned in Subsection 21.4.1, these bands are considered to be due to the one-electron transition from the perturbed bonding π orbital of the ethylenic bond, ϕ^{p}_{+1} , to the perturbed antibonding π orbital of the carbonyl bond, $\phi^{\mathrm{p}}_{-1'}$, which is accompanied with a partial electron transfer from the ethylenic bond to the carbonyl bond. From the energy level diagram in Fig. 21.3 it is expected that the energies of $\phi^{\mathrm{p}}_{-1'}$ and

[1] H. Labhart and G. Wagniere, *Helv. Chim. Acta* **42,** 2219 (1959).

[2] C. J. Norton, Ph.D. Thesis, Harvard University, 1955, quoted in the paper cited in Reference 3.

[3] R. C. Cookson, R. R. Hill, and J. Hudec, *Chem. & Ind. (London)* **1961,** 589.

ϕ^{p}_{+1} will be comparatively insensitive to change in the extent of the interaction between the carbonyl and ethylenic bonds. Therefore, a similar intramolecular CT transition may occur with a similar transition energy but with a reduced probability in $\beta\gamma$- and even in $\gamma\delta$-unsaturated ketones if the π orbitals of the carbonyl and ethylenic bonds overlap appreciably.

Labhart and Wagniere[1] evaluated the overlap integrals between the $2p\pi$ atomic orbitals of the carbonyl and ethylenic bonds and the oxygen nonbonding $2p$ orbital in compound I*a*. These are shown in Table 24.1. It is

TABLE 24.1

THE OVERLAP MATRIX OF THE $2p\pi$ AND $2pn$ ORBITALS OF BICYCLO-[2:2:2]-OCTENONE I*a*

	$\chi_{n'}$	$\chi_{2'}$	$\chi_{1'}$	χ_1	χ_2
$\chi_{n'}$	1	0	0	0.0021	0.0002
$\chi_{2'}$		1	0.214	0.0025	0
$\chi_{1'}$			1	0.0570	0.033
χ_1				1	0.262
χ_2					1

seen that there is quite a large overlap between χ_1 and $\chi_{1'}$, which is about one quarter of the overlap between χ_1 and $\chi_{1'}$ in sterically unhindered $\alpha\beta$-unsaturated ketones. The transannular interaction between the π orbitals of the carbonyl and ethylenic bonds occurring mainly through the overlap of χ_1 and $\chi_{1'}$ allows the one-electron CT transition from ϕ^{p}_{+1} to $\phi^{p}_{-1'}$ to occur with some probability.

The comparatively high intensity of the n–π^* band of I*a* ($\epsilon_{max} = 110$) is explained in terms of mixing of the n–π^* transition with the π–π^* CT transition. As is seen in Table 24.1, the oxygen nonbonding $2p$ orbital, $\chi_{n'}$, appreciably overlaps with the β-carbon $2p\pi$ orbital, χ_1, and hence the one-electron transition from the perturbed n orbital, $\chi^{p}_{n'}$, to $\phi^{p}_{-1'}$ has a contribution of the allowed one-electron transition from ϕ^{p}_{+1} to $\phi^{p}_{-1'}$. Such intensified n–π^* bands of asymmetric $\beta\gamma$-unsaturated ketones are in general associated with very large optical rotatory dispersions.[4]

The n–π^* band of bicycloheptenone I*c* has a normal intensity, although, as mentioned above, this compound shows the transannular CT band at 225 mμ. From the consideration of the molecular geometry it is inferred that in this compound the oxygen nonbonding $2p$ orbital does not appreciably overlap with the carbon $2p\pi$ orbitals of the ethylenic bond, while the $2p\pi$

[4] R. C. Cookson and S. MacKenzie, *Proc. Chem. Soc.* **1961,** 423.

orbitals of the carbonyl bond overlap appreciably with the $2p\pi$ orbitals of the ethylenic bond.

The spectrum of 3-methylenecyclobutanone in ethanol shows a weak n–π^* band at $\lambda_{max} = 285$ mμ ($\epsilon_{max} = 24$) and a moderately intense band at $\lambda_{max} = 214$ mμ ($\epsilon_{max} = 1500$).[5] There is probably an attenuated conjugative interaction between the carbonyl and ethylenic bonds in this compound too, and the 214-mμ band is interpreted as a transannular CT band due to the one-electron transition from ϕ^{p}_{+1} to $\phi^{p}_{-1'}$. The exciton interaction of the π–π^* transitions in the carbonyl and ethylenic bonds may play some role in appearance of the 214-mμ band. However, since the carbonyl and ethylenic bonds are coplanar, there is no overlap between the oxygen nonbonding $2p$ orbital and π orbitals, and hence there is no mixing of the π–π^* transition with the n–π^* transition, so that the n–π^* band intensity is normal.

The n–π^* band of 2,2-diphenylcyclohexanone has an enhanced intensity ($\epsilon_{max} = 125$), probably because the oxygen nonbonding $2p$ orbital and carbonyl π orbitals overlap with the π orbitals of one of the benzene rings.[6] In this connection, it is of interest that in a series of phenylcholestanones only compounds containing an axial phenyl group attached to the carbon atom in the α position to the carbonyl group show abnormally intense n–π^* bands.[7]

24.2.3. Interaction between a Carbonyl Group and an Ethylenic Bond Separated by More than Two Saturated Carbon Atoms

Interaction can occur even between a carbonyl group and an ethylenic bond separated by two or more saturated carbon atoms when the geometry is favorable for overlap of the orbitals concerned. For example, the spectrum of 5-cyclodecenone in cyclohexane is markedly different from that expected on the basis of noninteracting ethylenic and carbonyl chromophores, having two bands at 302 mμ ($\epsilon_{max} = 72$) and at 260 mμ ($\epsilon_{max} = 423$).[8] These bands are interpreted as the intensified n–π^* band and the transannular CT band, respectively. Scale models show that the ethylenic and carbonyl bonds in this $\delta\epsilon$-unsaturated ketone can readily approach each other in parallel planes. The enhanced intensity of the n–π^* band of this compound as compared with that of cyclodecanone ($\lambda_{max} = 288$ mμ, $\epsilon_{max} = 16$ in cyclohexane) is conceivably due to mixing of the n–π^* transition with the π–π^* transition.

There is a striking difference between the solvent effects on the n–π^* bands

[5] F. F. Caserio, Jr., and J. D. Roberts, *J. Am. Chem. Soc.* **80**, 5837 (1958).
[6] W. B. Bennet and A. Burger, *J. Am. Chem. Soc.* **75**, 84 (1953).
[7] R. C. Cookson and J. Hudec, *J. Chem. Soc.* **1962**, 429.
[8] N. J. Leonard and F. H. Owens, *J. Am. Chem. Soc.* **80**, 6039 (1958).

of 5-cyclodecenone and cyclodecanone. The n–π^* band of the simple ketone undergoes the expected hypsochromic shift of 5 mμ on changing the solvent from cyclohexane to ethanol, but the solvent change has no effect on the n–π^* band of the unsaturated ketone. No explanation has been advanced for this lack of solvent effect in the unsaturated ketone. Perhaps a folded structure made preferable by the interaction of the carbonyl bond with the ethylenic bond makes the oxygen lone-pair electrons inaccessible to solvent.

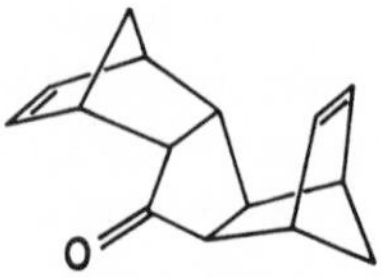

II

The spectrum of a complicated polycyclic $\gamma\delta$-unsaturated ketone II in cyclohexane has a moderately intense band at $\lambda_{max} = 218$ mμ ($\epsilon_{max} = 3200$), which is absent in the spectra of the derivatives lacking either the carbonyl group or the ethylenic bond.[3] This band is attributed to the transannular CT transition from ϕ^{p}_{+1} to $\phi^{p}_{-1'}$. The change of the solvent from cyclohexane to ethanol results in a reduction in intensity from $\epsilon_{max} = 3200$ to $\epsilon_{max} = 2200$ and a bathochromic shift of about 7 mμ. As will be expected, the n–π^* band of this compound has an enhanced intensity (in cyclohexane, $\lambda_{max} = 314$ mμ, $\epsilon_{max} = 110$; in ethanol, $\lambda_{max} = 313$ mμ, $\epsilon_{max} = 126$). The solvent effect on the position of the n–π^* band is also very small in this compound.

Some 4-methylenecyclohexanone derivatives which have a boat structure in which the carbonyl and ethylenic bonds form the stem and stern also show the transannular CT band at about 220 mμ, but the n–π^* bands of these compounds have normal intensities.[9] In these compounds, the overlap between the oxygen n orbital and the ethylenic bond π orbitals must be zero, but that between the carbonyl π orbitals and the ethylenic bond π orbitals must be nonzero.

24.2.4. Interaction between a Carbonyl Group and a Nonconjugated Heteroatom Bearing Lone-Pair Electrons

The n–π^* band of an acyclic simple carbonyl compound is shifted to longer wavelength and intensified by introduction of a halogen or oxygen (that is, hydroxyl or alkoxyl) substituent into the α position to the carbonyl group,

[9] S. Winstein, L. de Vries, and R. Orloski, *J. Am. Chem. Soc.* **83,** 2020 (1961).

TABLE 24.2

THE EFFECT OF A BROMINE SUBSTITUENT ON THE n–π^* BANDS OF ACETALDEHYDE AND ACETONE

Compound	Solvent	λ_{max}, mμ	ϵ_{max}
CH_3—CHO	Hexane	293.5	12
CH_3—CO—Br	Heptane	250	93
CH_3—CO—CH_3	Cyclohexane	275	22
CH_3—CO—CH_2Br	Hexane	310.5	83

while it is shifted to shorter wavelength and intensified by the same substituent at the carbonyl carbon atom (see Table 24.2).[10] The hypsochromic effect of such a substituent attached to the carbonyl carbon atom has been attributed to the conjugative interaction between the n orbital of the substituent and the antibonding π orbital of the carbonyl bond (see Section 21.2).

In cyclic ketones, the effect of such a substituent at the α position on the n–π^* band depends on the configuration of the substituent. Thus, an α-halogen or oxygen substituent in the axial configuration shifts the n–π^* band some 4–30 mμ to longer wavelength and brings about a larger intensity increase of the band than does the corresponding equatorial substituent, which shifts the band to shorter wavelength by 4–13 mμ.[11] For example, in the cyclohexanone series and in the steroid series, an axial bromine and an axial chlorine substituent at the α position to the carbonyl group produce bathochromic shifts of 20–30 mμ and of about 11 mμ, respectively, of the n–π^* band of the unsubstituted ketone, accompanied with large increases in intensity, and an equatorial bromine or chlorine substituent at the α position produces a small hypsochromic shift of about 5 mμ of the n–π^* band. Furthermore, such a halogen substituent at the γ position of a cyclic $\alpha\beta$-unsaturated ketone causes a small bathochromic shift (of about 2–5 mμ) of the n–π^* band of the conjugated enone system if it is in the equatorial configuration, and it causes a much greater bathochromic shift (of about 7–20 mμ) of the band if it is in the axial configuration.

These observations suggest that the important interaction between the carbonyl group and the equatorial halogen substituent at the α position is an attenuated conjugative interaction between the antibonding π orbital of the carbonyl bond and the nonbonding orbital of the substituent, which is weaker than, but similar in nature to, the interaction between the antibonding

[10] S. F. Mason, *Quart. Rev.* (*London*) **15,** 287 (1961).

[11] R. C. Cookson, *J. Chem. Soc.* **1954,** 282; R. C. Cookson and S. H. Dandegaonker, *ibid.* **1955,** 352, 1651; C. W. Bird, R. C. Cookson, and S. H. Dandegaonker, *ibid.* **1956,** 3675.

π orbital of the carbonyl bond and the nonbonding orbital of the halogen in acyl halides.

The strong bathochromic and hyperchromic effect of an axial halogen substituent at the α position on the n–π^* band of cyclic ketones is similar to the effect of a halogen substituent at the α position on the n–π^* band of acyclic ketones. In acyclic ketones having a halogen substituent at the α position, as in cyclic ketones having an axial halogen substituent at the α position, the carbonyl bond and the carbon–halogen bond are probably in a staggered configuration. No satisfactory explanation has been advanced for these effects. Perhaps the interaction of hyperconjugation type between the carbonyl bond and the carbon–halogen bond is an important factor contributing to these effects.

If a trivalent nitrogen atom possessing a lone pair of electrons is located in a cyclic ketone at a position suitable for transannular interaction with the carbonyl group in the same molecule, the compound shows a new absorption band, absent both in the spectra of the corresponding simple carbonyl compounds lacking the nitrogen atom and in the spectra of the corresponding simple amines lacking the carbonyl group. The correlation between ring size and nitrogen–carbonyl transannular interaction has been extensively studied by Leonard and co-workers,[12] both in the series of aza cyclic acyloins, such as III, and in the series of aza cyclanones, such as IV.

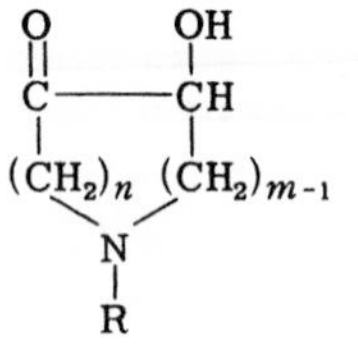

O=C—CH(OH), $(CH_2)_n$ $(CH_2)_{m-1}$, N—R

III

O=C, $(CH_2)_n$ $(CH_2)_m$, N—R

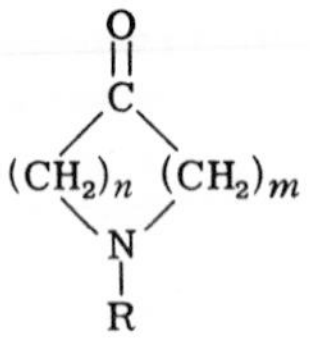

IV

The interaction apparently occurs for eight-, nine-, and ten-membered rings in which m and n are 3 or 4. For example, while the spectrum of cyclononanone has the n–π^* band at 264 mμ (log $\epsilon_{max} = 2.13$), the spectrum of 6-methyl-6-aza-2-hydroxycyclononanone (III, R = CH_3, $m = 4$, $n = 3$) has a moderately intense band at 228 mμ (log $\epsilon_{max} = 3.77$) and only weak absorption at 264 mμ. The 228-mμ band is probably due to some type of nitrogen–carbonyl interaction, the nature of which has not been determined. This band is possibly the intramolecular transannular CT band due to the

[12] N. J. Leonard and M. Oki, *J. Am. Chem. Soc.* **77**, 6241 (1955); N. J. Leonard, *Record Chem. Progr.* (*Kresge–Hooker Sci. Lib.*) **17**, 243 (1956).

one-electron transition from the perturbed nonbonding orbital of the nitrogen atom to the perturbed antibonding π orbital of the carbonyl group, which is of course accompanied with a partial electron transfer from the nitrogen atom to the carbonyl group. The spectrum of 5-methyl-5-azacyclooctanone (IV, R $= CH_3$, $m = n = 3$) also shows a similar moderately intense band at 225 mμ (log $\epsilon_{max} = 3.80$) and again only a weak band at 264 mμ.

24.3. Nonneighbor Interactions in Unsaturated Hydrocarbons

24.3.1. Interaction between Nonconjugated Ethylenic Bonds

The spectrum of bicycloheptadiene V is appreciably different from the spectra of compounds containing completely insulated ethylenic double bonds, for example, bicycloheptene VI. Thus, while the spectrum of the mono-olefin VI has an intense band at 195 mμ, the spectrum of the di-olefin V in ethanol shows a moderately intense band with vibrational structure in the 200–230 mμ region, a region in which the first intense bands of conjugated dienes appear. This moderately intense band has peaks at 205 mμ ($\epsilon =$ 2100), 214 mμ ($\epsilon = 1480$), and 220 mμ ($\epsilon = 870$) and a shoulder at 230 mμ ($\epsilon = 200$). This band in the vapor spectrum has well-resolved vibrational structure with at least seventeen sharp subbands, the strongest absorption occurring at 211 mμ. A sharp rise in intensity occurs below 198 mμ, but the band maximum has not been observed.[13]

1′ 1 2′ 2

V VI

In the diene V probably interaction of exciton type occurs effectively between the two ethylenic bonds, since these bonds are parallel to each other. The carbon atoms of these bonds are numbered as shown in the appended structural formula. The antibonding and bonding π orbitals of the 1—2 bond are denoted by ϕ_{-1} and ϕ_{+1}, respectively, and those of the 1′—2′ bond are denoted by $\phi_{-1'}$ and $\phi_{+1'}$, respectively:

$$\phi_{-1} = 2^{-1/2}(\chi_1 - \chi_2), \qquad \phi_{+1} = 2^{-1/2}(\chi_1 + \chi_2);$$
$$\phi_{-1'} = 2^{-1/2}(\chi_{1'} - \chi_{2'}), \qquad \phi_{+1'} = 2^{-1/2}(\chi_{1'} + \chi_{2'}).$$

[13] C. F. Wilcox, S. Winstein, and W. G. McMillan, *J. Am. Chem. Soc.* **82,** 5450 (1960).

Of course, the S—S one-electron transition from ϕ_{+1} to ϕ_{-1} and that from $\phi_{+1'}$ to $\phi_{-1'}$ require the same energy (E_V), and their transition moments are in the same direction and have the same magnitude (μ). The singlet LE configurations arising from these transitions are denoted by V_1^{11} and $V_1^{1'1'}$, respectively. Then, the wavefunctions and energies of the states resulting from the exciton interaction between these two electron configurations can be expressed as follows:

Wavefunction	Energy
$2^{-1/2}(V_1^{11} + V_1^{1'1'}) \equiv {}^1Ex^{(+)}$	$E_V + \mu^2/r^3$
$2^{-1/2}(V_1^{11} - V_1^{1'1'}) \equiv {}^1Ex^{(-)}$	$E_V - \mu^2/r^3$

Here, r represents the distance between the two ethylenic bonds. The transition to the upper exciton state, ${}^1Ex^{(+)}$, is allowed, the magnitude of its transition moment being $2^{1/2}\mu$, and the transition to the lower exciton state, ${}^1Ex^{(-)}$, is forbidden. When the values of μ and r are taken to be 3.4×10^{-18} e.s.u. and 2.37 Å, respectively, the molecule is expected to have a forbidden band 4300 cm^{-1} below and an allowed band 4300 cm^{-1} above the S–S π–π^* band of the isolated ethylenic bond.[14] As mentioned already, the corresponding monoethylenic compound VI has an intense band at 195 mμ, which is almost certainly the S–S π–π^* band. Hence it is expected that the diene V has a weak band at about 213 mμ and an intense band at about 180 mμ. Thus the exciton theory gives a rather satisfactory interpretation of the spectrum of this compound.

The π orbitals on different ethylenic bonds in the diene V may overlap considerably with one another. Through the overlap, the minus exciton state, ${}^1Ex^{(-)}$, may interact with the minus charge-resonance state, $2^{-1/2}(V_1^{11'} - V_1^{1'1}) \equiv {}^1CR^{(-)}$, with the interaction matrix element of $2\beta_{11'}$, and the plus exciton state, ${}^1Ex^{(+)}$, may interact with the plus charge-resonance state, $2^{-1/2}(V_1^{11'} + V_1^{1'1}) \equiv {}^1CR^{(+)}$, with the interaction matrix element of $-2\beta_{12'}$. However, the overlap effect does not appear to be very important in this case.

The diene V is treated by the simple MO method, by emphasizing the overlap effect, in the following way. The π orbitals of this system is considered as arising from the transannular conjugative interactions between the π

[14] J. N. Murrell, "The Theory of the Electronic Spectra of Organic Molecules," Sect. 7.1. Methuen, London, 1963.

orbitals on the two ethylenic bonds. It is evident that interactions can occur only between ϕ_{+1} and $\phi_{+1'}$ and between ϕ_{-1} and $\phi_{-1'}$, with the interaction matrix elements of $(k_{11'} + k_{12'})\beta$ and $(k_{11'} - k_{12'})\beta$, respectively. The extra-resonance energy of the ground state of this system relative to the two isolated ethylenic bonds is obviously exactly zero in this approximation. The highest bonding and lowest antibonding π orbitals, ψ_{+1} and ψ_{-1}, are given by $2^{-1/2}(\phi_{+1} - \phi_{+1'})$ and $2^{-1/2}(\phi_{-1} + \phi_{-1'})$, respectively. The one-electron transition from ψ_{+1} to ψ_{-1} is forbidden. The excited configuration of this

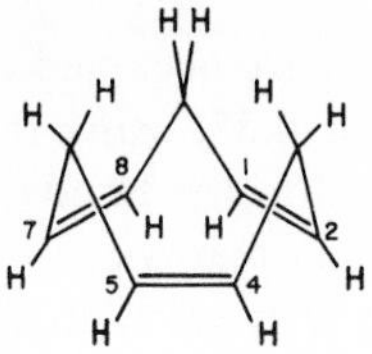

FIG. 24.1. The "crown" form of *cis,cis,cis*-1,4,7-cyclononatriene (VII).

transition corresponds to an equal mixture of the minus exciton state and the minus charge-resonance state, $2^{-1/2}(Ex^{(-)} + CR^{(-)})$. This transition requires the energy of $2(k_{12} - k_{11'})(-\beta)$, which is to be compared with the energy of $2k_{12}(-\beta)$ for the one-electron π–π^* transition in the isolated ethylenic bond. In this way the simple MO theory predicts that this compound should have a forbidden band $2k_{11'}(-\beta)$ below the S–S π–π^* band of the corresponding monoethylenic compound VI, and thus it can interpret at least qualitatively the spectrum of this compound.

The nonneighbor conjugative interaction between two mesomeric groups separated by one methylene group, such as the conjugative interaction between the two ethylenic bonds in compound V, is sometimes called the homoconjugation. It was expected that in *cis,cis,cis,*-1,4,7-cyclononatriene (VII) there should be homoconjugations between the three ethylenic bonds, and hence it was expected that this compound, having a homoconjugated cyclic six π-electron system, should have an aromaticity somewhat similar to that of benzene, that is, the so-called homoaromaticity. Recently this compound was synthesized by Radlick and Winstein[15] and by Untch,[16] and its most stable structure was found to be a "crown" form as illustrated in Fig. 24.1.

[15] P. Radlick and S. Winstein, *J. Am. Chem. Soc.* **85,** 344 (1963); see also R. S. Boikess and S. Winstein, *ibid.* **85,** 343 (1963).

[16] K. G. Untch, *J. Am. Chem. Soc.* **85,** 345 (1963); see also K. G. Untch and R. J. Kurland, *ibid.* **85,** 346 (1963).

The homoconjugated cyclic π-electron system in the crown form of VII can be considered to belong to point group C_{3v}. The ratio of resonance integrals β_{24}/β_{12} is denoted by a, and the orbital energies are expressed as $\alpha - w\beta_{12}$. The simple LCAO MO treatment of this system leads to six delocalized π orbitals with the w values of $\pm(1 + a)$, $\pm(1 - a + a^2)^{1/2}$, and $\pm(1 - a + a^2)^{1/2}$. These w values are to be compared with the w values of ± 2, ± 1, and ± 1 for the π orbitals of benzene. The highest orbital [with $w = +(1 + a)$] belongs to symmetry species A_2, the lowest orbital [with $w = -(1 + a)$] to symmetry species A_1, and the two doubly degenerate pairs of orbitals [with $w = \pm(1 - a + a^2)^{1/2}$] to symmetry species E. Of course, the lower three orbitals are occupied each by two electrons. Untch[16] calculated the value of a to be 0.25, using the valence angle values of 120° (for sp^2 bonds) and 109°28′ (for sp^3 bonds), and stated that the possible steric repulsion of the three inner methylene hydrogen atoms would cause the value of a to be greater than that calculated. By the use of the a value of 0.25, the w values of the highest bonding degenerate orbitals and of the lowest bonding orbital are calculated to be -0.9014 and -1.25, respectively. Accordingly, the delocalization energy of the ground state is calculated to be $0.1056(-\beta_{12})$. This value is to be compared with the corresponding value of $2(-\beta)$ for benzene.

The calculated delocalization energy of VII is so small that it probably cannot be said that this compound will have an aromaticity. However, the delocalization of π orbitals is expected to affect appreciably the electronic spectrum. Similarly to the case of benzene, when the electronic repulsion is taken into account, the four degenerate singlet electron configurations arising from one-electron transitions from the highest bonding degenerate π orbitals [with $w = -(1 - a + a^2)^{1/2}$] to the lowest antibonding degenerate π orbitals [with $w = +(1 - a + a^2)^{1/2}$] should split into two nondegenerate states (belonging to symmetry species A_1 and A_2) and a degenerate pair of states (belonging to symmetry species E). The transitions from the ground state to the A_1 and A_2 states are forbidden, and the transitions to the E states are allowed.

As expected, the spectrum of VII is considerably different from that expected from the assumption that the three ethylenic bonds were completely insulated. According to Radlick and Winstein,[15] the spectrum of this compound in heptane has a band at 198 mμ ($\epsilon_{max} = 11{,}600$) with an apparent shoulder at 200 mμ ($\epsilon = 11{,}200$) and a prominent shoulder at 212 mμ ($\epsilon =$

5000), and attempts at analytic separation of the 198-mμ band from that at longer wavelength indicate an absorption band at about 216 mμ with ε above 10^3.

24.3.2. Interaction between Nonconjugated Aromatic Rings

The spectrum of triptycene [9,10-(*o*-phenylene)-9,10-dihydroanthracene] somewhat resembles that of triphenylmethane with bathochromic and hyperchromic shifts.[17] The bridgehead hydrogen atoms cannot readily participate in hyperconjugation, but the three benzene rings in triptycene are suitably situated for transannular interaction, which is probably responsible for the bathochromic and hyperchromic shifts.

An interesting example of transannular interaction of nonconjugated aromatic rings is found in the series of *p*-cyclophanes. As mentioned in Section 18.3, the spectra of small ring *p*-cyclophanes such as di-*p*-xylylene are markedly different from the spectra of larger ring homologs as well as open-chain analogs. The difference has been interpreted as arising from a combination of transannular conjugative interaction (or perhaps exciton interaction) between the two closely faced benzene rings and nonplanarity of the rings in the small ring cyclophanes.

24.3.3. Effects of Transannular Interaction and of Steric Distortion in Cyclic Acetylenic Compounds

A phenomenon considerably similar to that found in the series of *p*-cyclophanes is found in some series of cyclic acetylenic compounds.[18]

The spectrum of the compound of structure VIII with $n = 5$ resembles that of an open-chain conjugated alkadiyne, for example, 6,8-tetradecadiyne $[CH_3—(CH_2)_4—C{\equiv}C—]_2$, which exhibits a band consisting of three well-developed vibrational peaks in the 220–260 mμ region. As the n becomes smaller from 5 to 4, and to 3 in the series of cycloalkatetraynes VIII, the vibrational structure rapidly becomes much more blurred.[19] This effect has

[17] P. D. Bartlett and E. S. Lewis, *J. Am. Chem. Soc.* **72,** 1005 (1950).
[18] M. Nakagawa, *Kagaku To Kogyo* (*Tokyo*) **15,** 844 (1962).
[19] F. Sondheimer, Y. Amiel, and R. Wolovsky, *J. Am. Chem. Soc.* **79,** 6263 (1957).

been attributed to increasing transannular interaction between the two diacetylene groups in smaller ring compounds.

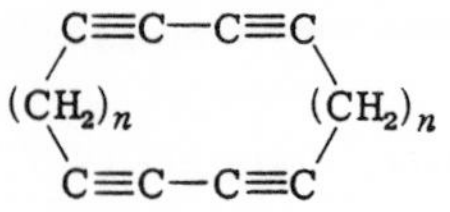

$$\begin{array}{ccc} & C\equiv C-C\equiv C & \\ (CH_2)_n & & (CH_2)_n \\ & C\equiv C-C\equiv C & \end{array}$$

VIII

Toda and Nakagawa[20] compared the spectra of compounds IX*a*, IX*b* (*trans* and *cis*), and IX*c* with the spectra of open-chain analogs X*a* and X*b*. The spectrum of IX*a* shows a considerably intense band in the 300–360 mμ region (log $\epsilon_{max} = 4.3$). This band is less intense than the corresponding band of X*a*, an open-chain analog of IX*a*. Furthermore, the vibrational structure of the former is much more blurred than that of the latter. The spectrum of the *cis* isomer of IX*b* resembles that of X*b*, showing a considerably well-developed vibrational structure, but the spectrum of the *trans* isomer shows a much more blurred vibrational structure. The blurring of the vibrational structure in the spectra of IX*a* and *trans*-IX*b* is attributed to the effect of transannular interaction of the diacetylene group with the benzene ring or the ethylenic bond.

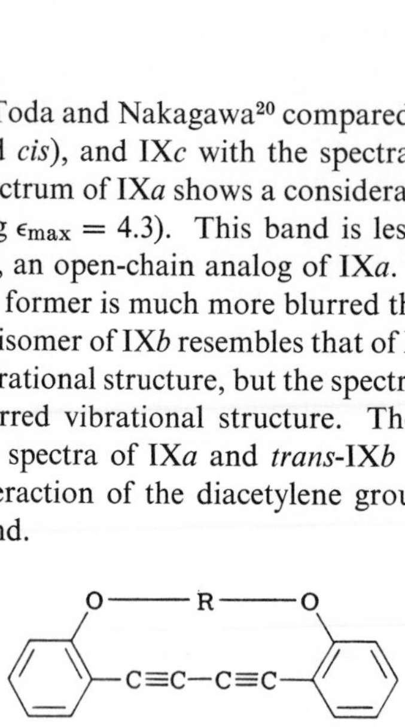

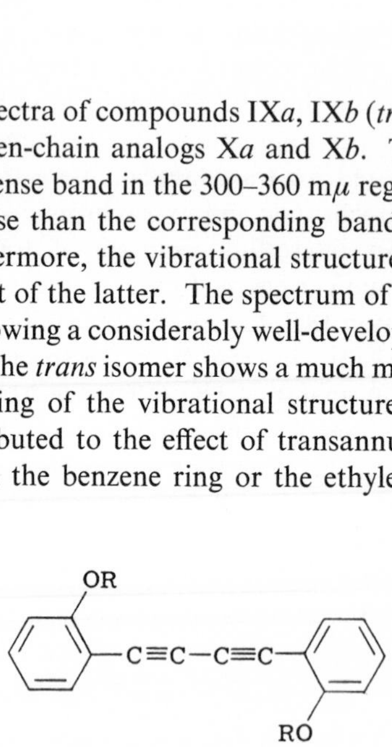

O—R—O (bridging two benzene rings linked by —C≡C—C≡C—)   OR … —C≡C—C≡C— … RO

IX*a* : R = —CH_2—$C_6H_4(p)$—CH_2— (*p*-xylylene)   X*a* : R = C_6H_5—CH_2— (benzyl)

IX*b* : R = —CH_2—CH═CH—CH_2—   X*b* : R = CH_3

IX*c* : R = —CH_2—$C_6H_4(o)$—CH_2— (*o*-xylylene)

IX*d* : R = —CO—$(CH_2)_n$—CO—   X*d* : R = CH_3—CO—

IX*e* : R = —$(CH_2)_n$—

[20] F. Toda and M. Nakagawa, *Bull. Chem. Soc. Japan* **34,** 874 (1961).

Compound IX*a* is considered to have no appreciable steric strain in the molecule. On the other hand, *trans*-IX*b*, *cis*-IX*b*, and IX*c* are considered to have considerable steric strain in the diacetylene group, owing to shorter bridge chain lengths in these compounds. It is of interest that the bands appearing in the 290–350 mμ region of the conceivably sterically strained compounds, *trans*-IX*b*, *cis*-IX*b*, and IX*c*, are at shorter wavelength by about 10 mμ than the corresponding bands of the conceivably sterically unstrained compound, IX*a*, and open-chain compounds X*a* and X*b*.

TABLE 24.3

THE LONGEST-WAVELENGTH ABSORPTION MAXIMA OF BRIDGED DIPHENYLDIACETYLENES IX*d* AND IX*e*

	IX*d*		IX*e*	
n	λ_{max}, mμ	$\epsilon_{max} \times 10^{-2}$	λ_{max}, mμ	$\epsilon_{max} \times 10^{-2}$
3	331	292	337	159
4	332	541	353	199
5	331	174	353	365
6	—	—	349	336
7	331	200	—	—
	(X*d*) 331	398	(X*b*) 349	286

Toda and Nakagawa correlated the spectra of compounds of structure IX*d*[21] and of structure IX*e*[22] with the molecular geometries. Data on the longest wavelength absorption maxima of compounds of these series and their open-chain analogs, X*d* and X*b*, are shown in Table 24.3. The values of ϵ_{max} for compound IX*d* with $n = 4$ and compound IX*e* with $n = 5$ are considerably greater than those for other members of the respective series. In addition, while the values of λ_{max} for compounds of the IX*d* series are almost the same, the values of λ_{max} for compounds IX*e* with $n = 4$ and $n = 5$ are appreciably greater than those for other members of the same series. In compound IX*d* with $n = 4$ and compound IX*e* with $n = 5$ the diphenyldiacetylene structure is probably fixed in a strainless planar and linear geometry. In the lower homologs the diacetylene structure is probably distorted from the linear

[21] F. Toda and M. Nakagawa, *Bull. Chem. Soc. Japan* **33,** 223 (1960).
[22] F. Toda and M. Nakagawa, *Bull. Chem. Soc. Japan* **34,** 862 (1961).

geometry. On the other hand, in the higher homologs, the two benzene rings can be noncoplanar with each other, and the probability that the two benzene rings are coplanar with each other will decrease with increasing ring size. Thus it is concluded that both the deviation of the two benzene rings from the coplanar configuration and the distortion of the diacetylene structure from the normal linear geometry cause a decrease in intensity of the band in both the IX*d* and IX*e* series, and that in the IX*e* series they cause a hypsochromic shift of the band.

Appendix. Notation for Electronic Spectral Bands

A large number of systems of notation for electronic spectral bands have been proposed and some of them are currently in use. The systems used in this book are summarized in Section A.1, and some other systems are given a brief explanation in Section A.2.

A.1. The Systems of Spectral Notation and Conventions Used in this Book

A.1.1. Empirical Notation

From a theoretical viewpoint it is most desirable to describe individual bands in terms of the electronic transitions responsible for the respective bands. However, when the transitions have not been assigned for bands, bands in the spectrum of a compound are named, for example, the A, B, C, . . . bands or the 1st, 2nd, 3rd, . . . bands in the order of decreasing wavelength. In this book the longest wavelength intense band, which can usually be attributed approximately to the singlet–singlet one-electron transition from the highest occupied π orbital to the lowest vacant one, has been referred to as the first (intense) band or the A band and sometimes as the conjugation band. Such bands correspond to the *p* bands of aromatic compounds in Clar's notation, which will be described later.

A.1.2. Notation for Orbitals and Description of Electronic Transitions in Terms of Orbitals

The simplest systematic notation describes an electronic transition in terms of the orbitals between which the electron jumps. The initial orbital is given first, the final last, and the two are connected by an arrow or a bar.

The orbitals may be referred to by the σ, π notation. In this case, a nonbonding orbital containing two electrons (that is, a lone pair of electrons) is named as an n orbital, which has also been referred to as a lone-pair orbital. According to this system of notation, which was proposed originally by Kasha,[1] one-electron transitions from a nonbonding orbital to an antibonding π orbital, from a bonding π orbital to an antibonding π orbital, and from a nonbonding orbital to an antibonding σ orbital are called n–π^*, π–π^*, and n–σ^* transitions, respectively.

The orbitals may be specified by the use of the Schönflies symbols of the symmetry species to which they belong. For this purpose the symbols are written in lower-case letters.

Any orbital is usually given by a lower-case Greek letter. Atomic orbitals (AO's), especially $p\pi$ atomic orbitals, have been indicated by the lower-case letter chi (χ) throughout this book. Molecular orbitals (MO's), especially molecular π orbitals, have been indicated by the lower-case letter psi (ψ). When a π system is considered as a composite system, that is, a system composed of two or more mesomeric fragments, the π orbitals of the fragments have been indicated by the lower-case letter phi (ϕ).

When π MO's of a whole π system are expressed as linear combinations of $p\pi$ AO's, the coefficients of the individual AO's have been expressed as c's:

$$\psi_m = \sum_i c_{m,i}\chi_i, \tag{A.1}$$

where the subscript i is used to define the individual AO's, and the subscript m identifies the different MO's. The orbital energies have been denoted by an old-style epsilon (ε) and expressed as follows:

$$\varepsilon_m = \alpha - w_m\beta. \tag{A.2}$$

By the way, molar extinction coefficients have been indicated by a Porson epsilon (ϵ).

When π MO's of fragments of a composite π system are expressed as linear combinations of $p\pi$ AO's, the coefficients of the individual AO's have been expressed as b's:

$$\phi_m = \sum_i b_{m,i}\chi_i. \tag{A.3}$$

The orbital energies have been denoted by a lower-case eta (η) and expressed as follows:

$$\eta_m = \alpha - v_m\beta. \tag{A.4}$$

[1] M. Kasha, *Discussions Faraday Soc.* **9**, 14 (1950).

The subscript that identifies the different MO's in a molecule may be chosen in a number of ways. In this book the following notation, called provisionally the two-directional notation, has been used. The highest bonding π MO is designated as $+1$, and lower bonding π MO's are designated as $+2, +3, \ldots$, in the order of decreasing energy; the lowest antibonding π MO is designated as -1, and higher antibonding π MO's are designated as $-2, -3, \ldots$, in the order of increasing energy. This notation has the advantage that the lowest-energy one-electron transition is always expressed by the same symbol, $\psi_{+1} \rightarrow \psi_{-1}$, independently of the kind of molecule and that analogous transitions in related molecules are conveniently correlated. In this notation nonbonding π MO's, which occur in π systems consisting of odd numbers of π centers, are designated as 0. The energy of the one-electron transition from ψ_{+k} to ψ_{-r}, which is the difference between the energies of ψ_{-r} and ψ_{+k}, has been expressed as ΔE_{kr} or Δw_{kr} (in units of $-\beta$), and the electric dipole moment of that transition has been expressed as $\mathbf{M}_{kr}$.

When a π system is considered as a composite system consisting of two fragments R and S, the π orbitals of fragment R are expressed as ϕ_m's or $\phi_{m(\mathrm{R})}$'s, and those of fragment S as $\phi_{n'}$'s or $\phi_{n(\mathrm{S})}$'s. When the two fragments are identical groups, the symmetry combinations of corresponding orbitals of different fragments are expressed as follows:

$$2^{-1/2}(\phi_m \pm \phi_{m'}) \equiv \phi_{m(\pm)} . \tag{A.5}$$

The π orbitals of a composite π system R–S can be considered as arising from interaction of the π orbitals localized on the two fragments R and S. When the π orbitals of the composite π system are expressed approximately by application of perturbation theory, the orbital that is considered as arising from perturbation of a localized orbital, say ϕ_m, is expressed as $\phi_m{}^{\mathrm{p}}$, and its energy as $\eta_m{}^{\mathrm{p}} = \alpha - v_m{}^{\mathrm{p}}\beta$. The superscript p indicates perturbed quantities.

A.1.3. Notation for Many-Electron Functions and Description of Electronic Transitions in Terms of Many-Electron Functions

Electronic transitions are classified according to the changes of the spin multiplicity accompanying the transitions. Since the ground electronic states of ordinary molecules are singlet, most electronic transitions of interest to us are singlet–singlet and singlet–triplet transitions. These transitions have frequently been referred to as S–S and S–T transitions, respectively.

Any wavefunction which is a function of coordinates of many (more than one) electrons is usually indicated by a capital letter. Electron configuration functions, formed as antisymmetrized products of occupied orbitals, have been indicated by the capital letter psi (Ψ), and electronic state functions, formed as linear combinations of electron configuration functions, have been indicated by the capital letter phi (Φ), prefixed by the superscript indicating the spin multiplicity, 1 for singlet and 3 for triplet.

For electron configurations the following notation has been used. The ground electron configuration is denoted by ${}^1\Psi_0$, and the singlet and triplet singly-excited electron configurations arising from the ground electron configuration by excitation of one electron in an occupied orbital ψ_{+k} to an unoccupied orbital ψ_{-r} are denoted by ${}^1\Psi_1^{kr}$ and ${}^3\Psi_1^{kr}$, respectively, the subscript 1 indicating the one-electron excitation. Frequently, the symbols ${}^1\Psi$ and ${}^3\Psi$ have been replaced by symbols V and T, respectively. According to this convention, the ground electron configuration is denoted by V_0, and the foregoing singlet and triplet singly-excited electron configurations are denoted by V_1^{kr} and T_1^{kr}, respectively.

In even alternant hydrocarbons, when $k \neq r$, V_1^{kr} and V_1^{rk} form a degenerate pair, and T_1^{kr} and T_1^{rk} also form a degenerate pair. The symmetry combinations of such paired electron configurations are expressed as follows:

$$2^{-1/2}(V_1^{kr} \pm V_1^{rk}) \equiv {}^{\pm}V_1^{kr}\,; \tag{A.6}$$

$$2^{-1/2}(T_1^{kr} \pm T_1^{rk}) \equiv {}^{\pm}T_1^{kr}\,. \tag{A.7}$$

Symmetry combinations of type ${}^+V_1^{kr}$ and ${}^+T_1^{kr}$ are called plus electron configurations, and symmetry combinations of type ${}^-V_1^{kr}$ and ${}^-T_1^{kr}$ are called minus electron configurations. Electron configurations of type V_1^{kk} and T_1^{kk}, which are called symmetrically-excited configurations, can interact only with electron configurations of the same type and plus electron configurations. Hence such electron configurations, say V_1^{kk} and T_1^{kk}, have frequently been expressed as ${}^+V_1^{kk}$ and ${}^+T_1^{kk}$, respectively. The ground electron configuration can interact only with singlet minus electron configurations. Hence the symbol V_0 has frequently been replaced by ${}^-V_0$.

When a π system is considered as a composite system consisting of two fragments R and S, the following notation has been used. The ground electron configuration is denoted by V_0 or G. Excited electron configurations arising from electronic excitations within one fragment, either R or S, are called locally excited (abbreviated as LE) configurations; excited electron configurations arising from electronic excitation accompanied with electron transfer from one fragment to another are called electron-transfer or charge-transfer (abbreviated as CT) configurations. Sometimes LE configurations arising from electronic excitation within fragment R have been referred to as LE(R) configurations, and LE configurations arising from electronic excitation within fragment S as LE(S) configurations; CT configurations arising from electronic excitation accompanied with electron transfer from R to S have been referred to as CT(R $\rightarrow$ S) configurations, and CT configurations arising from electronic excitation accompanied with electron transfer from S to R as CT(S $\rightarrow$ R) configurations.

The singlet and triplet LE(R) configurations arising from one-electron transition from ϕ_{+k} to ϕ_{-r} are denoted by V_1^{kr} or $^1LE_1^{kr}$ and T_1^{kr} or $^3LE_1^{kr}$, respectively; the singlet and triplet LE(S) configurations arising from one-electron transition from $\phi_{+k'}$ to $\phi_{-r'}$ are denoted by $V_1^{k'r'}$ or $^1LE_1^{k'r'}$ and $T_1^{k'r'}$ or $^3LE_1^{k'r'}$, respectively. The singlet and triplet CT(R $\rightarrow$ S) configurations arising from one-electron transition from ϕ_{+k} to $\phi_{-r'}$ are denoted by $V_1^{kr'}$ or $^1CT_1^{kr'}$ and $T_1^{kr'}$ or $^3CT_1^{kr'}$, respectively; the singlet and triplet CT(S $\rightarrow$ R) configurations arising from one-electron transition from $\phi_{+k'}$ to ϕ_{-r} are denoted by $V_1^{k'r}$ or $^1CT_1^{k'r}$ and $T_1^{k'r}$ or $^3CT_1^{k'r}$, respectively.

When the fragments R and S are identical, each of the following pairs of electron configurations is degenerate: $^1LE_1^{kr}$ and $^1LE_1^{k'r'}$; $^3LE_1^{kr}$ and $^3LE_1^{k'r'}$; $^1CT_1^{kr'}$ and $^1CT_1^{k'r}$; $^3CT_1^{kr'}$ and $^3CT_1^{k'r}$. We can construct the symmetry combinations of paired configurations as follows:

$$2^{-1/2}(^1LE_1^{kr} \pm {}^1LE_1^{k'r'}) \equiv {}^1LE_1^{kr(\pm)} \equiv {}^1Ex_1^{kr(\pm)}; \tag{A.8}$$

$$2^{-1/2}(^3LE_1^{kr} \pm {}^3LE_1^{k'r'}) \equiv {}^3LE_1^{kr(\pm)} \equiv {}^3Ex_1^{kr(\pm)}; \tag{A.9}$$

$$2^{-1/2}(^1CT_1^{kr'} \pm {}^1CT_1^{k'r}) \equiv {}^1CT_1^{kr'(\pm)} \equiv {}^1CR_1^{kr'(\pm)}; \tag{A.10}$$

$$2^{-1/2}(^3CT_1^{kr'} \pm {}^3CT_1^{k'r}) \equiv {}^3CT_1^{kr'(\pm)} \equiv {}^3CR_1^{kr'(\pm)}. \tag{A.11}$$

Electron configurations of type of $^1Ex_1^{kr(+)}$ and $^3Ex_1^{kr(+)}$ are called plus exciton configurations, and electron configurations of type of $^1Ex_1^{kr(-)}$ and $^3Ex_1^{kr(-)}$ are called minus exciton configurations; electron configurations of type of $^1CR_1^{kr'(+)}$ and $^3CR_1^{kr'(+)}$ are called plus charge-resonance configurations, and electron configurations of type of $^1CR_1^{kr'(-)}$ and $^3CR_1^{kr'(-)}$ are called minus charge-resonance configurations.

As mentioned above, electronic state functions have been indicated by the capital Greek letter phi (Φ). The ground electronic state is denoted by $^1\Phi_0$, singlet excited electronic states are denoted by $^1\Phi_1$, $^1\Phi_2$, $^1\Phi_3$, . . . , in the order of increasing energy, and triplet excited electronic states are denoted by $^3\Phi_1$, $^3\Phi_2$, $^3\Phi_3$, . . . , in the order of increasing energy.

Frequently, the following notation has been used for electronic states.

If the electron configuration that makes the largest contribution to a state is, for example, V_1^{kr}, the state is denoted by $[V]_1^{kr}$. Similarly, when the method of composite molecule is used, the state to which an LE configuration $^1LE_1^{kr}$ makes the largest contribution is denoted by $^1[LE]_1^{kr}$, and the state to which a CT configuration $^1CT_1^{kr'}$ makes the largest contribution is denoted by $^1[CT]_1^{kr'}$.

A state to which an LE configuration makes the largest contribution is called a perturbed LE state. Transitions from the ground state to perturbed LE states are called perturbed LE transitions, and bands due to perturbed LE transitions are called perturbed LE bands, or more simply, LE bands. A state to which a CT configuration makes the largest contribution is called a perturbed CT state. Transitions to perturbed CT states are called perturbed CT transitions, and bands due to perturbed CT transitions are called perturbed CT bands, or more simply, CT bands. Similarly, a state to which an exciton configuration makes the largest contribution is called an exciton state, transitions to exciton states are called exciton transitions, and bands due to exciton transitions are called exciton bands; a state to which a charge-resonance configuration makes the largest contribution is called a charge-resonance state, transitions to charge-resonance states are called charge-resonance transitions, and bands due to charge-resonance transitions are called charge-resonance bands.

Electron configurations and electronic states may be designated by the Schönflies symbols of the symmetry species to which they belong, sometimes prefixed by the superscript indicating the spin multiplicity. In this case, the symmetry species symbols are written in capital letters. For example, benzene belongs to point group D_{6h}, and the ground state, lowest singlet excited state, and lowest triplet excited state of this molecule are written as $^1A_{1g}$, $^1B_{2u}$, and $^3B_{1u}$, respectively.

It is internationally recognized that in describing a given electronic transition in terms of electronic states the higher energy state should be written first, the lower energy state last, with an arrow pointing from the starting state toward the final state. Thus, $B \rightarrow A$ denotes the transition from a higher energy state B to a lower energy state A, which will be accompanied with emission of light, and $B \leftarrow A$ denotes the excitation from a lower energy state A to a higher energy state B, which will be accompanied with absorption of light. However, according to this convention, the "unnatural" sequence such as $B \leftarrow A$ would be exclusively used in discussion on absorption spectra. Therefore, in this book, as in many other books which are predominantly concerned with absorption spectra, the "natural" sequence, $A \rightarrow B$, has been used for the excitation from a lower energy state A to a higher energy state B. For example, the lowest energy S–S transition in benzene has been referred to as ${}^1A_{1g} \rightarrow {}^1B_{2u}$.

A.1.4. Notation for Electronic Absorption Bands and Excited Electronic States of Aromatic Hydrocarbons

Clar[2] classified absorption bands of aromatic hydrocarbons on the basis of their experimentally observed characteristics, especially their intensity, into three types designated by symbols α, p, and β. The α bands are very weak ($\epsilon_{max} = 10^2$–10^3, $f = 10^{-3}$–10^{-2}), the p bands are moderately intense ($\epsilon_{max} \simeq 10^4$, $f \simeq 10^{-1}$), and the β bands are highly intense ($\epsilon_{max} \simeq 10^5$, $f \simeq 1$). Different bands of the same type in the spectrum of a compound were distinguished by primed superscripts. In the spectrum of an aromatic hydrocarbon usually four S–S bands appear. The longest wavelength band is the α band or the p band. In the spectra of benzene, naphthalene, and phenanthrene, the longest wavelength band is the α band, and the second longest is the p band. In the spectra of anthracene and higher polyacenes the first band is the p band, and the α band, which appears to be absent, is considered to underlie the p band or other intense bands. In shorter wavelength region, two intense bands appear, which are designated as the β and β' bands.

In this book the excited states of the transitions responsible for the p, α, β, and β' bands have been defined as the p, α, β, and β' states, respectively. When the notation for electronic states mentioned in the preceding subsection is used, the p, α, β, and β' bands of most aromatic hydrocarbons are interpreted as being due to the transitions from the ground state, $[V]_0$, to excited states $[V]_1^{11}$, ${}^-[V]_1^{12}$, ${}^+[V]_1^{12}$, and $[V]_1^{22}$, respectively. Accordingly, excited states $[V]_1^{11}$, ${}^-[V]_1^{12}$, ${}^+[V]_1^{12}$, and $[V]_1^{22}$ have been defined as the p, α, β, and β' states, respectively. Furthermore, the corresponding triplet states, that is, $[T]_1^{11}$, ${}^-[T]_1^{12}$, ${}^+[T]_1^{12}$, and $[T]_1^{22}$, have been defined as the t_p, t_α, t_β, and $t_{\beta'}$ states, respectively.

Benzene is a special case. In this molecule, because of its high symmetry, V_1^{11} and V_1^{22} form a degenerate pair, and T_1^{11} and T_1^{22} also form a degenerate pair. In this molecule the p and β' states are the symmetric and antisymmetric combinations of V_1^{11} and V_1^{22}, that is, $[V]_1^{11+22}$ and $[V]_1^{11-22}$, respectively, and the t_p and $t_{\beta'}$ states are the symmetric and antisymmetric combinations of T_1^{11} and T_1^{22}, that is, $[T]_1^{11+22}$ and $[T]_1^{11-22}$, respectively. Furthermore, the β and β' states form a degenerate pair of states, the ${}^1E_{1u}$

[2] E. Clar, "Aromatische Kohlenwasserstoffe," Springer Verlag, Berlin, 1952.

state, and the t_β and $t_{\beta'}$ states also form a degenerate pair of states, the $^3E_{1u}$ state.

A.2. Some Other Systems of Spectral Notation

A.2.1. Burawoy's Notation

Formerly Burawoy[3] proposed a system of notation for electronic absorption bands on the basis of his investigation of the solvent effect on the spectra of azo, carbonyl, and thiocarbonyl compounds. The low-intensity long-wavelength bands of these compounds which underwent blue-shift on changing the solvent from hexane to ethanol were named R bands [from the German *radikalartig* (radical-like)], and the high-intensity shorter wavelength bands which underwent red-shift on the same solvent change were named K bands [from the German *konjugierte* (conjugated)]. In addition, the medium-intensity band appearing in the 230–260 mμ region with vibrational fine structure in the spectrum of benzene and similar bands of benzene derivatives and benzenoid condensed hydrocarbons were named benzenoid bands or B bands. These three types of bands are characterized by their intensities as follows: R bands, $\epsilon_{max} = 10–100$; B bands, $\epsilon_{max} = 200–3000$; K bands, $\epsilon_{max} \simeq 10^4$.

Burawoy's nomenclature has little theoretical justification, and hence it is largely of historical value. It is now commonly accepted that most of Burawoy's R bands are S–S n–π^* bands, most of his K bands are S–S π–π^* bands which have been called the A bands in this book and which are intramolecular CT π–π^* bands in nature, and his B bands are Clar's α bands.

A.2.2. Mulliken's Notation

Mulliken[4] classified electronic transitions in molecules into the following several types. In his notation the ground (normal) electronic state is denoted by symbol N.

$N \rightarrow T$ transition : transition from the ground state to a triplet excited state.

$N \rightarrow V$ transition : transition of an electron without change of spin from a bonding molecular orbital to an antibonding molecular orbitals formed from the same atomic orbitals. Transitions of this type in a molecule are called successively $N \rightarrow V_1$, $N \rightarrow V_2$, $N \rightarrow V_3$, . . . transitions in the order

[3] A. Burawoy, *Ber.* **63,** 3155 (1930); *J. Chem. Soc.* **1939,** 1177; *ibid.* **1941,** 20.

[4] R. S. Mulliken, *J. Chem. Phys.* **7,** 20 (1939).

of increasing transition energy. S–S π–π^* and σ–σ^* transitions are of this type.

$N \rightarrow Q$ transition : transition of one of lone-pair electrons from a non-bonding orbital to an antibonding molecular orbital. The so-called n–π^* and n–σ^* transitions belong to this type.

$N \rightarrow R$ (Rydberg) transition : transition of an electron from a bonding orbital in the ground state to a higher energy orbital formed from atomic orbitals of higher quantum numbers. In such a transition, the molecule ion core appears to the electron as an atomic ion, and such a transition appears usually in the far ultraviolet region and is often the member of a series which has systematically narrowing spacings toward higher frequency and terminates in ionization, analogously to the Rydberg series in atomic spectroscopy.

A.2.3. Free-Electron Theory and Platt's Notation

Platt[5] proposed a system of notation for electronic states and electronic absorption bands of *cata*-condensed aromatic hydrocarbons [aromatic hydrocarbons of molecular formula $C_{4n+2}H_{2n+4}$ ($n = 1, 2, 3, \ldots$) which have no carbon atom belonging to more than two rings, that is, in which all the carbon atoms lie on a perimeter ring] on the basis of the free-electron method. In this method, which is a special case of the molecular orbital method, π MO's of a molecule are treated as analogous to the wavefunctions of an electron rotating around the periphery of the conjugated system. The periphery is considered to be distorted into a circle of the same length, and the potential on the circle is assumed to be constant. Then, the wavefunctions are given by

$$\psi_{m_q} = (2\pi)^{-1/2} e^{im_q \varphi}, \tag{A.12}$$

where $m_q = \pm q$, $q = 0, 1, 2, \ldots$, $i = \sqrt{-1}$, and φ is the polar angle about the center of the circle. The m_q has a meaning of the component of the angular momentum (in units of $h/2\pi$) of an electron in the orbital along the rotational axis (the z axis). If we require the orbitals in real form, we may use the equivalent functions

$$2^{-1/2}(\psi_{+q} + \psi_{-q}) = \pi^{-1/2} \cos (q\varphi); \tag{A.13}$$

$$(-2)^{-1/2}(\psi_{+q} - \psi_{-q}) = \pi^{-1/2} \sin (q\varphi). \tag{A.14}$$

The energies of the orbitals are given by

$$E_q = q^2h^2/2ml^2 \doteqdot 1{,}210{,}000q^2/l^2 \quad (\text{cm}^{-1}), \tag{A.15}$$

where h is Planck's constant, m is the mass of the electron, l is the length of the perimeter (in Å), and E_q is the energy (in cm^{-1}) of the orbital ψ_{+q} as well as

[5] J. R. Platt, *J. Chem. Phys.* **17**, 484 (1949).

ψ_{-q} measured upward from the constant potential. The energy levels are quadratically spaced. Orbitals having the same q value are said to form a shell. All shells except the lowest ($q = 0$) are doubly degenerate.

The quantum number q, which measures the angular momentum of an electron in the orbital ψ_{+q} as well as ψ_{-q}, determines the number of nodes of the orbital. When the assumption of the constant potential is removed and the periodic potential due to the atoms is introduced, q no longer describes accurately any momentum, but it remains a good quantum number because it still determines the number of nodes of the orbital around the perimeter.

In a *cata*-condensed system consisting of n benzene rings, there are $4n + 2$ carbon atoms and the equal number of π electrons. Since the lowest shell ($q = 0$) is nondegenerate and all others doubly degenerate, the $4n + 2$ π electrons just completely fill the first $n + 1$ shells in the ground state, so that the highest filled shell has $q = n$. This highest filled shell is called the f shell, and shells below this are called successively by the next preceding letters of the alphabet: for example, the shell with $q = n - 1$ is designated as e, and the shell with $q = n - 2$ as d. The lowest vacant shell ($q = n + 1$) is called the g shell, and higher shells are called successively by the next following letters of the alphabet: for example, the shell with $q = n + 2$ is designated as h.

A molecular electronic state is characterized by a quantum number Q, which measures the total angular momentum of the π electrons. The z component of the total angular momentum (in units of $h/2\pi$), $M_Q = \pm Q$, is obtained by summing the m_q for each electron. Platt has assigned A to states with $Q = 0$, and successive capital letters of the alphabet to states with successive values of Q: for states with $Q = 1$, B; for states with $Q = 2$, C; etc. Furthermore, states with $Q = 2n$, $2n + 1$, $2n + 2, \ldots,$ have been designated as K, L, $M, \ldots,$ respectively. These symbols are chosen to be independent of the value of n. As usual, the spin multiplicity is indicated by a superscript prefixed to the letter designating the state. In the ground state, which is of course a closed-shell system, the m_q's always occur in pairs of equal magnitude but opposite sign, so that their sum, M_Q, is zero and hence Q is zero. Therefore, the ground state is a ^1A state.

The lowest energy excitation from the ground state is described in one-electron notation by $f \rightarrow g$. For the f shell, q is n, and m_q is either $+n$ or $-n$; for the g shell, q is $n + 1$, and m_q is either $+(n + 1)$ or $-(n + 1)$. The one-electron transition from the orbital with $m_q = +n$ to the orbital with $m_q = +(n + 1)$ or from the orbital with $m_q = -n$ to the orbital with $m_q = -(n + 1)$ is accompanied with a change in M_Q (ΔM_Q) of $+1$ or -1, and hence it leads to an excited state with $Q = 1$, that is, a B state. On the other hand, for the one-electron transition from the orbital with $m_q = +n$ to the orbital with $m_q = -(n + 1)$ or from the orbital with $m_q = -n$ to the orbital with $m_q = +(n + 1)$, ΔM_Q is $-(2n + 1)$ or $+(2n + 1)$, and the final state is an L state ($Q = 2n + 1$). Thus the one-electron transitions $f \rightarrow g$ give rise to excited states B and L. Of course, these excited states may be either singlet or triplet. Quite analogously, the one-electron transitions $e \rightarrow g$ give rise to excited states C and K; the one-electron transitions $f \rightarrow h$ give rise to excited states C and M.

A states ($Q = 0$) are nondegenerate. On the other hand, if the potential were strictly axially symmetric, all states with $Q \neq 0$ would be doubly degenerate: the state with $M_Q = +Q$ and the state with $M_Q = -Q$ would have the same energy. As mentioned above, the lowest energy one-electron transitions, $f \rightarrow g$, give rise to B and L states. The L states have lower energy than the B states. As long as the potential is assumed to be axially symmetric, the transitions from the ground state ^1A to the ^1B states are allowed, and the transitions to the ^1L states are symmetry forbidden, or as Platt says, momentum forbidden: transitions of an electron on a circular ring are allowed only if there is an accompanying change of just one unit of angular momentum. The degeneracies of states with $Q \neq 0$ and the selection rule are removed in actual molecules except in some states of highly symmetrical molecules, because the periphery is not circular, because the potential is not constant around the periphery, and because the cross links in molecules exert an effect.

The two components of an otherwise doubly degenerate state are specified by subscripts a and b. Thus, for example, an otherwise doubly degenerate ^1B state is split to give two states ^1B_a and ^1B_b, and an otherwise doubly degenerate ^1L state is split to give two states ^1L_a and ^1L_b.

Each state characterized by the quantum number Q has Q nodal planes perpendicular to the molecular plane, or in other words, $2Q$ nodal cuts across the perimeter. Since states with $Q \neq 0$ are primarily doubly degenerate, for a value of Q there are two independent sets of nodal planes corresponding to the two components a and b. In the model in which the periphery is distorted into a circle, the nodal planes of each set are turned by an angle φ of π/Q about the rotational axis (the z axis) with respect to one another, and the nodal planes of one set are turned out of the nodal planes of the other set by an angle φ of $\pi/2Q$. Thus the nodes of one set lie at antinodes of the other set.

An L state has $2n + 1$ nodal planes, or $4n + 2$ nodal cuts across the perimeter. On the perimeter there are $4n + 2$ carbon atoms. By symmetry the nodes of an L state must pass through atoms or through the centers of bonds. The L state having nodes through the centers of all the bonds is denoted by L_a ; the L state having nodes through all the atoms is denoted by L_b. If the actual molecule possesses a twofold symmetry axis, states which have the same symmetry as the L_a state with respect to that axis are designated as a, and states which have the same symmetry as the L_b state with respect to that axis are designated as b. The nodes of the L and B states of benzene (D_{6h}) are shown below:

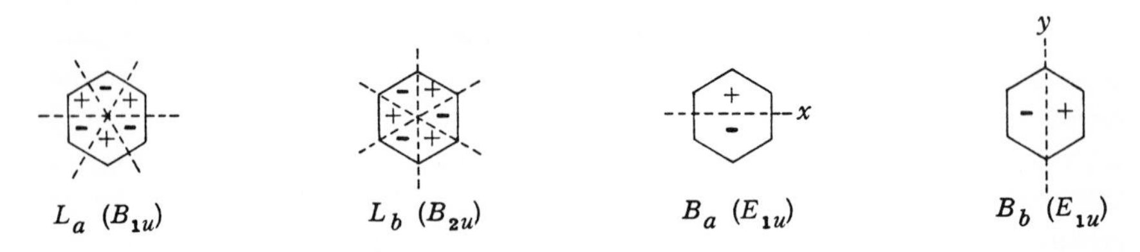

In benzene, because of its high symmetry, the B_a and B_b states have the same energy. In molecules having no twofold symmetry axis, the unambiguous choice of subscripts a and b is not possible for states other than the L states.

Since the ground state is totally symmetric, the transition density of the transition from the ground state to an excited state has the symmetry of the excited state. As is obvious from the above nodal-plane diagrams for excited states of benzene, in benzene the transitions to the L_a and L_b states are forbidden, the transition to the B_a state is allowed and polarized along the y axis, and the transition to the B_b state is allowed and polarized along the x axis. In general, the polarizations of transitions from the ground state are easily determined by means of the nodal-plane diagrams for the excited states.

The α, p, β, and β' states in the notation mentioned in Subsection A.1.4 correspond to the ^1L_b, ^1L_a, ^1B_b, and ^1B_a states in Platt's notation, respectively.

Platt[6] proposed further a system of notation for $N \rightarrow Q$ transitions in Mulliken's notation, that is, n–π^* and n–σ^* transitions. In Platt's notation, the upper state of a symmetry-forbidden $N \rightarrow Q$ transition is designated as U, prefixed by the usual superscript indicating the spin multiplicity, and the transition is called an $A \rightarrow U$ transition; the upper state of a symmetry-allowed $N \rightarrow Q$ transition is designated as W, and the transition is called an $A \rightarrow W$ transition. Transitions from nonbonding orbitals having no s component to antibonding π orbitals are usually forbidden by local symmetry. According to Platt's notation, such n–π^* transitions are $A \rightarrow U$ transitions. On the other hand, transitions from nonbonding orbitals having some s component to antibonding π orbitals are allowed by local symmetry. Such n–π^* transitions are $A \rightarrow W$ transitions.

[6] J. R. Platt, *J. Chem. Phys.* **18,** 1168 (1950); *J. Opt. Soc. Am.* **43,** 252 (1953).

General References

Electronic Absorption Spectroscopy

H. Baba, *in* "Jikken Kagaku Koza," (edited by the Chemical Society of Japan), Vol. 3, Chap. 7. Maruzen, Tokyo, 1957.

A. E. Gillam and E. S. Stern, "An Introduction to Electronic Absorption Spectroscopy in Organic Chemistry," 2nd ed. Arnold, London, 1958.

C. N. R. Rao, "Ultraviolet and Visible Spectroscopy, Chemical Applications." Butterworths, London, 1961.

The Chemical Society of Japan (ed.), "Jikken Kagaku Koza (Zoku)," Vol. 11. Maruzen, Tokyo, 1965.

Collections of Electronic Absorption Spectra

R. A. Friedel and M. Orchin, "Ultraviolet Spectra of Aromatic Compounds." Wiley, London and New York, 1951.

Organic Electronic Spectral Data, Inc., "Organic Electronic Spectral Data," Vol. I (covering 1946–1952, M. J. Kamlet, ed., 1960); Vol. II (covering 1953–1955, H. E. Ungnade, ed., 1960); Vol. III (covering 1956–1957, L. A. Kaplan and O. H. Wheeler, eds., 1966); Vol. IV (covering 1958–1959, J. P. Phillips and F. C. Nachod, eds., 1963). Wiley (Interscience), New York.

L. Láng (ed.), "Absorption Spectra in the Ultraviolet and Visible Region," Vols. I (2nd ed.) and II. Publishing Hause of the Hungarian Academy of Science, Budapest, 1961.

K. Hirayama, *in* "Jikken Kagaku Koza," (edited by the Chemical Society of Japan), Vol. 1, Chap. 1. Maruzen, Tokyo, 1957.

A. E. Gillam and E. S. Stern, "An Introduction to Electronic Absorption Spectroscopy in Organic Chemistry," 2nd ed. Arnold, London, 1958.

S. F. Mason, *Quart. Rev. (London)* **15,** 287 (1961).

H. M. Hershenson, "Ultraviolet and Visible Absorption Spectra," Index for 1930–54 (1956); Index for 1955–59 (1961); Index for 1958–62 (1964). Academic Press, New York.

Theory of Electronic Absorption Spectra

G. Herzberg, "Molecular Spectra and Molecular Structure I. Spectra of Diatomic Molecules," 2nd ed. Van Nostrand, Princeton, New Jersey, 1950.

A. B. F. Duncan and F. A. Matsen, *in* "Chemical Application of Spectroscopy" (W. West, ed.) ("Technique of Organic Chemistry," Vol. IX), Chap. V. Wiley (Interscience), London and New York, 1956.

C. Sandorfy, "Les Spectres Ēlectroniques en Chimie Théorique." Édition de la Revue d'optique théorique et instrumentale, Paris, 1959.

H. H. Jaffé and M. Orchin, "Theory and Applications of Ultraviolet Spectroscopy." Wiley, New York, 1962.

J. N. Murrell, "The Theory of the Electronic Spectra of Organic Molecules." Methuen, London, 1963.

J. R. Platt and co-workers at the Laboratory of Molecular Structure and Spectra, Department of Physics, University of Chicago, "Systematics of the Electronic Spectra of Conjugated Molecules: A Source Book." Wiley, New York, 1964.

Theory of Light Absorption

H. Eyring, J. Walter, and G. E. Kimball, "Quantum Chemistry," Chap. 8. Wiley, London and New York, 1944.

W. Heitler, "The Quantum Theory of Radiation," 3rd ed. Oxford Univ. Press, London and New York, 1954.

Quantum Chemistry

L. Pauling and E. B. Wilson, Jr., "Introduction to Quantum Mechanics with Applications to Chemistry." McGraw-Hill, New York, 1935.

H. Eyring, J. Walter, and G. E. Kimball, "Quantum Chemistry." Wiley, London and New York, 1944.

K. Higasi and H. Baba, "Ryoshi Yuki Kagaku (Quantum Organic Chemistry)." Asakura, Tokyo, 1956.

W. Kauzmann, "Quantum Chemistry: An Introduction." Academic Press, New York, 1957.

R. Daudel, R. Lefebvre, and C. Moser, "Quantum Chemistry: Methods and Applications." Wiley (Interscience), London and New York, 1959.

C. A. Coulson, "Valence," 2nd ed. Oxford Univ. Press, London and New York, 1961.

T. Yonezawa, C. Nagata, H. Kato, A. Imamura, and K. Morokuma, "Ryoshi-Kagaku Nyumon (An Introduction to Quantum Chemistry)," Vols. I and II. Kagakudojin, Kyoto, 1963–1964.

Simple LCAO MO Method

A. Streitwieser, Jr., "Molecular Orbital Theory for Organic Chemists." Wiley, London and New York, 1961.

J. D. Roberts, "Notes on Molecular Orbital Calculations." Benjamin, New York, 1961.

Free-Electron MO Method

J. R. Platt, K. Ruedenberg, C. W. Scherr, J. S. Ham, H. Labhart, and W. Lichten, "Free-Electron Theory of Conjugated Molecules: A Source Book." Wiley, London and New York, 1964.

Group Theory

H. Eyring, J. Walter, and G. E. Kimball, "Quantum Chemistry," Chap. 10 and Appendix 7. Wiley, London and New York, 1944.

A. Streitwieser, Jr., "Molecular Orbital Theory for Organic Chemists," Chap. 3. Wiley, New York, 1961.

H. H. Jaffé and M. Orchin, "Theory and Applications of Ultraviolet Spectroscopy," Chap. 4 and Appendix 3. Wiley, New York, 1962.

P. J. Wheatley, "The Determination of Molecular Structure," Chap. 1. Oxford Univ. Press, London and New York, 1959.

V. Heine, "Group Theory in Quantum Mechanics." Macmillan (Pergamon), London, 1960.

F. A. Cotton, "Chemical Applications of Group Theory." Wiley (Interscience), London and New York, 1963.

H. H. Jaffé and M. Orchin, "Symmetry in Chemistry." Wiley, London and New York, 1965.

Advanced MO Methods

R. Daudel, R. Lefebvre, and C. Moser, "Quantum Chemistry: Methods and Applications," Part II. Wiley (Interscience), London and New York, 1959.

R. G. Parr, "Quantum Theory of Molecular Electronic Structure." Benjamin, New York, 1963.

Collections of Data on Molecular Structure

"Tables of Interatomic Distances and Configuration in Molecules and Ions." The Chemical Society, London, 1958.

G. W. Wheland, "Resonance in Organic Chemistry," Appendix. Wiley, London and New York, 1955.

The Chemical Society of Japan (ed.), "Jikken Kagaku Koza," Vol. 3, Chap. 6, Appendix. Maruzen, Tokyo, 1957.

The Intensity and Shape of Electronic Absorption Bands

G. Herzberg, "Molecular Spectra and Molecular Structure I. Spectra of Diatomic Molecules," 2nd ed., Chap. IV. Van Nostrand, Princeton, New Jersey, 1950.

Geometry of Molecules in Excited Electronic States

D. A. Ramsay, *in* "Determination of Organic Structures by Physical Methods" (F. C. Nachod and W. D. Phillips, eds.), Vol. 2, Chap. 4. Academic Press, New York, 1962.

J. C. D. Brand and D. G. Williamson, *in* "Advances in Physical Organic Chemistry" (V. Gold, ed.), Vol. 1. Academic Press, New York, 1963.

Far-Ultraviolet Absorption Spectroscopy

D. W. Turner, *in* "Determination of Organic Structures by Physical Methods" (F. C. Nachod and W. D. Phillips, eds.), Vol. 2, Chap. 5. Academic Press, New York, 1962.

Electronic Absorption Spectra of Molecules in the Crystalline State

D. S. McClure, *in* "Solid State Physics" (F. Seitz and D. Turnbull, eds.), Vols. 8 and 9. Academic Press, New York, 1959.

H. C. Wolf, *in* "Solid State Physics" (F. Seitz and D. Turnbull, eds.), Vol. 9. Academic Press, New York, 1959.

A. S. Davydov, "Theory of Molecular Excitons" (translated by M. Kasha and M. Oppenheimer, Jr.). McGraw-Hill, New York, 1962.

Electronic Absorption Spectra of Molecular Complexes

L. J. Andrews, *Chem. Rev.* **54,** 713 (1954).

L. E. Orgel, *Quart. Rev.* (*London*) **8,** 422 (1954).

S. P. McGlynn, *Chem. Rev.* **58,** 1113 (1958).
G. Briegleb and J. Czekalla, *Angew. Chem.* **72,** 401 (1960).
G. Briegleb, "Elektronen-Donator-Acceptor-Komplexe," Springer-Verlag, Berlin, 1961.
J. N. Murrell, *Quart. Rev.* (*London*) **15,** 191 (1961).
H. Tsubomura and A. Kuboyama, *Kagaku To Kogyo* (*Tokyo*) **14,** 537 (1961).
R. S. Mulliken and W. B. Person, *Ann. Rev. Phys. Chem.* **13,** 107 (1962).
J. N. Murrell, "The Theory of the Electronic Spectra of Organic Molecules," Chap. 13. Methuen, London, 1963.
H. Tsubomura, *in* "Denshiron No Kiso To Oyo (Bases and Applications of Electronic Theory)" (S. Nagakura, ed.), Chap. 1. Iwanami, Tokyo, 1964.

Steric Effects

W. Klyne (ed.), "Progress in Stereochemistry," Vol. 1. Butterworths, London, 1954.
E. A. Braude, *in* "Determination of Organic Structures by Physical Methods" (E. A. Braude and F. C. Nachod, eds.), Vol. 1, Chap. 4. Academic Press, New York, 1955.
M. S. Newman (ed.), "Steric Effects in Organic Chemistry." Wiley, London and New York, 1956.
W. Klyne and P. B. D. de la Mare (eds.), "Progress in Stereochemistry," Vol. 2. Butterworths, London, 1958.
G. W. Gray (ed.), "Steric Effects in Conjugated Systems." Butterworths, London, 1958.
H. H. Jaffé and M. Orchin, "Theory and Applications of Ultraviolet Spectroscopy," Chap. 15. Wiley, New York, 1962.
J. N. Murrell, "The Theory of the Electronic Spectra of Organic Molecules," Chap. 11. Methuen, London, 1963.

Author Index

Numbers in parentheses are reference numbers and indicate that an author's work is referred to although his name is not cited in the text.

Subject Index